D1676075

High-Performance Materials from Bio-based Feedstocks

Wiley Series in Renewable Resources

Series Editor:

Christian V. Stevens, Faculty of Bioscience Engineering, Ghent University, Belgium

Titles in the Series:

Wood Modification: Chemical, Thermal and Other Processes
Callum A. S. Hill

Renewables-Based Technology: Sustainability Assessment
Jo Dewulf, Herman Van Langenhove

Biofuels
Wim Soetaert, Erik Vandamme

Handbook of Natural Colorants
Thomas Bechtold, Rita Mussak

Surfactants from Renewable Resources
Mikael Kjellin, Ingegärd Johansson

Industrial Applications of Natural Fibres: Structure, Properties and Technical Applications
Jörg Müssig

Thermochemical Processing of Biomass: Conversion into Fuels, Chemicals and Power
Robert C. Brown

Biorefinery Co-Products: Phytochemicals, Primary Metabolites and Value-Added Biomass Processing
Chantal Bergeron, Danielle Julie Carrier, Shri Ramaswamy

Aqueous Pretreatment of Plant Biomass for Biological and Chemical Conversion to Fuels and Chemicals
Charles E. Wyman

Bio-Based Plastics: Materials and Applications
Stephan Kabasci

Introduction toWood and Natural Fiber Composites
Douglas D. Stokke, Qinglin Wu, Guangping Han

Cellulosic Energy Cropping Systems
Douglas L. Karlen

Introduction to Chemicals from Biomass, 2nd Edition
James H. Clark, Fabien Deswarte

Lignin and Lignans as Renewable Raw Materials: Chemistry, Technology and Applications
Francisco G. Calvo-Flores, Jose A. Dobado, Joaquín Isac-García, Francisco J. Martín-Martínez

Sustainability Assessment of Renewables-Based Products: Methods and Case Studies
Jo Dewulf, Steven De Meester, Rodrigo A. F. Alvarenga

Cellulose Nanocrystals: Properties, Production and Applications
Wadood Hamad

Fuels, Chemicals and Materials from the Oceans and Aquatic Sources
Francesca M. Kerton, Ning Yan

Bio-Based Solvents
François Jérôme and Rafael Luque

Nanoporous Catalysts for Biomass Conversion
Feng-Shou Xiao and Liang Wang

Thermochemical Processing of Biomass: Conversion into Fuels, Chemicals and Power 2nd Edition
Robert Brown

Chitin and Chitosan: Properties and Applications
Lambertus A.M. van den Broek and Carmen G. Boeriu

The Chemical Biology of Plant Biostimulants
Danny Geelen, Lin Xu

Biorefinery of Inorganics: Recovering Mineral Nutrients from Biomass and Organic Waste
Erik Meers, Evi Michels, René Rietra, Gerard Velthof

Process Systems Engineering for Biofuels Development
Adrián Bonilla-Petriciolet, Gade P. Rangaiah

Waste Valorisation: Waste Streams in a Circular Economy
Carol Sze Ki Lin, Chong Li, Guneet Kaur, Xiaofeng Yang

Forthcoming Titles:

High-Performance Materials from Bio-based Feedstocks
Andrew J. Hunt, Nontipa Supanchaiyamat, Kaewta Jetsrisuparb, Jesper T. Knijnenburg

Handbook of Natural Colorants 2nd Edition
Thomas Bechtold, Avinash P. Manian and Tung Pham

Biogas Plants: Waste Management, Energy Production and Carbon Footprint Reduction
Wojciech Czekała

High-Performance Materials from Bio-based Feedstocks

ANDREW J. HUNT
Materials Chemistry Research Center
Department of Chemistry and Center of Excellence for Innovation in Chemistry
Khon Kaen University
Khon Kaen, Thailand

NONTIPA SUPANCHAIYAMAT
Materials Chemistry Research Center
Department of Chemistry and Center of Excellence for Innovation in Chemistry
Khon Kaen University
Khon Kaen, Thailand

KAEWTA JETSRISUPARB
Department of Chemical Engineering
Khon Kaen University
Khon Kaen, Thailand

JESPER T.N. KNIJNENBURG
International College
Khon Kaen University
Khon Kaen, Thailand

This edition first published 2022

Registered Offices
John Wiley & Sons, Inc., 111 River Street, Hoboken, NJ 07030, USA
John Wiley & Sons Ltd, The Atrium, Southern Gate, Chichester, West Sussex, PO19 8SQ, UK

Editorial Office
The Atrium, Southern Gate, Chichester, West Sussex, PO19 8SQ, UK

For details of our global editorial offices, customer services, and more information about Wiley products visit us at www.wiley.com.

Wiley also publishes its books in a variety of electronic formats and by print-on-demand. Some content that appears in standard print versions of this book may not be available in other formats.

For general information on our other products and services or for technical support, please contact our Customer Care Department within the United States at (800) 762-2974, outside the United States at (317) 572-3993 or fax (317) 572-4002.
Wiley also publishes its books in a variety of electronic formats. Some content that appears in print may not be available in electronic formats. For more information about Wiley products, visit our web site at ww.wiley.com.

Library of Congress Cataloging-in-Publication Data applied for

Hardback ISBN: 978-1-119-65572-5

Cover Design: Wiley
Cover Image: © Fietzfotos/Pixabay

Set in 10/12 TimesLTStd by Straive, Pondicherry, India
Printed and bound by CPI Group (UK) Ltd, Croydon, CR0 4YY

C9781119655725_210322

In memory of our friend and colleague Professor Janet L. Scott.

Contents

List of Contributors

Andrew P. Abbott School of Chemistry, University of Leicester, Leicester, UK

Mohammed Aqil Materials Science, Energy and Nanoengineering Department (MSN), Mohammed VI Polytechnic University (UM6P), Benguerir, Morocco

Hicham Ben Youcef High Throughput Multidisciplinary Research Laboratory (HTMR-Lab), Mohammed VI Polytechnic University (UM6P), Benguerir, Morocco

Bernardo Castro-Dominguez Department of Chemical Engineering, University of Bath, Bath, UK

Chaiyan Chaiya Department of Chemical Engineering and Materials, Rajamangala University of Technology Thanyaburi, Pathum Thani, Thailand

Prinya Chindaprasirt Sustainable Infrastructure Research and Development Center, Department of Civil Engineering, Khon Kaen University, Khon Kaen, Thailand

and

Academy of Science, Royal Society of Thailand, Bangkok, Thailand

Oranat Chuchuen Department of Chemical Engineering, Khon Kaen University, Khon Kaen, Thailand

Alan Connolly DSM Nutritional Products Ltd., Nutrition R&D Center Forms and Application, Basel, Switzerland

Doungporn Yiamsawas National Nanotechnology Center, National Science and Technology Development Agency, Pathum Thani, Thailand

Pradip K. Dutta Polymer Research Group, Department of Chemistry, Motilal Nehru National Institute of Technology, Allahabad, Prayagraj, India

Emile R. Engel Department of Chemistry, University of Bath, Bath, UK

Xiaotong Feng College of Chemistry and Chemical Engineering, Lanzhou University, Lanzhou, Gansu, China

Salim Hiziroglu Department of Natural Resource Ecology and Management, Oklahoma State University, Stillwater, OK, USA

Andrew J. Hunt Materials Chemistry Research Center (MCRC), Department of Chemistry, Centre of Excellence for Innovation in Chemistry, Khon Kaen University, Khon Kaen, Thailand

Itziar Iraola-Arregui High Throughput Multidisciplinary Research Laboratory (HTMR-Lab), Mohammed VI Polytechnic University (UM6P), Benguerir, Morocco

Shefali Jaiswal Polymer Research Group, Department of Chemistry, Motilal Nehru National Institute of Technology, Allahabad, Prayagraj, India

Sasiradee Jantasee Department of Chemical Engineering and Materials, Rajamangala University of Technology Thanyaburi, Pathum Thani, Thailand

Kaewta Jetsrisuparb Department of Chemical Engineering, Khon Kaen University, Khon Kaen, Thailand

Wiyong Kangwansupamonkon National Nanotechnology Center, National Science and Technology Development Agency, Pathum Thani, Thailand

Pornnapa Kasemsiri Department of Chemical Engineering, Khon Kaen University, Khon Kaen, Thailand

David F. Katz Departments of Biomedical Engineering & Obstetrics and Gynecology, Duke University, Durham, NC, USA

Sarah Key School of Chemistry, University of Leicester, Leicester, UK

Poramate Klanrit Department of Biochemistry, Faculty of Medicine, Khon Kaen University, Khon Kaen, Thailand

Jesper T.N. Knijnenburg International College, Khon Kaen University, Khon Kaen, Thailand

Wilaiporn Kraisuwan Department of Chemical Engineering, Srinakharinwirot University, Nakhon Nayok, Thailand

Santosh Kumar Department of Organic and Nano System Engineering, Konkuk University, Seoul, South Korea

Kritapas Laohhasurayotin National Nanotechnology Center, National Science and Technology Development Agency, Pathum Thani, Thailand

Duncan J. Macquarrie Green Chemistry Centre of Excellence, Department of Chemistry, University of York, York, UK

Cinthia J. Meña Duran Green Chemistry Centre of Excellence, Department of Chemistry, University of York, York, UK

Manunya Okhawilai Metallurgy and Materials Science Research Institute, Chulalongkorn University, Bangkok, Thailand

Tabitha H.M. Petchey Green Chemistry Centre of Excellence, Department of Chemistry, University of York, York, UK

Nawadon Petchwattana Department of Chemical Engineering, Srinakharinwirot University, Nakhon Nayok, Thailand

Uraiwan Pongsa Division of Industrial Engineering Technology, Rajamangala University of Technology Rattanakosin Wang Klai Kang Won Campus, Prachuap Khiri Khan, Thailand

Patcharapol Posi Department of Civil Engineering, Rajamangala University of Technology Isan Khon Kaen Campus, Khon Kaen, Thailand

Qiaosheng Pu College of Chemistry and Chemical Engineering, Lanzhou University, Lanzhou, Gansu, China

Wanwan Qu School of Chemistry, University of Leicester, Leicester, UK

Ubolluk Rattanasak Department of Chemistry, Faculty of Science, Burapha University, Chonburi, Thailand

Ismael Saadoune Technology Development Cell (TechCell), Mohammed VI Polytechnic University (UM6P), Benguerir, Morocco

and

IMED-Lab, Faculty of Science and Technology, Cadi Ayyad University, Marrakesh, Morocco

Chadamas Sakonsinsiri Department of Biochemistry, Khon Kaen University, Khon Kaen, Thailand

Janet L. Scott Department of Chemistry, University of Bath, Bath, UK

Anu Singh Polymer Research Group, Department of Chemistry, Motilal Nehru National Institute of Technology, Allahabad, Prayagraj, India

Benjatham Sukkaneewat Divison of Chemistry, Udon Thani Rajabhat University, Udon Thani, Thailand

Nontipa Supanchaiyamat Materials Chemistry Research Center (MCRC), Department of Chemistry, Centre of Excellence for Innovation in Chemistry, Faculty of Science, Khon Kaen University, Khon Kaen, Thailand

Alexandra Teleki Science for Life Laboratory, Department of Pharmacy, Uppsala University, Uppsala, Sweden

Vera Trabadelo High Throughput Multidisciplinary Research Laboratory (HTMR-Lab), Mohammed VI Polytechnic University (UM6P), Benguerir, Morocco

Christos Tsekou DSM Nutritional Products Ltd., Nutrition R&D Center Forms and Application, Basel, Switzerland

Series Preface

Renewable resources, their use and modification, are involved in a multitude of important processes with a major influence on our everyday lives. Applications can be found in the energy sector, paints and coatings, and the chemical, pharmaceutical, and textile industries, to name but a few.

The area interconnects several scientific disciplines (agriculture, biochemistry, chemistry, technology, environmental sciences, forestry, etc.), which makes it very difficult to have an expert view on the complicated interactions. Therefore, the idea to create a series of scientific books, focusing on specific topics concerning renewable resources, has been very opportune and can help to clarify some of the underlying connections in this area.

In a very fast-changing world, trends are not only characteristic of fashion and political standpoints; science too is not free from hypes and buzzwords. The use of renewable resources is again more important nowadays; however, it is not part of a hype or a fashion. As the lively discussions among scientists continue about how many years we will still be able to use fossil fuels – opinions ranging from 50 to 500 years – they do agree that the reserve is limited and that it is essential not only to search for new energy carriers but also for new material sources.

In this respect, the field of renewable resources is a crucial area in the search for alternatives for fossil-based raw materials and energy. In the field of energy supply, biomass- and renewables-based resources will be part of the solution alongside other alternatives such as solar energy, wind energy, hydraulic power, hydrogen technology, and nuclear energy. In the field of material sciences, the impact of renewable resources will probably be even bigger. Integral utilization of crops and the use of waste streams in certain industries will grow in importance, leading to a more sustainable way of producing materials. Although our society was much more (almost exclusively) based on renewable resources centuries ago, this disappeared in the Western world in the nineteenth century. Now it is time to focus again on this field of research. However, it should not mean a "retour à la nature," but should be a multidisciplinary effort on a highly technological level to perform research towards new opportunities, and to develop new crops and products from renewable resources. This will be essential to guarantee an acceptable level of comfort for the growing number of people living on our planet. It is "the" challenge for the coming generations of scientists to develop more sustainable ways to create prosperity and to fight poverty and hunger in the world. A global approach is certainly favored.

This challenge can only be dealt with if scientists are attracted to this area and are recognized for their efforts in this interdisciplinary field. It is, therefore, also essential that consumers recognize the fate of renewable resources in a number of products. Furthermore, scientists do need to communicate and discuss the relevance of their work. The use and modification of renewable resources may not follow the path of the genetic engineering concept in view of consumer acceptance in Europe. Related to this aspect, the series will certainly help to increase the visibility of the importance of renewable resources. Being convinced of the value of the renewables approach for the industrial world, as well as for developing countries, I was myself delighted to collaborate on this series of books focusing on the different aspects of renewable resources. I hope that readers become aware of the complexity, the interaction, and interconnections, and the challenges of this field, and that they will help to communicate on the importance of renewable resources.

I certainly want to thank the people of Wiley's Chichester office, especially David Hughes, Jenny Cossham, and Lyn Roberts, in seeing the need for such a series of books on renewable resources, for initiating and supporting it, and for helping to carry the project to the end.

Last, but not least, I want to thank my family, especially my wife Hilde and children Paulien and Pieter-Jan, for their patience, and for giving me the time to work on the series when other activities seemed to be more inviting.

Christian V. Stevens
Faculty of Bioscience Engineering, Ghent University, Belgium
Series Editor, "Renewable Resources"
June 2005

1

High-performance Materials from Bio-based Feedstocks: Introduction and Structure of the Book

Kaewta Jetsrisuparb[1], Jesper T.N. Knijnenburg[2], Nontipa Supanchaiyamat[3] and Andrew J. Hunt[3]

[1] *Department of Chemical Engineering, Khon Kaen University, Khon Kaen, Thailand*
[2] *International College, Khon Kaen University, Khon Kaen, Thailand*
[3] *Materials Chemistry Research Center (MCRC), Department of Chemistry, Centre of Excellence for Innovation in Chemistry, Faculty of Science, Khon Kaen University, Khon Kaen, Thailand*

1.1 Introduction

The overexploitation of the Earth's resources over the last century has led to a decrease in natural resources, a loss of natural habitat, climate change, and degradation of the environment, resulting in the extinction of several species [1]. The recovery of global economics after COVID-19 is also driving lifestyle changes, leading to increased high-performance materials production. As a result, a large number of nonrenewable resources are being utilized, which inevitably contributes to the generation of waste and may lead to detrimental effects to both environment and health. In addition, the scarcity of fossil resources and finite elements with potential global supply chain vulnerabilities are global concerns. Concerns over the supply of natural resources and potential damage to the environment have compelled governments to implement policies that mitigate the risk of further damage. The formation of the World Commission on Environment and Development (WCED) in 1983 and their report called "Our Common Future" in 1987 (also

High-Performance Materials from Bio-based Feedstocks, First Edition. Edited by Andrew J. Hunt, Nontipa Supanchaiyamat, Kaewta Jetsrisuparb and Jesper T.N. Knijnenburg.

called "Brundtland report") was one of the catalysts for the move toward a sustainable future for humankind [2]. The definition of sustainable development is the development that "meets the needs of the present without compromising the ability of future generations to meet their own needs" [3]. Importantly, sustainability is a complex balance between societal, economic, and environmental needs, where this must be achieved in unison [4]. Implementation of a bio-based circular economy including minimizing the waste by recycling materials and utilization of replenishable resources is key to sustainable development.

Historically, chemistry goes hand in hand with innovation, thus promoting a positive image of this industry. However, the perception of the industry can be tarnished with media reports of life-threatening accidents and environmental pollution [5]. Anastas and Warner pioneered the concept of green chemistry, "the invention, design and application of chemical products and processes to reduce or to eliminate the use and generation of hazardous substances" [6]. Today, green chemistry is recognized and widely accepted to pursue sustainable development. The 12 principles of green chemistry (stated next) are regarded as a blueprint for achieving the aims of green chemistry. Moreover, green chemistry can aid in the development of sustainable bio-based chemicals and importantly also high-performance materials.

The 12 principles of green chemistry as stated by Anastas and Warner [6] are:

1. **Prevention**
   It is better to prevent waste than to treat or clean up waste after it has been created.
2. **Atom Economy**
   Synthetic methods should be designed to maximize the incorporation of all materials used in the process into the final product.
3. **Less Hazardous Chemical Syntheses**
   Wherever practicable, synthetic methods should be designed to use and generate substances that possess little or no toxicity to human health and the environment.
4. **Designing Safer Chemicals**
   Chemical products should be designed to effect their desired function while minimizing their toxicity.
5. **Safer Solvents and Auxiliaries**
   The use of auxiliary substances (e.g. solvents, separation agents, etc.) should be made unnecessary wherever possible and innocuous when used.
6. **Design for Energy Efficiency**
   Energy requirements of chemical processes should be recognized for their environmental and economic impacts and should be minimized. If possible, synthetic methods should be conducted at ambient temperature and pressure.
7. **Use of Renewable Feedstocks**
   A raw material or feedstock should be renewable rather than depleting whenever technically and economically practicable.
8. **Reduce Derivatives**
   Unnecessary derivatization (use of blocking groups, protection/deprotection, temporary modification of physical/chemical processes) should be minimized or avoided if possible because such steps require additional reagents and can generate waste.
9. **Catalysis**
   Catalytic reagents (as selective as possible) are superior to stoichiometric reagents.
10. **Design for Degradation**
    Chemical products should be designed so that at the end of their function they break down into innocuous degradation products and do not persist in the environment.

11. **Real-time Analysis for Pollution Prevention**
    Analytical methodologies need to be further developed to allow for real-time, in-process monitoring and control prior to the formation of hazardous substances.
12. **Inherently Safer Chemistry for Accident Prevention**
    Substances and the form of a substance used in a chemical process should be chosen to minimize the potential for chemical accidents, including releases, explosions, and fires. [6]

By examining the 12 principles of green chemistry, the use of waste biomass and bio-based products to produce high-performance materials is in agreement with the seventh principle, which encourages the use of renewable feedstocks. The utilization of renewable resources has an added benefit as they can potentially lead to the development of carbon-neutral products.

According to the Kirk-Othmer Encyclopedia of Chemical Technology, bio-based materials refer to "products that mainly consist of a substance (or substances) derived from living matter (biomass) and either occur naturally or are synthesized, or it may refer to products made by processes that use biomass" [7]. Strictly speaking, this also includes traditional materials such as paper, leather, and wood, but these traditional uses are outside the scope of this book. It is important to note that bio-based materials are different from biomaterials (which involve biocompatibility), and being bio-based does not always mean the material will be biodegradable or safe.

The use of bio-based materials seems to be an appropriate approach to minimize the negative impact on the environment while harnessing the unique properties they offer. The development application of high-performance advanced bio-based materials through green synthetic approaches (i.e. application of the 12 principles of green chemistry) can aid in developing sustainable circular economies, while still minimizing environmental impacts. High-performance bio-based materials can be applied in catalysis, energy materials, polymers, medical devices, and even construction materials to name but a few.

A significant source of biomass which is ripe for exploitation into high-performance materials comes in the form of waste or agricultural residues. These include residues from food (e.g. corncob, sugarcane bagasse, rice husk, rice straw, and wheat straw) and non-food production (e.g. cellulose and lignin), forest residues, industrial by-products (e.g. ashes from biomass power generation), animal wastes (e.g. manure) as well as municipal wastes [7, 8]. These resources offer a complex mixture of polymers, inorganics, and chemicals, which can include but are not limited to polysaccharides, lignin, proteins, and ash, all of which are attractive alternative feedstocks to replace nonrenewable fossil-based resources. Exhaustion of fossil fuels and other finite resources is a driver for bio-based materials for high-performance applications. The structural diversity of biomass constituents and their unique properties are also promising for new applications including high-performance products.

Despite the great potential, some of the biggest challenges in using biomass as feedstock for high-performance applications are its heterogeneity, seasonal variation, and complexity regarding separation. In many cases, biomass needs to undergo some form of processing prior to being used as high-performance materials. Typically, the processing of biomass can be performed using chemical, biochemical, and thermochemical processes. These challenges are being tackled as part of the growth in holistic biorefineries, and such approaches that generate no waste are vital for maximizing the value of biomass.

However, unlike petrochemical feedstocks that require significant functionalization, bio-based feedstocks are blessed with an abundance of functionalities. As such, the development of high-performance materials from biomass requires different chemistries compared to those from fossil resources. The benefits of biomass utilization for industrial-scale production of high-performance materials are that they can potentially reduce waste and production costs, in addition to being carbon neutral, low cost, versatile, and renewable.

1.2 High-performance Bio-based Materials and Their Applications

Bio-based materials can be synthesized directly from biomass constituents (such as polysaccharides and other polymers, proteins and amino acids, and active biological compounds) that are directly extracted from biomass, but also from biomass-derived materials (e.g. bio-derived polymers, porous carbons, or ashes) that require additional processing steps such as polymerization, carbonization, or combustion.

1.2.1 Biomass Constituents

1.2.1.1 Polysaccharides

Polysaccharides are biopolymers that are made up of monosaccharide units connected by glycosidic linkages. Polysaccharides can be obtained from plants (e.g. cellulose, starch, and pectin), algae (alginate), animals (chitin/chitosan), bacteria (bacterial cellulose), and fungi (pullulan). In contrast to synthetic polymers, polysaccharides are abundant in nature, renewable and biodegradable, and are therefore considered as promising replacements of nonrenewable fossil fuel-based materials in a wide range of applications [9]. Yet, polysaccharides alone frequently present insufficient physicochemical properties, necessitating physical and/or chemical modifications to meet the required product specifications.

The hydrophilic nature of many polysaccharides presents a poor mechanical strength, which is a major hurdle for their widespread use. For this reason, polysaccharides are often blended with other polymers and/or chemically modified to improve their properties. In addition, inorganic additives can also play a key role to enhance materials properties from boosting strength to improving oxygen barrier and antibacterial properties. For example, esterification, etherification, acetylation, hydroxylation, and oxidation of starch, as well as the blending of starch with other biodegradable polymers can improve its physical and mechanical properties for food-packaging applications (Chapter 15). Similarly, the substitution of hydrogen by alkyl groups to form cellulose ethers including ethyl cellulose (EC) and carboxymethyl cellulose (CMC) increases the hydrophobicity of cellulose. The addition of CMC into food packaging can improve mechanical, thermal, and barrier properties (Chapter 15). Although biopolymers account for less than 1% of the total plastic production [9], there is a strong growth potential toward their wider application driven by the circular economy trend. Novel biopolymers with improved properties and new functionalities for various applications such as packaging films and coatings as well as textile applications have been developed for commercialization and are reviewed in Chapter 15 with the emphasis on the packaging.

Similar modifications to the hydrophobicity of biopolymers have been used in coating materials in controlled release fertilizers. In coated fertilizers, the hydrophilicity of native biopolymers results in the buildup of osmotic pressure inside the fertilizer beads. This causes the coatings to break prematurely, resulting in burst release of the nutrients. Naturally obtained biopolymers are thus not ideal fertilizer-coating materials owing to the lack of hydrophobicity and modifications are necessary to help control the nutrient release. By replacing hydroxy groups with ester groups, cellulose acetate (CA), which is a common cellulose derivative, can be prepared. The presence of ester introduces its pH sensitivity property to the materials and can be beneficial for applications not only in slow-release fertilizers (Chapter 16) but also in drug delivery (Chapter 8).

In other applications, the hydrophilicity and swelling behavior of biopolymers are used as an advantage. A different class of fertilizers is that of hydrogel-based fertilizers, which absorb and retain high amounts of water. Hydrogel composites with plant nutrients embedded in their network structure using alginate or chitosan have been developed to reduce the frequency of irrigation as well as controlling the rate of nutrient release (Chapter 16). Hydrogels also find applications in targeted drug delivery. The biocompatibility, biodegradability, and non-toxicity as well as high affinity for water make biopolymers like alginate, hyaluronic acid, pectin, and carrageenan attractive in controlled release and targeted drug delivery systems for HIV prophylaxis (Chapter 8). Alginate microparticles and films have also been used for anti-HIV drug delivery due to their excellent biocompatibility and biodegradability (Chapter 8), and alginate has been blended with other polymers to adjust the hydrophobicity of polyelectrolyte films for food-packaging applications (Chapter 15).

An interesting group of oligosaccharides with both hydrophilic and hydrophobic properties is that of cyclodextrins. The truncated cone structure of these circular oligosaccharides exhibits a hydrophobic inner cavity, while the upper and lower rims are hydrophilic. These unique properties enable cyclodextrins to contain hydrophobic molecules, but the high costs limit the applicability in foods (Chapter 14). Cyclodextrins can also be used to form metal–organic frameworks (MOFs) that have been investigated for a variety of potential applications including molecular separations, drug delivery, and biomedicine. Organic ligands are the main factors that determine if an MOF is bio-based, but the sustainability and safety of the metal ions should also be considered (Chapter 12).

Biodegradability is preferred in some applications such as packaging and fertilizers. The biodegradability needs to be tuned, however, to meet the desired product specifications. For example, in fertilizers, polysaccharide-based coatings based on, e.g. starch or cellulose, are too biodegradable, leading to premature nutrient release into the soil. To reduce the rate of degradation in soil and extend its service time, the biopolymer can be grafted with rubber or a different polymer with a lower biodegradability and higher hydrophobicity (Chapter 16). In food packaging, many packaging materials are based on blends of biodegradable polysaccharides such as starch, cellulose, and chitosan. In addition, cellulose nanocrystals, nanofibers, and bacterial cellulose have been used as biodegradable reinforcing fillers in various packaging films (Chapter 15).

Due to their biological nature, the biocompatibility of polysaccharides can be taken advantage of in applications where the polymer requires intimate contact with cells. For example, mucoadhesive films based on derivatives of cellulose, alginate, or chitosan can provide sustained release of several antiretroviral agents (Chapter 8). Moreover, biomaterials are critical to success in tissue engineering, act as a scaffold for tissue and cells to grow on. Such scaffolds can be derived from protein or carbohydrate biopolymers, including silk, collagen, fibrin, chitosan, alginate, and agarose. Biocompatibility and the ability to contribute to biological functions with the cells are necessary properties. Chapter 11 highlights bio-based feedstock that can be used in scaffold manufacturing for tissue engineering.

Carbohydrate-based materials play an essential role in cellular recognition processes. Carbohydrate-protein interactions can be probed by synthetic glycomaterials to diagnose viral and bacterial infections. In Chapter 10, bio-based glycomaterials and carbohydrate-functionalized materials are discussed including their application in drug/gene delivery, wound healing, biorecognition, and sensing.

Chitosan and chitin/glucan complexes have been used in a wide range of applications such as anticancer, antibacterial, antioxidant as well as gene delivery due to their unique biochemical properties (Chapter 9). The major limitation of these biopolymers is their insolubility in water and, as such, steps have been taken to improve their solubility by various modification methods. Other applications of chitosan include fertilizers (Chapter 16) and antibacterial food packaging (Chapter 15). For food-packaging applications, chitosan is limited by its low mechanical strength, rigid structure, and low thermal stability, which can be overcome by blending chitosan with other (bio)polymers (Chapter 15).

Due to their safety and biocompatibility, polysaccharide-based materials find applications in the food industry as emulsion stabilizers. Starch modified with octenyl succinic anhydride (OSA) generates an amphiphilic character of the starch, making it an effective stabilizer of oil-in-water emulsions. Gums such as gum acacia, gum tragacanth, xanthan gum, and guar gum are used as surfactants to stabilize emulsions and liposomes in the food industry, as well as the formation of coacervates. Additionally, cellulose nanocrystals find applications in food applications in the stabilization of Pickering emulsions (Chapter 14).

Polysaccharides have a high abundance of functional groups, e.g. hydroxyl groups of cellulose, amine of chitosan, and carboxylic acid of alginate. These functional groups have a good affinity for complexation of metal ions through mechanisms such as reduction, chelation, and complexation. The polysaccharides alone are poor sorbents due to their low stability and need to be improved by chemical (e.g. introduction of other functional groups) as well as physical modification (e.g. forming composites with other polymers as well as inorganic materials). The use of bio-based composites for the recovery of precious and heavy metals is discussed in detail in Chapter 7.

In electrochemical storage devices, synthetic binders such as polytetrafluoroethylene (PTFE) and polyvinylidene fluoride (PVDF) are traditionally used, which suffer from environmental and safety issues. As a green and renewable replacement, various biopolymer-based binders have been used in either pristine or modified forms. For example, cellulose and its modified forms such as CMC, EC, and CA, but also chitosan, alginate, and gums have shown great potential as environmentally friendly binders (Chapter 5). Biopolymers such as chitosan, cellulose, and carrageenan can also be used as components of proton exchange membranes in fuel cells. Especially, chitosan is a highly promising biopolymer for fuel cell applications due to its low cost, environmental friendliness, hydrophilicity, ease of modification, and low methanol permeability (Chapter 5).

As mentioned in Section 1.1, a major challenge in the use of biomass feedstocks is the complex processing to obtain sufficient purity and quantity for the use in high-performance products. Deep eutectic solvents (DES) comprise a class of solvents formed by eutectic mixtures of Lewis or Brønsted acids and bases, with physical properties similar to ionic liquids but different chemical properties [10]. These green solvents have the advantage over traditional ionic liquids since they are derived from biological resources and can be designed to have low toxicity and good biocompatibility. Chapter 6 of this book highlights a versatile approach of how biocompatible DES can be used as synthetic modifiers to achieve materials with a wide range of properties.

1.2.1.2 Other Biopolymers

Lignin is a three-dimensional biopolymer composed of coniferyl alcohol, sinapyl alcohol, and p-coumaryl alcohol monomers, but its specific structure and composition depend on the biomass [11]. A large number of hydroxyl groups in lignin provide a template for modification with specific functional groups. The ease and variability of lignin modification enable tailoring of the lignin structure to recover precious metals and rare earth metals (Chapter 7). Lignin has been used in fertilizer formulations, with the addition of plasticizers to improve the film-forming properties (Chapter 16). Other applications of lignin are as binders in Li-ion batteries (Chapter 5).

1.2.1.3 Proteins and Amino Acids

Proteins are biomolecules that are composed of amino acids and can be divided into plant-based (e.g. soy protein, zein, and gliadin) and animal-based proteins (e.g. gelatin, casein, and whey proteins). Plant proteins such as zein (extracted from maize) and gliadin (a component of gluten, extracted from wheat) are applied in the food industry as stabilizers of Pickering emulsions. However, pure protein emulsions are sensitive to changes in pH, which can be overcome by the formation of polymer-biopolymer complexes. Zein nanoparticles can also be used to protect a hydrophobic core (such as vitamins) against oxidation and hydrolysis. Soy protein isolate, whey protein isolate, and casein are used in the formation of coacervates to encapsulate lipophilic-active ingredients (Chapter 14). Gelatin, a natural water-soluble protein obtained from collagen, is used in food-packaging applications (Chapter 15).

Amino acids, the building blocks of proteins, and aliphatic diacids, like fumaric, glutaric, azelaic, and itaconic acids, have been used for the synthesis of MOFs with good water sorption and heat transfer applications (Chapter 12).

1.2.1.4 Active Biological Compounds

Apart from biopolymers and proteins, bio-based resources may contain various active compounds that possess unique properties. Such high-value chemicals may include fragrances, flavoring agents, and nutraceuticals like vitamins and antioxidants, and are generally extracted first before further processing of the biomass [12]. Naturally occurring fat-soluble (A, D, E, and K) and water-soluble vitamins (B and C) find applications in food. Here the key challenge is to retain the stability of the vitamins during storage and processing, which can be done by encapsulating the vitamin into delivery vesicles such as liposomes or coacervates (Chapter 14). Lecithins (such as phosphatidylcholine) and saponins (such as Quillaja saponins) are molecules with amphiphilic properties, which are used as natural surfactants to stabilize various vitamin-containing emulsions (Chapter 14). Other active compounds include antioxidants, essential oils (e.g. oregano, thyme, clove, and cinnamon), and various extracts (e.g. from spent coffee grounds). Their incorporation into food packaging can prolong the shelf life of foods (Chapter 15).

1.2.2 Bioderived Materials

1.2.2.1 Polymers Derived from Biological Monomers

Polylactic acid (PLA) is a bio-based thermoplastic composed of lactic acid monomers that are derived from renewable resources such as corn, sugarcane, or cassava. Due to its outstanding properties (biocompatible, biodegradable, and good mechanical strength), PLA and poly(lactic-co-glycolic) acid (PLGA) in biomedical applications such as anti-HIV drug delivery, in the form of drug-loaded nanoparticle formulations, topical products, and long-acting inserts (Chapter 8). The good mechanical strength, optical transparency, biodegradability, and processability make PLA attractive for food-packaging applications. A limitation of PLA is its poor impact force, which can be improved by addition of toughening agents [13]. Further modification of PLA can also improve properties such as antibacterial, antifogging, and gas barrier properties (Chapter 15).

Another interesting group of biomolecules is vegetable oils (VOs), which are composed of triglycerides from fatty acids. The VOs can be modified and used as monomers for the synthesis of alkyd resins and polyurethanes that find applications as coating materials in fertilizers (Chapter 16).

1.2.2.2 Carbon-based Materials Derived from Biomass

After the extraction of valuable compounds, a significant proportion of the biomass remains with inferior composition and value. Or, in some cases, it may not be possible to extract valuables from the biomass due to, e.g. technical, financial, compositional, or hygienic reasons. Instead of leaving them to decay, these biomass feedstocks can be used as a carbon source to produce various carbon-based materials. Carbon materials are inert, possess high mechanical stability, high surface area, and are chemically stable in the absence of oxygen under high or low pH. These properties make carbon materials appealing and suitable for various applications ranging from chemical sorbents, electrode materials, catalysts, and catalyst supports to slow-release fertilizers [14–16].

Conversion of carbon-rich biomasses can be carried out through thermochemical processes such as carbonization, pyrolysis, or similar techniques to produce porous carbons with tunable porosity. The porous structure can vary greatly depending on the cellulose content and source of the biomass as well as the pyrolysis conditions (e.g. temperature, atmosphere, duration, and presence of additives). The simplest form of porous carbon is biochar, which can directly be used as heterogeneous catalyst or catalyst support (Chapter 2). In soil, the porous biochar structure serves as a nutrient host to improve nutrient use efficiency. The water retention capacity, porosity, and high porous area of biochars can help slow down the nutrient release and thereby increase fertilizer use efficiency (Chapter 16).

Porous carbons can be categorized according to their pore size: microporous (<2 nm), mesoporous (2–50 nm), and microporous (>50 nm) [17]. High surface area and proper surface chemistry play a key role in catalysis and adsorption, and especially mesoporous carbons offer a wide range of applications. Most porous carbons, however, are microporous, allowing only small molecules or atoms to diffuse into the pore structure. Various activation techniques involving gases such as steam or CO_2 (physical activation) or compounds such as KOH, H_3PO_4, or $ZnCl_2$ (chemical activation) can be used to open the structure. The synthesis of porous carbons from various biomasses and the relations between synthesis conditions and material properties are discussed in detail in Chapter 2 for applications as catalyst or catalyst support, and in Chapter 4 for electrochemical applications.

Surface modification of porous carbons can be carried out to incorporate heteroatoms and functional groups through different methods including surface oxidation, halogenation, sulfonation, grafting, and impregnation [18]. Oxygen-containing moieties such as carboxylic acids and phenolic hydroxyl groups have been successfully introduced onto the carbon surface via oxidative modification [18]. Apart from oxygen, also the incorporation of N-, S-, and P-containing groups has been widely investigated. Sulfonation has been carried out to introduce $-SO_3H$ group to the porous structure, offering enhanced catalytic activity (Chapters 2 and 3). The incorporation of nitrogen-containing pyridine-N and pyrrole-N into carbon electrodes improves the pseudocapacitance as well as catalytic activity for the oxygen reduction reaction in fuel cells (Chapter 4). The presence of phosphorus groups in biowastes can improve the pseudocapacitance of the carbon electrodes (Chapter 4).

Examples of other carbon materials that have been applied in energy storage applications include hard carbon, carbon nanofibers (CNFs), carbon nanotubes (CNTs), graphene, and carbon aerogels. Hard carbon can be produced from pyrolysis of biowastes such as fruit waste, sucrose, or glucose, and possesses a highly irregular and disordered carbon structure that make it a promising anode material for Na-ion batteries [19]. The preparation of these carbons from biomass feedstocks and electrochemical applications are presented in Chapters 4 and 5. Production of biomass-derived CNT and graphene and their catalytic performance are discussed in Chapter 2.

Starbon® is a special type of porous carbon with high mesoporosity, produced from polysaccharide hydrogels as structural templates. The relatively large pores (compared to the more traditional activated carbons) can effectively adsorb and desorb bulky molecules, showing promising applications as catalysts, catalyst supports, adsorbents, and recently as battery materials. Surface modification through sulfonation of Starbon® is very promising for catalysis applications. Chapter 3 highlights the synthesis and properties of (modified) Starbon® and its applications in adsorption and catalysis.

The versatility, environmental compatibility, and high availability are the driving forces for using bio- and agricultural wastes as starting materials of bio-based carbon materials. For high-performance applications, (modified) porous carbons have shown tremendous benefits and will continue developing to meet new application demands. Surface modification is necessary to fully exploit the potential of carbon materials. The challenges include the large-scale yet well-controlled and cost-effective synthesis approaches. In addition, the inconsistency of the feedstocks and contamination may adversely affect the performance. Another concern is that biomass feedstock should not compete with food supply. Lignocellulosic biomass is particularly interesting for material development and plays an important role, both now and in the future.

1.2.2.3 Inorganic Materials Derived from Biomass

The elements C, O, H, and N generally present the major organic constituents of biomass in the form of polysaccharides, proteins, lipids, or other molecules. In addition to these, biomasses may also contain a significant fraction of inorganic material. The inorganic fraction can be separated from the organic matrix by combustion in the form of ash, of which the amount and elemental composition vary greatly between biomass sources and combustion conditions [20, 21]. Most commonly, inorganic elements consist of Si, Ca, K, P, Al, Mg, Fe, S, Na, and Ti with different proportions depending on the biomass source [20].

Biomasses such as rice husk contain relatively high amounts of Si, which can be extracted and used in various applications from energy storage to construction materials [22]. The successful use of Si derived from biomass is shown as anode material for Li-ion batteries. A major challenge for Si-based anodes is the low capacity retention, large volume change during lithium insertion/deinsertion process, and poor electrical conductivity, which can be partly overcome by using nanostructured materials [23]. Chapter 4 provides some critical perspectives on the application of biomass-derived Si in Li-ion batteries.

Another potential use of biomass ashes is demonstrated in the production of construction materials. Removal of the organic matrix by controlled combustion produces Si-rich ash that can be used to produce geopolymer materials as an alternative to traditional Portland cement [24]. These biomass ashes can substitute aluminosilicate sources instead of traditional fly ash from coal burning and leads to lower greenhouse gas production. Although large-scale application of bio-based inorganic materials is not available at the present time, the utilization of inorganic materials from biomasses is expected to increase with increasing environmental awareness. Challenges include the reduction of transportation cost of bulky biomass, and the inhomogeneity of ashes needs to be overcome to promote their usage. Various treatment options and applications of biomass ashes for geopolymer applications are discussed and assessed in Chapter 13.

1.3 Structure of the Book

This book covers a wide range of bio-based materials for high-performance applications, including their processing and comparison to state-of-the-art materials. First, Chapters 2–5 focus on the synthesis and applications of biomass-derived carbons. Chapter 2 starts with presenting the characteristics of biomasses and their thermochemical conversion into carbon-based catalysts and catalyst supports, with examples of their application in various reactions. Chapter 3 focuses on Starbon®, a mesoporous carbonaceous material derived from waste polysaccharides. The unique properties of pristine and modified Starbon® are highlighted with selected applications in adsorption and catalysis. Chapter 4 presents the conversion of biowastes into carbon electrodes through carbonization and activation, emphasizing the importance of biowaste selection, structure control, and heteroatom doping for optimizing electrochemical performance. The chapter also presents selected applications of carbon electrodes in various energy devices. Chapter 5 continues with applications of bio-derived materials in electrochemical energy storage and conversion devices such as fuel cells, capacitors, batteries, and as alternative binders therein.

Chapters 6–11 put emphasis on the extraction, modification, and applications of polysaccharides and other biopolymers for various applications in the environment and health. Chapter 6 presents the recent developments in biocompatible DES used to modify the mechanical, morphological, and chemical properties of bio-derived materials. In addition, the use of DES in the formation of biocomposites and gels is discussed. In Chapter 7, biopolymer composites prepared from cellulose, alginate, chitosan, and lignin for the recovery of precious and heavy metals are presented. The adsorption performance of biopolymer composites with magnetic materials, polymers, and other materials are reviewed.

Health-related applications of high-performance bio-based materials are presented in Chapters 8–11. Chapter 8 reviews the recent developments of bio-based materials in anti-HIV drug delivery systems. A wide variety of pre-exposure prophylaxis products based on bio-based polymers and their derivatives is presented. Chapter 9 follows with the synthesis and modification of chitin, chitin-glucan complexes, and chitosan from biological feedstocks for anticancer, antibacterial, antioxidant, and gene delivery applications. In Chapter 10, bio-based glycomaterials and carbohydrate-functionalized materials are discussed including their application in drug/gene delivery, wound healing, biorecognition, and sensing. Chapter 11 highlights bio-based feedstocks that can be used in scaffold manufacturing for tissue engineering.

The next two chapters make a transition toward inorganic materials. Chapter 12 discusses the green synthesis of bio-based MOFs and the wide range of applications. Applications of geopolymers prepared from biomass ashes with applications in the construction industry are presented in Chapter 13.

Finally, the last three chapters present the use of bio-based materials in food, packaging, and fertilizers. Chapter 14 highlights the use of various bio-based materials as functional ingredients used in the formulation of lipophilic nutraceuticals. Key developments on the role of bio-based materials in emulsions, colloidal delivery vehicles, as well as in drying technologies are reviewed. The use of bio-based materials like polysaccharides and bio-derived polymers in advanced packaging materials is discussed in Chapter 15. Finally, Chapter 16 presents the recent developments in controlled fertilizer applications using bio-based materials such as polysaccharides, alkyd resins, polyurethanes, and biochars.

References

1. Munier, N. (2005). *Introduction to Sustainability: Road to a Better Future*. Dordrecht: Springer.
2. Blewitt, J. (2008). *Understanding Sustainable Development*. Padstow: Earthscan.
3. United Nations World Commission on Environment and Development (1987). *Our Common Futute (Brundtland Report)*. Oxford: Oxford University Press.
4. The United Nations (1993). *Report of the United Nations Conference on Environment and Development*. New York: United Nations.
5. Schummer, J., Bensaude-Vincent, B., and Van Tiggelen, B. (2007). *The Public Image of Chemistry*. Singapore: World Scientific Publishing.
6. Anastas, P.T. and Warner, J.C. (1998). *Green Chemistry: Theory and Practice*. New York: Oxford University Press.
7. Curran, M.A. (2010). Biobased materials. In: *Kirk-Othmer Encyclopedia of Chemical Technology* (ed. C. Ley), 1–19. Hoboken, NJ: Wiley.
8. Taherzadeh, M., Bolton, K., Wong, J., and Pandey, A. (2019). *Sustainable Resource Recovery and Zero Waste Approaches*. Elsevier.
9. Babu, R.P., O'connor, K., and Seeram, R. (2013). Current progress on bio-based polymers and their future trends. *Progress in Biomaterials* 2 (1): 1–16.
10. Smith, E.L., Abbott, A.P., and Ryder, K.S. (2014). Deep eutectic solvents (DESs) and their applications. *Chemical Reviews* 114 (21): 11060–11082.
11. Supanchaiyamat, N., Jetsrisuparb, K., Knijnenburg, J.T.N. et al. (2019). Lignin materials for adsorption: current trend, perspectives and opportunities. *Bioresource Technology* 272: 570–581.
12. Ragauskas, A.J., Williams, C.K., Davison, B.H. et al. (2006). The path forward for biofuels and biomaterials. *Science* 311 (5760): 484–489.
13. Hamad, K., Kaseem, M., Ayyoob, M. et al. (2018). Polylactic acid blends: the future of green, light and tough. *Progress in Polymer Science* 85: 83–127.
14. González-García, P. (2018). Activated carbon from lignocellulosics precursors: a review of the synthesis methods, characterization techniques and applications. *Renewable and Sustainable Energy Reviews* 82: 1393–1414.
15. Sim, D.H.H., Tan, I.A.W., Lim, L.L.P., and Hameed, B.H. (2021). Encapsulated biochar-based sustained release fertilizer for precision agriculture: a review. *Journal of Cleaner Production* 303, 127018.
16. Xiong, X., Yu, I.K.M., Cao, L. et al. (2017). A review of biochar-based catalysts for chemical synthesis, biofuel production, and pollution control. *Bioresource Technology* 246: 254–270.
17. Thommes, M., Kaneko, K., Neimark, A.V. et al. (2015). Physisorption of gases, with special reference to the evaluation of surface area and pore size distribution (IUPAC Technical Report). *Pure and Applied Chemistry* 87 (9–10): 1051–1069.
18. Stein, A., Wang, Z., and Fierke, M.A. (2009). Functionalization of porous carbon materials with designed pore architecture. *Advanced Materials* 21 (3): 265–293.
19. Zhu, J., Yan, C., Zhang, X. et al. (2020). A sustainable platform of lignin: from bioresources to materials and their applications in rechargeable batteries and supercapacitors. *Progress in Energy and Combustion Science* 76: 100788.
20. Vassilev, S.V., Baxter, D., Andersen, L.K., and Vassileva, C.G. (2013). An overview of the composition and application of biomass ash. Part 1. Phase-mineral and chemical composition and classification. *Fuel* 105: 40–76.
21. Zając, G., Szyszlak-Bargłowicz, J., Gołębiowski, W., and Szczepanik, M. (2018). Chemical characteristics of biomass ashes. *Energies* 11 (11): 2885.
22. Pode, R. (2016). Potential applications of rice husk ash waste from rice husk biomass power plant. *Renewable and Sustainable Energy Reviews* 53: 1468–1485.
23. Goutam, S., Omar, N., Van Den Bossche, P., and Van Mierlo, J. (2017). Review of nanotechnology for anode materials in batteries. In: *Emerging Nanotechnologies in Rechargeable Energy Storage Systems* (ed. L.M. Rodriguez-Martinez and N. Omar), 45–82. Elsevier.
24. Thomas, B.S., Yang, J., Mo, K.H. et al. (2021). Biomass ashes from agricultural wastes as supplementary cementitious materials or aggregate replacement in cement/geopolymer concrete: a comprehensive review. *Journal of Building Engineering* 40: 1–12.

2

Bio-based Carbon Materials for Catalysis

Chaiyan Chaiya and Sasiradee Jantasee

Department of Chemical Engineering and Materials, Rajamangala University of Technology Thanyaburi, Pathum Thani, Thailand

2.1 Introduction

Developments of inexpensive heterogeneous catalysts to replace conventional homogeneous catalysts for extensive applications are currently investigated by many research studies. Inorganic materials such as silica (SiO_2), alumina (Al_2O_3), zirconia (ZrO_2), titania (TiO_2), magnesium oxide (MgO), zeolite, and clays have received much attention for their use in catalysis [1]. Carbon-based catalysts are among the promising heterogeneous catalysts used in a wide range of applications, which can be derived from both inorganic and organic resources. These materials have high stability in several reaction media and highly versatile chemical and physical properties. A great number of attempts have been made to synthesize carbon-based materials from various raw biomasses and biomass-derived compounds such as agricultural products and residues, sugar derivatives, and non-lignocellulosic materials [2, 3]. The bio-based carbon materials have been applied as a catalyst or as catalyst support in processes involving biorefinery, hydrogenation, bio-oil upgrading, and biomass conversions into chemicals, to name a few [4–9]. Benefits of these catalysts include their high stability and simple preparation from a low-cost carbon source. Carbon-based catalysts can be formed in diverse physical structures and their chemical properties can be easily tailored by modification and functionalization processes. The bio-based carbon materials synthesized from biomass mostly present amorphous structures and several chemical functional groups depending on parameters such as the composition of the biomass resource, biomass conversion process and conditions, and modification technique.

High-Performance Materials from Bio-based Feedstocks, First Edition. Edited by Andrew J. Hunt, Nontipa Supanchaiyamat, Kaewta Jetsrisuparb and Jesper T.N. Knijnenburg.

The effects of these parameters on the bio-based carbon properties are reviewed in this chapter for their utilization in catalysis. This chapter also provides an overview of catalysis applications of the carbon-based materials exclusively originated from biomass resources, especially the most commonly used bio-based carbon materials that show superior performance in Figure 2.1.

2.2 Biomass Resources for Carbon Materials

Carbon materials are one of many constituent substances on Earth. Their structures depend on the raw materials, processes, and preparation conditions. The biomass resources that are transformed into carbon materials for their use in catalytic applications are discussed in this section.

Biomass is a natural material grown by solar energy and comprises the world's natural resources such as plants, animals, vegetables, and animal waste or organic waste. Biomass can be classified into two main types as illustrated in Figure 2.2. First, lignocellulosic materials are found in wood from natural forests and agricultural residues. The chemical compositions of general lignocellulosic materials are displayed in Table 2.1. These structures are primarily derived from carbon (C), hydrogen (H), and oxygen (O) atoms. The proper thermal processes can convert lignocellulose into carbon-based materials that possess strong structures, which are suitable as a catalyst or catalyst support. Lignin is the hardest fraction in biomass and requires a high temperature for destruction. Second, non-lignocellulosic materials, which can be obtained from algae, animal manure, sewage sludge, and others, contain a distinct proportion of proteins and lipids instead of lignocellulose. The non-lignocellulosic carbon material generally has low strength and a nonrigid structure. According to the literature review, carbon-based materials can be derived from several methods depending on the applications. In catalysis, the carbon material can play two roles, namely as a catalyst and as catalyst support. The required strength and rigidity under various reaction conditions mean that such carbon materials are typically obtained from lignocellulosic biomass resources rather than non-lignocellulosic ones.

2.2.1 Wood from Natural Forests

Many types of wood can be exploited as a carbon source since their major components are carbon-rich cellulose, hemicellulose, and lignin. Cellulose is a basic component of wood cell walls consisting of a long chain linear polysaccharide of glucose. Hemicellulose, on the other hand, is a lower-molecular-weight polysaccharide, and its structure is weaker than cellulose. Finally, lignin, which acts as a binder of wood cells, is a highly branched polyphenolic molecule with a complex structure and high molecular weight resulting in the relatively high hardness and rigidity of wood. The ratio of the major components in wood depends on the geographical location and growth conditions. Red oak or *Quercus rubra* was analyzed by decomposition using thermogravimetric analysis (TGA) in the absence of oxygen. The weight loss started at 200 °C because of hemicellulose decomposition, and the weight loss increased again at 290 °C due to the decomposition of cellulose. The lignin decomposed at a maximum temperature of 360 °C [11]. A suitable temperature to convert lignocellulosic materials into a biochar or activated carbon should therefore be more than 360 °C. To confirm this, when birchwood was carbonized at 280 °C, the cellulose fraction was close to zero, while the lignin fraction still appeared [12]. The cellulose content should thus be considered when selecting lignocellulosic materials for carbon material production.

The major components of several wood types are presented in Table 2.2. Some wood types that have a low cellulosic percentage conversely have high lignin content, e.g. Japanese cedar and eucalyptus. Although each type of wood displays differences in the amount of cellulose

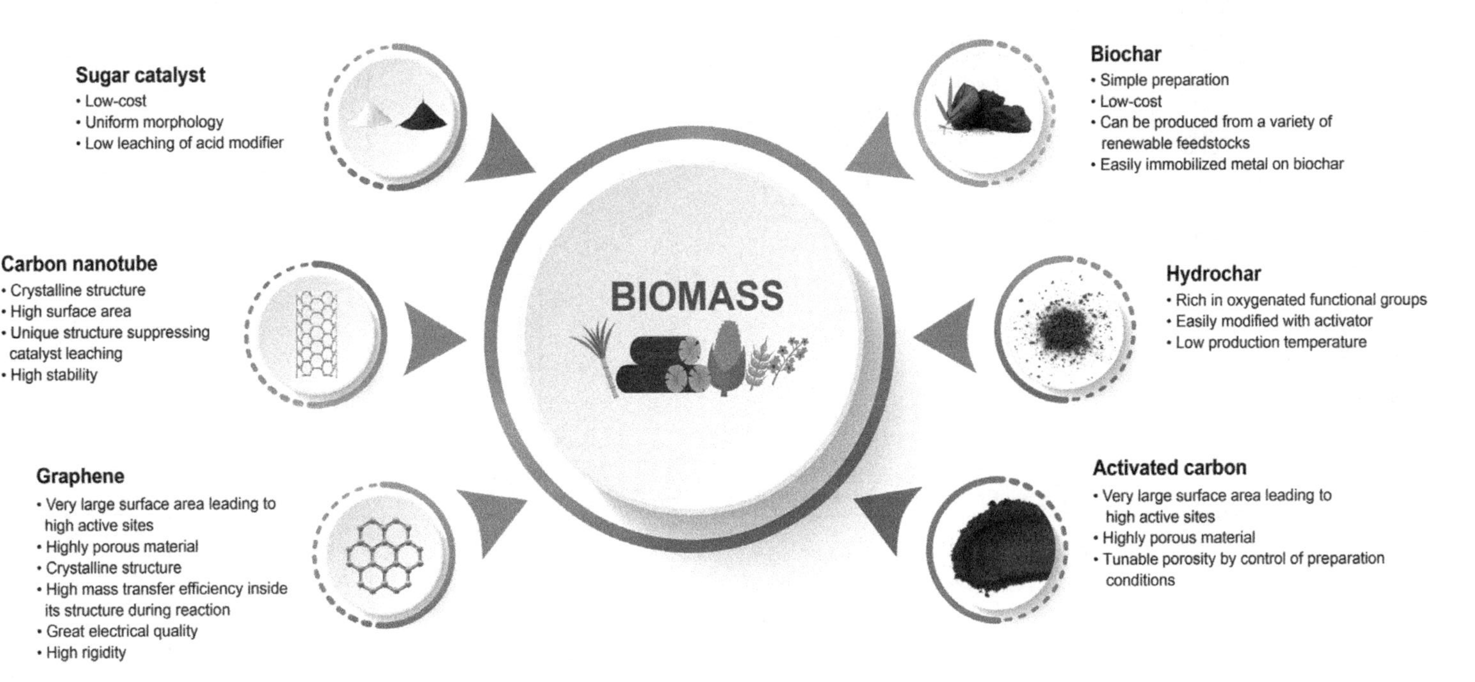

Figure 2.1 Some advantages of bio-based carbon materials.

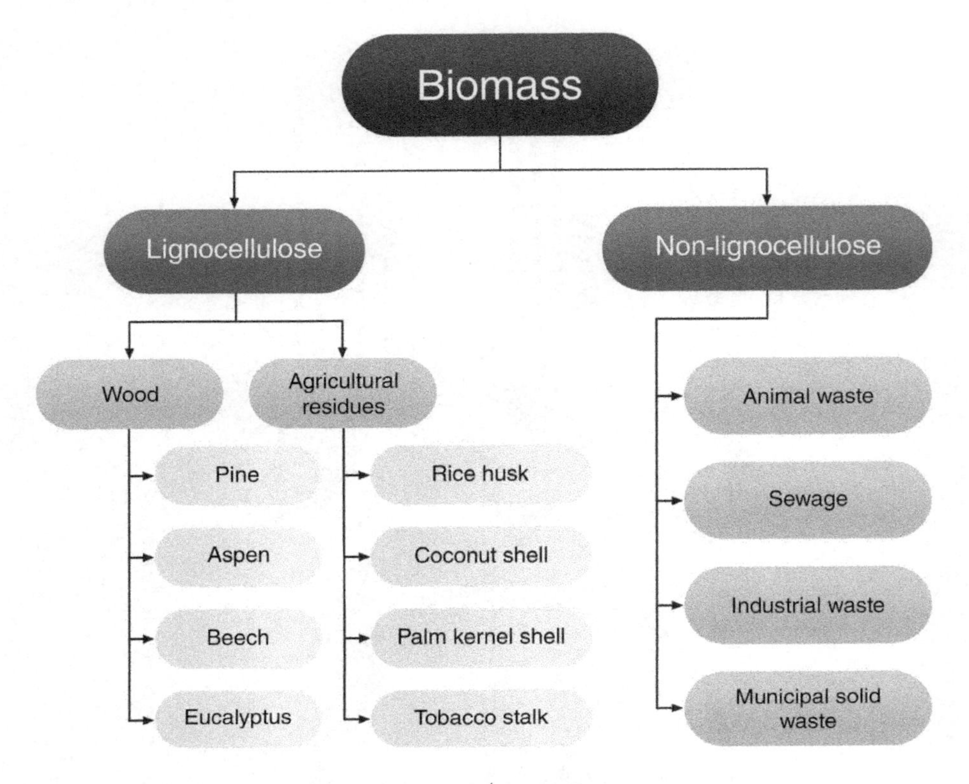

Figure 2.2 *Classification of biomass.*

Table 2.1 *Weight ratio of chemical compositions in lignocellulosic materials Based on [10].*

Chemical compositions	Content (wt%)
Cellulose $(C_6H_{10}O_5)_n$	25–50
Hemicellulose	15–40
Lignin	10–40
Extractives and minerals	1–15

Table 2.2 *Chemical components in various wood categories [13–18].*

	Contents (%)			
Feedstock	Cellulose	Hemicellulose	Lignin	Others
Pine	46.90	20.30	27.30	5.50
Fir	45.00	22.00	30.00	3.00
Aspen	47.14	19.64	22.11	11.11
Beech	46.40	31.70	21.90	0.00
Japanese cedar	38.60	23.10	33.80	4.50
Eucalyptus	34.70	27.30	35.80	2.20

and lignin, the temperatures for their conversion into carbon-based materials are similar when the combined cellulose and lignin content is more than 70%. The temperature for wood transformation was evaluated from the total amount of cellulose and lignin, which should be in excess of 360 °C. Pragmatically, the thermal conversion of wood usually operates at much higher than 360 °C to ensure that the volatile matter, which is an unwanted matter in this process, is removed from the feedstock structure. The volatile matter that is generated during carbonization usually consists of condensable (i.e. water vapor and light hydrocarbon) and non-condensable gases (i.e. nitrogen, hydrogen, and carbon monoxide). When eucalyptus was transformed by flash pyrolysis at 400 °C, the weight loss was only 9% while the resulting surface area was 37.52 m^2 g^{-1}. The majority of components were still present in the wood structure due to its incomplete carbonization [19]. Until the temperature reached 600 °C, the weight loss increased to 30%, and the surface area increased up to 367.25 m^2 g^{-1}. This indicated that the decomposition of the lignocellulosic structure was complete at 600 °C. Thus, the temperature for wood decomposition should be higher than the experimental results. The temperature widely used for wood decomposition is higher than 500 °C [20–22]. The ultimate analysis of wood shows the same range of carbon content of 40–50%, 5–7% of hydrogen, and 1% of nitrogen and oxygen [22, 23], which implies that the amount of each element inside the material is not directly related to the composition of lignocellulosic materials. On the other hand, proximate analysis can give more interesting properties such as fixed carbon, volatile matter, ash, and moisture content. The fixed carbon is the carbon structure that remains as a residue after all the volatile matter has been removed, and is essential to evaluate the suitability of a feedstock for the production of carbon material. Fir consisted of high cellulose of 45.00% and carbon content of 47.20%, but it contained only 10.50% fixed carbon [24]. This means that after undergoing the same thermal process, the carbon product yield from Eastern red cedar was higher than that from fir trees. Moreover, the carbon product yield can be evaluated from the volatile matter content since these are inversely related. The moisture content should be considered as well because it can assess the energy of the thermal process. Fixed carbon and moisture content of Japanese cedar were 17.60% and 15.50%, respectively, while those of eucalyptus were 18.82% and 9.40%, respectively. Seeing that the carbon product yield obtained from eucalyptus and Japanese cedar was almost the same, the thermal conversion of eucalyptus required less energy [14, 25].

2.2.2 Agricultural Residues

Agricultural residues are cellulosic materials that are discarded from agricultural or agro-industrial processes. Much literature has reported the production of carbon-based materials from several kinds of agricultural residues, of which rice husks and rice straw are the most extensively used [26, 27]. Rice husk contains 37.00% cellulose, 23.43% hemicellulose, and 24.77% lignin, while rice straw contains 34.53% cellulose, 18.42% hemicellulose, and 20.22% lignin [25]. The rice husks and rice straw had a lower hardness than wood because the total amounts of cellulose and lignin present in rice husks and rice straw were lower than 70%. After thermal treatment, the carbon material obtained from rice husks and rice straw were powders with low hardness [28]. Additionally, the rice husks and rice straw still had a substantial total carbon and fixed carbon content. When comparing rice husks to rice straw, the carbon content was 41.92% and 36.07%, with a fixed carbon content of 11.44% and 6.93%, respectively [29]. The rice husks were probably a more suitable feedstock for the production of carbonaceous material.

Many works also refer to the potential of biomass stones and shells for the production of carbonaceous materials [30, 31]. Olive stone was converted into a bio-based carbon material, with a carbon content of 47.50% and fixed carbon of 13.80% [32]. Another effective feedstock, peanut shell, possessed a high carbon content of 52.73% and fixed carbon of 22.60% [33]. Carbonized peanut shells at 500 °C resulted in a biochar with little ash and volatile matter. Cocopeat presented a high carbon content of 61.57% and fixed carbon of 25.30% [34]. The cocopeat that was pyrolyzed at 500 °C generated a biochar containing up to 84.44% carbon content with up to 67.25% fixed carbon. However, it contained volatile matter of about 14.30%, volatile matter of about 14.30%, moisture 2.55% and ash 15.90%. Only temperatures over 800 °C can produce pure fixed carbon without any volatile matter. In European countries, a great deal of research is done on the utilization of walnut shells (45.10% carbon and 15.90% fixed carbon) and almond shells (50.50% total carbon and 9.10% fixed carbon) for the production of hard biochars with high yield [35]. In Southeast Asia, palm kernel shells (50.49% carbon) and coconut shells (49.20% carbon), which are local plants, have been converted into carbonaceous materials in the industry [28].

The thermochemical conversion of petioles, leaves, branches, leaflets, and stalk from various plants has also been studied. These feedstocks have a moderate carbon content, low lignin, and fixed carbon that affect the hardness and yield of biochar [36]. Although both date palm leaflets and tobacco stalks (the biomass waste after tobacco harvesting) had a carbon content of up to 50%, the produced biochars were rather fragile [37, 38]. These feedstocks may thus not be applicable to produce carbon material for catalysis, but can be used as additives for soil improvement. Camphor leaves and camellia leaves contained a carbon content of over 70% but they produced a brittle biochar owing to the small amount of lignin and fixed carbon. Such materials are usually applied as an adsorbent instead of a catalyst or catalyst support. Nevertheless, the empty fruit bunch of oil palm containing only 47.78% carbon produced a biochar with high hardness [39].

2.3 Thermochemical Conversion Processes

Lignocellulosic biomass can be chemically transformed into carbon materials by diverse thermochemical processes as demonstrated in Figure 2.3. Each process presents different influences on the properties of the obtained carbon materials, both chemically and physically. The selection of the conversion process mainly depends on the application.

2.3.1 Carbonization and Pyrolysis

Carbonization is a thermal process of biomass conversion in the absence of oxygen with charcoal or biochar as the main product. Similarly, the pyrolysis process results in a condensed liquid product from the gas called bio-oil, as well as biochar as a by-product. The equipment used in carbonization and pyrolysis of biomass are seemingly identical; both processes essentially require a furnace for heating and creating an oxygen-free environment [33]. Some researchers carried out these processes in a horizontal tube furnace under nitrogen flow to remove oxygen from their surroundings [37]. Nevertheless, the pyrolysis process needs additional condensation units to compress the produced gas into the bio-oil [40]. Both carbonization and pyrolysis processes require an appropriate temperature. As discussed in Section 2.2.1, the destruction of the lignocellulosic structure occurs at around 360 °C. However, many researchers usually operate the carbonization process of biomass at

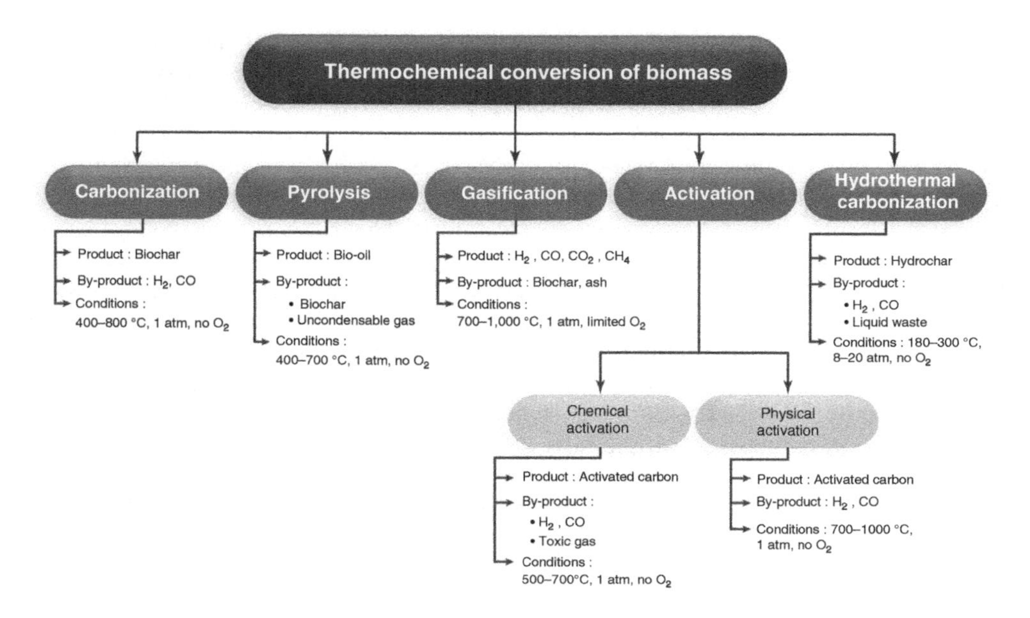

Figure 2.3 *Thermochemical processes for biomass conversion.*

temperatures above 500 °C under atmospheric pressure and oxygen-free conditions in order to achieve the complete conversion of lignocellulosic material [33, 35]. Nitrogen may be supplied into the processes, where the flow rate is typically around 100–300 cm^3 min^{-1} depending on the process scale. From the fixed carbon analysis, highly crystalline biochar without any impurities can be produced at a temperature higher than 800 °C.

Considering the cellulose conversion steps, cellulose is first transformed into anhydrocellulose. Subsequently, the charcoal or biochar is formed. The overall chemical reaction of the cellulose transformation is shown in Eq. (2.1).

$$C_6H_{10}O_5 \rightarrow 3.74C + 2.65H_2O + 1.17CO_2 + 1.08CH_4 \quad (2.1)$$

From the preceding equation, the number of carbon atoms can vary according to the reaction conditions such as temperature and pressure. For instance, the use of a reaction temperature of 400 °C, which is slightly higher than the theoretically required 360 °C, presents the incomplete decomposition of cellulose through Eq. (2.2).

$$8C_6H_{10}O_5 \rightarrow C_{30}H_{18}O_4 + 23H_2O + 4CO_2 + 2CO + C_{12}H_{16}O_3 \quad (2.2)$$

Previous research showed that with the aim to produce a biochar with high porosity and surface area, the process should be operated at temperatures in the range of 600–800 °C [41]. A type of pinewood, *Pinus sylvestris*, was pyrolyzed at various reaction temperatures, and increasing temperature enhanced the porosity and surface area of the obtained carbonaceous product. Pyrolysis at 500 °C generated a biochar with a surface area of 19 m^2 g^{-1}, while increasing the temperature to 600 and 700 °C greatly improved the surface area to 254 and 470 m^2 g^{-1}, respectively. Conversely, the pyrolysis of non-lignocellulosic material at temperatures above 700 °C resulted in a decrease in surface area due to the collapse of the biochar structure [42]. The change in pore diameter depends on several parameters, e.g. type of raw material and heating rate. An increase in carbonization temperature strongly enhanced the number of pores but it slightly affected the pore diameter. However, with the biomass having low lignin content or low hardness, increasing the carbonization temperature could significantly increase the pore diameter of biochar. This clearly shows that the carbonization and pyrolysis processes benefit the porosity of biochar, which is an interesting characteristic for catalysis application. However, the production of biochar via these thermochemical processes has some disadvantages as it reduces the amount of valuable chemical groups. Biomass commonly consists of oxygen-containing functional groups with carbon–oxygen bonds (C–O), carbonyl groups (C=O), and hydroxyl groups (–OH), which are useful in many catalytic processes. An increase in carbonization temperature led to a decrease in such chemical functionalities [43]. For this reason, the physical and chemical properties of the biochar should be modified prior to its use in specific applications including catalysis. For example, mixtures of rapeseed oil cake and walnut shells were carbonized at 400 and 750 °C. Only a few functional groups appeared on the surface of the produced biochar, and either chemical (using agents such as ammonium and monoethanolamide) or thermal treatment were required before use in CO_2 capture [44].

2.3.2 Activation

The activation process is an oxidation reaction at elevated temperatures. The objective of the activation process is to enhance the surface area, pore size, and pore volume of the bio-based carbon material during carbonization, as the highly porous structure of carbon material is advantageous in several applications such as adsorption and catalysis. In order to clarify the distinction between carbonization and activation, the latter process requires both an activating agent and heat. Activation is performed under an oxygen-free or oxygen-poor atmosphere depending on the type of activating agent. The product obtained from the activation process is a so-called activated carbon, which is a highly porous carbon-rich material. Lignocellulosic materials are promising raw materials for the production of activated carbons since they can provide a great yield of activated carbon together with high porosity and hardness compared to non-lignocellulosic materials. Previous research showed that activated carbon prepared from lignocellulosic biomass had a surface area in the range of 300–2700 $m^2\ g^{-1}$ and various pore sizes [21, 45]. Conventionally, there are two routes in the activation process. First, biomass is carbonized to reduce moisture, volatile matter, and contaminants, and the obtained biochar is subsequently activated with an oxidizing agent and carbonized again. This route can be called a two-step activation process [46]. A different route is where the biomass is directly activated during the carbonization step, which is a one-step activation process [28, 39]. The activation method is typically classified into two types depending upon the kind of activating agent employed, and these are chemical and physical activation.

2.3.2.1 Chemical Activation

The utilization of agricultural material as a raw material for the production of activated carbon is usually carried out by chemical activation. The proper temperature for chemical activation should be 400–600 °C. As the chemical activating agent should penetrate and erode the cellulose and lignin and destroy their structures during carbonization, the property of the activating agent should be strongly corrosive. Not only strong acidic and basic agents but also other strong oxidizing agents have been exploited. The most widely used chemical activating agents are potassium hydroxide (KOH), phosphoric acid (H_3PO_4) and zinc chloride ($ZnCl_2$) [45, 47], but chemical agents such as sodium chloride (NaCl), sodium hydroxide (NaOH), hydrogen peroxide (H_2O_2), and potassium carbonate (K_2CO_3) have also been used [48–50]. The chemical activating agents in an aqueous form are typically applied onto the raw materials via impregnation for a desired period of time. The duration of the chemical activation is often around 60–120 minutes, and the ratio of chemical activating agent and raw material is usually between 1 : 1 and 5 : 1. The reaction pathways when using KOH as an activating agent are presented in Eqs. (2.3)–(2.6) as follows [26]:

$$6KOH + 2C \rightarrow 2K + 2K_2CO_3 + 3H_2 \tag{2.3}$$

$$K_2CO_3 + 2C \rightarrow 2K + 3CO \tag{2.4}$$

$$K_2CO_3 \rightarrow K_2O + CO_2 \tag{2.5}$$

$$K_2O + C \rightarrow 2K + CO \tag{2.6}$$

The activated carbon produced from soybean pods activated by KOH showed the highest surface area of 2245 $m^2\ g^{-1}$ [47]. An interesting shortcut in the reaction pathways was taken by employing K_2CO_3 as an activating agent instead of KOH [48]. The K_2CO_3 is an initial substance in Eq. (2.4), thus Eq. (2.3) could be eliminated. The activated carbon produced by using K_2CO_3 also showed an extremely high surface area of 2613 $m^2\ g^{-1}$ along with a high pore volume of 1.66 $cm^3\ g^{-1}$. However, the KOH showed a higher activation potential than the K_2CO_3 at low carbonization temperatures with regard to the porosity and surface area of the resulting activated carbon [51], which was ascribed to the difference in melting point. The K_2CO_3 has a higher melting temperature of 890 °C while the melting point of KOH is 360 °C. When using NaOH as an activating agent, the reaction pathways are presented in Eqs. (2.7) and (2.8) [52].

$$6NaOH + 2C \rightarrow 2Na + 2Na_2CO_3 + 3H_2 \tag{2.7}$$

$$4NaOH + C \rightarrow 4Na + CO_2 + 2H_2O \tag{2.8}$$

The reaction steps in heterogeneous catalysis will be discussed in detail in Section 2.4. Since most of the reactant adsorption onto the catalyst surface takes place within the micropores of activated carbon, the number and the dimension of micropores should be carefully considered. Macroporous and mesoporous structures of activated carbon provide the passageway of the reactant to the micropores, hence they impact the overall activity of a heterogeneous catalyst as well. Activated carbon obtained from the chemical activation process regularly presents a microporous or sub-mesoporous structure with pore diameters in the range of 1–3 nm. Pretreatment of the raw material before activation showed a surprising result. The palm shell was soaked in 5 H_2SO_4 for 24 hours and the excess H_2SO_4 was removed by washing with deionized water. Finally, the treated palm shell was activated with H_3PO_4 at 650 °C for two hours. The pore diameter of the obtained activated carbon was around 20 nm, which corresponds to a mesoporous structure. Furthermore, the proportion of each pore size can be controlled by the combination of several activating agents in one process [50]. Activation with

NaCl resulted in activated carbon with a macroporous structure while NaOH provided micropores. The combination of NaCl and NaOH in the appropriate ratio thus enables control of the pore diameter.

Apart from the pore size and structure, also the chemical properties of activated carbon can be modified by chemical activation. The functional groups of activated carbon, especially oxygen-containing groups, play an important role in catalysis. A variety of chemical activating agents affect the chemical properties of activated carbon. For example, NaOH activation of durian shell produced an activated carbon that contained a large amount of oxygen-containing groups such as OH, C=O (ketone, aldehyde, lactones, and carboxyl), and C–O (anhydrides) [53]. In contrast, the oxygen-containing functional groups disappeared from the surface of the activated carbon produced from *Euphorbia rigida*, which is an oil-rich biomass [54]. Although $ZnCl_2$, K_2CO_3, NaOH, and H_3PO_4 were used as activating agents, the produced activated carbon only contained hydrophobic groups (C–H, C–C, and C=C) [48]. As the hydrocarbon groups in the oil-rich biomass could be cracked during the carbonization process, it generated more aromaticity of the activated carbon. Therefore, not only the type of chemical activating agent but also the category of biomass feedstock influence the chemical properties of activated carbon.

2.3.2.2 Physical Activation

For the development of the specific surface area and porous structure of activated carbon, the physical activation process is a favorable approach. Physical activation is achieved by carbonization in the presence of oxidizing gases such as steam, carbon dioxide (CO_2), or a mixture of these. This process resembles a partial gasification process which is generally carried out at temperatures between 700 and 1000 °C at atmospheric pressure. The furnace for the physical activation process is constructed with a gas flow unit, and a horizontal tube furnace or a fluidization system in a vertical tube are frequently selected [55, 56]. The oxygen in the activator gas molecule reacts with the carbon atom of biomass resulting in the generation of carbon monoxide (CO). The precise reaction pathways depend on the type of oxidizing gas and the activation temperature. Physical activation by steam and CO_2 during carbonization occurs through the endothermic reactions as shown in Eqs. (2.9) and (2.10).

$$C + H_2O \rightarrow CO + H_2 \quad -29\,\text{kcal mol}^{-1} \tag{2.9}$$

$$C + CO_2 \rightarrow 2CO \quad -39\,\text{kcal mol}^{-1} \tag{2.10}$$

As previously mentioned, the physical activation is similar to partial gasification because the oxygen atoms of the oxidizing gas molecules can react with the carbon atom inside the biochar during activation. From the steam activation in Eq. (2.9), the carbon atom reacts with water molecule yielding CO, which can increase the porosity and surface area of activated carbon. Other prominent points of physical activation are to provide less contamination of activated carbon than chemical activation [57]. Physical activation does not require any neutralization procedure for the removal of the oxidizing agent. So, this process is more environmentally friendly. However, the use of fresh biomass or cellulosic material tends to produce ash during the physical activation process, which results in low activated carbon yields [3].

The activation conditions are one of the important factors considered in promoting a high surface area and high product yield. The proper carbonization period is more influential on the final properties of activated carbon than the activation period. In the work of Rezma et al. [36], the biomass was first carbonized at 1000 °C and the resulting product was activated by CO_2 at 750–900 °C for 30 minutes. The activated carbon presented a very low surface area and product yield [36]. As shown by Grima-Olmedo et al. [19], the surface area of carbonized eucalyptus was improved when the temperature was raised from 600 to 800 °C. The addition of CO_2 during carbonization strongly enhanced the surface area of the produced activated carbon [19].

Both micro- and mesopores can be generated through physical activation. For example, the micropores inside carbonized rubber wood sawdust could be changed into mesopores via steam activation, resulting in a high surface area of 1134 $m^2\ g^{-1}$ [58]. The surface functional groups of physically activated carbons correspond to those produced by chemical activation [39]. The factors that affect the physical and chemical properties of activated carbon can be prioritized as: (i) carbonization and activation temperatures, (ii) type of feedstock, (iii) particle size of feedstock, (iv) heating rate, (v) gas flow rate, and (vi) activation time [28]. This information is significant for the design of the physical activation process.

2.3.3 Hydrothermal Carbonization

Hydrothermal carbonization (HTC) is an environmentally friendly thermochemical process for biomass conversion. This process operates at temperatures of 180–300 °C under the pressure of 8–20 bar in water [59]. The thermal processing at high pressure in water destructs the biomass structure and the resulting carbon material is called a hydrochar. The HTC is proposed as a productive process for fine biomass particles because it presents a higher yield of the bio-based carbon product compared to regular carbonization [60]. Moreover, carbonization under high vapor pressure of water can produce hydrochar with uniform particles.

According to the hydrothermal reaction of the lignocellulosic material, the cellulose and hemicellulose are hydrolyzed to obtain CO_2 as a by-product according to Eq. (2.11).

$$C_6H_{10}O_5 \rightarrow C_{5.25}H_4O_{0.5} + 3H_2O + 0.75CO_2 \tag{2.11}$$

Carbon atoms do not appear in Eq. (2.11), which differs from the ordinary carbonization process shown in Eq. (2.1), as the transformation of biomass into hydrochar is essentially associated with three reactions: hydrolysis, dehydration, and decarboxylation [61]. However, the operating conditions of the HTC process are not able to destroy the lignin structure. Fourier-transform infrared (FTIR) spectra showed the amount of lignin that was still present in the carbon material produced from the HTC process [62]. The HTC process is thus not appropriate for the conversion of lignin-rich biomasses. Interestingly, the hydrothermal conversion of glucose ($C_6H_{12}O_6$) or other monosaccharides as seen in Eq. (2.12) could directly affect the carbon yield at 200 °C and 20 bar [27].

$$C_6H_{12}O_6 \rightarrow 6C + 6H_2O \tag{2.12}$$

Even though hydrochars usually have a lower surface area than biochars, hydrochars possess a 1–4 times higher oxygen content than the carbon materials derived from carbonization and pyrolysis processes [33]. The hydrochars prepared from pinewood, peanut shells, or bamboo showed several oxygen-containing groups such as C–O, –OH, and C=O, which were absent in the biochar [30]. The superior oxygen content in hydrochar makes it advantageous for many industrial sections such as catalysts or catalyst support. Nevertheless, an increase in hydrothermal temperature decreased the amount of oxygen and, conversely, the carbon content increased [11, 38]. Hydrochar is frequently used as a starting material for the production of activated carbon. The oxidizing agent can react with the oxygenated groups on the hydrochar surface resulting in its excellent catalytic behavior [63]. Moreover, the surface of hydrochar could be modified for specific applications. For example, hydrochar treated with sulfonic groups was effective in the production of biofuels [64].

2.3.4 Graphene Preparation from Biomass

Graphite is an infinite three-dimensional (3D) material comprising several stacked layers anchored by weak Van der Waals forces. Each stacked layer is a two-dimensional (2D) carbon sheet structured in a hexagonal honeycomb lattice called graphene [65, 66]. Graphene was discovered in 2004 and presents astonishing properties. Its appearance is like a soft film, and it has a high Young's modulus and excellent thermal and electrical properties. Previous research reveals that graphene has a very high surface area of over 2000 $m^2\ g^{-1}$ and can be easily chemically modified [67]. Within the past decade, numerous investigations have used graphene in various applications along with exploring the practical synthesis procedures of graphene from biomass [68, 69]. Graphene can be synthesized through several methods including micromechanical exfoliation of graphite, chemical vapor deposition (CVD) of graphite, epitaxial graphene growth on silicon carbide, chemical graphitization of graphene oxide, and carbonization of biomass and waste materials [70–74]. Micromechanical exfoliation and CVD of graphite produced good-quality graphene. The transformation of biomass into graphene can be accomplished via the carbonization of biomass under specific conditions. For example, pyrolysis of camphor leaves at an ultra-high temperature of 1200 °C under nitrogen or argon atmosphere produced a biochar, which was subsequently reacted with D-tyrosine as a bio-based dispersion agent under sonication to obtain graphene [75]. The graphene derived from biomass contained many impurities, whereas the graphene synthesized from graphite is usually purer. To avoid contamination, some researchers used pure lignin to produce fine graphene through a stepwise pyrolysis process [76]. The biomass type may affect the properties of graphene as well. Populus wood was mixed with KOH before carbonization under nitrogen. Surprisingly, the graphene-like carbon material was found on the structure of the obtained biochar [20]. Sucrose, xylitol, and glucose were also used as starting materials for graphene synthesis. The carbonization of sugar mixed with $FeCl_3$ generated graphene-like carbon materials, which presented excellent catalytic activity in the hydrogenation of nitrobenzene [77]. Even though graphene is one of the beneficial carbon materials, the production of graphene from biomass without any impurities remains a challenge.

To meet the requirement to be a heterogeneous catalyst or support, the physicochemical properties and thermal stability of the material are crucial as these dictate the catalytic efficiency. As previously described, bio-based carbon materials have predominant characteristics that benefit their application as a direct catalyst or catalyst support. Biomass-derived porous carbon materials with high porosity provide a greater number of active sites. Moreover, the pore size and specific surface area can be tailored via the process parameters (temperature, time, heating rate, type of biomass, activating agent, etc.). Several forms of bio-based carbon materials including granules, pellets, tubes, and sheets can be produced. A variety of natural oxygenated surface functional groups in bio-based carbon materials is useful for metal adsorption when used as a support. These functional groups can also enhance reactant adsorption in heterogeneous catalysis. Some bio-based carbon materials are cheaper than commercial catalysts and supports, e.g. zeolite, silica, and alumina. Before going into the details on the applications of promising bio-based carbon materials in a field of catalysis, some of their properties are summarized in Table 2.3.

Table 2.3 Comparison of typical physical and chemical properties of bio-based carbon materials [30, 39, 43, 48].

Bio-based carbon materials	Surface area (m^2 g^{-1})	Pore volume (cm^3 g^{-1})	Functional groups
Biochar	100–500	0.03–0.25	–COOH, –OH, C–H, C=C, C=O
Hydrochar	1–43	0.007–0.2	–OH, C=C, C=O, C–C, $-CH_2$, $-CH_3$
Activated carbon (chemical treatment)	400–2700	0.3–1.53	–OH, C–H, C=C, C–O
Activated carbon (physical treatment)	300–1000	0.1–0.99	–OH, C=O, C–O
Graphene	800–1800	0.3–0.9	C–H, C–OH, C=C
Sugar-derived carbon	1–5	—	—
Carbon nanotube	300–600	2.5	–COOH, –OH

2.4 Fundamentals of Heterogeneous Catalysis

Over 90% of chemical industries (e.g. petroleum, renewable energy, fine chemicals, polymers, and food and pharmaceutical industries) are associated with catalytic processes. In 2019, approximately 72% of industrial catalytic processes used heterogeneous (solid) catalysts [1]. In order to develop more industrially friendly catalysts, many investigations have advanced the use of different bio-based carbon materials for a wide range of catalytic applications. The studied bio-based carbon catalysts and supports initially started with biochar, activated carbon, carbon nanotubes (CNTs), mesoporous carbons, and sugar catalysts, but have now been extended to include graphene and its derivatives. The crucial target for these developments is the upgrading of bio-based carbon materials into direct catalysts or as catalyst supports to maximize catalytic efficiency. A simple process for the preparation of bio-based carbon materials as well as achieving high conversion and highly desired product yield under mild reaction conditions are aimed for.

An understanding of the mechanisms of heterogeneous catalysis could address the appropriate characteristics of bio-based carbonaceous catalysts. Generally, a heterogeneous catalytic reaction takes place through the following steps: (1) dispersion of the substrate from the bulk fluid to the pore entrance on the external catalyst surface; (2) diffusion of the substrate from the pore entrance into the internal catalyst pore; (3) adsorption of the substrate on the active catalyst site; (4) reaction of the substrate on the active site to generate a product; (5) desorption of the product from the active site; (6) diffusion of the product from the internal catalyst pore to the external surface of the catalyst; and (7) dispersion of the product from the external surface of catalyst into the bulk fluid [78].

The adsorption of the reactant on the active sites dispersed throughout the catalyst pores is the most important reaction step in heterogeneous catalysis. In case there is more than one reactant, competition in reactant adsorption on the active sites may occur. The concentration of each reactant on the active sites of the solid catalyst should be appropriate. The catalytic activity of the catalyst is based on the proportion of the reactants adsorbed on the active sites, and for each individual reaction, the chemical properties and functional groups of the bio-based carbon catalyst should be designed to enable proper interactions with the adsorbed reactants. Diffusion steps 2 and 6 depend on the molecular size of the reactant and product as well as the pore morphology and catalyst particle size. Large catalyst particles diminish the reactant approaching the active sites inside the pores. So, the use of small catalyst particles can enhance the opportunity of a reactant to reach active positions within the pores. However, small catalyst particles cause a high pressure drop when used in a fixed-bed reactor. To overcome this, the morphology of bio-based carbon catalyst should be designed to have a higher external surface area such as a hollow cylindrical shape. Such information enables us to specifically control the catalyst characteristics and operating conditions of bio-based carbon synthesis for catalysis applications.

2.5 Catalysis Applications of Selected Bio-based Carbon Materials

An overview of high-performance bio-based carbon materials as catalysts and as carbon-supported catalysts in various reactions are discussed here to present state-of-the-art bio-based carbon materials in a wide range of catalysis applications.

Chemical reactions can be categorized into various groups, and some important types of chemical reactions in which the performance of biomass-derived carbon catalysts has been studied are exemplified in Figure 2.4. The details of each reaction will be presented based on the types of bio-based carbon materials in the following sections.

2.5.1 Biochar

Biochar is a carbonaceous solid product created by thermochemical conversion of biomass in an oxygen-free or oxygen-poor atmosphere by carbonization, pyrolysis and gasification. Among the various bio-based carbon materials, biochar has promising characteristics allowing it to be used as a heterogeneous catalyst and as catalyst support in numerous reactions. As discussed in Section 2.3.1, biochars have been gaining increased attention in catalysis applications due to their low cost, high porosity, stability, easy regeneration, and being more environmentally safe than other synthetic carbon materials. Biochars were reported to exhibit good catalytic performance in biodiesel production, steam reforming, pyrolysis, photocatalysis, bio-oil upgrading processes, and biomass conversions into fuels and chemicals [4, 6–9, 79].

Some outstanding physical and chemical features of an unmodified biochar enable its direct use as a catalyst. Although unmodified biochar received directly from the thermochemical

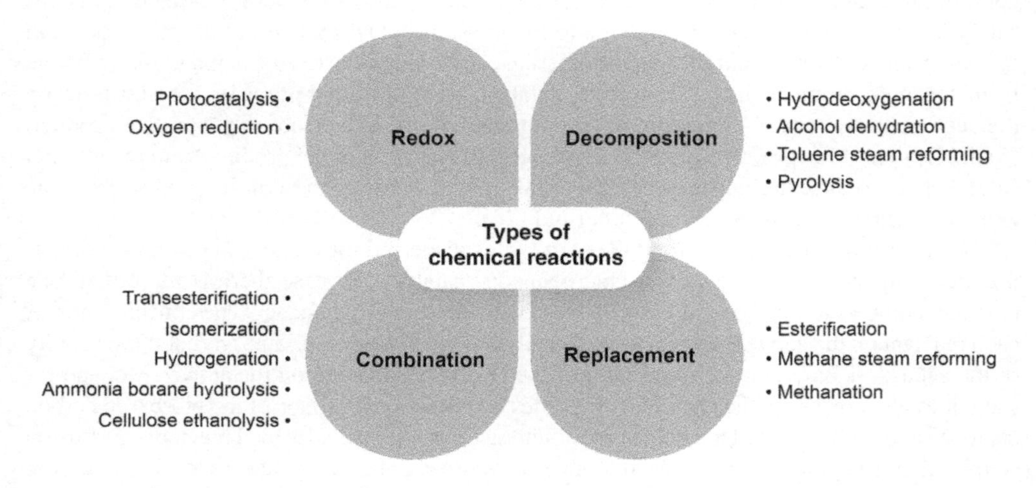

Figure 2.4 *Examples of chemical reactions catalyzed by biomass-derived carbons.*

conversion of biomass presents a modest surface area and porosity, it contains interesting surface functional groups and some mineral matters [80]. Functional groups such as C–O bonds as well as hydroxyl, carbonyl, carboxylic, and phenolic hydroxyl groups and also certain inorganic species significantly contribute to catalyst performance by promoting adsorption of reactants onto the active sites as well as the catalytic activity [81–83]. The catalytic performance of an unmodified biochar obtained from the gasification of poplar wood chips at 850 °C was investigated in the pyrolysis of low-density polyethylene (LDPE) and high-density polyethylene (HDPE) [8]. The main product in the gas phase from the catalytic pyrolysis of LDPE was propane instead of hydrogen since the surface functional groups on the unmodified biochar possibly reacted with the generated hydrogen radicals and hydrocarbon radicals. Besides, a great deal of wax was simultaneously produced. Addition of the unmodified biochar catalyst not only increased the gas yield from 18.3 to 25.4 wt% but also enhanced the light tar yield from 3.3 to 5.9 wt%, while the wax yield was decreased from 37.5 to 25.8 wt%. For the catalytic pyrolysis of HDPE, the gas yield was decreased from 20.1 to 15.7 wt% and the heavy tar yield also reduced from 35.5 to 29.5 wt%, but wax and light tar yields were increased in the presence of the catalyst. The unmodified biochar catalyst could not productively promote the cracking of heavy tar into light tar and gas products. However, it effectively promoted the polymerization during HDPE pyrolysis to form more wax with a combined ring structure. Surprisingly, the spent biochar catalyst in the pyrolysis of LDPE and HDPE had higher contents of hydroxyl groups and carbonyls in aldehydes since the oxygen species appeared to relocate from the reaction intermediates to the surface of the unmodified biochar catalyst. Crystallinity and minerals of the spent biochar catalyst were also enhanced compared to the fresh biochar. Considering biochar properties after the catalytic pyrolysis of polyolefin, the spent biochar catalyst may be applied as a soil amendment. Hence, it could economize on solid waste treatment cost. Vidal et al. revealed that the surface functional groups in the unmodified biochar had high thermal stability for the production of cyclic carbonates from CO_2 and epoxides [84]. TGA showed that the unmodified biochar had a low decomposition temperature of 250 °C, while the oxidized biochar with nitric acid (HNO_3) rapidly decomposed. The appearance of a lower decomposition temperature was attributed to the functional groups produced entirely on the surface of the unmodified biochar during its production. The oxidized biochar retained more residuals from the oxidation step, which were lost at high temperature. In addition, the carboxyl groups on the surface of the unmodified biochar could boost the polarization of the C–O bond resulting in catalyzing the epoxide ring opening followed by the formation of cyclic carbonates.

The existence of organic compounds in the renewable resource matrix, particularly animal wastes, brings about occurring minerals and inorganic alkalis in the structure of the produced biochar, such as K, Ca, Mg, N, P, and S [85–88]. These elements may be present in the form of chemical compounds such as $CaCO_3$, KCl, or $SiCl_4$ [8]. These minerals and inorganic alkalis can behave like a natural promoter of biochar activity in some catalyzed reactions. For example, the alkali and alkali earth metallic species could markedly promote the catalytic activity of biochar in tar reforming during biomass gasification [89]. An increasing biomass pyrolysis temperature resulted in enhanced fixed carbon and mineral contents in the produced biochar [83]. However, the number of surface functional groups within the biochar decreased with increasing carbonization temperatures because hetero-atomic functional groups containing such oxygen and nitrogen atoms were volatilized from the biomass structure [90]. The remaining elements in the biomass after the volatile compounds' detachment rearranged themselves to form more aromatic structures, causing a reduction in the number of active sites.

2.5.2 Modified Biochar

To develop the physicochemical properties and the number of active species on biochar toward a higher activity and maximum catalytic performance, a large number of modification techniques have been investigated including metal modifications as well as chemical and physical treatments.

2.5.2.1 Tar-reforming Processes

The production of syngas via biomass gasification has attracted a great deal of interest. However, this technology faces some challenges, the biggest one of which is excessive tar formation resulting in clogged up equipment. Consequently, the total cost of biomass-derived syngas production is increased making it difficult to develop this process into industrial manufacturing. To overcome this drawback, tar removal technologies have been intensively researched with regard to their economic and environmental impacts. The catalytic thermochemical conversion using biochar catalysts has been reported as a capable technique for tar reforming, but currently still shows inferior performance compared to conventional metal-supported catalysts. Particularly steam and CO_2-treated biochars are very powerful catalysts for tar reforming. The biochar prepared by pyrolysis from several biomass sources was activated with 15 vol% H_2O mixed with argon or with CO_2 at 800 °C for a short time. The treated biochar catalysts increased the catalytic activity in the steam or CO_2 reforming of tar compared to a regular biochar. Treatment of the biochar resulted in an increased surface area and pore volume (both microporous and mesoporous), and high content of oxygenated functional groups [91, 92]. The CO_2 generated more micropores in the biochar, whereas conversely, steam created more mesopores, which adds further importance for tar reforming. Even though micropores showed a greater initial tar conversion, these are rapidly deactivated due to coke deposition in the pores [4, 91, 93]. The steam-activated biochar had more oxygenated functional groups in the aromatic C–O forms, which are more active sites for tar reforming, than those treated by CO_2 [89, 92, 94]. Tar molecules were probably absorbed onto the unstable aromatic C–O structures on the biochar surface bringing about the transformation of tar into the gas products. When comparing the reforming gases, the treated biochar catalyzing the tar steam reforming was more effective than that in the CO_2 reforming [95]. The steam fed during the reforming reaction could produce additional oxygenated functional groups and aromatic C–O structures [92], which resulted in a gradual increase in catalytic performance. Therefore, steam was a promising agent for boosting and preserving the catalytic activity of biochar in tar reforming.

As discussed previously, raw biochar had slower rates of tar elimination than metal-supported catalysts. Doping an active metal such as Ni and Fe onto the biochar surface is another favored approach for the development of catalytic biochar [96]. Kastner et al. prepared biochar by pyrolysis of pine bark, which was then impregnated with Fe [97]. Compared to the regular biochar, the Fe-modified biochar catalyst possessed a lower surface area and pore volume as the Fe particles could hinder and shrink the biochar pores. Although the physical properties of the Fe-modified biochar catalyst were not remarkable, this catalyst showed that the tar decomposition rate was increased and the activation energy was decreased by 47%. The tar conversion was significantly altered by the metal loading and the tar decomposition increased with higher metal content on the biochar. At a decomposition temperature of 900 °C, the biochar modified with 13% Fe had a tar conversion of 100%. The addition of alkali along with the metal impregnation on biochar was developed in order to further enhance the catalytic activity. The raw biomass material was impregnated with potassium ferrate prior to carbonization at 900 °C for two hours to obtain a K–Fe bimetallic catalyst-supported biochar [98]. The K_2CO_3 was formed as an active site of this catalyst and showed a superior catalytic activity in cracking of biomass pyrolysis tar at a relatively low reaction temperature (600–700 °C). The active metal sites and pore structure were still maintained after the reaction. This catalyst also showed a high stability as no significant change in tar conversion was observed after five cycles.

2.5.2.2 Biodiesel Production Processes

Biodiesel, a fuel derived from renewable sources such as vegetable oils and animal fats, has received much attention due to the continuous reduction in petroleum reserves and environmental issues. Biodiesel production via transesterification (Figure 2.5), also known as alcoholysis, is currently the most attractive approach, and can be divided into non-catalytic, biocatalytic, and chemical catalytic processes. The non-catalytic process or supercritical alcohol process is carried out in conditions above critical temperature and pressure of the reaction mixture determined from the critical properties of alcohols and triglycerides. This process requires a relatively high reaction temperature of 230–450 °C and high pressure of 19–60 MPa, as well as excess methanol (molar ratio of oil to methanol approximately 1 : 40). These requirements are a limitation to the non-catalytic process for biodiesel production on an industrial scale [99]. A biocatalytic or enzymatic process produces biofuel with a low environmental impact and can be performed at mild temperature and pressure. Such processes are also not sensitive to the free fatty acid and water content in the feedstock [100, 101]. However, enzyme stability, enzyme reuse, and the high cost of enzyme immobilization are the major drawbacks of this process. In conventional biodiesel production, chemical catalysts (both acidic and basic) are usually used. Solid catalysts have been widely applied in biodiesel production owing to their ease of separation from products and excess reactants. A number of research studies have been conducted on biodiesel production using various types of acidic and basic solid catalysts. Most common is the solid base catalyst or alkali catalyst that can catalyze transesterification reaction in even milder reaction conditions and shorter reaction time than acidic solid catalysts. Calcium oxide (CaO) solid base catalyst can be derived from calcium carbonate-rich materials such as horn shell and eggshell. With the basic catalyst produced from calcined eggshell, the biodiesel yield reached 97% in transesterification of waste cooking oil and methanol at the ratio of 1 : 6 [102]. Nevertheless, the preparation of a solid base catalyst from calcium carbonate-rich feedstocks requires high temperatures of 800–900 °C. In addition, the free fatty acid and moisture contents in the oil feedstock should be considered for alkali catalysts. Water molecules in the feedstock can hydrolyze triglycerides into diglycerides and monoglycerides, which yield a greater amount of free fatty acids.

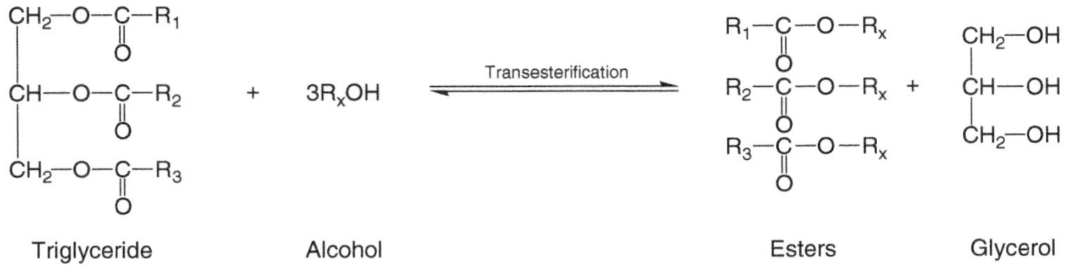

$$\underset{\text{Triglyceride}}{\begin{matrix} CH_2{-}O{-}C(=O){-}R_1 \\ CH{-}O{-}C(=O){-}R_2 \\ CH_2{-}O{-}C(=O){-}R_3 \end{matrix}} + \underset{\text{Alcohol}}{3R_xOH} \xrightleftharpoons{\text{Transesterification}} \underset{\text{Esters}}{\begin{matrix} R_1{-}C(=O){-}O{-}R_x \\ R_2{-}C(=O){-}O{-}R_x \\ R_3{-}C(=O){-}O{-}R_x \end{matrix}} + \underset{\text{Glycerol}}{\begin{matrix} CH_2{-}OH \\ CH{-}OH \\ CH_2{-}OH \end{matrix}}$$

Figure 2.5 *Transesterification of triglyceride.*

The alkali catalyst is able to convert free fatty acids into soap via the saponification side-reaction. The free fatty acid in the feedstock should be lower than 2% for transesterification.

For feedstocks containing high free fatty acid content, the use of an acid solid catalyst is an alternative to produce biodiesel via esterification reaction (Figure 2.6) [103]. Several solid acids such as metal oxides (particularly ZrO_2, TiO_2, and SnO_2), sulfonic acid-modified materials, and heteropolyacids have been used in esterification [104–106]. Considering the economic aspect of the biodiesel production process, the price of raw materials and the catalyst are two major determinants of the overall biodiesel production cost [107]. Therefore, much attention has been paid to the utilization of low-cost bio-based carbonaceous solids in biodiesel production because of their inexpensive carbon sources, convenient preparation, and easily tunable physical and chemical properties [108]. Furthermore, covalent attachment of active groups such as SO_3H, COOH, and phenolic OH groups on the surface of biochar improve the biochar characteristics. Of the various functionalized catalysts, the sulfonated carbon-based catalyst is among the most competent for biodiesel production [109]. Effects of the different forms of sulfonating precursors modified on raw biochar support were investigated. Biochar sulfonated with fuming sulfuric acid had a superior acid density over that sulfonated with concentrated sulfuric acid, resulting in a higher activity in the transesterification of vegetable oil. Further modification of biochar with KOH before sulfonating with fuming sulfuric acid could maximize the catalytic activity for transesterification as KOH ascribed to an enhanced porosity, surface area, and acid density of the catalyst [6]. Carbonization of biomass followed by sulfonation resulted in catalysts containing SO_3H, COOH, and OH groups, in which the number of SO_3H groups significantly influence the catalytic activity for biodiesel production [110]. Leaching of SO_3H groups caused the deactivation of the catalyst. The efficiency of esterification between oleic acid and ethanol calculated from the reduction of the reactant acidity reached 98.40% when the time to introduce the SO_3H groups into the bamboo-derived catalyst was increased from 60 to 120 minutes [111]. Thus, not only the type of sulfonating precursor but also the sulfonating conditions affect the catalytic efficiency.

More information on biodiesel yield produced in the presence of various heterogeneous catalysts is summarized in Table 2.4. In comparison to non-biomass-derived catalysts mentioned in this discussion, not only the catalysts produced from carbon-rich biomass but also from calcium carbonate-rich biomass have the potential to be competitive with commercial alkali and metal oxide catalysts in biodiesel production. Even though biomass-derived catalysts showed a slightly lower biodiesel yield, paying more attention to research in biodiesel production with bio-based catalysts will shortly achieve better results in terms of product yield, catalyst stability, environmental friendliness and cost-efficiency.

2.5.3 Biomass-Derived Activated Carbon

Biomass-derived activated carbons are conventionally produced from diverse renewable sources via carbonization and activation. The sequence of processes in activated carbon production from

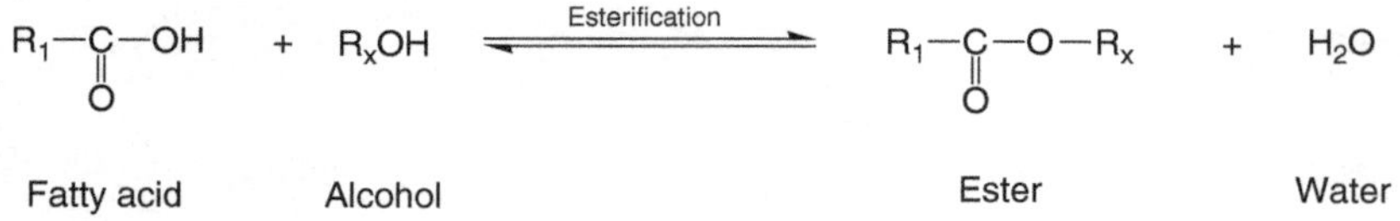

$$\underset{\text{Fatty acid}}{R_1{-}C({=}O){-}OH} + \underset{\text{Alcohol}}{R_xOH} \xrightarrow{\text{Esterification}} \underset{\text{Ester}}{R_1{-}C({=}O){-}O{-}R_x} + \underset{\text{Water}}{H_2O}$$

Figure 2.6 Esterification of fatty acid.

Table 2.4 Comparison of various catalysts for biodiesel production.

Catalyst	Feedstock	Reaction conditions				Yield (%)	References
		Alcohol:Oil	Catalyst amount (wt%)	Temperature (°C)	Time		
La_2O_3/CaO	Jatropha oil	25 : 1	3	160	3 h	98.76	[112]
Li/CaO	Jatropha oil	12 : 1	5	65	1–2 h	>99	[113]
$ZnO\text{-}TiO_2\text{-}Nd_2O_3/ZrO_2$	Soybean oil	5.7 : 1	—	195	44 min	99	[104]
SO_4^{2-}/SnO_2	Waste cooking oil	15 : 1	3	150	3 h	92.3	[105]
Bio-based carbon modified with Ni and Na_2SiO_3	Soybean oil	9 : 1	7	65	100 min	98.1	[114]
Sulfonated modified coconut meal residual	Waste palm oil	12 : 1	5	65–70	10 h	92.7	[115]
Sulfonated modified carbonized coconut shell	Palm oil	30 : 1	6	60	6 h	88.25	[116]
Sulfonated woody biochar	Canola oil	30 : 1	7	315	3 h	48.1	[117]
Sulfonated carbonized bamboo	Oleic acid	7 : 1	6	90	2 h	98.4	[111]
Calcium oxide from eggshell	Waste cooking oil	6 : 1	5.8	Room temperature	11 h	97	[118]

biomass causes a considerable increase in the porosity and specific surface area. Other unique properties of activated carbon comprise thermal resistance, stability in both acidic and basic environment, and the possibility to tailor both its physical and surface chemical properties. To satisfy the requirements for catalyzing a specific reaction, activated carbons can be treated with KOH or CaO via wet impregnation. The stability of such immobilized catalysts, however, is relatively low. To improve the performance of activated carbon, chemical treatment with oxidizing acids such as SO_3H, H_3PO_4, or HNO_3 is an interesting choice [119–121]. In view of the catalytic potential, sulfonated carbon catalysts are widely attractive in reactions such as hydrolysis, esterification, and transesterification. Sulfonated bio-based carbon catalysts can be prepared by direct sulfonation or sulfonation via reductive alkylation/arylation reactions.

Sulfonated activated carbons from various biomass resources were tested for the production of biodiesel through esterification or transesterification. Various kinds of biomass wastes were activated with H_3PO_4 prior to carbonization and subsequently sulfonated through arylation of 4-benzenediazonium. The resulting bio-based activated carbon materials were highly porous with surface areas of 640–970 $m^2\ g^{-1}$ depending on the biowaste source. Under optimal conditions, conversion of up to 94% could be achieved for esterification [120]. Slow reaction rates, the requirements of high methanol-to-oil molar ratios, and high reaction temperatures are, however, weaknesses of the sulfonated carbon catalyst for biodiesel production [122]. A further application with regard to the upgrading of bio-oil via reactive distillation including water removal, catalytic esterification, and neutralization steps was carried out with p-toluene sulfonic acid embedded on an activated carbon obtained by activation of sawdust pellets [123]. The catalyst was applied to the catalytic esterification in the reactive distillation procedure. After water removal from the oil, the esterification reaction took place at 80 °C under atmospheric pressure with the acid-modified activated carbon catalyst and zeolite. The heating value of the upgraded bio-oil obtained from fast pyrolysis was improved from 14.80 to 23.13 $MJ\,kg^{-1}$ and was ascribed to the presence of higher ester and acetic acid contents in the oil according to the reaction in Figure 2.7. The properties of the upgraded oil were consistent with the diesel standard. Interestingly, sulfonated activated carbon catalysts have the potential for bio-jet fuel production through co-pyrolysis of biomass and plastic waste [124]. Prior to the sulfonation step, an activated corncob was converted by microwave-assisted carbonization. The sulfonation temperature was a major factor that influenced the number of SO_3H groups on the catalyst. These active groups benefited the co-pyrolysis oil yield of nearly 98% at 500 °C for 15 minutes. The obtained oil consisted of aromatic and C_9–C_{16} alkanes, which was qualified as a bio-jet fuel.

A drawback of the sulfonated bio-based activated carbon catalyst is that its preparation process is environmentally unfriendly owing to the heating of carbon materials in concentrated sulfuric acid for an extended time. Apart from sulfuric acid, tribasic potassium phosphate (K_3PO_4) was recently used to activate carbon materials. The K_3PO_4 impregnated on the activated carbon catalyst benefited the biodiesel production from waste cooking oil. With 5% catalyst loading, the biodiesel yield was 98% in a 12 : 1 methanol-to-oil molar ratio at 60 °C for four hours. Only 20% reduction in biodiesel yield was observed when the catalyst was used for five reaction cycles [125].

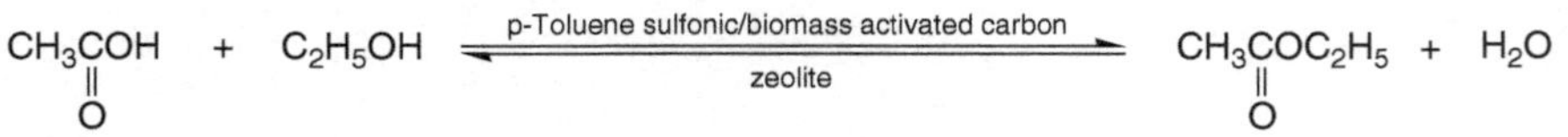

$$CH_3C(=O)OH + C_2H_5OH \xrightleftharpoons[\text{zeolite}]{\text{p-Toluene sulfonic/biomass activated carbon}} CH_3C(=O)OC_2H_5 + H_2O$$

Figure 2.7 Esterification of acetic acid.

One way to further enhance the catalytic activity of bio-based activated carbon is via impregnation with various metal ions. The Fe_3O_4 embedded on beetroot-derived activated carbon catalyst (Fe_3O_4/BRAC) was examined for its performance in the reduction of nitroarenes as illustrated in Figure 2.8 [126]. The process yielded a high aromatic amines product (99%) even at 85 °C for 15 minutes. The reaction pathway as described in Veerakumar et al. [126] can be briefly summarized as follows: first, the nitrobenzene and isopropanol were chemisorbed on the surface of Fe_3O_4/BRAC. Hydrogen in the form of hydride ion was transferred from the isopropanol to the substrate. Subsequently, the existence of KOH and the Fe_3O_4/BRAC catalyst promoted the transformation of alcohol into alkoxide species. The nitro groups on the catalyst surface were reduced to nitrosobenzene, which reacted with hydrogen ion from isopropanol to form aniline. The product desorbed from the catalyst surface in the final step. The immobilization of Fe_3O_4 on the large surface area activated carbon facilitated the reaction under mild conditions in a short reaction time. The Fe_3O_4/BRAC catalyst boosted the hydrogen relocation from interacting isopropanol and KOH. Moreover, it presented high stability with less than 10% decrease in aniline yield after five repeated cycles. Importantly, separation of the Fe_3O_4 impregnated carbon catalyst from the products was readily achieved by an external magnetic field owing to the magnetic properties of Fe_3O_4.

Lignocellulosic waste-derived activated carbons impregnated with $ZnCl_2$ (giving Lewis acid sites) or H_2SO_4 (giving Brønsted acid sites) were very effective for the conversion of glucose to 5-hydroxymethylfurfural (HMF) through isomerization and dehydration according to pathway 1 presented in Figure 2.9. These catalysts were more productive than homogeneous catalysts (H_2SO_4 and $ZnCl_2$), even though these contained higher Lewis and Brønsted acid sites than the impregnated activated carbon [127]. This implies that the activated carbon itself had specific catalytic behavior for this reaction. Activated carbon possessed some Lewis and Brønsted acid sites but it had very high specific surface area and high amount of oxygenated functional groups. Therefore, the two latter predominant properties of the activated carbon strongly affected its catalytic efficiency. Its large specific surface area might increase the opportunity for adsorption of reactants on the active site. The large amount of oxygen functional groups on its surface might cooperate in the fructose dehydration to HMF.

It is well known that KOH activation before metal impregnation increases the surface area and pore volume of the carbon, particularly the micropore volume. The metallic K interposes itself between the carbon structure, leading to permanent enlargement of the carbon lattice after the removal of potassium metal. Thus, it is capable of increasing both the catalytic activity and the stability of the metal-impregnated carbon catalysts in many reactions, e.g. Mizoroki–Heck, Suzuki, and Sonogashira coupling reactions, toluene steam reforming, and methanation [128–130]. Even if the KOH activator can intensively improve the surface area and porosity of carbon, it requires a relatively high activation temperature. Alternatively, H_3PO_4 is an interesting activator that requires a lower activation temperature [131]. The H_3PO_4 was used to activate olive stone at room temperature before carbonization, followed by wet

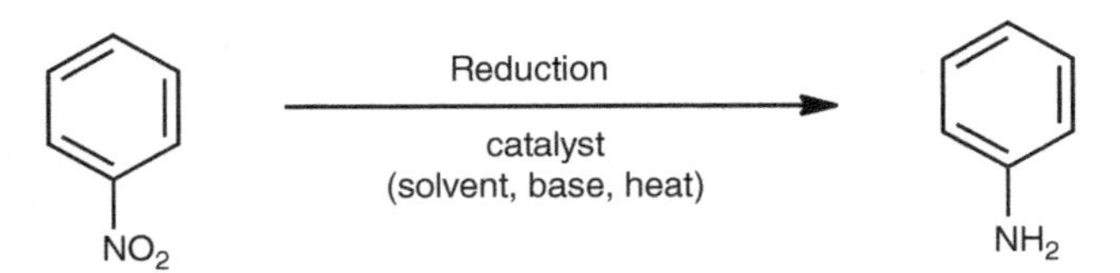

Figure 2.8 *Reduction of nitrobenzene.*

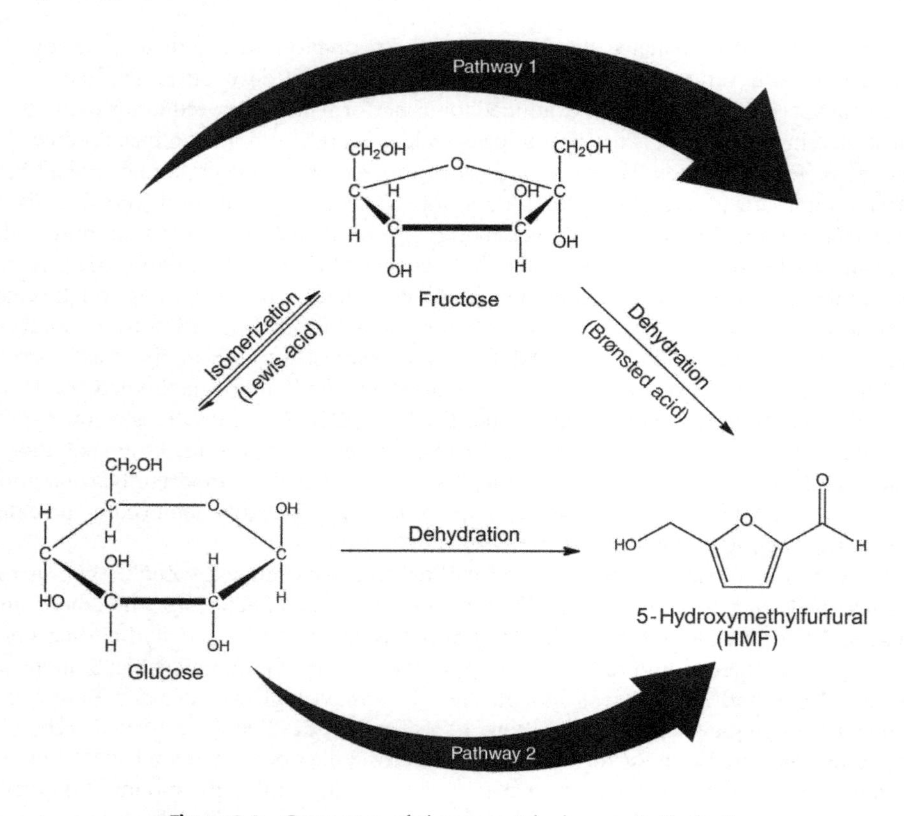

Figure 2.9 Conversion of glucose to 5-hydroxymethylfurfural.

impregnation with zirconium, and the resulting catalyst was employed in methanol dehydration to dimethyl ether [132]. This preparation method produced a superb mesoporous structure and large surface area (1190 m^2 g^{-1}) despite the activation occurring at room temperature. Additionally, the extensive formation of active zirconium phosphate functional groups over the surface area of carbon was observed. The catalyst presented both a high conversion in methanol dehydration and high selectivity to dimethyl ether. An increase in the H_3PO_4 activation temperature along with the addition of $NaBH_4$ promoted the microporous area and the total surface area of the activated carbon (2559 m^2 g^{-1}) [133]. Activated carbon was widely employed as a support for several metal catalysts such as nanoparticles of Co, Ni, Cu, and noble metals. These catalysts were effective in bio-oil hydrodeoxygenation and hydrogen production via both biomass hydrolysate reforming and ammonia borane hydrolysis [131, 134, 135]

2.5.4 Hydrothermal Bio-based Carbons

A green hydrothermal carbonaceous material was developed from by-products of sugar dehydration in hot compressed water at 150–250 °C [136, 137]. Hydrothermal carbon has been used not only in catalysis but also in agriculture, energy storage, and adsorption. Hydrothermal reaction mechanisms comprise dehydration and decarboxylation, and the process parameters such as reaction temperature and type of biomass need to be considered since these factors affect the catalytic property of the carbon. Originally, carbohydrate derivatives were utilized as a model structure for the preparation of hydrothermal carbons. Various researchers have prepared hydrothermal carbons from mono-, di-, and polysaccharides [138–140]. The hydrothermal carbon produced from glucose had a uniform spherical morphology with micron-sized particles and a smooth surface [141]. The hydrothermal temperature had a stronger influence than the reaction time, not merely on the yield but also on the elemental compositions (i.e. C, H, and O) of the hydrochar. An increase in temperature resulted in a decrease in oxygen and hydrogen contents, while the carbon content increased [142]. The hydrothermal carbon possessed a large amount of oxygen-containing functional groups including hydroxyl and aromatic C=O groups (carboxyl, carbonyl, ester, and quinone). These functional groups catalyze several reactions involved in the transformation of biomass to chemicals and organic pollutant degradation but the activity was not outstanding [143, 144]. Its porosity and surface area were also relatively restricted, which were less than $10\,cm^3\,g^{-1}$ and $40\,m^2\,g^{-1}$, respectively. Chemical modification with an oxidizing agent and metal immobilization are usually applied to enhance its catalytic activity and other properties. After modification with H_2SO_4, the hydrothermal carbon presented a rougher surface that was more accessible to the reactant, as well as a large amount of Brønsted acids and some Lewis acids [140]. The sulfonation created polycyclic aromatic carbon rings in an irregular form [145]. The improved properties of the hydrothermal carbon benefited the yield of ethyl levulinate and ethyl glucoside from the cellulose ethanolysis, and no tar or char was formed throughout the reaction.

Other than carbohydrates, raw biomass has been used as a possible feedstock for the production of hydrothermal carbon. The irregular shape and agglomerated particles of the hydrothermal carbon from biomass were observed [144, 145]. Its abundance of oxygen, especially in the form of ketonic groups, was similar to the hydrothermal carbon prepared from carbohydrate feedstocks. The appearance of several functional groups on the hydrothermal carbon confirmed that the hemicellulosic polysaccharide in biomass was destroyed by the hydrothermal treatment. However, its stability and reactivity in biomass conversion into chemicals were distinctly different from the monosaccharide-derived hydrothermal carbon due to the difference in the hydrothermal carbon formation mechanism. The hydrothermal pathways of glucose feedstock were mostly via glucose dehydration to HMF, condensation, and aromatization. Whereas for the biomass feedstock, there are probably two reaction pathways: (1) partially by biomass hydrolysis, glucose dehydration, condensation, and aromatization, and (2) partially by direct biomass condensation and aromatization. The higher stability and aromaticity in the hydrothermal carbon from biomass was ascribed to the direct reaction pathways [144].

2.5.5 Sugar-Derived Carbon Catalysts

Recently, sugars such as D-glucose, sucrose, and cellulose have been reported as promising carbon sources for the preparation of catalysts or catalyst supports through various biomass conversion pathways [146]. A solid acid catalyst synthesized from sugar, also known as a sugar catalyst, was shown to be stable and active in many reactions. In particular, a sulfonated carbon-based solid acid catalyst had outstanding physical properties and was very active in both esterification and hydrolysis reactions. The formation of water-soluble saccharides could be achieved via the hydrolysis of microcrystalline cellulose (see Figure 2.10) using the sugar-derived sulfonated solid catalyst. Despite its relatively low surface area of approximately $2\,m^2\,g^{-1}$, the obtained sulfonated solid catalyst was capable of producing soluble saccharides consisting of 64% β-1,4-glucan and 4% glucose even at a low reaction temperature of 100 °C. Its intrinsic adsorption proficiency to attach the β-1,4-glucan on the surface likely prompted the catalytic activity, an ability that other solid acid catalysts did not possess. Meanwhile,

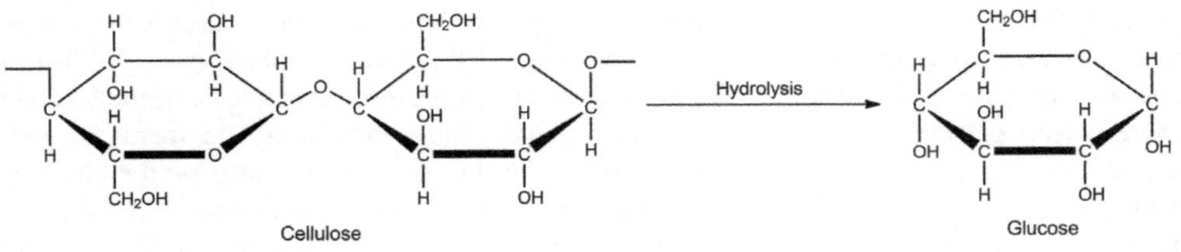

Figure 2.10 Hydrolysis of cellulose.

typical solid acid catalysts, for instance, Amberlyst-15, H-mordenite, Nafion, and niobic acid were inactive for this reaction [146].

Aside from the hydrolysis reaction, a sugar-derived solid catalyst was found to have potential in biodiesel production by esterification. The D-glucose-derived sugar acid catalyst was compared with conventional solid catalysts (sulfated zirconia, Amberlyst-15, and niobic acid) for its effectiveness to esterify a short-chain aliphatic alcohol with palmitic acid and oleic acid [147]. The sugar-derived acid catalyst presented superb thermal stability and the highest activity among other acid catalysts, and its activity was almost the same as a homogeneous acid catalyst. To contribute to the cost-competition, the sugar catalyst was used in biodiesel production from waste oil with high free fatty acid content (27.8%). It showed prominent performances in the reactivity with a yield of higher than 90%, in spite of the oil containing high amounts of free fatty acids.

2.5.6 Carbon Nanotubes from Biomass

CNTs, which are a 3D-structured material, have recently been developed for their application in catalysis due to the promising physical and chemical properties. Several kinds of carbon-rich biomasses and bio-resource residues were used as feedstock in the production of biomass-derived CNTs. Yao et al. utilized a biomass residue from the liquor industry to produce CNTs by activation and carbonization to receive an activated carbon followed by thermal condensation [148]. The activated carbon was reacted with dicyandiamide and Ni metal precursor and subsequently annealed at 700, 800, or 900 °C to embed the Ni nanoparticles in the N-doped CNTs supported on porous carbon (Ni-CNTs-C). The growth of CNTs on the surface of the porous carbon support was promoted by the increasing annealing temperature. The bamboo-like structure of CNTs annealed at 700 °C had a diameter of approximately 10–15 nm and a length of 100–300 nm. Conversely, the surface area was decreased from 558 to 283 $m^2\ g^{-1}$ when the annealing temperature was raised from 700 to 900 °C. The Ni-CNTs-C annealed at 700 °C revealed the highest catalytic activity with ~98.2% removal of organic dye (Orange II) within 90 minutes because its largest surface area provided more active sites. The dye removal percentage was reduced to 96.9 and 68.5% with annealing at 800 and 900 °C, respectively. Leaching and agglomeration of Ni^0 and CNTs were shielded by the carbon layer leading to high stability and activity of these catalysts in acidic media. The activity of the reused catalyst, however, decreased from 98.2 to 64.2% after five cycles. The great number of CNTs presented both a high number of active sites and promoted electron transfer on the carbon surface. The Ni-CNTs-C catalyst was also tested for its performance in the conversion of harmful Cr(VI) to less toxic Cr(III). Among catalysts tested, the catalyst annealed at 800 °C (Ni-CNTs-C-800) showed the highest activity and could completely reduce Cr(VI) to Cr(III) in 160 minutes as observed from the change in color from yellow to colorless [148].

The HTC of biomass including sugar, cellulose, and crude plants yields many intermediate products of liquid alkanes such as 5-hydroxymethyl-2-furaldehyde, furfural, 5-methylfurfural and 1,2,4-benezenetriol. When these intermediates have the appropriate concentration of oxygenated functional groups, they could self-transform into various structures of carbonaceous materials through various conversion pathways, i.e. dehydration, aromatization, and polymerization. The occurring carbonaceous particles could attach to the CNT matrix, and repeatedly bind layer-by-layer to obtain oxygenated functional groups anchored to the CNTs. The resulting material is suitable both as a direct catalyst and as support for metal catalysts as it contained numerous reactive functional groups. The metal immobilized on the functionalized CNT catalyst presented both high activity and selectivity for the hydrogenation of unsaturated aldehyde to unsaturated alcohol with 90% selectivity at a conversion of 55% [149].

2.5.7 Graphene and Its Derivatives

Graphene, a member of the graphite family, is a flat monolayer of carbon atoms arranged into a 2D honeycomb structure. Graphene and graphene-based materials (e.g. graphene oxides) have been extensively exploited because of their excellent thermal, mechanical, and electronic characteristics. Graphene, both with and without modifications, has been explored for its effectiveness in various reactions. Graphene produced from biomass commonly contains a high content of impurities that could either act as a poison or as a natural promoter in catalysis [150]. The imperfect graphene contaminated with nitrogen dopant was active and stable for the selective CO_2 hydrogenation to methane. For example, the natural N-doped graphene prepared by pyrolysis of chitosan at 500 °C showed a higher CO_2 conversion than other dopants and 99.2% selectivity toward CH_4 [151]. Alternatively, external N doping into graphene was employed as a metal-free catalyst which also showed high stability and selectivity in hydrogenation [77]. Huang et al. [152] synthesized N-doped graphene via the carbonization of biomass guanine along with colloidal silica at 1000 °C under N_2 atmosphere followed by the addition of hydrofluoric acid to remove the silica template. It was observed that its electrochemical activity in the oxygen reduction reaction under acid and base conditions was remarkable due to its porous structure and high N doping [152]. Other than N doping onto graphene, an Ru metal-supported graphene synthesized via carbonization and calcination of a mixture of glucose, $FeCl_3$, and $RuCl_3$ also demonstrated to be highly effective for levulinic acid hydrogenation to γ-valerolactone (see Figure 2.11) with 100% conversion and 96% yield [153]. In addition, a catalyst based on graphene-encapsulated Fe_3C embedded in CNTs was successfully produced using the inexpensive co-pyrolysis of a mixture of biomass (glucose, xylitol, or sucrose), melamine, and iron compound [77]. The functionalized CNTs were stable and efficient in the hydrogenation of nitrobenzene to anilines. The 100% conversion of nitrobenzene along with nearly 99% selectivity toward anilines was achieved with the catalyst that was carbonized at 700 °C. The catalytic activity of this catalyst was mainly based on the pyrolysis temperature; the Fe_3C, which represents the active site for this reaction, could be destroyed when the pyrolysis temperature reached over 700 °C.

2.6 Summary and Future Aspects

The exponentially growing development in the transformation of biomass into carbon-based materials is presented in a wide variety of publications for more than a decade. Various examples related to the utilization of biomass-derived carbon materials with applications in

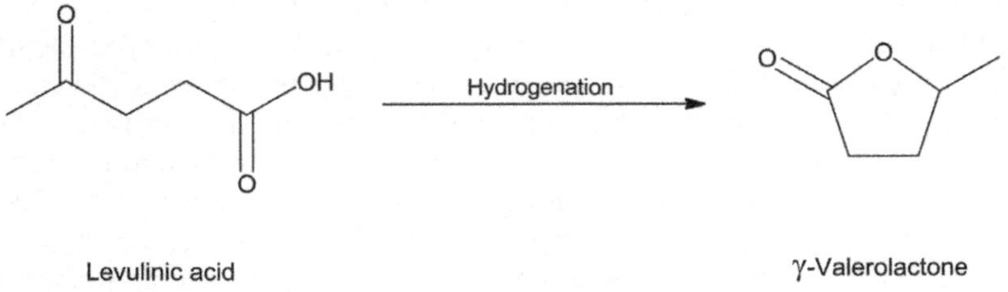

Hydrogenation

Levulinic acid

γ-Valerolactone

Figure 2.11 *Hydrogenation of levulinic acid.*

catalysis have been presented and discussed in this chapter. Several outstanding features of bio-based carbon materials such as various physical properties, chemical functionalities, facile activation and functionalization, and environmental friendliness allow them to straightforwardly act as a catalyst and catalyst support. It is important to consider issues that take into account the variety and complication of agricultural biomasses utilized as raw materials in carbon synthesis. The surface area and pore diameter of carbon materials as well as surface properties including hydrophobicity, hydrophilicity, acidity, basicity, and chemical functional groups could be accommodated by the selection of the biomass and control of the preparation conditions. Biochar is a simple bio-based carbon material commonly having diverse surface functional groups. In order to acquire a high catalytic activity and high stability of biochar, a physical or chemical modification step is required. Hydrochar also possesses a large amount of oxygen-containing functional groups that are beneficial for the production of bio-based chemicals and the degradation of organic pollutants. The abundance of mesopores and large surface area of activated carbon could lead to higher efficiency by providing more active sites. Novel crystalline bio-based carbon materials including graphene and CNTs are currently focused on as high-efficiency materials in catalysis. Even though current bio-based carbon materials have accomplished much in catalysis applications relating to product yield, selectivity, and stability, most are not quite ready for production on an industrial scale. To accomplish a high catalytic efficiency, the production cost of biomass-derived carbon catalysts is still higher than that of conventional catalysts. Thus, a more intensive investigation is required to decrease the production cost together with enhancing the catalytic efficiency of the bio-based carbon materials. With the intention of moving the production and utilization of biomass-derived carbon materials from laboratory scale to commercialization, techno-economic and environmental feasibility should also be analyzed.

References

1. Grand View Research, US (2020). Catalyst Market Size & Share Analysis Report, 2019–2025.
2. Idrees, M., Batool, S., Kalsoom, T. et al. (2018). Animal manure-derived biochars produced via fast pyrolysis for the removal of divalent copper from aqueous media. *Journal of Environmental Management* 213: 109–118. https://doi.org/10.1016/j.jenvman.2018.02.003.
3. Osman, N.B., Shamsuddin, N., and Uemura, Y. (2016). Activated carbon of oil palm empty fruit bunch (EFB); core and shaggy. *Procedia Engineering* 148: 758–764. https://doi.org/10.1016/j.proeng.2016.06.610.
4. Buentello-Montoya, D., Zhang, X., Li, J. et al. (2020). Performance of biochar as a catalyst for tar steam reforming: effect of the porous structure. *Applied Energy* 259: 114176. https://doi.org/10.1016/j.apenergy.2019.114176.
5. Chen, R.X. and Wang, W.C. (2019). The production of renewable aviation fuel from waste cooking oil. Part I: Bio-alkane conversion through hydro-processing of oil. *Renewable Energy* 135: 819–835. https://doi.org/10.1016/j.renene.2018.12.048.

6. Dehkhoda, A.M., West, A.H., and Ellis, N. (2010). Biochar based solid acid catalyst for biodiesel production. *Applied Catalysis A: General* 382 (2): 197–204. https://doi.org/10.1016/j.apcata.2010.04.051.
7. Lee, H.W., Lee, H., Kim, Y.M. et al. (2019). Recent application of biochar on the catalytic biorefinery and environmental processes. *Chinese Chemical Letters* 30 (12): 2147–2150. https://doi.org/10.1016/j.cclet.2019.05.002.
8. Li, C., Zhang, C., Gholizadeh, M. et al. (2020). Different reaction behaviours of light or heavy density polyethylene during the pyrolysis with biochar as the catalyst. *Journal of Hazardous Materials* 399: 123075. https://doi.org/10.1016/j.jhazmat.2020.123075.
9. Nejati, B., Adami, P., Bozorg, A. et al. (2020). Catalytic pyrolysis and bio-products upgrading derived from *Chlorella vulgaris* over its biochar and activated biochar-supported Fe catalysts. *Journal of Analytical and Applied Pyrolysis* 152: 104799. https://doi.org/10.1016/j.jaap.2020.104799.
10. Kan, T., Strezov, V., and Evans, T.J. (2016). Lignocellulosic biomass pyrolysis: a review of product properties and effects of pyrolysis parameters. *Renewable and Sustainable Energy Reviews* 57: 1126–1140. https://doi.org/10.1016/j.rser.2015.12.185.
11. Byrne, C.E. and Nagle, D.C. (1997). Carbonization of wood for advanced materials applications. *Carbon* 35 (2): 259–266. https://doi.org/10.1016/S0008-6223(96)00136-4.
12. Volpe, M., Messineo, A., Mäkelä, M. et al. (2020). Reactivity of cellulose during hydrothermal carbonization of lignocellulosic biomass. *Fuel Processing Technology* 206: 106456. https://doi.org/10.1016/j.fuproc.2020.106456.
13. Demirbas, A. (2005). Pyrolysis of ground beech wood in irregular heating rate conditions. *Journal of Analytical and Applied Pyrolysis* 73 (1): 39–43. https://doi.org/10.1016/j.jaap.2004.04.002.
14. Elyounssi, K., Collard, F.X., Mateke, J.N. et al. (2012). Improvement of charcoal yield by two-step pyrolysis on eucalyptus wood: a thermogravimetric study. *Fuel* 96: 161–167. https://doi.org/10.1016/j.fuel.2012.01.030.
15. Erçin, D. and Yürüm, Y. (2003). Carbonisation of Fir (*Abies bornmulleriana*) wood in an open pyrolysis system at 50–300°C. *Journal of Analytical and Applied Pyrolysis* 67 (1): 11–22. https://doi.org/10.1016/S0165-2370(02)00011-6.
16. Kato, Y., Enomoto, R., Akazawa, M. et al. (2016). Characterization of Japanese cedar bio-oil produced using a bench-scale auger pyrolyzer. *SpringerPlus* 5 (1): 177. https://doi.org/10.1186/s40064-016-1848-7.
17. Nanda, S., Mohanty, P., Pant, K.K. et al. (2013). Characterization of North American lignocellulosic biomass and biochars in terms of their candidacy for alternate renewable fuels. *BioEnergy Research* 6 (2): 663–677. https://doi.org/10.1007/s12155-012-9281-4.
18. Pedersen, T.H., Grigoras, I.F., Hoffmann, J. et al. (2016). Continuous hydrothermal co-liquefaction of aspen wood and glycerol with water phase recirculation. *Applied Energy* 162: 1034–1041. https://doi.org/10.1016/j.apenergy.2015.10.165.
19. Grima-Olmedo, C., Ramírez-Gómez, Á., Gómez-Limón, D. et al. (2016). Activated carbon from flash pyrolysis of eucalyptus residue. *Heliyon* 2 (9): E00155. http://dx.doi.org/10.1016/j.heliyon.2016.e00155.
20. Ekhlasi, L., Younesi, H., Rashidi, A. et al. (2018). Populus wood biomass-derived graphene for high CO_2 capture at atmospheric pressure and estimated cost of production. *Process Safety and Environmental Protection* 113: 97–108. https://doi.org/10.1016/j.psep.2017.09.017.
21. Ilnicka, A., Kamedulski, P., Aly, H.M. et al. (2020). Manufacture of activated carbons using Egyptian wood resources and its application in oligothiophene dye adsorption. *Arabian Journal of Chemistry* 13 (5): 5284–5291. https://doi.org/10.1016/j.arabjc.2020.03.007.
22. Wang, X., Liu, Y., Zhu, L. et al. (2019). Biomass derived N-doped biochar as efficient catalyst supports for CO_2 methanation. *Journal of CO2 Utilization* 34: 733–741. https://doi.org/10.1016/j.jcou.2019.09.003.
23. Chen, W., Fang, Y., Li, K. et al. (2020). Bamboo wastes catalytic pyrolysis with N-doped biochar catalyst for phenols products. *Applied Energy* 260: 114242. https://doi.org/10.1016/j.apenergy.2019.114242.
24. Qian, K., Kumar, A., Bellmer, D. et al. (2015). Physical properties and reactivity of char obtained from downdraft gasification of sorghum and eastern red cedar. *Fuel* 143: 383–389. https://doi.org/10.1016/j.fuel.2014.11.054.
25. Wang, S., Dai, G., Yang, H. et al. (2017). Lignocellulosic biomass pyrolysis mechanism: a state-of-the-art review. *Progress in Energy and Combustion Science* 62: 33–86. https://doi.org/10.1016/j.pecs.2017.05.004.
26. Shen, Y. (2018). Rice husk-derived activated carbons for adsorption of phenolic compounds in water. *Global Challenges* 2 (12): 1800043. https://doi.org/10.1002/gch2.201800043.

27. Steurer, E. and Ardissone, G. (2015). Hydrothermal carbonization and gasification technology for electricity production using biomass. *Energy Procedia* 79: 47–54. https://doi.org/10.1016/j.egypro.2015.11.473.
28. Rashidi, N.A., Yusup, S., Ahmad, M.M. et al. (2012). Activated carbon from the renewable agricultural residues using single step physical activation: a preliminary analysis. *APCBEE Procedia* 3: 84–92. https://doi.org/10.1016/j.apcbee.2012.06.051.
29. Biswas, B., Pandey, N., Bisht, Y. et al. (2017). Pyrolysis of agricultural biomass residues: comparative study of corn cob, wheat straw, rice straw and rice husk. *Bioresource Technology* 237: 57–63. https://doi.org/10.1016/j.biortech.2017.02.046.
30. Huff, M.D., Kumar, S., and Lee, J.W. (2014). Comparative analysis of pinewood, peanut shell, and bamboo biomass derived biochars produced via hydrothermal conversion and pyrolysis. *Journal of Environmental Management* 146: 303–308. https://doi.org/10.1016/j.jenvman.2014.07.016.
31. Rahman, M.M., Awang, M., Mohosina, B.S. et al. (2012). Waste palm shell converted to high efficient activated carbon by chemical activation method and its adsorption capacity tested by water filtration. *APCBEE Procedia* 1: 293–298. https://doi.org/10.1016/j.apcbee.2012.03.048.
32. Bader, N. and Ouederni, A. (2016). Optimization of biomass-based carbon materials for hydrogen storage. *Journal of Energy Storage* 5: 77–84. https://doi.org/10.1016/j.est.2015.12.009.
33. Apaydın-Varol, E. and Pütün, A.E. (2012). Preparation and characterization of pyrolytic chars from different biomass samples. *Journal of Analytical and Applied Pyrolysis* 98: 29–36. https://doi.org/10.1016/j.jaap.2012.07.001.
34. Lee, Y., Park, J., Ryu, C. et al. (2013). Comparison of biochar properties from biomass residues produced by slow pyrolysis at 500°C. *Bioresource Technology* 148: 196–201. https://doi.org/10.1016/j.biortech.2013.08.135.
35. González, J.F., Román, S., Encinar, J.M. et al. (2009). Pyrolysis of various biomass residues and char utilization for the production of activated carbons. *Journal of Analytical and Applied Pyrolysis* 85 (1): 134–141. https://doi.org/10.1016/j.jaap.2008.11.035.
36. Rezma, S., Birot, M., Hafiane, A. et al. (2017). Physically activated microporous carbon from a new biomass source: date palm petioles. *Comptes Rendus Chimie* 20 (9): 881–887. https://doi.org/10.1016/j.crci.2017.05.003.
37. Al-Wabel, M.I., Rafique, M.I., Ahmad, M. et al. (2019). Pyrolytic and hydrothermal carbonization of date palm leaflets: characteristics and ecotoxicological effects on seed germination of lettuce. *Saudi Journal of Biological Sciences* 26 (4): 665–672. https://doi.org/10.1016/j.sjbs.2018.05.017.
38. Cai, J., Li, B., Chen, C. et al. (2016). Hydrothermal carbonization of tobacco stalk for fuel application. *Bioresource Technology* 220: 305–311. https://doi.org/10.1016/j.biortech.2016.08.098.
39. Tobi, A.R., Dennis, J.O., Zaid, H.M. et al. (2019). Comparative analysis of physiochemical properties of physically activated carbon from palm bio-waste. *Journal of Materials Research and Technology* 8 (5): 3688–3695. https://doi.org/10.1016/j.jmrt.2019.06.015.
40. Chaiya, C. and Reubroycharoen, P. (2013). Production of bio oil from para rubber seed using pyrolysis process. *Energy Procedia* 34: 905–911. https://doi.org/10.1016/j.egypro.2013.06.828.
41. Fu, P., Hu, S., Xiang, J. et al. (2012). Evaluation of the porous structure development of chars from pyrolysis of rice straw: effects of pyrolysis temperature and heating rate. *Journal of Analytical and Applied Pyrolysis* 98: 177–183. https://doi.org/10.1016/j.jaap.2012.08.005.
42. Xu, Q., Tang, S., Wang, J. et al. (2018). Pyrolysis kinetics of sewage sludge and its biochar characteristics. *Process Safety and Environmental Protection* 115: 49–56. https://doi.org/10.1016/j.psep.2017.10.014.
43. Zhang, J., Gao, J., Chen, Y. et al. (2017). Characterization, preparation, and reaction mechanism of hemp stem based activated carbon. *Results in Physics* 7: 1628–1633. https://doi.org/10.1016/j.rinp.2017.04.028.
44. David, E. and Kopac, J. (2014). Activated carbons derived from residual biomass pyrolysis and their CO_2 adsorption capacity. *Journal of Analytical and Applied Pyrolysis* 110: 322–332. https://doi.org/10.1016/j.jaap.2014.09.021.
45. Kumar, A. and Jena, H.M. (2016). Preparation and characterization of high surface area activated carbon from Fox nut (*Euryale ferox*) shell by chemical activation with H_3PO_4. *Results in Physics* 6: 651–658. https://doi.org/10.1016/j.rinp.2016.09.012.
46. Mistar, E.M., Alfatah, T., and Supardan, M.D. (2020). Synthesis and characterization of activated carbon from *Bambusa vulgaris* striata using two-step KOH activation. *Journal of Materials Research and Technology* 9 (3): 6278–6286. https://doi.org/10.1016/j.jmrt.2020.03.041.

47. Liu, Z., Zhu, Z., Dai, J. et al. (2018). Waste biomass based-activated carbons derived from soybean pods as electrode materials for high-performance supercapacitors. *ChemistrySelect* 3 (21): 5726–5732. https://doi.org/10.1002/slct.201800609.
48. Kılıç, M., Apaydın-Varol, E., and Pütün, A.E. (2012). Preparation and surface characterization of activated carbons from *Euphorbia rigida* by chemical activation with $ZnCl_2$, K_2CO_3, NaOH and H_3PO_4. *Applied Surface Science* 261: 247–254. https://doi.org/10.1016/j.apsusc.2012.07.155.
49. Ncibi, M.C., Ranguin, R., Pintor, M.J. et al. (2014). Preparation and characterization of chemically activated carbons derived from Mediterranean *Posidonia oceanica* (L.) fibres. *Journal of Analytical and Applied Pyrolysis* 109: 205–214. https://doi.org/10.1016/j.jaap.2014.06.010.
50. Ponomarev, N. and Sillanpää, M. (2019). Combined chemical-templated activation of hydrolytic lignin for producing porous carbon. *Industrial Crops and Products* 135: 30–38. https://doi.org/10.1016/j.indcrop.2019.03.050.
51. Hayashi, J., Kazehaya, A., Muroyama, K. et al. (2000). Preparation of activated carbon from lignin by chemical activation. *Carbon* 38 (13): 1873–1878. https://doi.org/10.1016/S0008-6223(00)00027-0.
52. Lillo-Ródenas, M.A., Cazorla-Amorós, D., and Linares-Solano, A. (2003). Understanding chemical reactions between carbons and NaOH and KOH: an insight into the chemical activation mechanism. *Carbon* 41 (2): 267–275. https://doi.org/10.1016/S0008-6223(02)00279-8.
53. Foo, K.Y. and Hameed, B.H. (2012). Textural porosity, surface chemistry and adsorptive properties of durian shell derived activated carbon prepared by microwave assisted NaOH activation. *Chemical Engineering Journal* 187: 53–62. https://doi.org/10.1016/j.cej.2012.01.079.
54. Pütün, A.E., Gerçel, H.F., Koçkar, Ö.M. et al. (1996). Oil production from an arid-land plant: fixed-bed pyrolysis and hydropyrolysis of *Euphorbia rigida*. *Fuel* 75 (11): 1307–1312. https://doi.org/10.1016/0016-2361(96)00098-1.
55. Mazlan, M.A.F., Uemura, Y., Yusup, S. et al. (2016). Activated carbon from rubber wood sawdust by carbon dioxide activation. *Procedia Engineering* 148: 530–537. https://doi.org/10.1016/j.proeng.2016.06.549.
56. Rajgopal, S., Karthikeyan, T., Prakash Kumar, B.G. et al. (2006). Utilization of fluidized bed reactor for the production of adsorbents in removal of malachite green. *Chemical Engineering Journal* 116 (3): 211–217. https://doi.org/10.1016/j.cej.2005.09.026.
57. Prakash Kumar, B.G., Shivakamy, K., Miranda, L.R. et al. (2006). Preparation of steam activated carbon from rubberwood sawdust (*Hevea brasiliensis*) and its adsorption kinetics. *Journal of Hazardous Materials* 136 (3): 922–929. https://doi.org/10.1016/j.jhazmat.2006.01.037.
58. Im, U.S., Kim, J., Lee, S.H. et al. (2019). Preparation of activated carbon from needle coke via two-stage steam activation process. *Materials Letters* 237: 22–25. https://doi.org/10.1016/j.matlet.2018.09.171.
59. Funke, A. and Ziegler, F. (2010). Hydrothermal carbonization of biomass: a summary and discussion of chemical mechanisms for process engineering. *Biofuels, Bioproducts and Biorefining* 4 (2): 160–177. https://doi.org/10.1002/bbb.198.
60. Demirbas, A. (2004). Effects of temperature and particle size on bio-char yield from pyrolysis of agricultural residues. *Journal of Analytical and Applied Pyrolysis* 72 (2): 243–248. https://doi.org/10.1016/j.jaap.2004.07.003.
61. Titirici, M., Antonietti, M., and Baccile, N. (2008). Hydrothermal carbon from biomass: a comparison of the local structure from poly- to monosaccharides and pentoses/hexoses. *Green Chemistry* 10: 1204–1212. https://doi.org/10.1039/B807009A.
62. Rodríguez Correa, C., Stollovsky, M., Hehr, T. et al. (2017). Influence of the carbonization process on activated carbon properties from lignin and lignin-rich biomasses. *ACS Sustainable Chemistry & Engineering* 5 (9): 8222–8233. https://doi.org/10.1021/acssuschemeng.7b01895.
63. Bhat, V.V., Contescu, C.I., and Gallego, N.C. (2009). The role of destabilization of palladium hydride in the hydrogen uptake of Pd-containing activated carbons. *Nanotechnology* 20 (20): 204011. https://doi.org/10.1088/0957-4484/20/20/204011.
64. Fuertes, A.B., Arbestain, M.C., Sevilla, M. et al. (2010). Chemical and structural properties of carbonaceous products obtained by pyrolysis and hydrothermal carbonisation of corn stover. *Australian Journal of Soil Research* 48 (7): 618–626. https://doi.org/10.1071/SR10010.
65. Dai, L., Chang, D.W., Baek, J.-B. et al. (2012). Carbon nanomaterials for advanced energy conversion and storage. *Micro and Nano: No Small Matter* 8 (8): 1130–1166. https://doi.org/10.1002/smll.201101594.

66. Dreyer, D.R., Ruoff, R.S., and Bielawski, C.W. (2010). From conception to realization: an historial account of graphene and some perspectives for its future. *Angewandte Chemie International Edition* 49 (49): 9336–9344. https://doi.org/10.1002/anie.201003024.
67. Georgakilas, V., Otyepka, M., Bourlinos, A.B. et al. (2012). Functionalization of graphene: covalent and non-covalent approaches, derivatives and applications. *Chemical Reviews*. American Chemical Society 112 (11): 6156–6214. https://doi.org/10.1021/cr3000412.
68. Lee, H.C., Liu, W.W., Chai, S.P. et al. (2017). Review of the synthesis, transfer, characterization and growth mechanisms of single and multilayer graphene. *RSC Advances* 7 (26): 15644–15693. https://doi.org/10.1039/C7RA00392G.
69. Zhao, G., Wen, T., Chen, C. et al. (2012). Synthesis of graphene-based nanomaterials and their application in energy-related and environmental-related areas. *RSC Advances* 2 (25): 9286–9303. https://doi.org/10.1039/C2RA20990J.
70. Gao, M., Pan, Y., Huang, L. et al. (2011). Epitaxial growth and structural property of graphene on Pt(111). *Applied Physics Letters* 98 (3): 33101. https://doi.org/10.1063/1.3543624.
71. Hernandez, Y., Nicolosi, V., Lotya, M. et al. (2008). High-yield production of graphene by liquid-phase exfoliation of graphite. *Nature Nanotechnology* 3 (9): 563–568. https://doi.org/10.1038/nnano.2008.215.
72. Moon, I.K., Lee, J., Ruoff, R.S., and Lee, H. (2010). Reduced graphene oxide by chemical graphitization. *Nature Communications* 1 (1): 73. https://doi.org/10.1038/ncomms1067.
73. Wang, H., Li, Z., and Mitlin, D. (2014). Tailoring biomass-derived carbon nanoarchitectures for high-performance supercapacitors. *ChemElectroChem* 1 (2): 332–337. https://doi.org/10.1002/celc.201300127.
74. Wang, Y., Zheng, Y., Xu, X. et al. (2011). Electrochemical delamination of CVD-grown graphene film: toward the recyclable use of copper catalyst. *ACS Nano* 5 (12): 9927–9933. https://doi.org/10.1021/nn203700w.
75. Shams, S.S., Zhang, L.S., Hu, R. et al. (2015). Synthesis of graphene from biomass: a green chemistry approach. *Materials Letters* 161: 476–479. https://doi.org/10.1016/j.matlet.2015.09.022.
76. Du, Q.S., Li, D.P., Long, S.Y. et al. (2018). Graphene like porous carbon with wood-ear architecture prepared from specially pretreated lignin precursor. *Diamond and Related Materials* 90: 109–115. https://doi.org/10.1016/j.diamond.2018.10.011.
77. Shi, J., Wang, Y., Du, W. et al. (2016). Synthesis of graphene encapsulated Fe_3C in carbon nanotubes from biomass and its catalysis application. *Carbon* 99: 330–337. https://doi.org/10.1016/j.carbon.2015.12.049.
78. Dumesic, J.A., Huber, G.W., and Boudart, M. (2008). Principles of heterogeneous catalysis. In: *Handbook of Heterogeneous Catalysis* (ed. G. Ertl, H. Knözinger, F. Schüth and J. Weitkamp). Online. Wiley https://doi.org/10.1002/9783527610044.hetcat0001.
79. Chen, J., Wang, M., Wang, S. et al. (2018). Hydrogen production via steam reforming of acetic acid over biochar-supported nickel catalysts. *International Journal of Hydrogen Energy* 43 (39): 18160–18168. https://doi.org/10.1016/j.ijhydene.2018.08.048.
80. Liu, W.J., Jiang, H., and Yu, H.Q. (2015). Development of biochar-based functional materials: toward a sustainable platform carbon material. *Chemical Reviews* 115 (22): 12251–12285. https://doi.org/10.1021/acs.chemrev.5b00195.
81. Chellappan, S., Nair, V., Sajith, V. et al. (2018). Synthesis, optimization and characterization of biochar based catalyst from sawdust for simultaneous esterification and transesterification. *Chinese Journal of Chemical Engineering* 26 (12): 2654–2663. https://doi.org/10.1016/j.cjche.2018.02.034.
82. Weber, K. and Quicker, P. (2018). Properties of biochar. *Fuel* 217: 240–261. https://doi.org/10.1016/j.fuel.2017.12.054.
83. Zhao, S.X., Ta, N., and Wang, X.D. (2017). Effect of temperature on the structural and physicochemical properties of biochar with apple tree branches as feedstock material. *Energies* 10 (9): https://doi.org/10.3390/en10091293.
84. Vidal, J.L., Andrea, V.P., MacQuarrie, S.L., and Kerton, F.M. (2019). Oxidized biochar as a simple renewable catalyst for the production of cyclic carbonates from carbon dioxide and epoxides. *ChemCatChem* 11 (16): 4089–4095. https://doi.org/10.1002/cctc.201900290.
85. Cao, X., Sun, S., and Sun, R. (2017). Application of biochar-based catalysts in biomass upgrading: a review. *RSC Advances* 7 (77): 48793–48805. https://doi.org/10.1039/c7ra09307a.
86. Chia, C.H., Gong, B., Joseph, S.D. et al. (2012). Imaging of mineral-enriched biochar by FTIR, Raman and SEM-EDX. *Vibrational Spectroscopy* 62: 248–257. https://doi.org/10.1016/j.vibspec.2012.06.006.

87. Lin, Y., Munroe, P., Joseph, S. et al. (2012). Nanoscale organo-mineral reactions of biochars in ferrosol: an investigation using microscopy. *Plant and Soil* 357 (1–2): 369–380. https://doi.org/10.1007/s11104-012-1169-8.
88. Tasim, B., Masood, T., Shah, Z.A. et al. (2019). Quality evaluation of biochar prepared from different agricultural residues. *Sarhad Journal of Agriculture* 35 (1): 134–143. https://doi.org/10.17582/journal.sja/2019/35.1.134.143.
89. Feng, D., Zhao, Y., Zhang, Y. et al. (2016). Effects of K and Ca on reforming of model tar compounds with pyrolysis biochars under H_2O or CO_2. *Chemical Engineering Journal* 306: 422–432. https://doi.org/10.1016/j.cej.2016.07.065.
90. Lee, J., Kim, K.H., and Kwon, E.E. (2017). Biochar as a catalyst. *Renewable and Sustainable Energy Reviews* 77: 70–79. https://doi.org/10.1016/j.rser.2017.04.002.
91. Feng, D., Zhang, Y., Zhao, Y. et al. (2018). Improvement and maintenance of biochar catalytic activity for in-situ biomass tar reforming during pyrolysis and H_2O/CO_2 gasification. *Fuel Processing Technology* 172: 106–114. https://doi.org/10.1016/j.fuproc.2017.12.011.
92. Liu, Y., Paskevicius, M., Wang, H. et al. (2020). Difference in tar reforming activities between biochar catalysts activated in H_2O and CO_2. *Fuel* 271: 117636. https://doi.org/10.1016/j.fuel.2020.117636.
93. Sun, H., Feng, D., Zhao, Y. et al. (2020). Mechanism of catalytic tar reforming over biochar: description of volatile-H_2O-char interaction. *Fuel* 275: 117954. https://doi.org/10.1016/j.fuel.2020.117954.
94. Liu, Y., Paskevicius, M., Wang, H. et al. (2019). Role of O-containing functional groups in biochar during the catalytic steam reforming of tar using the biochar as a catalyst. *Fuel* 253: 441–448. https://doi.org/10.1016/j.fuel.2019.05.037.
95. Feng, D., Zhao, Y., Zhang, Y. et al. (2017). Effects of H_2O and CO_2 on the homogeneous conversion and heterogeneous reforming of biomass tar over biochar. *International Journal of Hydrogen Energy* 42 (18): 13070–13084. https://doi.org/10.1016/j.ijhydene.2017.04.018.
96. Ashok, J., Dewangan, N., Das, S. et al. (2020). Recent progress in the development of catalysts for steam reforming of biomass tar model reaction. *Fuel Processing Technology* 199: 106252. https://doi.org/10.1016/j.fuproc.2019.106252.
97. Kastner, J.R., Mani, S., and Juneja, A. (2015). Catalytic decomposition of tar using iron supported biochar. *Fuel Processing Technology* 130: 31–37. https://doi.org/10.1016/j.fuproc.2014.09.038.
98. Guo, F., Liang, S., Jia, X. et al. (2020). One-step synthesis of biochar-supported potassium-iron catalyst for catalytic cracking of biomass pyrolysis tar. *International Journal of Hydrogen Energy* 45 (33): 16398–16408. https://doi.org/10.1016/j.ijhydene.2020.04.084.
99. Saka, S. and Kusdiana, D. (2001). Biodiesel fuel from rapeseed oil as prepared in supercritical methanol. *Fuel* 80 (2): 225–231. https://doi.org/10.1016/S0016-2361(00)00083-1.
100. Hsu, A.F., Jones, K.C., Foglia, T.A. et al. (2004). Continuous production of ethyl esters of grease using an immobilized lipase. *Journal of the American Oil Chemists' Society* 81 (8): 749–752. https://doi.org/10.1007/s11746-004-0973-9.
101. Xiao, M., Mathew, S., and Obbard, J.P. (2009). Biodiesel fuel production via transesterification of oils using lipase biocatalyst. *GCB Bioenergy* 1 (2): 115–125. https://doi.org/10.1111/j.1757-1707.2009.01009.x.
102. Tang, Z.E., Lim, S., Pang, Y.L. et al. (2018). Synthesis of biomass as heterogeneous catalyst for application in biodiesel production: state of the art and fundamental review. *Renewable and Sustainable Energy Reviews* 92: 235–253. https://doi.org/10.1016/j.rser.2018.04.056.
103. Chang, B., Fu, J., Tian, Y. et al. (2012). Mesoporous solid acid catalysts of sulfated zirconia/SBA-15 derived from a vapor-induced hydrolysis route. *Applied Catalysis A: General* 437–438: 149–154. https://doi.org/10.1016/j.apcata.2012.06.031.
104. Kim, M., DiMaggio, C., Salley, S.O. et al. (2012). A new generation of zirconia supported metal oxide catalysts for converting low grade renewable feedstocks to biodiesel. *Bioresource Technology* 118: 37–42. https://doi.org/10.1016/j.biortech.2012.04.035.
105. Lam, M.K., Lee, K.T., and Mohamed, A.R. (2009). Sulfated tin oxide as solid superacid catalyst for transesterification of waste cooking oil: an optimization study. *Applied Catalysis B: Environmental* 93 (1–2): 134–139. https://doi.org/10.1016/j.apcatb.2009.09.022.
106. Li, Y., Zhang, X.D., Sun, L. et al. (2010). Fatty acid methyl ester synthesis catalyzed by solid superacid catalyst SO_4^{2-}/ZrO_2-TiO_2/La^{3+}. *Applied Energy* 87 (1): 156–159. https://doi.org/10.1016/j.apenergy.2009.06.030.
107. Mardhiah, H.H., Ong, H.C., Masjuki, H.H. et al. (2017). A review on latest developments and future prospects of heterogeneous catalyst in biodiesel production from non-edible oils. *Renewable and Sustainable Energy Reviews* 67: 1225–1236. https://doi.org/10.1016/j.rser.2016.09.036.

108. Mardhiah, H.H., Ong, H.C., Masjuki, H.H. et al. (2017). Investigation of carbon-based solid acid catalyst from *Jatropha curcas* biomass in biodiesel production. *Energy Conversion and Management* 144: 10–17. https://doi.org/10.1016/j.enconman.2017.04.038.
109. Hara, M. (2010). Biodiesel production by amorphous carbon bearing SO_3H, COOH and phenolic OH groups, a solid Brønsted acid catalyst. *Topics in Catalysis* 53 (11–12): 805–810. https://doi.org/10.1007/s11244-010-9458-z.
110. Dawodu, F.A., Ayodele, O.O., Xin, J. et al. (2014). Application of solid acid catalyst derived from low value biomass for a cheaper biodiesel production. *Journal of Chemical Technology and Biotechnology* 89 (12): 1898–1909. https://doi.org/10.1002/jctb.4274.
111. Zhou, Y., Niu, S., and Li, J. (2016). Activity of the carbon-based heterogeneous acid catalyst derived from bamboo in esterification of oleic acid with ethanol. *Energy Conversion and Management* 114: 188–196. https://doi.org/10.1016/j.enconman.2016.02.027.
112. Lee, H.V., Juan, J.C., and Taufiq-Yap, Y.H. (2015). Preparation and application of binary acid-base CaO-La_2O_3 catalyst for biodiesel production. *Renewable Energy* 74: 124–132. https://doi.org/10.1016/j.renene.2014.07.017.
113. Kaur, M. and Ali, A. (2011). Lithium ion impregnated calcium oxide as nano catalyst for the biodiesel production from karanja and jatropha oils. *Renewable Energy* 36 (11): 2866–2871. https://doi.org/10.1016/j.renene.2011.04.014.
114. Zhang, F., Wu, X.H., Yao, M. et al. (2016). Production of biodiesel and hydrogen from plant oil catalyzed by magnetic carbon-supported nickel and sodium silicate. *Green Chemistry* 18 (11): 3302–3314. https://doi.org/10.1039/c5gc02680f.
115. Thushari, I. and Babel, S. (2018). Sustainable utilization of waste palm oil and sulfonated carbon catalyst derived from coconut meal residue for biodiesel production. *Bioresource Technology* 248: 199–203. https://doi.org/10.1016/j.biortech.2017.06.106.
116. Endut, A., Abdullah, S.H.Y.S., Hanapi, N.H.M. et al. (2017). Optimization of biodiesel production by solid acid catalyst derived from coconut shell via response surface methodology. *International Biodeterioration and Biodegradation* 124: 250–257. https://doi.org/10.1016/j.ibiod.2017.06.008.
117. Dehkhoda, A.M. and Ellis, N. (2013). Biochar-based catalyst for simultaneous reactions of esterification and transesterification. *Catalysis Today* 207: 86–92. https://doi.org/10.1016/j.cattod.2012.05.034.
118. Piker, A., Tabah, B., Perkas, N. et al. (2016). A green and low-cost room temperature biodiesel production method from waste oil using egg shells as catalyst. *Fuel* 182: 34–41. https://doi.org/10.1016/j.fuel.2016.05.078.
119. Shu, Y., Zhang, F., Wang, F. et al. (2018). Catalytic reduction of NO_x by biomass-derived activated carbon supported metals. *Chinese Journal of Chemical Engineering* 26 (10): 2077–2083. https://doi.org/10.1016/j.cjche.2018.04.019.
120. Tang, Z.E., Lim, S., Pang, Y.L. et al. (2020). Utilisation of biomass wastes based activated carbon supported heterogeneous acid catalyst for biodiesel production. *Renewable Energy* 158: 91–102. https://doi.org/10.1016/j.renene.2020.05.119.
121. Zhang, Y., Lei, H., Yang, Z. et al. (2018). Renewable high-purity mono-phenol production from catalytic microwave-induced pyrolysis of cellulose over biomass-derived activated carbon catalyst. *ACS Sustainable Chemistry and Engineering* 6 (4): 5349–5357. https://doi.org/10.1021/acssuschemeng.8b00129.
122. Konwar, L.J., Das, R., Thakur, A.J. et al. (2014). Biodiesel production from acid oils using sulfonated carbon catalyst derived from oil-cake waste. *Journal of Molecular Catalysis A: Chemical* 388–389: 167–176. https://doi.org/10.1016/j.molcata.2013.09.031.
123. Wang, C., Hu, Y., Chen, Q. et al. (2013). Bio-oil upgrading by reactive distillation using p-toluene sulfonic acid catalyst loaded on biomass activated carbon. *Biomass and Bioenergy* 56: 405–411. https://doi.org/10.1016/j.biombioe.2013.04.026.
124. Mateo, W., Lei, H., Villota, E. et al. (2020). Synthesis and characterization of sulfonated activated carbon as a catalyst for bio-jet fuel production from biomass and waste plastics. *Bioresource Technology* 297: 122411. https://doi.org/10.1016/j.biortech.2019.122411.
125. Ahmad Farid, M.A., Hassan, M.A., Taufiq-Yap, Y.H. et al. (2017). Production of methyl esters from waste cooking oil using a heterogeneous biomass-based catalyst. *Renewable Energy* 114: 638–643. https://doi.org/10.1016/j.renene.2017.07.064.

126. Veerakumar, P., Panneer Muthuselvam, I., Hung, C. et al. (2016). Biomass-derived activated carbon supported Fe_3O_4 nanoparticles as recyclable catalysts for reduction of nitroarenes. *ACS Sustainable Chemistry and Engineering* 4 (12): 6772–6782. https://doi.org/10.1021/acssuschemeng.6b01727.
127. Rusanen, A., Lahti, R., Lappalainen, K. et al. (2019). Catalytic conversion of glucose to 5-hydroxymethylfurfural over biomass-based activated carbon catalyst. *Catalysis Today* 357: 94–101. https://doi.org/10.1016/j.cattod.2019.02.040.
128. Patel, A.R., Asatkar, A., Patel, G. et al. (2019). Synthesis of rice husk derived activated mesoporous carbon immobilized palladium hybrid nano-catalyst for ligand-free Mizoroki-Heck/Suzuki/Sonogashira cross-coupling reactions. *ChemistrySelect* 4 (19): 5577–5584. https://doi.org/10.1002/slct.201900384.
129. Quan, C., Wang, H., and Gao, N. (2020). Development of activated biochar supported Ni catalyst for enhancing toluene steam reforming. *International Journal of Energy Research* 44 (7): 5749–5764. https://doi.org/10.1002/er.5335.
130. Zhu, L., Yin, S., Yin, Q. et al. (2015). Biochar: a new promising catalyst support using methanation as a probe reaction. *Energy Science and Engineering* 3 (2): 126–134. https://doi.org/10.1002/ese3.58.
131. Tabak, A., Sevimli, K., Kaya, M. et al. (2019). Preparation and characterization of a novel activated carbon component via chemical activation of tea woody stem. *Journal of Thermal Analysis and Calorimetry* 138 (6): 3885–3895. https://doi.org/10.1007/s10973-019-08387-2.
132. Palomo, J., Rodríguez-Cano, M.A., Rodríguez-Mirasol, J. et al. (2019). On the kinetics of methanol dehydration to dimethyl ether on Zr-loaded P-containing mesoporous activated carbon catalyst. *Chemical Engineering Journal* 378: 122198. https://doi.org/10.1016/j.cej.2019.122198.
133. Akbayrak, S., Özçifçi, Z., and Tabak, A. (2020). Activated carbon derived from tea waste: a promising supporting material for metal nanoparticles used as catalysts in hydrolysis of ammonia borane. *Biomass and Bioenergy* 138: 105589. https://doi.org/10.1016/j.biombioe.2020.105589.
134. Cordero-Lanzac, T., Palos, R., Arandes, J.M. et al. (2017). Stability of an acid activated carbon based bifunctional catalyst for the raw bio-oil hydrodeoxygenation. *Applied Catalysis B: Environmental* 203: 389–399. https://doi.org/10.1016/j.apcatb.2016.10.018.
135. Meryemoglu, B., Irmak, S., and Hasanoglu, A. (2016). Production of activated carbon materials from kenaf biomass to be used as catalyst support in aqueous-phase reforming process. *Fuel Processing Technology* 151: 59–63. https://doi.org/10.1016/j.fuproc.2016.05.040.
136. Titirici, M.M., Thomas, A., Yu, S.H. et al. (2007). A direct synthesis of mesoporous carbons with bicontinuous pore morphology from crude plant material by hydrothermal carbonization. *Chemistry of Materials* 19 (17): 4205–4212. https://doi.org/10.1021/cm0707408.
137. Titirici, M.M., Thomas, A., and Antonietti, M. (2007). Back in the black: hydrothermal carbonization of plant material as an efficient chemical process to treat the CO_2 problem? *New Journal of Chemistry* 31 (6): 787–789. https://doi.org/10.1039/b616045j.
138. Joo, J.B., Kim, Y.J., Kim, W. et al. (2008). Simple synthesis of graphitic porous carbon by hydrothermal method for use as a catalyst support in methanol electro-oxidation. *Catalysis Communications* 10 (3): 267–271. https://doi.org/10.1016/j.catcom.2008.08.031.
139. Morais, R.G., Rey-Raap, N., Figueiredo, J.L. et al. (2020). Highly electroactive N–Fe hydrothermal carbons and carbon nanotubes for the oxygen reduction reaction. *Journal of Energy Chemistry* 50: 260–270. https://doi.org/10.1016/j.jechem.2020.03.039.
140. Wen, Z., Ma, Z., Mai, F. et al. (2019). Catalytic ethanolysis of microcrystalline cellulose over a sulfonated hydrothermal carbon catalyst. *Catalysis Today* 355: 272–279. https://doi.org/10.1016/j.cattod.2019.05.070.
141. Wu, Q., Zhang, G., Gao, M. et al. (2019). Clean production of 5-hydroxymethylfurfural from cellulose using a hydrothermal/biomass-based carbon catalyst. *Journal of Cleaner Production* 213: 1096–1102. https://doi.org/10.1016/j.jclepro.2018.12.276.
142. Wataniyakul, P., Boonnoun, P., Quitain, A.T. et al. (2018). Preparation of hydrothermal carbon as catalyst support for conversion of biomass to 5-hydroxymethylfurfural. *Catalysis Communications* 104: 41–47. https://doi.org/10.1016/j.catcom.2017.10.014.
143. Hu, W., Tong, W., Li, Y. et al. (2020). Hydrothermal route-enabled synthesis of sludge-derived carbon with oxygen functional groups for bisphenol A degradation through activation of peroxymonosulfate. *Journal of Hazardous Materials* 388: 121801. https://doi.org/10.1016/j.jhazmat.2019.121801.

144. Wataniyakul, P., Boonnoun, P., Quitain, A.T. et al. (2018). Preparation of hydrothermal carbon acid catalyst from defatted rice bran. *Industrial Crops and Products* 117: 286–294. https://doi.org/10.1016/j.indcrop.2018.03.002.
145. Ibrahim, S.F., Asikin-Mijan, N., Ibrahim, M.L. et al. (2020). Sulfonated functionalization of carbon derived corncob residue via hydrothermal synthesis route for esterification of palm fatty acid distillate. *Energy Conversion and Management* 210: 112698. https://doi.org/10.1016/j.enconman.2020.112698.
146. Yamaguchi, D., Kitano, M., Suganuma, S. et al. (2009). Hydrolysis of cellulose by a solid acid catalyst under optimal reaction conditions. *Journal of Physical Chemistry C* 113 (8): 3181–3188. https://doi.org/10.1021/jp808676d.
147. Zong, M.H., Duan, Z.Q., Lou, W.Y. et al. (2007). Preparation of a sugar catalyst and its use for highly efficient production of biodiesel. *Green Chemistry* 9 (5): 434–437. https://doi.org/10.1039/b615447f.
148. Yao, Y., Lian, C., Wu, G. et al. (2017). Synthesis of “sea urchin”-like carbon nanotubes/porous carbon superstructures derived from waste biomass for treatment of various contaminants. *Applied Catalysis B: Environmental* 219: 563–571. https://doi.org/10.1016/j.apcatb.2017.07.064.
149. Ming, J., Liu, R., Liang, G. et al. (2011). Knitting an oxygenated network-coat on carbon nanotubes from biomass and their applications in catalysis. *Journal of Materials Chemistry* 21 (29): 10929–10934. https://doi.org/10.1039/c1jm10989h.
150. Das, V.K., Shifrina, Z.B., and Bronstein, L.M. (2017). Graphene and graphene-like materials in biomass conversion: paving the way to the future. *Journal of Materials Chemistry A* 5 (48): 25131–25143. https://doi.org/10.1039/c7ta09418c.
151. Jurca, B., Bucur, C., Primo, A. et al. (2019). N-Doped defective graphene from biomass as catalyst for CO_2 hydrogenation to methane. *ChemCatChem* 11 (3): 985–990. https://doi.org/10.1002/cctc.201801984.
152. Huang, B., Xia, M., Qiu, J. et al. (2019). Biomass derived graphene-like carbons for electrocatalytic oxygen reduction reaction. *ChemNanoMat* 5 (5): 682–689. https://doi.org/10.1002/cnma.201900009.
153. Wu, L., Song, J., Zhou, B. et al. (2016). Preparation of Ru/graphene using glucose as carbon source and hydrogenation of levulinic acid to γ-valerolactone. *Chemistry – An Asian Journal* 11 (19): 2792–2796. https://doi.org/10.1002/asia.201600453.

3

Starbon®: Novel Template-Free Mesoporous Carbonaceous Materials from Biomass – Synthesis, Functionalisation and Applications in Adsorption, and Catalysis

Duncan J. Macquarrie, Tabitha H.M. Petchey and Cinthia J. Meña Duran

Green Chemistry Centre of Excellence, Department of Chemistry, University of York, York, UK

3.1 Introduction

Porous materials are of great importance in a wide range of applications, including adsorption, catalysis, energy applications and composite materials. The majority of these materials are microporous; from the International Union of Pure and Applied Chemistry(IUPAC) definition [1, 2], they contain pores of <2 nm. In reality, the pores are generally constrained to <1 nm, making them able to adsorb small molecules within their pore structure, whereas larger molecules can only adsorb externally, meaning that the majority of the surface area is unavailable. Long micropores also suffer from diffusional issues, and it is recognised that much of a pore system may indeed play little role in the material's performance.

These issues have driven research into the development of mesoporous materials (IUPAC definition encompasses pores from 2 to 50 nm) which allow much faster diffusion and thus utilise a greater proportion of the pore system. These larger pores are generally less easily

High-Performance Materials from Bio-based Feedstocks, First Edition. Edited by Andrew J. Hunt, Nontipa Supanchaiyamat, Kaewta Jetsrisuparb and Jesper T.N. Knijnenburg.

blocked than the much smaller micropores, but a proportion may have constrictions such as the classic 'ink bottle' shape with a narrow entrance to a much larger space. It is very much possible that the mesoporous materials also contain micropores, and the mesopores may play a role in improving access to the micropores.

While (amorphous unstructured) mesoporous materials such as silica and alumina have been known for decades, more regular materials took much longer to develop. It was the early 1990s that saw an explosion of methodologies that delivered a wide range of very controllable routes to highly regular structured materials, largely silicas, but also a few alumina systems and others were developed [3, 4]. This work served to illustrate the advantages of mesoporosity, with a range of applications being demonstrated.

While these strides were being made in inorganic materials, the field of carbon was still dominated by predominantly microporous activated carbons, generally prepared in two steps. The first step is the thermal decomposition of waste biomass, which is followed by a chemical 'activation' step. This second step usually involved reaction with metal species (KOH and $ZnCl_2$ in large quantities) or gases (oxidants such as oxygen or CO_2) in order to open up the structure and generate some mesoporosity. Nonetheless, these materials remain predominantly microporous.

The enormous activity in well-defined, highly mesoporous inorganic materials led to their use as templates, including for carbon, and various groups have published work showing that the mesopores could be filled with organic material which could then be carbonised [5, 6]. Removal of the inorganic template then gave an inverse replica carbonaceous material that often displayed mesoporosity.

These so-called hard templating methods provide a wide range of well-ordered materials [7] but have the drawback of complex and wasteful syntheses, destroying both the template for the silica and the silica itself. An elegant approach has recently been put forward by Jiang et al. [8] who coated the internal surface of mesoporous silicas with a layer of carbon, leading to carbon silica composites (CSCs). Such CSCs directly replicate the porosity of the silica but have a carbonaceous surface. This route is simpler, less wasteful, and retains positive features from the silica, including mechanical strength and the variety of pore architectures that silica can have.

Another approach, first published in 2006, utilises the inherent structure of certain polysaccharides to expand to give mesostructured hydrogels [9]. With suitable drying strategies, these gels can be dried and the resultant aerogel is pyrolysed to provide a highly mesoporous material with excellent pore volume, surface area, and importantly, tunable surface properties ranging from polysaccharide-like at lower pyrolysis temperatures to carbon-like at higher temperatures. While the process consists of three stages, it does not involve templating, instead relying on the inherent properties of the starch to assemble into naturally mesoporous structures. These materials are known as Starbon®, from a combination of *star*ch and car*bon*, and will be the focus of this chapter. Starbon has been commercialised by Starbons Ltd.

3.2 Choice of Polysaccharide

The choice of polysaccharides is a key aspect of the synthesis of Starbon. Not all polysaccharides have been successfully expanded and carbonised using the Starbon approach, notable examples which have proved impossible being cellulose and chitosan. On the other hand, various starches, alginic acid [10], and pectin [11] all readily form mesoporous structures under the appropriate conditions, and these mesostructures can be successfully translated into carbonaceous materials. Recently, Attard et al. [12] have demonstrated that chitosans can be successfully incorporated into Starbon materials via mixing with alginic acid over a range of compositions, maintaining porosity while, at the same time, incorporating nitrogen.

3.2.1 Synthetic Procedure

3.2.1.1 Gelation

The formation of Starbon, irrespective of the nature of the polysaccharide, has three distinct stages (gelation, drying, and pyrolysis). First, gelation of the polysaccharide takes place in water to swell the material (starch, for example, exists in nature as densely packed granules and functions as an energy source) to produce a hydrogel. The gelation is promoted by a combination of temperature and shear forces, which aid in breaking up the tightly packed grain, allowing some of its contents to move outwards, creating porosity. A subsequent step is a retrogradation, where the gel is allowed to 'rest' at 4 °C for approximately 24 hours. This causes slight structural changes and stabilisation.

3.2.1.2 Drying of the Hydrogel

A critical stage is the subsequent drying step, as direct drying of the hydrogel leads to extensive collapse. This is due to the material being relatively soft and structurally weak compared to the significant capillary forces generated on the evaporation of water from the surface of a meniscus (Figure 3.1). These forces act to pull the walls of the pore together, causing collapse if the material is mechanically weak, which is the case with soft polysaccharide gels.

The fact that water has high surface tension and that it forms menisci in nanopores means that direct drying of the hydrogel puts the structure under great stress as the evaporation causes the pore walls to be pulled together – these forces are too great for the porous starch network to survive and structural collapse ensues. Roundabout methods, therefore, need to be utilised in order to avoid these forces and maintain the porosity.

The first methodology utilised solvent exchange – water has a much higher surface tension (72.74 mN m^{-1} at 20 °C [13]) than most organic liquids (e.g. acetone 23.70 mN m^{-1}, ethanol 22.27 mN m^{-1}, and methanol 22.60 mN m^{-1} [14]).

Therefore, exchanging water first for ethanol by simple mixing and filtration (×3) led to approximately 12% water in ethanol-filled gel, which had to be further exchanged with acetone using the

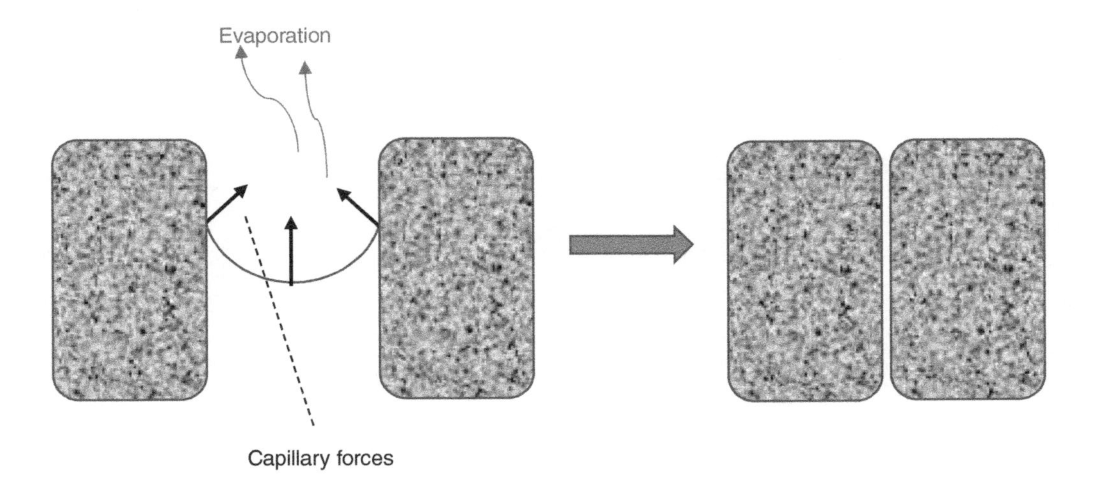

Figure 3.1 *Role of capillary forces in the collapse of soft porous materials.*

same methodology in order to provide an aerogel with porosity intact. What is interesting here is that the gel produced after the ethanol exchanges contains a liquid (c. 12% water in ethanol) which has a surface tension (22.6 mN m^{-1}) [15] very close to that of acetone (23.70 mN m^{-1}) [14], yet fails to lead to a satisfactory material. While this may be due to the evaporation of an ethanol-rich vapour (meaning that the remaining liquid is enriched in water, leading to the surface tension of the remaining liquid increasing and causing damage), more extensive ethanol washing does not lead to better materials. Similarly, direct exchange with acetone does not provide acceptable materials. Thus, both solvents must be used for a successful material.

A major downside to this approach (apart from cost) is the large volume of mixed solvents that are very difficult to purify and reuse. While the materials obtained are very good, the process lacks environmental credentials.

In order to develop a more efficient, less solvent-intensive process, Borisova et al. developed a route involving freeze-drying of hydrogels doped with t-butanol, a molecule often utilised in freeze-drying to control the freezing process [16]. This approach avoids surface tension issues as the phase transition is from solid to gas, and the capillary forces that plague direct drying are thus avoided.

The optimal porosity characteristics were achieved at the eutectic points of the water – t-BuOH system, with a more distinct optimum found at the water-rich eutectic (c. 23% t-butanol). At this solvent composition, the macroporosity that dominates in freeze-drying of the pure water hydrogels is strongly reduced and considerable mesoporosity is developed. While all three polysaccharides display qualitatively similar behaviour, the values of mesopore volume (pectin: 3.1 cm^3 g^{-1}; alginic acid: 1.8 cm^3 g^{-1}; starch: 1.1 cm^3 g^{-1}) are quite different, but nonetheless impressive, in each case (Figure 3.2).

Therefore, compared with a solvent exchange, the quality of the materials (in terms of mesopore content) is excellent, and much less solvent is required as the t-butanol can be recovered and reused.

3.2.1.3 Pyrolysis of the Expanded Aerogel

The final stage of pyrolysis involves a slow temperature ramp from room temperature to the desired temperature. This drives off water, a complex bio-oil (typically formed between 250 and 400 °C) and smaller gases (CO_2, CO, etc.) which are the dominant species at higher temperatures.

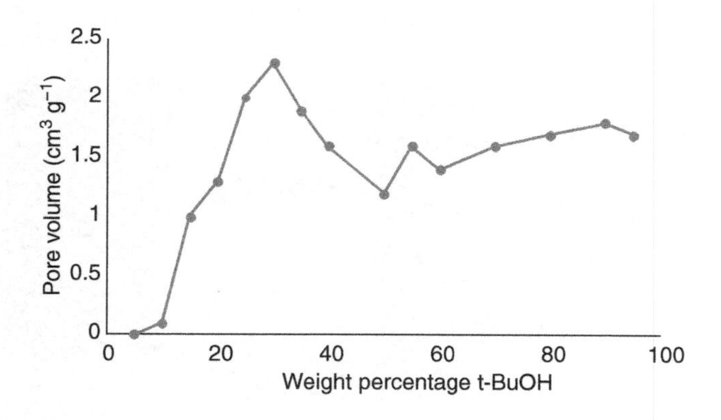

Figure 3.2 Evolution of porosity as a function of water : butanol composition. Eutectic points are at 25 wt% t-BuOH and at 95 wt% t-BuOH. Source: Original data from Borisova et al. [16].

With increasing temperature, the nature of the surface changes from polysaccharide-like, gradually increasing the C : O ratio from 1 : 1 to around 30 : 1 at 800 °C. The increasingly carbon-like nature of the material underlies profound changes in chemistry. While these changes are complex and not fully understood, globally it can be said that, for starch at least, initial dehydrations occur which lead to unsaturation (C=C and C=O functionality) leading to an enhanced reactivity. These groups react further, cross-linking and releasing small organic molecules into the gas phase to form a broadly aromatic matrix [9]. This is somewhat of an oversimplification as significant aliphatic functionality is always seen by ^{13}C NMR. This may be due, in part at least, to the early dehydration of sugars, which can lead initially to $-CH_2-C(=O)-$ units, although the reactivity of such groups suggests that more structural features of greater stability are initially formed. After the initial substantial mass loss, further heating causes much more gradual mass loss, predominantly in the form of small molecules such as water, CO, and CO_2.

Thus, the chemical functionality of the surface of the Starbon materials represents a continuum of changing functionality ranging from hydroxylic to highly functional (hydroxylic, unsaturated) to a more aromatic, low oxygen structure.

The mesoporosity of the materials compared to the total (meso + micro) porosity are shown in Figure 3.3 for the three materials derived from alginic acid, starch, and pectin [9–11].

What can be seen from a comparison of the three types of materials is that, overall, the pore volumes remain fairly constant over a wide temperature range. However, in the case of alginic-acid-derived materials, there is an increase at lower temperatures followed by a drop and then relative constancy. For room temperature pectin, there is evidence of increasing porosity. The total volumes are broadly constant over all three material types. The most variation is in the difference between total and mesopore volumes, indicating the extent of microporosity. Alginic-acid-derived materials have virtually no microporosity at any temperature and pectin a modest amount. In contrast, starch-derived materials display very little microporosity at low pyrolysis temperatures, but from 300 °C onwards, the materials develop a considerable amount (up to c. 30%).

Critical to maintaining the porosity of the materials is that the pyrolysis and cross-linking reactions occur before the aerogel melts or softens considerably. For alginic acid and pectin, the more reactive nature of the polysaccharides' structure – acid and ester groups – means that this is not an issue, while for starch, a strong Brønsted acid (typically p-toluene sulphonic acid) must be added.

3.2.2 Derivatisation

Since the development of Starbon as a high-surface-area mesoporous carbon, multiple reports have been published on its use as solid catalyst support. Starbon itself has a very modest catalytic activity – as will be seen next, there is little more than some weakly acidic groups on the surface. Therefore, the addition of more powerfully active groups (e.g. sulphonic acid sites or basic amines) is likely to provide a much more active material for catalytic applications, while N functionalised materials may also improve adsorbency of metals and function as ligands for catalytically active surfaces.

3.2.2.1 Sulphonation

3.2.2.1.1 Method of Sulphonation

The first published method of Starbon sulphonation was described by Budarin et al. [17]. In this synthesis, sulphonation was carried out on Starbon having been pre-carbonised at 400 °C (Starbon-400) as this was considered to have the best ratio of hydrophilic and hydrophobic

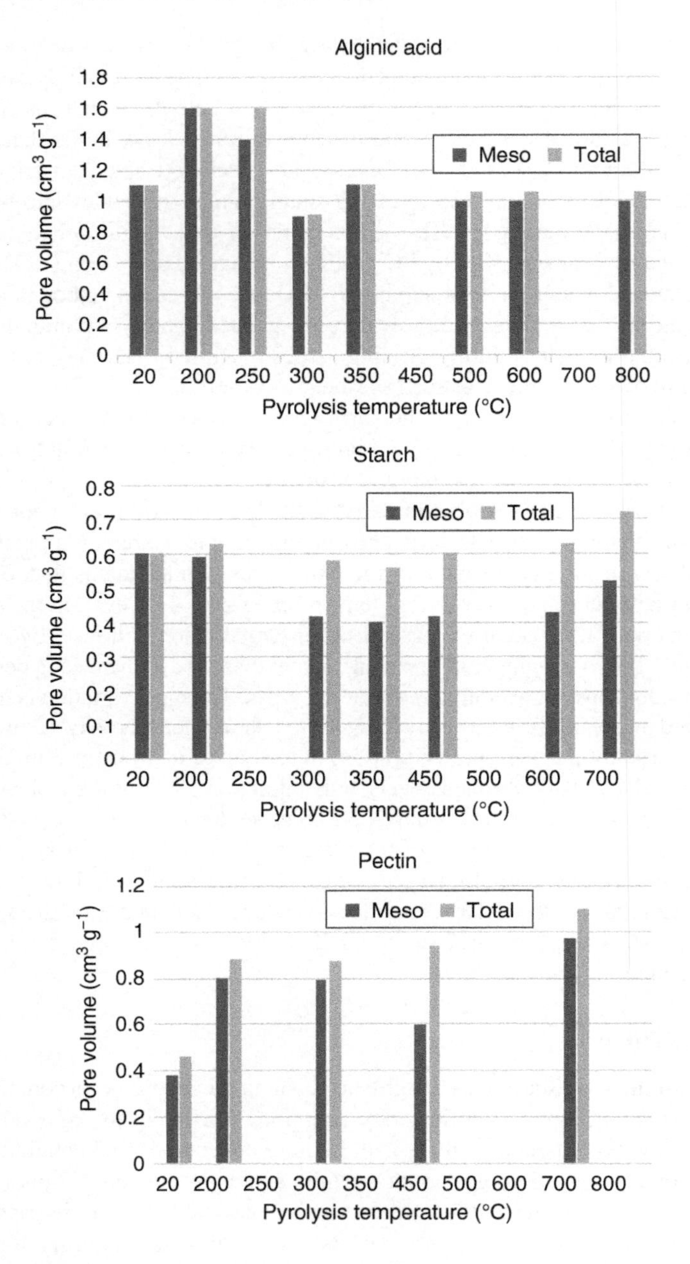

Figure 3.3 *Comparison of porosity of the three major Starbon types. Source: Original data adapted from Budarin et al. [9], White et al. [10], White et al. [11].*

functional groups. The Starbon-400 was heated for four hours at 80 °C in a suspension of H_2SO_4 (99.999% purity) at a ratio of 10 ml acid to 1 g Starbon, and the sulphonated material was subsequently washed with distilled water until no further acid was leached. This was followed by conditioning in toluene for four hours at 150 °C and then in water for three hours at 100 °C. The resultant solid acid was dried overnight in an oven at 100 °C. Subsequent publications from the same authors noted a reduction in the aqueous conditioning temperature to 80 °C [18, 19].

A more recent publication expanded on this synthesis by variation of the sulphonating agent and suspension times [20]. In this report, Starbon-300 was chosen as the base material. The first synthesis was similar to that previously reported, except that a ratio of 0.2 g Starbon to 10 ml H_2SO_4 (96%) was used, and a suspension time of 15 hours. Further syntheses were reported using a mixture of $ClSO_3H/H_2SO_4$ as the sulphonating agent, in ratios of 2 : 10 and 3 : 10 by volume, respectively. In these cases, sulphonation was carried out at reflux for five hours.

3.2.2.1.2 Characterisation of Sulphonated Material

Sulphonated Starbon has been analysed thoroughly in terms of porosity, elemental composition, and SO_3H loading: all characteristics affecting the validity of these materials as solid acid catalysts. In the original publication, the SO_3H loading on Starbon-400 was reported as 0.5 mmol g^{-1} as determined by thermogravimetry coupled to infrared spectroscopy (thermogravimetry-infra red [TGIR]) [17]. Further investigation showed that the loading remained in this region for preparation temperatures of 600 °C and below. Starbon-650 and above resulted in a lower concentration of active acid sites (0.3–0.4 mmol g^{-1}) [18, 21]. This was further corroborated by pyridine absorption experiments showing that the acidity of sulphonated Starbon did not change considerably with preparation temperatures of 600 °C or less. However, stronger Lewis acidity was observed at lower temperatures [18, 19]. Later, Aldana-Pérez et al. reported a $-SO_3H$ content of 1.2, 1.8, and 2.3 mmol g^{-1}, and a total number of acid sites as 8.0, 8.2, and 10 mmol g^{-1}, respectively, for Starbon-300 sulphonated by H_2SO_4 for 15 hours, a mixture of 2 : 10 $ClSO_3H/H_2SO_4$ for five hours, and a mixture of 3 : 10 $ClSO_3H/H_2SO_4$ for five hours [20]. In this case, acidity was measured by potentiometric titration.

Transmission electron microscopy (s) and scanning electron microscopy (SEM) imaging of Starbon have shown that sulphonation did not considerably change the particle size or morphology of the material, which is amorphous after functionalisation, but the structure was prone to cracking at higher carbonisation temperatures as is true for the unsulphonated Starbon [18]. Starbon-400 was shown to remain predominantly mesoporous (pore size 5–15 nm) after sulphonation, although the mean pore diameter and surface area decreased. As with the original Starbon, microporosity increased with preparation temperatures above 500 °C [19, 21]. A homogeneous distribution of elements was also observed [20]. Additional N_2 adsorption studies showed that the average pore diameter of sulphonated Starbon was 8–12 nm [21]. Sulphonated Starbon-400 had a Brunauer-Emmett-Teller (BET) surface area of 386 $m^2\ g^{-1}$ and pore volume of 0.62 $cm^3\ g^{-1}$. Aldana-Pérez et al. reported a decrease in BET surface area on sulphonation from 163 $m^2\ g^{-1}$ for Starbon-300 to 66, 75, and 77 $m^2\ g^{-1}$, respectively, for Starbon-300 sulphonated by H_2SO_4 for 15 hours, a mixture of 2 : 10 $ClSO_3H/H_2SO_4$ for five hours, and a mixture of 3 : 10 $ClSO_3H/H_2SO_4$ for five hours [20].

Diffuse reflectance infra red (DRIFT IR) analysis of the original sulphonated Starbon-400 showed little change in surface structures before and after sulphonation [18, 21]. Peaks in the region of 1300–600 cm^{-1} were attributed to $-SO_3H$ groups confirming the incorporation of sulphur into the Starbon structure, although differentiation between $O-SO_3H$ and $C-SO_3H$ was not possible. Aldana-Pérez et al. performed fourier Transform infra red (FTIRP) photoacoustic spectra – an infrared technique particularly well suited to study dark samples – of the sulphonated and presulphonation Starbon, as shown in Table 3.1 [20]. In particular, this confirmed the presence of $-SO_3$ groups in the sulphonated Starbon by the appearance of the S=O band at approximately 1040 cm^{-1}. Additionally, there is a reduction in the CH_2/CH_3 signal at approximately 3000 cm^{-1}, which was attributed to oxidation of aliphatic carbons to carboxylic acids, although oxidation to ketones should also not be excluded in Starbon-300, where significant carbohydrate character (in particular –CH(OH)– groups) should remain.

***Table 3.1** Summarised IR data of a range of sulphonated and unsulphonated Starbon.*

Wavenumber (cm^{-1})	Vibration	Appearance
1040	S=O symmetric stretch	Sulphonated
1610	Aromatic C=C stretch	Both
1719	C=O stretch	Both
3000	CH_2/CH_3	Unsulphonated
3440	C(O)*OH*/PhO*H* stretch	Both

Source: From Aldana-Pérez et al. [20].

Solid and liquid ^{13}C NMR data were also collected for the sulphonated catalyst, Starbons-300-2–H_2SO_4–$ClSO_3H$–TEPO, using triethylphosphine oxide (TEPO) as a probe molecule to find the relative acid strength [20]. This was done by dissolving 0.05 g of material in 5 ml of 1 M TEPO in hexane, stirring for one hour, followed by drying at 40 °C under vacuum. The resulting solid was dissolved in deuterated chloroform. Having managed to get better resolution of signals in the liquid state, bands at 128.94, 131.05, and 132.54 ppm were assigned to aromatic carbons while band 167.94 ppm was assigned to carboxylic carbons. The ^{31}P MAS NMR (magic angle spinning nuclear magnetic resonance) of the same material showed the shift of a signal from 53.89 ppm in the unsulphonated Starbon to 85.30 ppm in the sulphonated material, which was attributed to the change from hydroxyl to sulphonic groups, whose interaction with TEPO confirmed strong Brønsted acid character.

Elemental analysis of the original sulphonated Starbon gave a sulphur content of 1.9% (Starbon-400-SO_3H), 1.4% (Starbon-650-SO_3H), and 1.3% (Starbon-750-SO_3H) [21], whereas Aldana-Pérez et al. reported a higher sulphur content of 3.30% by X-ray photoelectron spectroscopy (XPS), possibly due to the much longer duration of sulphonation in their method [20].

3.2.2.2 N-Starbons

3.2.2.2.1 Methods of N Incorporation

Attard et al. first reported a method for the N-doping of Starbon in 2018 [12] by using chitosan (a natural N-containing polysaccharide) or ammonia as nitrogen sources. Three methods were described. Route A involved the gelation of chitosan and alginic acid together in the formation of Starbon. Routes B and C involved injecting ammonia into the Starbon at the aerogel stage and into the final product, respectively, where ammonia was adsorbed onto the surface of the material. These N-doped materials are collectively referred to as N-Starbon.

Another study used the impregnation of Starbon with monoethanolamine [22]. This involved mixing Starbon derived from both corn and potato starch with a known amount of amine in a solution of ethanol and water. The product was subsequently dried at 105 °C.

3.2.2.2.2 Characterisation

In the account by Attard et al. [12], N-doped Starbon produced by either the inclusion of chitosan or ammonia displayed similar characteristics. A significant amount of nitrile was observed in these materials, something that was known to be unusual in other forms of N-doped carbons [12]. The TGIR results showed that chitosan released ammonia upon heating to low temperatures, whereas N-Starbon did not. It was concluded that the amine groups in chitosan may become trapped in the form of nitriles, the presence of which was confirmed by a DRIFT absorbance band at 2210 cm^{-1}. The nitrile was assumed chemically bonded to the Starbon as it remained present after generous washing with ethanol and water.

The presence of nitriles was further explored by preparing a range of materials and it was found that nitriles only form at carbonisation temperatures above 300 °C. It was also shown that the combination of chitosan and polysaccharide precursor to Starbon was necessary to retain the mesoporous structure of the N-doped materials, as chitosan alone collapsed to a microporous structure on carbonisation.

Porosimetry was carried out on N-Starbon from each synthetic route, showing that N-doping did not negatively influence mesoporosity. Route A, using a chitosan/polysaccharide combination, resulted in larger pore volumes and surface areas than the ammonia adsorption method. A summary of BET surface area studies and pore volume is shown in Table 3.2.

A large amount of nitrogen was detected in all samples, as confirmed by elemental analysis and XPS, and displayed in Table 3.3.

In a study to produce N-Starbon for the application of carbon capture, Sreedhar et al. found that higher carbonisation temperatures of 750 °C for a duration of six hours were required to achieve the high surface areas and pore volumes needed [22]. Different weight loadings of monoethanolamine (10, 20, and 30%) were characterised by FTIR, with amine signals being present at 1650–1550 cm^{-1}. Further studies by X ray diffraction (XRD) confirmed that N-Starbons from both corn and potato starch were amorphous materials with very little difference between them.

Table 3.2 *Textural properties of N-doped Starbon compared with N-free analogues.*

Carbonisation temp. (°C)	Material	Mesoporosity (%)	Total pore volume (cm^3 g^{-1})	BET surface area (m^2 g^{-1})
300	Starbon	88.3	0.627	174.2
	N-Starbon-A	86.7	0.663	412.6
	N-Starbon-B	98.7	0.645	240.5
450	Starbon	84.5	0.515	339.5
	N-Starbon-A	81.9	0.615	448.3
600	Starbon	79.8	0.686	519.5
	N-Starbon-A	83.5	0.750	519.2
	N-Starbon-C	92.8	0.324	249.5

Source: Data from Attard et al. [12].

Table 3.3 *Nitrogen content of N-doped Starbons.*

	Nitrogen content (wt%)		C/N ratio	
Carbonisation temp. (°C)	CHN	XPS	CHN	XPS
300-A	6.4	4.9	9.2	15.3
300-B	6.8	6.1	8.8	12.4
450-A	10.6	11.6	5.7	5.8
600-A	6.5	7.0	11.1	11.3
600-C	12.4	12.9	5.8	5.9

Source: Data from Attard et al. [12].

3.2.2.3 Derivatisation via Bromination and Activation of Hydroxyl Functionality

According to the functionality of the Starbon materials, particularly (but not exclusively) the lower temperature materials should have a significant number of –OH functionalities as well as several alkene and similar sites. Such materials clearly have the potential for functionalisation too, although less work has been published in utilising these functional groups as anchor points for groups (e.g. epoxides which could be further elaborated). Sreedhar's approach is likely to bind the amines to the surface via aldehydes and other active groups on the surface, reactions that should be readily achieved under mild conditions [22]. Two other articles indicate the potential for further functionalisation of the Starbon surface, the results of which will be discussed in more detail in the catalysis section. The first approach, from Matharu et al. [23], utilises the surface hydroxyls of the uncarbonised material and of a Starbon-350 material to attach a complex ligand to the Starbon surface via succinimidyl carbonate functionalisation. This approach resulted in significant loadings (a degree of substitution of 0.33 was achieved for the uncarbonised material), representing an approximately 10% functionalisation of the total hydroxyl population (and therefore significantly more in terms of accessible surface groups).

In an alternative approach, Carneiro et al. [24] used the unsaturation present in a Starbon-700 to first brominate and then displace the bromine atoms to build a bis-oxazolidine structure onto the surface. The lability of the bromines suggests that alkene functionality is common on such materials, as the alternative bromination of (electron-rich) aromatics, which are also present, might occur even under the very mild conditions used, but this would lead to less active sites for functionalisation. A 10% drop in the C content of the brominated material suggests a substantial amount of bromination took place, and loadings of the catalyst (0.5 wt% Cu) are significant.

3.2.3 Applications

3.2.3.1 Catalysis

A range of catalytic applications of Starbon materials has been investigated, mainly with sulphonated Starbons, where strong acid functionality has been included via reaction with sulphuric acid.

3.2.3.1.1 Sulphonated Starbon in Esterifications

As discussed earlier, reaction of Starbon with concentrated sulphuric acid has been achieved [18, 19] and good loadings of sulphonic acid/sulphuric acid esters have been obtained. These materials have proved themselves to be very promising solid acids in a range of reactions as described next.

Esterification of succinic acid was shown to be possible, even in water-rich environments. Succinic acid, an important platform molecule, can be produced in high concentration (up to c. 12 wt%) by fermentation of waste polysaccharides [25]. Clark and Budarin focused on valorising such fermentation broths containing succinic acid, as recovery of the acid itself from such complex media is extremely difficult and generates considerable waste. Conversion to di-ethyl succinate, a liquid with low miscibility with the broths, is an elegant solution, but esterifications in water are theoretically likely to be unsuccessful due to reversibility. Interestingly, Clark et al. showed that sulphonated Starbon was a very effective esterification catalyst, giving the diester in excellent yields (95% after eight hours' reaction), and under conditions where a range of other acid catalysts failed to give more than modest quantities of diester, even after considerably longer periods of reaction [18, 19].

As can be seen from Figure 3.4, there is a significant rate dependence on the Starbon carbonisation temperature, and the optimum activity is not a simple function of acid site density. Textural properties are relatively similar throughout, and therefore there may be a significant role being played by the nature of the surface beyond the active sites themselves. One potential factor is hydrophobicity – pores that are prone to exclude water (even partially) will allow the esterification equilibrium to shift toward the ester product, whereas hydrophilic pores will do the opposite. Low acidity in the bulk liquid will ensure that related esters will hydrolyse only very slowly.

Further bio-derived acids (itaconic, fumaric, and levulinic) were also successfully converted to their esters/diesters under similar conditions [17]. In the same paper, a range of benzylic alcohols were also esterified with acetic acid under microwave irradiation within one minute, with phenol also reacting, albeit 10 times more slowly.

Oleic acid was successfully esterified to give ethyl oleate using sulphonated Starbon-300 [20]. The blank reaction and Starbon-300 gave essentially no conversion, but Starbon-300 sulphonated with sulphuric acid gave a 45% conversion to the ester after 24 hours. Starbon-300 sulphonated with a mixture of $ClSO_3H$ and sulphuric acid shows considerably better activity and gave conversions of 90% after 10 hours. The rationale for the enhanced activity was that the latter catalysts had a significantly higher loading of acidic groups, which lead to approximately 20-fold greater turnover frequencies. However, the comparisons shown in Figure 3.5 suggest that, as mentioned earlier, there are other factors at play. The nonaqueous reaction medium (in contrast to the succinic acid system as mentioned earlier) means that hydrophobicity/hydrophilicity is unlikely to be as relevant in this case. It may well be that the strongly acidic nature of the sulphonation medium alters the surface by, e.g. dehydration or cross-linking as well as via sulphonation, and that this also plays a role in the activity of the materials.

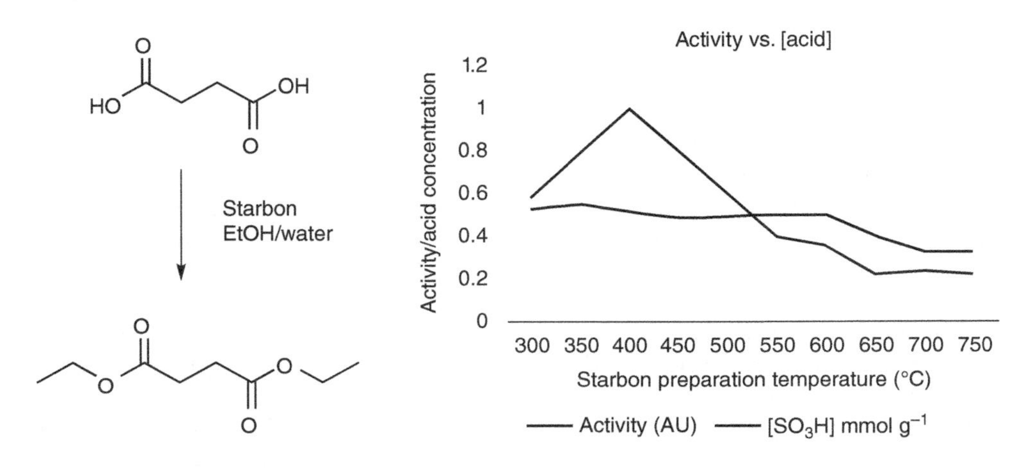

Figure 3.4 *Diesterification of succinic acid in aqueous ethanol as a function of Starbon preparation temperature. Source: Data from Clark et al. [19].*

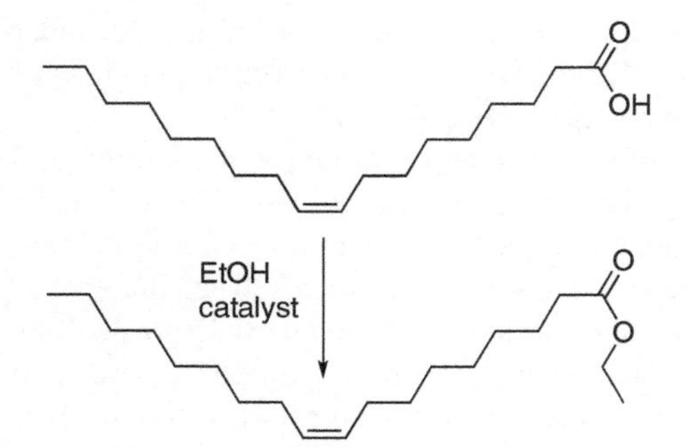

Catalyst	mmol H⁺/g	TOF
S300-S	8.0	23
S300-ClA	8.2	509
S300-ClB	10.0	471

Catalyst: S300-S – Starbon-300 treated with sulfuric acid
S300-ClA – Starbon-300 treated with sulfuric acid and chlorosulfonic acid at a 5 : 1 ratio
S300-ClB – Starbon-300 treated with sulfuric acid and chlorosulfonic acid at a 5 : 2 ratio

TOF = conversion at three hours per mmol SO_3H

Figure 3.5 *Esterification of oleic acid with sulphonated Starbon materials. Source: Data from Aldana-Pérez et al. [20].*

3.2.3.1.2 Sulphonated Starbon in Dehydrations

Millán et al. [26] have demonstrated the use of Starbon-450-SO_3H in the dehydration of xylose to furfural in biphasic conditions. They found that, in a single aqueous phase, conversion of xylose was readily achieved, but furfural was further degraded, meaning selectivity and yield of furfural was low (38% yield and 65% selectivity at best). The addition of cyclopentyl methyl ether (CPME) as a water-immiscible co-solvent was found to aid the reaction considerably by rapidly removing furfural from the aqueous solution, leading to optimal performance of 96% conversion and 72.5% selectivity after one hour at 200 °C (Figure 3.6). Interestingly, given the rather challenging conditions, reusability (measured at 175 °C for 18 hours) was excellent, with no change in performance over 3 runs.

3.2.3.1.3 Sulphonated Starbon in Amide Synthesis

Starbon-400-SO_3H has been used to form amides in excellent yields from a range of anilines and acetic acid under microwave irradiation [27].

After 15 minutes, at a maximum temperature of 130 °C, yields approaching quantitative were achieved for a range of aliphatic amines and anilines. Interestingly, the simplest primary amines (C_5, C_6, and C_{10}) tended to give slightly poorer yields, despite them being typically more active than aromatic systems. As expected in amide formation, selectivity was excellent. Other aliphatic acids were also screened with very good results, with secondary acids being slightly less reactive on steric grounds. Other strong solid acids were investigated with significantly lower conversions.

The same catalyst was utilised by Mesquita et al. in the acid-catalysed Ritter reaction of steroids, again forming amide products [28]. Starting from epoxy steroids and acetonitrile, the authors found that Starbon-400-SO_3H catalysed the ring opening, and trapping of the resultant carbocation via hydration gave the resultant product as shown in Figure 3.7.

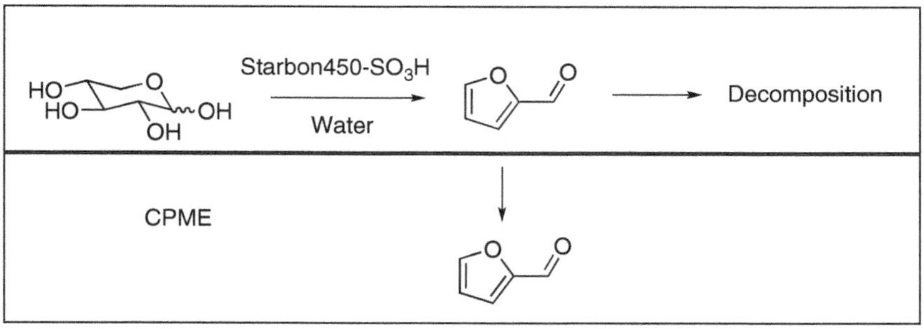

Figure 3.6 Conversion of xylose to furfural and extraction of furfural.

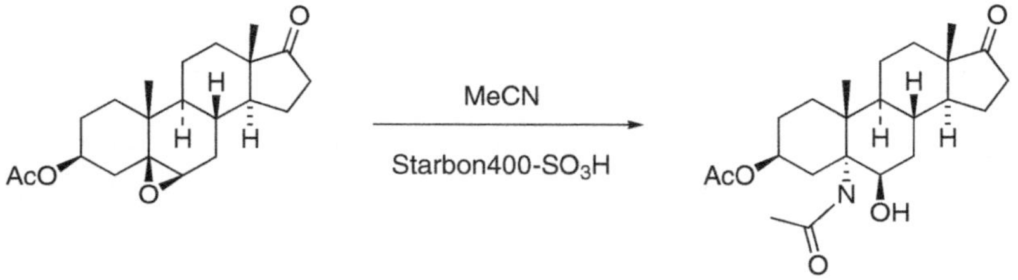

Figure 3.7 Functionalisation of steroids via Starbon acid-catalysed Ritter reaction.

A series of different steroids was successfully functionalised in yields of 88–96%, with a lower yield of 60% being obtained with a steroidal epoxide of the opposite configuration to that shown in Figure 3.7. Similar yields were obtained when methylthioacetonitrile ($MeSCH_2CN$) was used, with the product being the $MeSCH_2C(O)N$-product. With $MeSO_2CH_2CN$ as the amide, oxazoline was formed by trapping of the intermediate carbocation (formed by attack of the nitrile on the protonated epoxide) by the vicinal hydroxy group.

3.2.3.1.4 Sulphonated Starbon in Acylations and Alkylations

Luque et al. have also published research on a range of different Starbon acids, with sulphonic and carboxylic groups attached, as well as $ZnCl_2$ and BF_3 adsorbed onto the surface [29]. They characterised the nature of the acidity by pyridine titration and found that the sulphonic acid systems had both Brønsted and Lewis acidity. In contrast, the two Lewis acid-doped materials had a greater Lewis acidity (as expected) but an unexpectedly low quantity of acid sites. The various materials were tested in the acetylation of 5-acetyl methyl salicylate, and were found to be very active in the O-acetylation, with Friedel Crafts acylation being absent despite the presence of Lewis acidic sites (Figure 3.8a). Alkylation of phenol with cyclohexene was also investigated (Figure 3.8b). While O-alkylation was very dominant, especially after short reaction times, C alkylation (mainly in the ortho position) increased upon prolonged reaction. The reasons for this are not discussed in the paper, but similar behaviour is seen with some other catalysts. This may be due to a change in the nature of the catalyst with time, or more likely, to a slow Fries rearrangement reaction following a rapid O-alkylation.

Further alkylations (of toluene and xylenes) with benzyl chloride have also been successfully carried out using Starbon-400-SO_3H under microwave conditions. High yields (70–95%)

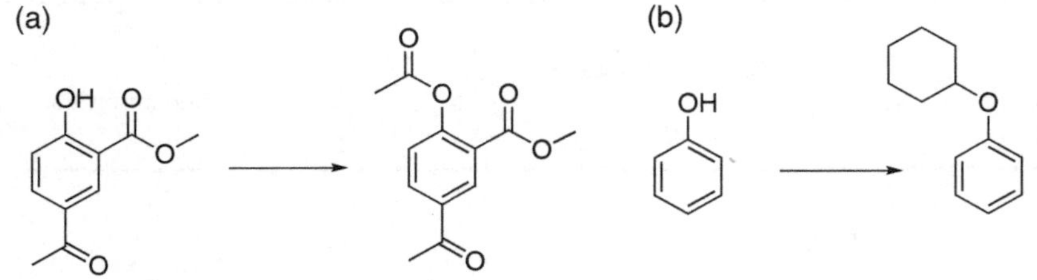

Figure 3.8 Friedel Crafts reactions catalysed by a range of Starbon acids. (a) O-Acylation of a phenol and (b) O-alkylation of a phenol.

were achieved after 15 minutes for toluene and p-xylene, while longer times were required for m-xylene and, surprisingly, anisole [17].

3.2.3.1.5 Supported Metal Complexes

Only a very few examples of supported metal complexes and their catalytic activity have been reported, perhaps surprising, given the large numbers that exist with silica as support. Interestingly, little has been done to extend the well-established organo-silica chemistry beyond the initial studies by Doi et al. [30] who used expanded starch (i.e. non-pyrolysed Starbon) as a support. Two approaches that successfully attach complex catalytic species to the surface of Starbon are available, and are discussed next.

Matharu et al. [23] published details of an Fe-NHC catalytic system (NHC = N-heterocyclic carbene), which they tethered to the surface of a starch-based Starbon-350 and also to the surface of the precursor-expanded starch (Figure 3.9). They utilised a multistage ligand synthesis, anchored the ligand to the surface, and then finally attached the metal species [23].

The synthesis of the heterocyclic ligand containing amine functionality as an anchor was carried out. Separately, Starbon surface was functionalised with a succinimyl carbonate group, following an adapted literature procedure, where the toxic dimethyl formamide (DMF) solvent was replaced with propylene carbonate, a safer alternative to dipolar aprotics [31]. Finally, the ligand was bound to the functionalised Starbon surface. Anchoring of the ligand system was achieved by reaction of the functionalised Starbon with the amine pendant on the ligand moiety. The degree of substitution achieved for the succinimidyl carbonate grafting was 0.33, approximating to 1 in 10 hydroxyls being substituted (in the case of the expanded starch, this is likely to be somewhat higher for the Starbon-350, although the complexity of the structure is much greater with a wider range of functionalities). Given the extensive H-bonding and steric hindrance pertinent to the majority of the hydroxyls in such polysaccharides, this is a reasonably significant degree of substitution, and led to metal centre loadings of 0.26 and 0.3 $mmol\,g^{-1}$, well within the range of loadings achieved for highly porous silicas.

Catalytic activity was very promising in the dehydration of fructose to 5-hydroxymethyl-2-furaldehyde (HMF), with the expanded starch catalyst slightly outperforming the Starbon-350 material (86% vs. 81% yield after 0.5 hour at 100 °C). Reuse was also very good, with consistent performance over 5 runs, and no discernible leaching of iron. Given the simpler route to the expanded starch material, it is clear that this is the catalyst of choice here.

While Matharu's approach utilised the abundant hydroxyl groups on the surface of the material, Silva and co-workers have described the synthesis of a chiral bis-oxazoline catalyst

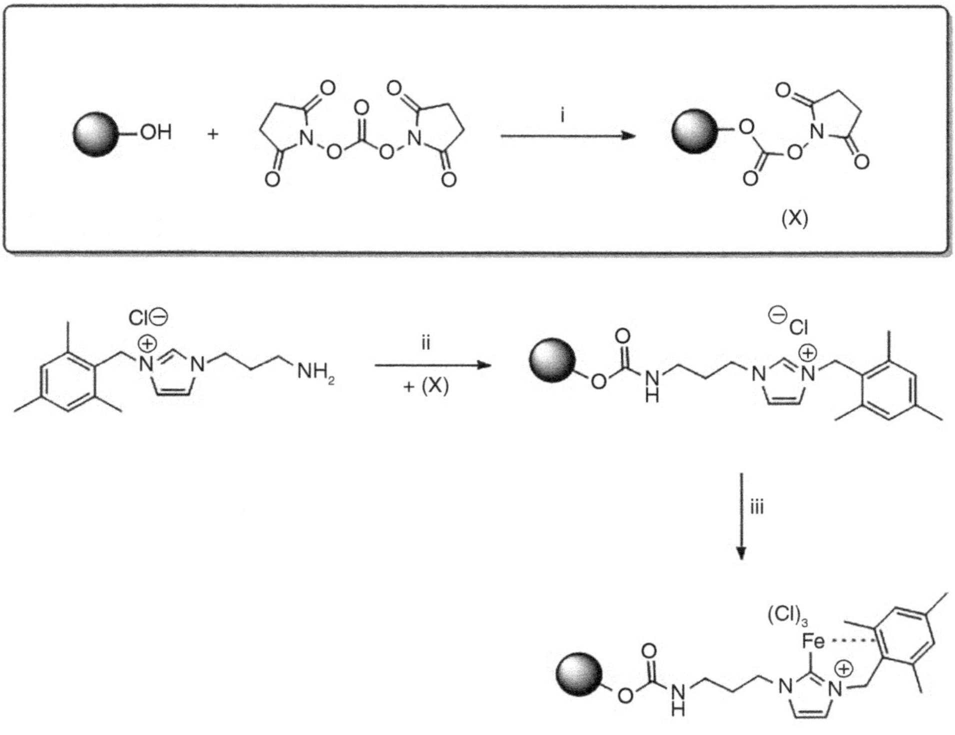

Figure 3.9 *Synthesis of an N-heterocyclic carbine-based catalyst on the Starbon surface.*

on the surface of a Starbon-700 material utilising the abundant unsaturation present in the higher temperature materials [24]. Given the low O content of these higher temperature materials, the typical functionalisation routes involving surface hydroxyls (i.e. reaction with silane esters such as $(RO)_3SiR'$ [30] and the succinimidyl carbonate route described earlier) are unsuitable. To circumvent this problem, Carneiro et al. [24] utilised a bromine functionalisation, reacting bromine with double bonds on the surface, to give a 1,2-dibromo-functionalised surface (Figure 3.10). As the ligand to be attached is a diol, the formation of pairs of anchor sites close together is particularly attractive.

The resultant materials were used in the kinetic resolution of 1,2-diphenylethane-1,2-diol. While activity was high, the enantiomeric excess of the process was significantly lower than that obtained by a solution-phase process (Figure 3.11). This was attributed to a low concentration of catalytic groups on the surface of the catalyst, and the high conversion suggests non-chiral-active sites may also be present.

3.2.3.1.6 Photocatalytic Processes

A further aspect of Starbon catalysis relates to the photocatalytic decomposition of water pollutants. Colmenares and colleagues have recently published work [32] that illustrates the excellent activity of Starbon-based titanium dioxide photocatalysts for the total oxidation of phenol, one of the more toxic of water-borne pollutants.

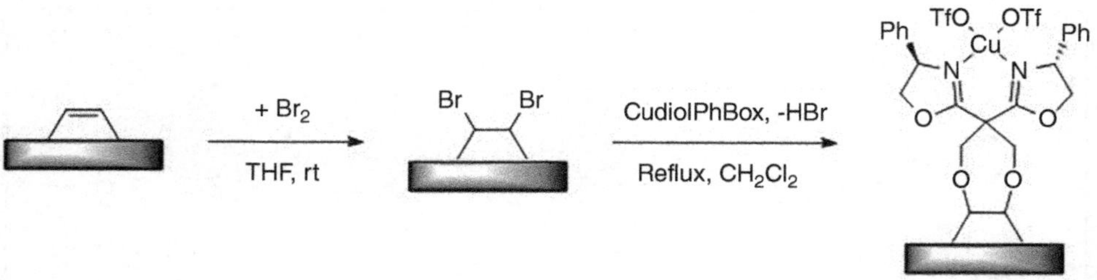

Figure 3.10 Preparation of a chiral bis-oxalolidinone catalyst via surface bromination/displacement.

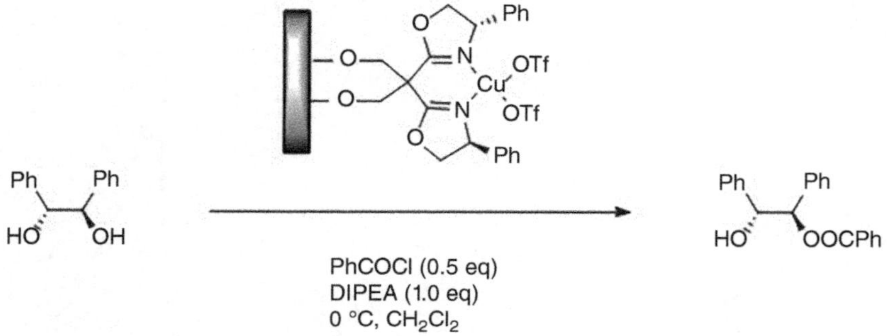

Figure 3.11 Selective acylation of a diol by a supported Cu-bis-oxalodinone/Starbon catalyst.

Part of the rationale for choosing Starbon as a support for the titania catalyst is that the TiO_2 particles need to be nanoscale in order to have sufficient surface area to function efficiently. This raises two issues. First, separation of the nanoparticles from water is a major challenge, and second, agglomeration of the nanoparticles is a significant issue, which reduces the activity and efficiency of the catalytic system.

In their work, Colmenares et al. [32] utilised ultrasound-assisted formation and deposition of titania onto thermally pre-treated Starbon, avoiding the use of surfactants, and adding to the green credentials of the process. The materials were then dried and re-calcined to produce a hybrid material containing 25 wt% of TiO_2. The authors suggest that carboxylic groups on the surface act as nucleation sites for the titania particles, which is further discussed in a second paper [33]. Interestingly, only the anatase form of titania was formed on Starbon, whereas similar amounts of anatase and rutile were produced on Norit, a standard activated carbon. This is important because anatase is the catalytically active form. The crystallinity was also found to be excellent on Starbon, meaning few defects are present – defects reduce the efficiency of the photocatalytic process by facilitating electron-hole recombination. Graphene oxide seemed to produce almost no discernible crystalline material. Titania particle sizes for Starbon- and Norit-derived materials were largely similar and in the range 20–30 nm. Loading of the inorganic phase resulted in a reduction in surface area by a factor of 1.6 for Starbon, and 1.3 for graphene oxide, but the microporous Norit suffered a sevenfold decrease in surface area.

Photochemical mineralisation of phenol was carried out using 50 ppm solutions of phenol in water, a Hg lamp (365 nm) and without bubbling of air through the solution (i.e. passive diffusion of oxygen into the solution). The Starbon materials had substantially greater activity than either Norit or graphene oxide-derived materials, with the mineralisation rate constant being approximately 3 times higher for Starbon-titania than for either of the other two systems. This was ascribed mainly to the pure, highly crystalline anatase phase formed on Starbon, as well as the mesoporosity of the system that allowed much better transport of phenol. Runs utilising Starbon itself also indicated that Starbon's ability to adsorb phenol from the aqueous solutions was very good, meaning that it can pick up phenol from water and deliver it to the catalytic sites efficiently. Physical mixtures of Starbon and commercial anatase (Evonik P25) likewise showed excellent results, having approximately 50% higher rate constant than the ultrasound synthesised materials. Rate curves for the various systems indicated that Starbon-based systems reduced phenol levels to 0.3% of the initial (i.e. to 0.3 ppm, well below the 1 ppm target) compared to 0.5 ppm for Norit and 0.7 ppm for graphene oxide. Not only were the final levels lower, but the rate of removal was faster as well.

3.2.4 Adsorption Processes

One of the classic uses of high-surface-area materials is the adsorption of dissolved or gas-phase species, either with a view to concentrating and winning high-value products, or to decontaminating waste streams, polluted air or environments. Starbon has been used widely in all these application areas, showing considerable promise in several instances, which are discussed next.

3.2.4.1 Adsorption of Gases

Milescu et al. [34] have published data on the adsorption of three toxic gases on starch, alginic acid, and pectin-based materials and found some very interesting behaviour. The authors studied the room temperature adsorption of the gases from 5000 ppm in air mixtures, and looked at ammonia, hydrogen sulphide, and sulphur dioxide (separately). They found excellent adsorption in many cases, with materials outperforming standard activated carbons such as Norit, and also giving impressive results compared with a range of other adsorbents discussed in the literature.

For ammonia, adsorption was best with low-temperature-activated alginic acid material (in particular, the uncarbonised A000) with performance dropping to very low levels with the higher temperature materials. This correlates very well with the availability of acidic functionality (via the alginic acid-repeating units) which is highest at carbonisation temperatures <200 °C. This functionality is lost above this temperature, and with it, the ability to adsorb ammonia drops dramatically. For both the starch and the pectin-derived materials, there is no inherent acidity and the best performance for these materials is seen at 300 °C activation. This activity likely correlates well with a high level of functionality, including aldehydic/ketonic groups formed by dehydration of vicinal diols, and also possibly by ring scission. Such groups will readily form imines, trapping the nitrogen. Evidence for nitrile formation is also strong, possibly via dehydration of primary amides, from acidic functionality that can also develop during carbonisation. Further carbonisation leads to the gradual loss of this rich functionality, leading to a reduced ability to adsorb ammonia.

For both the sulphur-containing gases, adsorption ability was dominated by the highest temperature materials (in this case, 800 °C was the maximum used). Indeed, for hydrogen sulphide, the pectin-derived materials were exceptionally active, adsorbing 20 times as much as any other material tested, and approximately 4 times as much as a pectin-derived P550 material.

Interestingly, XPS studies on the loaded samples showed that, while most of the sulphur loaded was in the original oxidation state, there were significant quantities of oxidised sulphur species present (S(IV) and S(VI)). Small amounts of reduced sulphur species were noted on SO_2 adsorption, along with more prevalent oxidised species. The presence of high oxidation state N species in very low quantities on the surface of the materials was suggested as a potential catalytic route to oxidation of the S species. The significant amounts of inorganics in the materials may also play a role, especially in the high-temperature materials, where the inorganics are concentrated via loss of organics – S800 has only 2.6% inorganics, A800 a significant 8.8%, and P800 (from pectin) a huge 28.6%, much of it potassium.

Dura et al. published results relating to the impressive ability of Starbon to adsorb carbon dioxide, which is an important topic for control of CO_2 levels and also to concentrate CO_2 prior to its conversion into valuable chemicals such as cyclic carbonates [35]. The authors compared the predominantly mesoporous Starbon (68–92% of the total pore volume was in the mesopore range) with Norit-activated carbon which is c. 75% microporous. They utilised both starch-based and alginic-acid-derived materials in their study, and these were prepared over a wide temperature range from 300 to 1200 °C.

Pressure swing adsorption data were collected over 5 cycles at 10 bar pressure. This demonstrated that, in both the starch- and alginic-acid-derived materials, CO_2 adsorption increased from low levels for the S300 and A300, reaching a maximum at the S800 and A800 materials. Beyond the 800 materials, there was a slight drop in the case of the starch materials, but a substantial reduction in adsorption in the alginic acid series. The optimal adsorption capacity was 40% (S800) and 50% greater (A800) than for Norit. Adsorption kinetics were the same for all three adsorbents, with saturation after 30 minutes at 5 bar pressure or after 10 minutes for 10 bar pressure. Desorption under atmospheric pressure took 20 minutes in all cases.

Simultaneous thermal analysis was used to measure adsorption/desorption under flowing conditions alternating between CO_2 and nitrogen. This indicated rapid and complete reversibility of adsorption over many cycles, as well as allowing estimation of the enthalpy of adsorption. The values (between −14 and −18 kJ mol^{-1}) indicate that physisorption is the predominant mechanism. Regression analysis was used to develop an empirical equation to correlate the adsorption to textural properties and led to the equation:

$$\mathrm{Mass}\left(\mathrm{CO_2\ adsorbed}\right) = 0.095 + 2.10\,\mathrm{V}\,micro + 3.51\,\mathrm{V}\,micro \times \mathrm{V}\,meso1$$

In other words, the adsorption behaviour is strongly linked both to micropore adsorption and to a combination of micropores and mesopores, suggesting that the combination of mesopores and micropores allows excellent transport to the adsorption sites as well as some mesopore adsorption.

Very importantly, CO_2 to nitrogen selectivity was measured for Norit, S800 and A800 using mixed gas streams at 298 and 323 K. What was striking here was that selectivity to CO_2 was much higher for the Starbon materials (14.0–20.3) than for Norit (5.4 at 298 K and 4.0 at 323 K) meaning that adsorption in real situations would benefit from the use of the Starbon materials.

3.2.4.2 Adsorption or Organics from Solution

Parker et al. [36] demonstrated that Starbon materials have a very impressive ability to adsorb a range of small phenolic molecules from water. They chose the relatively hydrophobic and low oxygen content Starbon-800 materials as their adsorbents, and compared starch-derived systems with alginic-acid-derived systems. Both materials exhibited equally good adsorbency

Table 3.4 *Collected data for the adsorption of phenols on alginic-acid-derived Starbon (A) and on starch-derived Starbon (S).*

	ΔG (kJ mol^{-1})		ΔH (kJ mol^{-1})		ΔS (J mol^{-1}K^{-1})		Q_o (mg g^{-1})		n	
Phenol	A	S	A	S	A	S	A	S	A	S
H-	−0.13	−0.91	1.7	−3.8	6.0	−9.7	40.3	100	0.03	5.7
2-Me	−2.8	−1.6	−16	4.2	−4.3	20	0.08	94.6	3.6	5.8
2-F	−2.0	−2.5	−11	−4.1	−32	−5.4	206	131.8	0.01	6
3-NH_2	−0.42	−0.57	8.9	−1.8	32	−4.0	241	101.2	0.01	6.6
4-OMe	−1.2	−2.8	37	−4.9	128	−6.9	146	118.6	3.8	14.1

Source: Data from Parker et al. [36].

behaviour over a range of phenols in an aqueous solution (Table 3.4). The starch-derived S800 had around twice the surface area of the alginate A800, but much of the 'extra' surface area was present as micropores and ultramicropores, which likely contribute less to the adsorption behaviour. It is tempting to think that the process is therefore as simple as being directly related to a similar mesopore surface area; however, a more detailed analysis of the relevant adsorption isotherms indicated that somewhat more complexity is involved.

First, using Langmuir and Freundlich isotherms, it was postulated that the surfaces have significant heterogeneity, with some residual oxygen functionality being responsible for the heterogeneous surface energy. Physisorption was also indicated by the isotherms (via heterogeneity factor and adsorption energy values). Perhaps, unsurprisingly, for systems where mesoporosity is important, multilayer adsorption is evident, with all systems going well beyond surface coverage. What is surprising is that, despite the similar capacities of the two materials, the monolayer adsorption capacities are similar for all phenols for S800, but vary significantly for the same molecules in the case of A800. Similarly, the thermodynamic parameters for the adsorption of S800 and A800 are very different, suggesting a quite different mode of adsorption within the group of phenolics, with some being entropy-driven (approximating to release of waters of hydration at the surface/around the phenol) and others enthalpic (stabilising interactions between surface and adsorbate). A positive contribution to the adsorption via entropy seems more prevalent in the alginic acid series of adsorbents.

The same group, in collaboration with a group at Menoufia University in Egypt, also examined the potential of Starbon materials in decontaminating grey water from laundrettes. This is the post-wash waters that are expelled from washing machines and which are contaminated with suspended solids, minerals such as potassium and phosphate, as well as detergent molecules and aggregates. Re-use of the very substantial quantities of grey water would go a long way to reducing the water requirements of large conurbations.

The first paper [37] focused on a combination of Fentons reagent (Fe(III) and hydrogen peroxide) with Starbon-300 to clean up grey water by a combination of oxidation and adsorption. Chemical oxygen demand (COD) reduction was used as a determinant of success. A series of operating parameters were investigated for each of the two components separately, and then the optimum combination of the two was investigated. For the Fenton's reagent and hydrogen peroxide system, optimal iron and hydrogen peroxide concentrations were determined beyond which efficiency dropped significantly. The influence of pH was also investigated, and a definite trend was seen with lower pH values giving significantly better results. This is attributed to the role of pH in guiding the relatively complex Fenton's chemistry to the highest production of hydroxyl radicals, the most active species. For adsorption on Starbon, similarly acidic conditions were optimal for adsorption of materials and reduction in COD.

A separate series of experiments using dual treatment (Fentons reagent and Starbon together) led to a different set of optimum conditions, indicating that there is a significant interaction between the two systems, or that the products of Fenton's treatment adsorb under different conditions than the untreated components. This allowed a significant improvement of performance with around half the Fenton's reagent (half the Fe concentration and half the hydrogen peroxide). Importantly too, neutral pH (pH 7.4) was found to be ideal, closer to real conditions and avoiding acidification. Up to 93% reduction in COD was demonstrated.

The second paper focused purely on adsorbency of a range of Starbon materials [38]. In this case, alginic acid-based materials were used, with pyrolysis temperatures from 300 to 800 °C, whereas the earlier paper used corn starch-derived materials. A direct comparison of adsorbency between the A300 used in the 2019 paper and the corn starch-derived S300 used in the original paper indicates contrasting results in terms of pH dependence (Figure 3.12). This may possibly indicate that the surface chemistry of the S300 and A300 materials is significantly different – perhaps not surprising as there are significant amounts of acidic groups remaining in the A300 material [39] which should not be present in the starch-derived materials.

What can also be seen is that increasing the pyrolysis temperature of the alginic acid material gives a dramatic increase in adsorbency, with A800 not only being very effective at neutral and slightly alkaline pH values but also losing very little adsorbency over the entire pH range evaluated. This is likely related to the much lower oxygen content and hydrophobicity of the material, meaning that its surface charge is much less affected by pH, but also that hydrophobic effects may be important in adsorbing at least some of the components of the grey water. It was found that the adsorption capacity for A800 was 0.92 g g^{-1}, outperforming the other adsorbents tested (and also silica gel and Norit-activated carbon) in terms of capacity, and also reaching maximal adsorbency in the fastest time (c. 10 minutes cf. 60 minutes for the other adsorbents tested). The high capacity and rapid adsorption make the A800 material a very promising material for laundry waste clean-up, giving potential to the reuse of grey laundry water in domestic situations.

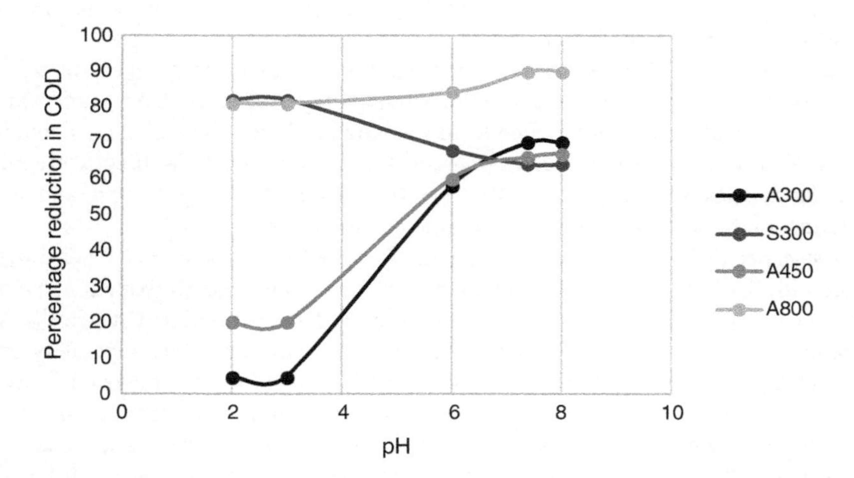

Figure 3.12 Comparison of pH-dependent adsorption behaviour (via COD reduction) of A300 and S300 and of A450 and A800. Source: Data from Shannon et al. [39].

3.2.4.3 Metal Recovery

Muñoz Garcia et al. [40] have shown that Starbon is an excellent adsorbent for precious metals, in particular, gold. Stirring S800 with an aqueous solution of a range of metals designed to mimic a model waste stream from a platinum group metals mine – Au(III), Pt(II), Pd(II), Ni(II), Cu(II), Ir(II), and Zn(II) – under mildly acidic (HCl) conditions gave remarkable results. Gold (99% adsorption), followed by Pd (>90%) and Pt (>80%) were adsorbed very well, whereas iridium (31%) followed by the others (<10%) were less well adsorbed. This selectivity was also observed even when the solution contained far higher concentrations of the poor adsorbers. Adsorption ability appeared to follow the reduction potential of the metals, with gold being the most easily reduced. This fits with the XPS data that confirmed that the predominant species adsorbed was the M(0) oxidation state, as well as demonstrating an increase in oxidised forms of carbon on the Starbon material. The highest capacity observed for gold was an astonishing 3.8 g Au/g Starbon, well in excess of other adsorbents [41]. Selectivity was explained on the basis of reduction potentials, and data from previous work (albeit under different conditions of solvent and precursor (chloride ions can play a major role in metal speciation)) indicate that the reductive adsorption of other metals with similar reduction potentials is also relatively facile on a Starbon surface [42].

The recovery of arsenic (As) from wastewater has been demonstrated by Baikousi et al. [43] using iron oxide/hydroxide-modified Starbon. They conducted an in-depth study of the surface of the Starbon-S700 material before and after iron adsorption, providing details on the nature of surface sites, and a mechanistic understanding of the As uptake. Potentiometric titration of the S700 shows the presence of two sets of functional groups, one (considered to be carboxylic acids) with a pKa of 2.75. This is slightly stronger than most carboxylic acids, but is in line with carboxylic acid strength, as measured on other carbonaceous surfaces. The second set of functionalities had a pKa of 10.3, and were considered to be phenolic in nature. Such an analysis is consistent with that of Shannon et al. [39] who carried out similar analyses on alginic-acid-derived materials (see Section 3.2.4.4), suggesting a reasonable degree of similarity in surface functionality between the two material classes.

Loading iron onto the material involved the adsorption of $FeCl_3$ from ethanol, evaporation, and reduction with aqueous sodium borohydride. This gave nanoparticles of iron on the surface of the Starbon. This led to additional protonatable functionalities at pKa = 3.8 and pKa = 7.3, attributed to carboxylic groups and Fe–OH groups, respectively. The latter are thought to be formed by aerial oxidation of the surface of the Fe nanoparticles.

Adsorption of As from water was then carried out at pH 7, at which the As is present as the neutral $HAsO_3$ species. Maximum loading was found to be 26.8 mg As/g material at pH 7. This is attributed to adsorption solely on the FeOH sites, and accounts for 75% of these sites adsorbing 1 As unit. The Starbon itself adsorbs no As, and the pH dependence of adsorption (which drops off rapidly at higher pH, where both the adsorption sites and the As develop negative charges) is consistent with the suggested mechanism. The adsorption capacity is significantly higher than that found for other metal oxide/clay systems.

3.2.4.4 Adsorption and Release of Bioactives

Shannon et al. [39] investigated the adsorption and desorption of bioactive molecules from alginic-acid-derived Starbon compared to activated carbon. The set of molecules studied included two plant growth promoters and two plant growth inhibitors, with a view to developing controlled release technology for horticultural applications (Figure 3.13).

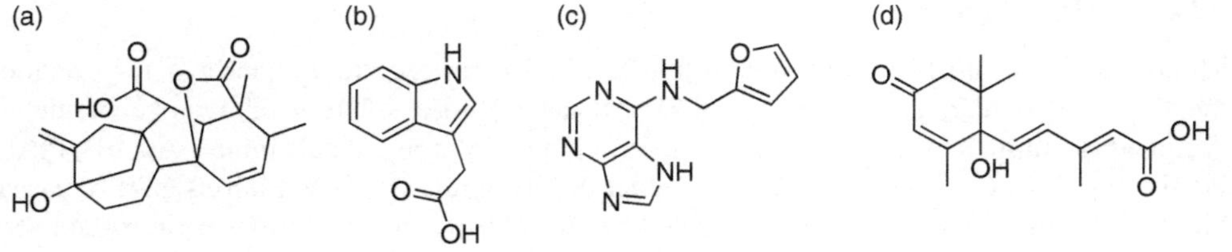

(a) (b) (c) (d)

Figure 3.13 Four bioactive molecules studied for adsorption/desorption behaviour. (a) Gibberellic acid, (b) indole-3-acetic acid, (c) kinetin, (d) abietic acid. The first three are growth promoters, and the fourth is an inhibitor.

Alginic-acid-derived materials carbonised at 300, 500, and 800 °C were investigated alongside an activated carbon. The authors studied the physicochemical nature of the four sorbents using pH drift and Boehm titrations to determine surface functionality and acidity/basicity. They found that pH_{pzc}, the pH at which the surface has net zero charge, was 7.9 for the activated carbon, but started at 6.1 for the A300 material, with the value shifting significantly to higher values for the A500 and A800 materials (8.7 and 9.2, respectively). Combining the adsorption results with Boehm titration indicated that the changes in pH_{pzc} are predominantly due to the partial loss of carboxylic functionality in the alginic acid structure at 300 °C, a process that is complete by 500 °C, where more basic, oxygen-based, functionality predominates. As discussed in Section 3.2.4.1, the alginate materials contain significant inorganic content, which will increase by a factor of 2 from 300 to 800 °C – this is likely to increase the basic character of the materials as a function of temperature. The other major differences between the sorbents were in porosity, with the activated carbon being 85% microporous and the alginate materials predominantly mesoporous (66–93%), albeit with some microporosity developing at higher temperatures.

Adsorption capacities are given in Table 3.5. It can be seen that all four adsorbates are taken up by all of the adsorbents without dramatic differences in capacity. Adsorption of the four molecules followed second-order kinetics, and it appeared that surface area was an important factor, and that diffusion limitations were important in systems with high microporosity.

Where the Starbons really shine is in their ability to desorb the adsorbed materials. While activated carbon adsorbs all four materials well, it does not desorb them – the highest desorption in aqueous systems was for indole acetic acid, and that was as low as 1.2% of the total adsorbed. For the Starbons, the situation is much more variable, but up to 46% release was observed. This perhaps indicates that for the essentially microporous activated carbons, once

Table 3.5 Adsorption capacities for various alginic-acid-derived Starbons and an activated carbon.

	Adsorption capacity (mg g^{-1})			
Bioactive	AC	A300	A500	A800
Gibberelic acid (A)	72	98	76	118
Indole-acetic acid (B)	210	115	150	157
Kinetin (C)	205	120	125	121
Abietic acid (D)	314	282	239	370

Source: Data from Shannon et al. [39].

desorption has taken place, desorption out of the micropores is extremely difficult, whereas, for the mixed micro-/mesoporous systems, there may be a proportion of the material adsorbed in larger, more open pores where diffusion (out) is easier. Alternatively, it may be that the micropores are shallower and 'decorate' the walls of the mesopores in the case of Starbons, but in activated carbons, the pores are longer and thus retain material better. Whatever the reason, the Starbon materials offer excellent potential for adsorption and release of key bioactive molecules.

3.2.5 Conclusion

What this chapter has attempted to show is that highly functional and functioning advanced materials can be produced relatively simply from biomass residues, utilising the inherent functionality and structural properties of the polysaccharide materials. This aids in the important aim of transforming our mindset as a society from a linear economy model (extract – make – use – discard) to a more circular structure, where the discarded elements are treated as a resource to be valorised rather than to be disposed of. Such models are crucial if we are to attain a sustainable society which can support the needs of all of us.

Due to the conversion of the 'waste' polysaccharides into a range of tunable high-surface area, highly mesoporous materials can now be carried out at scale and the understanding of the processes has developed rapidly over the last few years, such that these materials have moved from the lab to production and commercialisation. This chapter has focused on two major areas of activity: adsorption and catalysis. We describe the various adsorption/desorption processes at which the materials excel, partly due to their relatively large pore size (with respect to the more traditional activated carbons) which allows them to effectively adsorb (and desorb) relatively large molecules, which are excluded from micropores. The tunable surface functionality also plays a significant role here, with the evolution of the surface being controlled by thermal treatments. However, it is not only small molecules that are adsorbed in impressive amounts – a wide range of acidic and basic gases are also taken up by the material, including ammonia, sulphur dioxide, hydrogen sulphide, and carbon dioxide, making these materials capable of air purification as well as water treatment.

The catalytic aspects of the materials have also been explored, and in this part of the review, a range of surface functionalisation methodologies are presented. Again, these rely on the various surface functionalities for their success (e.g. bromination of unsaturated functionality to provide anchor points for further functionalisation, attachment of activating groups via hydroxyl functionalities and sulphonation to generate strongly acidic sites) and complex structures can be built up on the surface of the materials, again aided by large enough pores to allow ingress and egress of relatively bulky substrates.

For purposes of space, the chapter does not include the development of nanoparticulate metal – Starbon composites, readily carried out, generally without the need for an additional reductant, making the process particularly green. Such materials have proved their use as catalysts, and related metal oxide nanoparticle – Starbon systems have recently been shown to be highly performing in battery applications. Similarly, the functionalisation of the materials with elements such as nitrogen is in its very early stages, with promising initial results presented here. However, much remains to be done to develop, control, and understand these materials, and to make the most of the perturbation that the nitrogen centres provide to the material.

References

1. Sing, K.S.W., Everett, D.H., Haul, R.A.W. et al. (1985). Reporting physisorption data for gas/solid systems with special reference to the determination of surface area and porosity. *Pure and Applied Chemistry* 57: 603–619.
2. Thommes, M., Kaneiko, K., Neimark, A.V. et al. (2015). Physisorption of gases, with special reference to the evaluation of surface area and pore size distribution (IUPAC Technical Report). *Pure and Applied Chemistry* 87: 1051–1061.
3. Morris, S.M., Fulvio, P.F., and Jaroniec, M. (2008). Ordered mesoporous alumina-supported metal oxides. *Journal of the American Chemical Society* 45: 15210–15216.
4. Zhao, R.-H., Li, C.P., Guo, F., and Chen, J.-F. (2007). Scale-up preparation of organized mesoporous alumina in a rotating packed bed. *Industrial & Engineering Chemistry Research* 46: 3317–3320.
5. Lee, J., Yoon, S., Oh, S. et al. (2000). Development of a new mesoporous carbon using an HMS aluminosilicate template. *Advanced Materials* 12: 359–362.
6. Ryoo, R., Joo, S.H., and Jun, S. (1999). Synthesis of highly ordered carbon molecular sieves via template-mediated structural transformation. *The Journal of Physical Chemistry B* 103: 7743–7746.
7. Liu, A.H. and Schüth, F. (2006). Nanocasting: a versatile strategy for creating nanostructured porous materials. *Advanced Materials* 18: 793–1805.
8. Jiang, T., Budarin, V.L., Shuttleworth, P. et al. (2015). Green preparation of tuneable carbon–silica composite materials from wastes. *Journal of Materials Chemistry A* 3: 14148–14156.
9. Budarin, V., Clark, J.H., Hardy, J.J.E. et al. (2006). Starbons: new starch-derived mesoporous carbonaceous materials with tunable properties. *Angewandte Chemie International Edition* 45: 3782–3786.
10. White, R.J., Antonio, C.A., Budarin, V.L. et al. (2010a). Polysaccharide-derived carbons for polar analyte separations. *Advanced Functional Materials* 20: 1834–1841.
11. White, R.J., Budarin, V.L., and Clark, J.H. (2010b). Pectin-derived porous materials. *Chemistry – A European Journal* 16: 1326–1335.
12. Attard, J., Milescu, R., Budarin, V. et al. (2018). Unexpected nitrile formation in bio-based mesoporous materials (Starbons®). *Chemical Communications* 54: 686–688.
13. International Association for the Properties of Water and Steam (IAPWS) (2014). Revised Release on Surface Tension of Ordinary Water Substance. http://www.iapws.org/relguide/Surf-H2O-2014.pdf (accessed 10 November 2021).
14. Lange, N.A. and Dean, J.A. (1967). *Handbook of Chemistry*, 10ee, 1661–1665. New York: McGraw Hill.
15. Vasquez, G., Alvares, E., and Navaza, J.M. (1995). Surface tension of alcohol water + water from 20 to 50 °C. *Journal of Chemical and Engineering Data* 40: 611–614.
16. Borisova, A., De Bruyn, M., Budarin, V.L. et al. (2015). A sustainable freeze-drying route to porous polysaccharides with tailored hierarchical meso-and macroporosity. *Macromolecular Rapid Communications* 36: 774–779.
17. Budarin, V.L., Luque, R., Clark, J.H., and Macquarrie, D.J. (2007a). Versatile mesoporous carbonaceous materials for acid catalysis. *Chemical Communications* 634–636. https://doi.org/10.1039/b614537j.
18. Budarin, V., Clark, J.H., Luque, R. et al. (2007b). Tunable mesoporous materials optimised for aqueous phase esterifications. *Green Chemistry* 9: 992–995.
19. Clark, J.H., Budarin, V., Dugmore, T. et al. (2008). Catalytic performance of carbonaceous materials in the esterification of succinic acid. *Catalysis Communications* 9 (8): 1709–1714.
20. Aldana-Pérez, A., Lartundo-Rojas, L., Gómez, R., and Niño-Gómez, M.E. (2012). Sulfonic groups anchored on mesoporous carbon Starbons-300 and its use for the esterification of oleic acid. *Fuel* 100: 128–138.
21. Budarin, V., Clark, J.H., Luque, R., and Macquarrie, D.J. (2007c). Towards a bio-based industry: benign catalytic esterifications of succinic acid in the presence of water. *Chemistry – A European Journal* 13: 6914–6919.
22. Sreedhar, I., Aniruddha, R., and Malik, S. (2019). Carbon capture using amine modified porous carbons derived from starch (Starbons®). *SN Applied Sciences* 1: 463.
23. Matharu, A.S., Ahmed, S., Almonthery, B. et al. (2018). Starbon/high-amylose corn starch-supported N-heterocyclic carbene–iron(III) catalyst for conversion of fructose into 5-hydroxymethylfurfural. *ChemSusChem* 11: 716–725.

24. Carneiro, L., Silva, A.R., Shuttleworth, P.S. et al. (2014). Synthesis, immobilization and catalytic activity of a copper(II) complex with a chiral bis(oxazoline). *Molecules* 19: 11988–11998.
25. Zeikus, J.G., Jain, M.K., and Elankovan, P. (1999). Biotechnology of succinic acid production and markets for derived industrial products. *Applied Microbiology and Biotechnology* 51: 545–552.
26. Gómez Millán, G., Phiri, J., Mäkelä, M. et al. (2019). Furfural production in a biphasic system using a carbonaceous solid acid catalyst. *Applied Catalysis A. General* 585: 117180.
27. Luque, R., Budarin, V., Clark, J.H., and Macquarrie, D.J. (2009). Microwave-assisted preparation of amides using a stable and reusable mesoporous carbonaceous solid acid. *Green Chemistry* 11: 459–461.
28. Mesquita, L.M.M., Pinto, R.M.A., Salvador, J.A.R. et al. (2015). Starbon® 400-HSO3: a green mesoporous carbonaceous solid acid catalyst for the Ritter reaction. *Catalysis Communications* 69: 170–173.
29. Luque, R., Budarin, V., Clark, J.H. et al. (2011). Starbon® acids in alkylation and acetylation reactions: effect of the Brönsted-Lewis acidity. *Catalysis Communications* 12: 1471–1476.
30. Doi, S., Clark, J.H., and Macquarrie, D.J. (2002). New materials based on renewable resources: chemically modified expanded corn starches as catalysts for liquid phase organic reactions. *Chemical Communications* 2: 2632–2633.
31. North, M., Pasquale, R., and Young, C. (2010). Synthesis of cyclic carbonates from epoxides and CO_2. *Green Chemistry* 12: 1514–1539.
32. Colmenares, J.C., Lisowski, P., and Łomot, D. (2013). A novel biomass-based support (Starbon) for TiO_2 hybrid photocatalysts: a versatile green tool for water purification. *RSC Advances* 3: 20186.
33. Colmenares, J.C., Lisowski, P., Mašek, O. et al. (2018). Design and fabrication of TiO_2/lignocellulosic carbon materials: relevance of low-temperature sonocrystallization to photocatalysts performance. *ChemCatChem* 10: 3469–3480.
34. Milescu, R.A., Dennis, M.R., McElroy, C.R. et al. (2020). The role of surface functionality of sustainable mesoporous materials Starbon® on the adsorption of toxic ammonia and sulphur gasses. *Sustainable Chemistry and Pharmacy* 15: 100230.
35. Durá, G., Budarin, V.L., Castro-Osma, J. et al. (2016). Importance of micropore–mesopore interfaces in carbon dioxide capture by carbon-based materials. *Angewandte Chemie International Edition* 55: 9173–9177.
36. Parker, H.L., Budarin, V.L., Clark, J.H., and Hunt, A.J. (2013). Use of Starbon for the adsorption and desorption of phenols. *ACS Sustainable Chemistry and Engineering* 1: 1311–1318.
37. Tony, M.A., Parker, H.L., and Clark, J.H. (2016). Treatment of laundrette wastewater using Starbon and Fenton's reagent. *Journal of Environmental Science Part A* 51: 974–979.
38. Tony, M.A., Parker, H.L., and Clark, J.H. (2019). Evaluating Algibon adsorbent and adsorption kinetics for launderette water treatment: towards sustainable water management. *Water and Environment Journal* 33: 401–408.
39. Shannon, J.M., Clark, J.H., de Heer, M.I. et al. (2018). Kinetic and desorption study of selected bioactive compounds on mesoporous Starbons: a comparison with microporous-activated carbon. *ACS Omega* 3: 18361–18369.
40. Muñoz Garcia, A., Hunt, A.J., Budarin, V.L. et al. (2015). Starch-derived carbonaceous mesoporous materials (Starbon®) for the selective adsorption and recovery of critical metals. *Green Chemistry* 17: 2146–2149.
41. Dodson, J.R., Parker, H.L., Munoz Garcia, A. et al. (2015). Bio-derived materials as a green route for precious & critical metal recovery and re-use. *Green Chemistry* 17: 1951–1965.
42. White, R.J., Budarin, V.L., Luque, R. et al. (2009). Supported metal nanoparticles on porous materials. Methods and applications. *Chemical Society Reviews* 38: 481–494.
43. Baikousi, M., Georgiou, Y., Daikopolous, C. et al. (2015). Synthesis and characterization of robust zero valent iron/mesoporous carbon composites and their applications in arsenic removal. *Carbon* 93: 636–647.

4

Conversion of Biowastes into Carbon-based Electrodes

Xiaotong Feng and Qiaosheng Pu

College of Chemistry and Chemical Engineering, Lanzhou University, Lanzhou, Gansu, China

4.1 Introduction

With the rapid development of human society, the demand for energy is increasing and energy sources have become the most essential foundation of modern life. Traditional fossil fuels such as coal, oil, and natural gas are facing the problems of resource depletion and serious environmental damage caused by their exploitation and consumption [1]. The development of clean energy technologies is the key to solving these problems. Great progress has been made on wind energy, solar energy, and geothermal energy, which all depend on certain types of energy storage and conversion devices. Highly efficient electrochemical devices such as lithium batteries, fuel cells, and supercapacitors are promising candidates and have been within the focus of researchers for a long time [2].

Among all components of electrochemical energy devices, high-performance electrodes are critical to the energy storage and conversion processes. Electrode materials such as carbon and its composites are extensively studied in electrochemical researches due to their good conductivity, excellent stability, and easy modification [3]. Pure carbon, doped carbon, and metal/carbon or polymer/carbon composites are widely used in electrochemical energy storage, catalysis, adsorption, and other fields, showing various unique advantages [4]. In industry, most carbon electrode materials are derived from coal or petrochemical products, which involve complex synthesis and environmental issues. To get rid of the dependence on fossil resources, the use of biowastes to produce carbon-based electrodes is an attractive route toward green utilization of natural resources.

High-Performance Materials from Bio-based Feedstocks, First Edition. Edited by Andrew J. Hunt, Nontipa Supanchaiyamat, Kaewta Jetsrisuparb and Jesper T.N. Knijnenburg.

Biowastes normally refer to various unwanted organic masses of biological origin, which include wood wastes, crop residues, kitchen and restaurant garbage, food-processing plant wastes, sewage sludges, animal effluents, and many others [5]. As a group of biomass, it may not always be appropriate to literally take them as "waste" because some of them have long been used as fertilizers or soil conditioners, and could be principal resources of biodevelopment. Biomass has huge reserves in nature and is highly reproducible. It is estimated that the biomass production by 2019 exceeded 220 billion dry tons, and a large portion of them are, or have finally been, transferred into biowastes. Disposition of biowastes, such as agricultural residues and forest byproducts through direct discarding or simple burning, has contributed to the greenhouse effect, environmental pollution, and low efficiency with regard to energy utilization. The reuse of biowastes may reduce environmental pollution and create a green circulation to mitigate global warming. It has been reported that 100–300 million tons of CO_2 in the atmosphere can be removed from the carbon cycle through storage in biomass-derived carbon materials [6]. The conversion of biowastes into carbon-based electrode materials is therefore an important way to achieving this ambitious goal.

Biowastes have some unique advantages to serve as sources of electrode materials. Over billions of years of evolution, plants, animals, and microbes have developed complex internal morphology and abundant element composition [7, 8]. Biowastes with original porous structures can act as self-template to construct the special skeletons during carbonization, and the inherited hierarchical pores make the resulting materials possess variable structures and pore distributions, which are extremely beneficial to improve the specific surface area (SSA) and mass transfer capacity [9]. In addition, various elements other than carbon in biowastes such as oxygen (O), nitrogen (N), chlorine (Cl), and sulfur (S) can form active sites in situ during the pyrolysis process. It is reported that more functional groups can be formed on the surface of the biowaste-derived carbon materials than industrial activated carbon, and both heteroatom doping and surface functional groups may effectively modulate the properties of the resulting carbon materials and enhance their performance [10]. More importantly, the highly organized distribution of heteroatoms in biomass may lead to distinctive dispersion of heteroatom doping in carbide materials [11]. Because the electrochemical energy storage and conversion mainly depend on physical interactions and chemical reactions on the surface of an electrode, it is of great significance to explore the effective ways to convert biowastes into high-quality electrode materials through the dedicated control of the final structure and element composition.

This chapter focuses on the applicability and advantages of preparing carbon-based electrode materials from biowastes, with emphasis on the biomass conversion processes of carbonization and activation as well as the effects on the resulting materials. The structure and doping of biowaste-derived materials are first introduced. Subsequently, we focus on the performance of biowaste-derived electrode materials in a variety of electrochemical applications, including supercapacitors, capacitive deionization (CDI) cells, electrocatalysis, fuel cells, and lithium batteries. Finally, based on the current problems in the conversion and application of biowastes, the future directions of improvement are discussed.

4.2 Conversion Techniques of Biowastes

A critical step for the utilization of biowastes as functional materials is their conversion (Figure 4.1). Different functional materials may be produced depending on the elements present in a specific biowaste. Because of their large carbon content, biowastes are usually

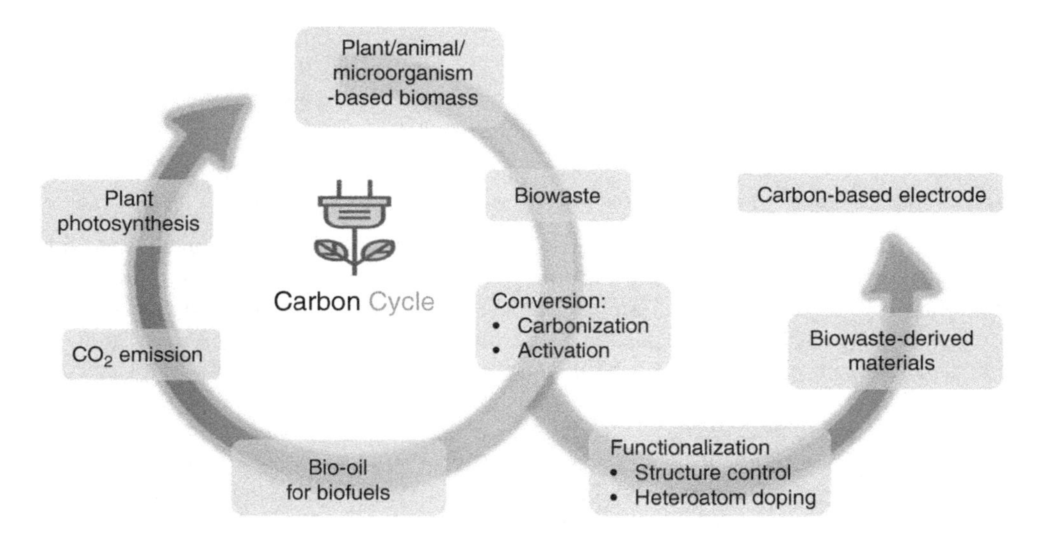

Figure 4.1 Schematic illustration of the carbon cycle and the production of biowaste-derived materials.

converted into carbon materials [12]. Biowaste-derived carbon materials are extremely suitable as electrodes due to their high conductivity, high porosity, and abundant doping of heteroatoms. Some biowastes can be used for preparing other functional materials due to their heteroatomic content and structural characteristics. For example, with 15–20% of silicon present in the form of nanoparticles, rice husk can be a promising precursor for silicon (Si)-containing composites, which have been used as electrode materials in various applications. Zhang et al. [13] reported the preparation of N-doped Si nanoparticle/carbon nanotube (CNT) composite materials using rice husk as the Si source by a simple electrospray method. The final material exhibited excellent electrochemical performance with a reversible specific capacity of 1380 mAh g^{-1} at 0.5 A g^{-1} for the next-generation rechargeable lithium-ion batteries (LIBs), due to the hierarchical hybrid structure of the composite spheres. Liu et al. [14] prepared porous Si materials using reed leaves. The 3D hierarchical structure of reed leaves avoided the formation of Si nanoparticles found in rice husk-derived materials, and the resulting products maintained hierarchical porous structures. As an anode material of lithium battery, the obtained porous Si with inside carbon coating showed excellent cycling behavior; the reversible capacity was 420 mAh g^{-1} after 4000 cycles at a rate of 10 C. For any kind of biowaste-derived material, the development of conversion technologies that can accelerate the conversion rate adjust the pore structure and heteroatom doping may effectively improve its performance as electrodes. There are two roles of conversion techniques: carbonization and activation [15].

4.2.1 Carbonization

The thermal reactions of a biowaste can be described as follows: the biowaste is first broken down into gases, tar, and solid residue, and then a secondary reaction occurs between the released gas and the solid residue [16]. In the presence of additives, continuous chemical reactions occur among the gas/solid products of biowastes and additives, resulting in activation (see Section 4.2.2). Low-temperature carbonization and high-temperature pyrolysis are both used to convert biowastes into carbon materials. However, for some biowastes, pretreatment may be necessary, such as torrefaction, which is a mild pyrolysis process carried out at 200–300 °C under the inert condition to remove moisture [17]. Other typical pretreatment methods include acidification, alkalization, leaching, steam explosion, and pre-oxidation [18–20].

4.2.1.1 Low-Temperature Carbonization

Hydrothermal treatment is a commonly used low-temperature carbonization method in which biowastes are processed in a high-pressure water environment at 120–300 °C. The product properties are affected by the catalyst, reactant concentration, reaction time, and reaction temperature [21]. The obtained materials from hydrothermal treatment usually show a high content of heteroatom doping and can be directly used as electrode materials [22]. However, oxygen-containing groups make these products poor in electrical conductivity, and the SSA and porosity are also low.

Microwave processing is also a low-temperature technology for biowaste-derived materials with the advantages of faster thermal conduction to the inner body of the reactor and shorter processing time compared to traditional heating methods [23]. Shang et al. [24] reported a microwave pyrolysis technology for treating biowastes, and studied in detail the influence of microwave power, heating time, moisture content as well as the wood slag used on the carbon products. Shi et al. [25] successfully prepared multiwalled CNTs with a diameter of 50 nm and a wall thickness of approximately 5 nm using a wood source with a microwave. At 500 °C, the mineral matter in biomass served as the catalyst, and the released volatiles acted as the carbon precursor gas. The formed amorphous carbon nanospheres subsequently self-assembled into multiwalled CNTs during the microwave pyrolysis.

4.2.1.2 High-Temperature Pyrolysis

High-temperature pyrolysis is a traditional way to produce carbon materials. Biowastes are treated at a temperature of 500–1000 °C in an inert atmosphere, and the porosity, SSA, and pore structure of the resulting material are greatly improved. Zhu et al. [26] compared the materials produced from fungi via a hydrothermal approach at 120 °C with high-temperature pyrolysis at 700 °C. After high-temperature pyrolysis, the morphology of the carbon material was greatly improved and the oxygen content decreased. During the process, the heating rate, carbonization temperature, dwell time, catalyst, and other conditions exhibit huge impacts on the SSA, pore distribution, and elemental composition of the products. Undoubtedly, control of the carbonization conditions is the fundamental way to ensure the performance of carbon materials.

4.2.1.3 Other Carbonization Methods

Other mechanisms were also proposed for the carbonization of biowastes. Due to the high thermal stability, negligible vapor pressure, and a low melting point of ionic liquids, ionothermal synthesis is potentially an effective, low-cost, and green reusable synthetic route. For example, Xie et al. [27] reported the ionothermal synthesis of porous carbon materials from a variety of carbohydrate precursors using [Bmim][$FeCl_4$] as the soft template, solvent, and catalyst, which resulted in enhanced carbon yields. After ionothermal processing, the SSA of the carbon materials could be significantly increased from 44 to 350 m^2 g^{-1} and led to a promising hierarchical pore structure. Molten salt carbonization (MSC) is another method for preparing carbon materials from biowastes through cracking the large molecules in biowastes. During MSC, biowastes are immersed into the molten salt to produce a porous structure at a higher temperature [28]. Deng et al. [29] reported the preparation of N-doped hierarchically porous carbon materials using chitosan as the N-containing carbon precursor through molten $ZnCl_2$ carbonization at 400–700 °C. The obtained hierarchically porous carbon could be used as a supercapacitor electrode material, which showed specific capacitances of 252 and 145 F g^{-1} at current densities of 0.5 and 50 A g^{-1}, respectively. The yield of the carbon is high and $ZnCl_2$ could be easily recovered.

4.2.2 Activation

Activation refers to an operation step to modify the porous structure and surface groups of porous carbon materials, which can be employed during or after the carbonization process. Activation is important because it enables fine-tuning of the physical/chemical properties of the product for a specific application. Based on the procedure, activation during the conversion of biowastes can be classified into physical and chemical activation.

4.2.2.1 Physical Activation

Physical activation is usually performed at a temperature higher than 1000 °C, with a gas such as O_2, air, water vapor, or CO_2 passing through the carbon products. This process removes volatile substances from biowaste-derived carbon and opens the closed pore channels to improve the SSA and porosity. The final porosity depends on factors such as the type of biowaste, activation gas, activation temperature, and reaction time [30]. Physical activation is relatively clean and does not require subsequent processing, but needs a higher carbonization temperature. On the other hand, O_2 or air activation requires lower activation energy, but the reaction is difficult to control and the use of these two atmospheres may lead to a serious decline in the yield of resulting carbon. The use of steam and CO_2 as activation gases is cheaper and easier to control, and the SSA and porosity of resulting products can be significantly increased [31, 32]. However, steam activation will introduce more oxygen-containing groups on the surface of the carbon materials, which is not conducive to conductivity. Qu et al. [33] carried out steam activation during the carbonization of corncob residue and coconut shell (CNS) and studied the electrochemical performance of the activated products. The corncob residues led to a high SSA of 1210 m^2 g^{-1}, and the proportion of mesopores in the products increased with the higher steam flow and longer activation time, so that the cyclic stability and high-rate capability were significantly improved. There is no specific range of pore sizes produced by each activation gas. It has been reported that carbon materials produced by steam activation show microporous structures with smaller pore size but wider distribution. The use of two activation gases simultaneously can be a way to improve the activation effect and modulate the reaction processes [34]. In addition to the physical activation agents as mentioned earlier, NH_3, Cl_2, and SO_2 can also be used for physical activation. For example, the cotton stalk was treated by CO_2/NH_3, resulting in N-doped carbon material with an SSA of 627.15 m^2 g^{-1} [35].

4.2.2.2 Chemical Activation

Chemical activation refers to the processes in the presence of chemicals such as alkaline hydroxide or decomposable activators such as carbonate, nitrate, zinc chloride ($ZnCl_2$), or phosphoric acid (H_3PO_4) during carbonization. Chemical activation generally takes place at lower temperatures than physical activation and is typically around 500–900 °C for better tuning of the structure and porosity with high SSA [36]. Another difference compared to physical activation is that chemical activation can be achieved simultaneously with carbonization.

The activator can be mixed with the starting materials through two typical ways: mechanical mixing and impregnation.

Potassium hydroxide (KOH) is one of the most commonly used chemical activators for porous carbon materials. KOH activation can not only obtain a very high SSA (values of higher than 3000 $m^2\ g^{-1}$ have been reported) but can also form diverse pore structures. Wang et al. [37] prepared porous carbon by direct pyrolysis of celtuce leaves and KOH at 600 °C, then raised to 800 °C for one hour. After activation with KOH, the SSA of the resulting material reached 3404 $m^2\ g^{-1}$, and the electrodes demonstrated a specific capacitance of 421 F g^{-1} (0.5 A g^{-1}). Lillo-Rodenas et al. [38] proposed that the following chemical reaction occurred in the gasification of graphite in the presence of alkali metals:

$$6KOH + 2C \rightarrow 2K + 3H_2 + 2K_2CO_3$$

KOH begins to react over 700 °C, and successively generates potassium carbonate (K_2CO_3), potassium oxide (K_2O), and is finally reduced to potassium metal (K). The produced K_2CO_3 can be further decomposed. So, K_2CO_3 is also an effective activator. The chemical reactions are:

$$K_2CO_3 + 2C \rightarrow 2K + 3CO$$
$$K_2CO_3 \rightarrow K_2O + CO_2$$

Generated CO_2 interacts with the surrounding carbon skeleton for physical activation. The generated metal K may be embedded in the carbon matrix or penetrate through it as vapor, which expands the spacing between carbon layers, resulting in an increase in the total pore volume [39].

H_3PO_4 and $ZnCl_2$ are also commonly used as activators in the synthesis of porous carbon [40, 41]. Unlike activators such as KOH and K_2CO_3 that cause oxidation of carbon matrix, these two activators often act as dehydrators. During dehydration, the starting materials may be cross-linked and form the carbon backbone. Donald et al. [42] proposed that the development of mesopores during H_3PO_4 activation was directly related to the cross-linked structure through phosphate esters encased in the carbon block. The authors also observed deoxidization and dehydrogenation caused by $ZnCl_2$ under high-temperature carbonization. These effects were believed to relate to the "swelling" effects of $ZnCl_2$ by breaking the lateral bonds in the cellulose molecules, resulting in increased inter- and intra-voids and creating interspaces between carbon layers to produce high microporosity. Senthilkumar et al. [43] prepared water hyacinth-derived porous carbon through $ZnCl_2$ activation and the material exhibited excellent capacitive performance in 1 M H_2SO_4 electrolyte; the specific capacitance and energy density were 472 F g^{-1} and 9.5 Wh kg^{-1}, respectively. When KI was added into the electrolyte, the capacitance and energy density increased to 912 F g^{-1} and 19.04 Wh kg^{-1}, facilitating the capacitive behavior of electric double layers. In addition, $ZnCl_2$ is a Lewis acid that can catalyze the aromatic condensation reaction and produce active sites on the adjacent molecules to undergo aromatization reactions [44].

Due to the diversity of biowastes and reaction conditions, it is important to carefully select proper activation methods, activation temperature, heating rate, dwell time, and activator : biowaste ratio to further improve the porosity, SSA, and pore size distribution, and finally improve the electrochemical performance of biowaste-derived carbon [45, 46]. In addition, the results of activation on porosity and pore size distribution can be improved by a combination of physical and chemical activation or a two-stage activation. For example, Fu et al. [47] compared a two-stage KOH activation with a one-stage process using rice husk as the precursor of carbonization. For the two-stage KOH activation, the rice husk was first treated at low temperature followed by KOH activator addition. At 750 °C, the product yield of the two-stage process was 19.4%, much higher than that of the one-stage process (2.5%). The two-stage KOH activation led to a higher SSA of 2138 $m^2\ g^{-1}$ and more abundant micropores compared to one-stage activation (589 $m^2\ g^{-1}$). As the pyrolysis time increases in a one-stage process, char can be greatly consumed, resulting in a low yield of activated carbon. With a two-stage process, char will be formed with a high yield during the first step and is partly gasified by activating agents to develop new pores in the second step.

Conversion and activation techniques are summarized in Table 4.1 with examples of biowastes used for producing carbon electrodes. SSA, doping element(s), and the application of the resulting carbon are given in parentheses.

4.3 Structure and Doping

4.3.1 Biowaste Selection

Biowastes may contain various functional groups and multiple elements other than carbon, making them ideal candidates for the design of materials with a wide range of morphologies and elemental compositions. A better understanding of the structure and composition of biowastes (precursors) is critical to select a suitable one to meet the requirements of a final carbon product. Lignin, cellulose, and hemicellulose are the main components of plant wastes [48]. Cagnon et al. [103] showed that lignin had good thermal stability and led to high carbon yield, while cellulose and hemicellulose had poor thermal stability, resulting in lower carbon yield. Teng et al. [104] demonstrated that the content of lignin and cellulose in biomass affected the porous structure of generated carbon materials. High content of lignin tends to form macropores while a high cellulose content gives micropores.

Chitin is abundant in the shell of some crustaceans and insects. It has a high N content, exhibits high thermal stability, and provides high carbon yield. Moreover, it is easy to form crosslinked networks or intermolecular hydrogen bonds between chitin and other molecules, making it a promising functional material. Zhao et al. [51] reported the preparation of 3D porous carbon materials by using chitin-rich ants as a precursor. The shell with high chitin content formed a porous skeleton during the carbonization process, while the proteins and fatty acids containing P and N atoms provided a synergistic effect. The resulting material had an SSA of 2650 $m^2\ g^{-1}$ and an excellent specific capacitance of 352 F g^{-1} at 0.1 A g^{-1}.

Animal and human wastes also show important use. The molting, hoof, or hair of animals contains fibrous keratin. Keratin is composed of α- and β-keratin, which form superhelical structures by disulfide bonds and hydrogen bonds with extremely high thermal and chemical stability. Qian et al. [52] reported the synthesis of heteroatom-doped porous carbon flakes using human hairs as a precursor. The resulting materials showed a lamellar structure with hierarchical pores and exhibited a high-charge storage capacity of 340 F g^{-1} in 6 M KOH and 126 F g^{-1} in 1 M $LiPF_6$ ethylene carbonate/diethyl carbonate (EC/DEC) organic electrolyte.

Because of the diversity of internal structures and element compositions of biowastes, products with different structures and doping status may be obtained even when the same activation and carbonization process is applied. Khalil et al. [105] compared porous carbon materials carbonized and activated with KOH using three kinds of biowastes, namely oil palm empty fruit bunch (PFB), bamboo strips (BS), and CNS. Mesoporosity in the product prepared from

Table 4.1 *Conversion techniques and activation procedures for the preparation of biowaste-derived carbon electrodes.*

	Activation		Biowastes (SSA m^2 g^{-1}/doping/application[a])
High temperature carbonization	Physical	Air, steam, CO_2	**Olive-tree wood** (4810)[32], **olive bagasse** (1106/N,O)[31], **corncob** (1210/N,O/SC)[33], **kraft lignin** (O/SC)[48], **banana peel** (217/N,O,Si,Cl/NIB)[49], **waste wood shavings** (3223/O/SC)[50]
	Chemical	KOH	**Celtuce leaves** (3404/O/SC)[37], **ant powder** (2650/N,O,S/SC)[51], **human hair** (1360/N,O,S/SC[52] or 1617/N,O/LIB[53]), **willow catkin** (1533/N,O,S/SC)[54], **lignin** (907/O/SC)[55], **corn husk** (928/N,O/SC)[56], **pomelo peel** (2725/O/SC)[57], **sugarcane bagasse** (1940/N,O/SC and LIB)[58], **corn straw** (2790/N,O/SC)[59], **garlic skin** (2818/O/SC)[60], **bagasse** (1892/SC)[61], **chitosan** (2435/N,O/SC)[62], **wheat straw** (916/N,O/LIB)[63], **dry elm samara** (1947/N,O/SC)[64], ***Typha angustifolia*** (3061/N,P/SC)[65], **pine needle** (2433/N,O/HER and SC)[66], **cattle bone** (1070/Co,N,O/Zn-air battery)[67], **bamboo carbon** (792/S/LSB)[68], **coconut shells** (682/LIB)[69]
		$ZnCl_2$	***Arundo donax*** (3298/N,O/CO_2 capture)[40],**water hyacinth** (580/SC[43] or 950/N,O/ORR[70]), **coconut shell** (1874/O/SC)[71], **chestnut shells** (1987/N,O/SC)[72], **fish scale** (850/N,O/ORR)[73]
		H_3PO_4	**Lignocelluloses** (1135/P,O/SC)[74], **coconut shell** (1216/N,O,P/ORR)[75], **pomelo peels** (1272/P,O/NIB)[76]
		$(NH_4)_2HPO_4$, $CaCl_2$, $KHCO_3$, $MgCl_2$	**Kapok fiber** (1600/N,O,P/SC)[77], **bagasse** (946/N,O/SC)[78], **soybean shell** (1036/N,O/CDI) [79], **peanut root nodules** (513/N,O,S/HER) [80]
	Combined	KOH/CO_2 $H_3PO_4/ZnCl_2$ $NH_4H_2PO_4/KHCO_3$	**Coffee endocarp** (1038/N,O/SC)[8], **oil palm empty fruit bunches** (1704/SC)[39], **peat slurry** (890)[42], **pomelo peel** (2726/N,P/CDI)[81]
Hydrothermal treatment + Carbonization	KOH		**Tobacco rods** (2115/N,O/SC)[82], **soybean meal** (2103/N,O/SC)([83]), **ginkgo leaves** (1132,/N,S/SC)[84], **stem bark of broussonetia papyrifera** (1212/N,O/SC) [85], **hemp bast fiber** (2287/N,O/SC)[86]
	—		**Kiwifruit** (379/O/LIB and SC)[87], **reed leaves** (101/Si,O/LIB)[14], **fungi** (80.08/O/SC) [26], **willow catkins** (155.7/O,Ni/SC)[88], **crab shells** (827.5/N,O,S/ORR)[89], **mollusc shell** (440/Ni,Co,O/SC)[90]), **tea waste** (CDI)[91], **bacterial cellulose** (118/Mo,S,N/HER)[92], **sunflower seeds** (401/Mo,S,N/HER)[93], **sheep-horn** (313/S,O,N/ORR and OER)[94], **cow dung** (Fe,Si,F,N,O/ORR)[95], **ginkgo leaves** (1436/N,O/ORR)[96] **egg** (970/Fe,N,O,P/ORR and OER) [97], **catkin** (461/Fe,N,O/ORR)[98], **cotton cloth** (163/O/LSB)[99], **disposable bamboo chopsticks** (808/O,F,Cl/LIB)[100]
Molten salt synthesis	Na_2CO_3–K_2CO_3, LiCl/KCl,Na metal		**Peanut shell** (408/N,O/SC)[28], **chitosan** (1582/N,O/SC)[29], **cotton** (1716/SC) [101]
Microwave induced pyrolysis	K_2CO_3		**Guimwood** (5.2/O,Si,Ca,Cu)[25]
Ionothermal synthesis	[Bmim][$FeCl_4$], KCl/$ZnCl_2$		**D-Glucose, D-fructose, D-xylose, starch** (>350/N,O)[27], **beech wood** (1589/N,O/ORR)[102]

[a] Application is presented as abbreviations: CDI, capacitive deionization; HER, hydrogen evolution reaction; LIB, lithium-ion battery; LSB, lithium-sulfur battery; NIB, sodium-ion battery; ORR, oxygen reduction reaction; SC, supercapacitors.

empty fruit bunch (EFB) was developed nicely compared to the other two products. The carbon material of CNS exhibited a better pore network with smooth walls and open channels, while the products of BS and PFB showed rough surface and distorted pores. Different activation methods may lead to different pore morphologies even for those from the same biowaste. Ramakrishnan and Namasivayam [106] synthesized a series of jatropha husk-derived carbon materials through different activators including H_2SO_4, HNO_3, HCl, H_3PO_4, H_2SO_4/H_2O_2, $(NH_4)_2S_2O_3$, and NaOH/$ZnCl_2$. All the products presented a faveolate structure, but the pore sizes were quite different. As a rule of thumb, biowastes with higher molecular weight, higher cross-linking, and higher thermal stability should be chosen first because they are conducive to the formation of aromatic structures and high carbon yields, while the presence of reasonably high content of heteroatoms in the original biowastes mostly results in heteroatom-doped material with better performance than undoped ones.

4.3.2 Structure Control

Taking advantage of the intrinsic characteristics of starting materials, the structure of the final materials is largely controllable. 1D, 2D, or 3D structures can be produced with different biowastes as precursors and carefully selected carbonization/activation procedures. The materials of different dimensions have their own unique advantages. The 1D materials are mostly fibrous or tubular. These structures show excellent mechanical strength with Young's modulus reaching 138 GPa, which is beneficial for the preparation of flexible electrodes through the construction of a cross-linked network to form a strong membrane or substrate [50, 107]. The surface of 1D materials is rich in active groups, which may be used in further reactions to form modified layers. In addition, if hollow tubular fibers are obtained, these may give a higher SSA that provides more accessible active sites for electrochemical reactions and acts as buffer containers for the electrolyte [77, 108]. Cellulose, lignin, and hemicellulose normally exhibit high mechanical strength and inherent fibrous structure, which are the ideal choice for 1D electrode materials. With a proper activation process, these precursors can be converted into activated carbon fibers, and by changing the ratio of activators such as KOH or NaOH, the pore distribution in the materials can be adjusted for better energy storage properties. Wang et al. [101] prepared hierarchical porous hollow carbon fibers from cotton through molten sodium activation. The molten Na metal was employed to restructure the natural cotton for obtaining hierarchical porous hollow graphitic carbon fibers, which greatly shortened the ion transfer distance and mass transfer resistance. Under a current density of 60 A g^{-1}, the specific capacitance of the materials was still maintained at 61%. Further construction of 1D composite materials can greatly improve the application performance. Li et al. [88] synthesized a $Ni(OH)_2$ nanosheet/hollow CNT composite material using catkins as the carbon source with increased surface area and additional pseudocapacitance. The specific capacitance of the resulting composite material was 1568 F g^{-1} (1.0 A g^{-1}).

2D materials tend to achieve a higher SSA than 1D materials. 2D carbon materials also show a high conductivity, which accelerates the electron transfer in the plane of carbon layers. The construction of a porous structure into 2D materials can shorten the ion transfer distance between carbon layers and slow down the volume change during charge and discharge processes, so as to improve the cycle stability of the materials [109]. On the other hand, heteroatoms doped in the 2D materials are more easily exposed to form effective active sites. These advantages ensure 2D materials exhibiting excellent performance in energy storage and conversion [110, 111]. The direct preparation of 2D materials by chemical synthesis requires complicated processes. Therefore, the simple synthesis of 2D carbon materials with graphene-like structure, using biowastes, is of interest [84, 112]. For example, Li et al. [54] prepared porous carbon nanosheets by one-step carbonization of catkins. The hollow multilayer structure of the resulting material could be controlled by adjusting the ratio of activator to biomass during the carbonization process, leaving rich N and S doping. The specific capacitance reached 298 F g^{-1} (0.5 A g^{-1}), and at the current density of 50 A g^{-1}, as much as 78% of specific capacitance could still be maintained. An assembled symmetric cell was prepared, which showed remarkable specific energy of 21.0 Wh kg^{-1} at 180 W kg^{-1} with a wide voltage range of 1.8 V. Sun et al. [71] synthesized a porous graphene-like sheet carbon material from CNS by a simultaneous activation-graphitization route. During the synthesis process, $FeCl_3$ and $ZnCl_2$ were used as the graphitization catalyst and activator, respectively. The obtained porous graphene-like sheet structures showed good electrical conductivity. In particular, the increase of the graphitization degree of the material significantly promoted the specific capacitance, which was 196 F g^{-1} at a current density of 1.0 A g^{-1} in organic electrolytes.

A 3D hierarchical porous structure refers to a structure composed of micropores, mesopores, and macropores with different pore sizes. Since pore size and SSA are critical to the electrochemical applications of carbon materials, it is important to understand the advantages of different types of pores. According to the definition of International Union of Pure and Applied Chemistry (IUPAC), pores that are classified as micropores are less than 2 nm, mesopores are between the size of 2 and 50 nm, and macropores are over 50 nm. Micropores contribute to a higher SSA and are of great value in gas adsorption and capacitance [113]. However, the twisty pores in microporous materials lead to a long ion transport distance and low ion adhesion in electrochemical applications. Mesopores can improve the mass transfer efficiency while ensuring a reasonably high SSA, which is why they are widely used in energy storage and energy conversion. Because the active sites are more easily accessed, mesoporous materials are also broadly used in electrocatalysis [114]. Macropores ensure the rapid entry and transfer of ions, making them important in the field of adsorption and separation of macromolecules [115]. Apparently, it is of great significance to combine different pore sizes to form a 3D hierarchical porous structure to take advantage of each pore size. The hierarchical porous structures can effectively reduce steric hindrance, greatly promote the transport of ions in the pore channels, and improve the mass transfer rate and contact efficiency between the electrolytes and the active sites. The construction of hierarchically porous structures has become a research hotspot in the development of electrode materials [82, 87, 116]. Biowastes with special internal porous structures may be beneficial to obtain hierarchical porous carbon (HPC). The pore distribution can be further optimized through different carbonization and activation procedures. Both cellulose and lignin have been used to form 3D structures through physical or chemical activation. Zhang et al. [55] converted lignin into an HPC material through a template-free method with KOH impregnation. The formed lignin crystals acted as activators and self-templates during the carbonization process. The resulting product showed a specific capacitance of 165 F g^{-1} (0.5 A g^{-1}). Song et al. [56] constructed a 3D hierarchical porous structure using corn husks as raw materials through KOH activation. KOH promoted the separation of cellulose from hemicellulose and lignin in corn husks, and cellulose was the main source of a hierarchical porous skeleton with a turbostratic carbon structure and uniform pore size. The specific capacitance of the final product as supercapacitor electrode material was 356 F g^{-1} at 1 A g^{-1}. Moreover, the material exhibited an ultra-high-rate capability with an 88% retention rate from 1 to 10 A g^{-1} and outstanding cycling stability with 95% capacitance retention after 2500 cycles. Biowastes such as grapefruit peel, bagasse, corn stalk, and garlic peel have been employed to obtain HPC materials by KOH activation, and have found application in different electrochemical operations [57–60].

Preparing porous carbon aerogel is another important aspect of the conversion from biowastes. Due to its large SSA, high porosity, hierarchical porous structure, and open channels, carbon aerogel ensures rapid ion transfer and high rate performance [117]. Hao et al. [61, 62] prepared an HPC aerogel by carbonizing the freeze-dried bagasse aerogel using NaOH activation. The product exhibited a high SSA and excellent mass transfer. The authors also prepared N-doped graphene/HPC aerogel composite materials from chitosan. Macropores and mesopores were formed during carbonization, while micropores were introduced during activation. The composite structure improved the stability of the whole material, and the resulting material showed excellent capacitance characteristics of 179 F g^{-1} at a current density of 0.2 A g^{-1}, which retained 92.1% after 10000 cycles.

4.3.3 Heteroatom Doping

Introducing heteroatoms into carbon materials is an effective way to improve electrochemical performance. Heteroatom doping includes the introduction of functional groups and the formation of composites with other materials. Dopants may largely change the nature of a porous material and directly affect the migration of ions in the pore channels and the interaction between ions and the surface. Through proper tuning, doping can improve the adsorption capacity, ion selectivity, charge efficiency, and long-term performance of the electrodes [85, 118]. Because of the abundance of heteroatoms in many biowastes, it is usually unnecessary to add additional dopants for this particular purpose. Carbon materials obtained from biowastes are often oxygen-rich, mainly in the form of functional groups such as hydroxyl, carboxyl, and ester groups. The amount and species of oxygen-containing groups depend on the biowaste precursors and the activation process. These groups can undergo redox reactions that introduce pseudocapacitance [119]. Beyond that, the presence of oxygen-containing groups can improve the wettability of the electrode materials. However, the strong polar oxygen-containing groups may hinder ion migration, resulting in an increase in resistance and a decrease in current density. Therefore, control of proper oxygen content is also important to guarantee the performance of electrode materials.

Nitrogen (N) doping plays a crucial role in the electrochemical application of carbon materials. Some studies have shown that N-doping can effectively improve electron mobility, regulate the charge distribution in the carbon material, and increase the pseudocapacitance of the material by a redox reaction. Lone pair electrons in N atoms are hybridized with p-electrons in carbon atoms to increase the electronegativity of adjacent carbon atoms, thus providing more active sites for Li-ion adsorption and electrocatalysis [120]. Biowastes often contain a large amount of N element and are therefore promising precursors of N-doped carbon materials. Chen et al. [63] synthesized highly N-doped porous carbon materials from wheat stalks. They proved that the pyridine N and pyrrole N had higher energy storage performance than NO_x and quaternary N groups in LIBs. Hulicova-Jurcakova and coworkers [121] prepared a membrane electrode material by the carbonizing eggshell film. The SSA of the material was only 221 m^2 g^{-1}, but its specific capacitance reached 284 and 297 F g^{-1} in acidic and alkaline media, respectively. With a relatively low SSA, the excellent capacitive properties of the material depended mainly on the high doping amounts of 8% N and 10% O. Among different N forms, pyridine and quaternary N are the main performance improvers because they can accelerate the electron transfer and introduce pseudocapacitance.

Other heteroatoms such as P, S, and B are also widely present in biowastes. These heteroatoms may cause synergistic effects with N atoms, which are also of great significance to improve the electrochemical performance of materials. Phosphorous groups in biowastes can effectively enhance the pseudocapacitance of the electrode materials. Hulicova-Jurcakova et al. [122] introduced P into the supercapacitor electrode materials for the first time to obtain better capacitance with a tolerance voltage higher than 1.3 V, confirming the effectiveness of P on the improvement of pseudocapacitance and stability of the electrode. Following this study, many biowaste-derived carbon materials rich in P have been used as supercapacitor electrodes [72, 74]. Li et al. [89] reported a method for preparing N, S co-doped carbon materials with hierarchical porous structure by simple pyrolysis of crab shells without template agent and activation process. Due to the good distribution of N and S atoms as active sites and honeycomb-like structures, the synthesized material presented high catalytic activity for oxygen reduction reaction (ORR) with an onset potential of 0.072 V and a half-potential of 0.110 V (vs Ag/AgCl) in alkaline electrolytes, and good methanol resistance and stability better than 20% Pt/C catalyst (Figure 4.2).

In addition to direct doping of heteroatoms, combining other functional materials to form composites can also improve electrochemical performance. Functional electrochemical materials such as metal oxides and conductive polymers can introduce high pseudocapacitance especially in the field of supercapacitors, but with poor stability. The formation of carbon composites can effectively improve their stability. Xiong et al. [90] prepared porous carbon from mollusk shells with honeycomb-like structure and hexagonal macroporous channels, which effectively prompted transmission and rapid infiltration of the electrolyte. Combining this carbon material with $NiCo_2O_4$ nanowires, the composite material exhibited anomalously high specific capacitance of 1696 F g^{-1}, and showed high power density and cycle stability (88% retention after 2000 cycles). Yu et al. [123] prepared PANI nanowire arrays/HPC composite from wheat stalks as symmetrical capacitor electrode material. Within a voltage window of 0–1.8 V in 1 M Na_2SO_4 aqueous electrolyte, the capacitor showed excellent energy densities, up to 60.3 Wh kg^{-1}. At a power density of 18 kW kg^{-1}, 91.6% of the capacitance remained after 5000 cycles.

4.4 Electrochemical Applications

The properties of carbon-based materials obtained from biowastes can be finely tuned to achieve high SSA, high conductivity, hierarchical porous structure, good wettability, improved mass transfer, and others. Their application, therefore, spreads in various electrochemical fields, including supercapacitors, CDI cells, hydrogen and oxygen evolution, fuel cells, and lithium-ion batteries.

4.4.1 Supercapacitors

As a type of energy storage device, supercapacitors are regarded as the bridge between traditional capacitors and batteries. Supercapacitors exert unique advantages, such as high power density (~10 kW kg^{-1}), long cycle life (>10^5 cycles), and the ability to achieve fast charging and discharging (within seconds) [124]. According to the storage mechanism, supercapacitors can be divided into two groups: double-layer capacitor and Faradic pseudocapacitor. The former achieves energy storage by electrostatic attraction between the ions in the electrolyte and the charged electrode surface to generate an electric double layer. The carbon material is representative of the electrode material of the double-layer capacitor [125]. High SSA and

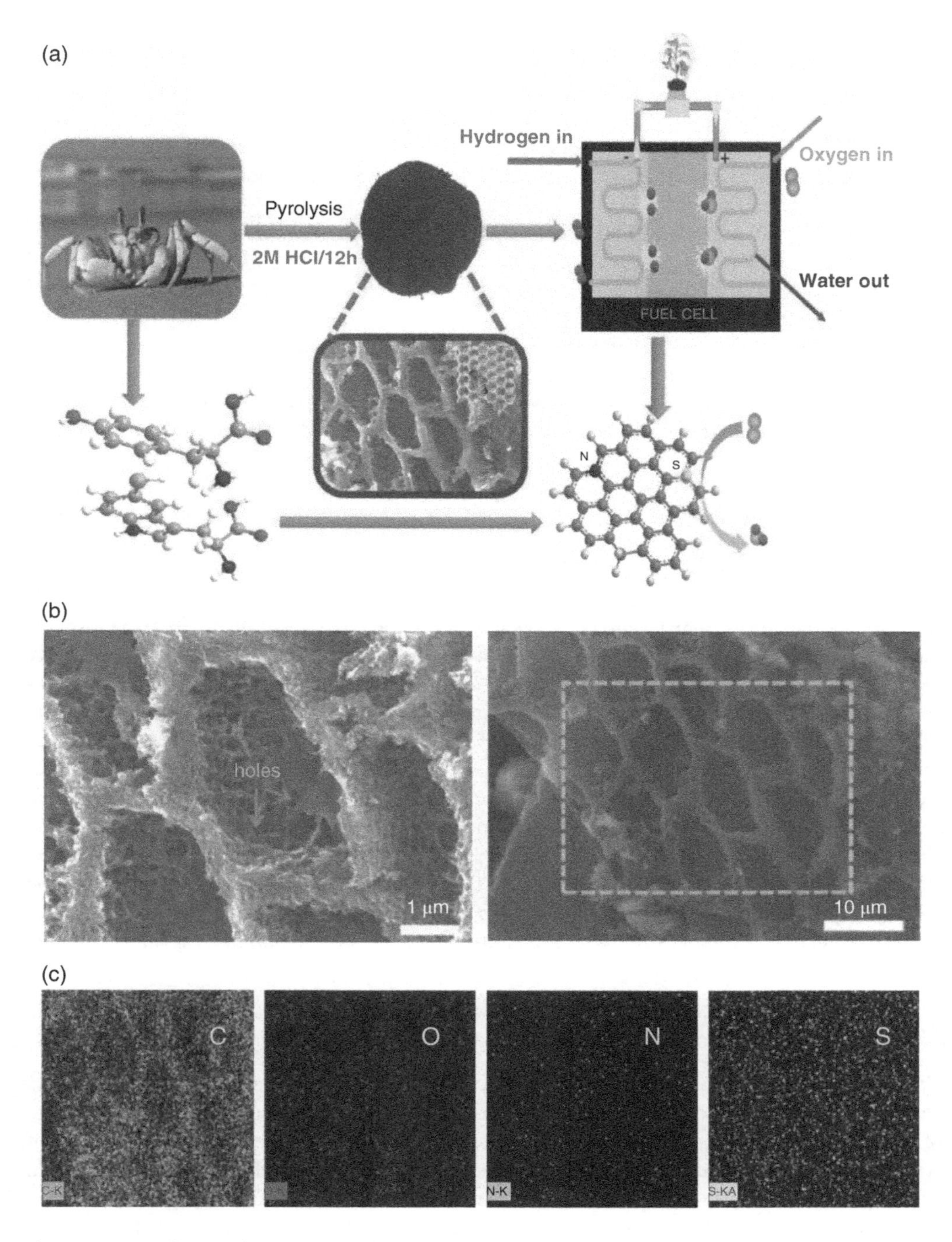

Figure 4.2 *(a) Illustrated procedure of fabricating N, S co-doped porous carbons from Chinese hairy crab shells, (b) scanning electron microscope (SEM) micrographs of the carbon material, (c) elemental mapping images of the carbon material. Source: Adapted from Li et al. [89]. Reprinted with permission from Elsevier.*

proper pore structure of the electrode materials are critical to improve the double-layer capacitance and realize the rapid charge and discharge. The latter depends on the active sites on the surface of the electrodes, such as conductive polymers, metal oxides, or doped active elements, which enable rapid reversible redox reactions or Faraday charge transfer to achieve energy storage [83, 126]. Heteroatom-doped carbon and metal oxide/carbon composites are widely studied for improving Faradic pseudocapacitance through the introduction of active sites for redox reactions.

It is generally believed that micropores can effectively increase the SSA of carbon materials. It was also shown that ultra-micropores are beneficial to the entry of non-solvated ions and greatly improve the capacitance performance [127]. However, the narrow and distorted microporous channels may limit ion transfer to the inner surface of the micropores especially under high current density. Mesopores can effectively improve the mass transfer of the material. Therefore, the key to improving the capacitance is to form a hierarchical porous structure with both micropores and mesopores, as well as control the size and distribution of these pores in the materials. Chen et al. [64] carbonized dry elm samara and obtained 3D porous carbon nanosheets with high microporosity and high SSA. The specific capacitance of the material reached 470 F g^{-1} in 6 M KOH solution, which is the highest reported so far for a supercapacitor based on biowastes. Wang et al. [86] reported the successful hydrothermal-based synthesis of highly interconnected carbon nanosheets using hemp bast fiber as the precursor (Figure 4.3). The middle layer of the hemp is a layered structure consisting of crystalline cellulose, which is sandwiched between a lignin layer and a hemicellulose layer. During carbonization, KOH penetrated the loose connection between the layers and the resulting carbon nanosheets showed a thickness of 10–30 nm and a hierarchical porous structure. This material was ideally suited for both low (down to 0 °C) and high (100 °C) temperature ionic liquid-based supercapacitor application. Even at 100 °C and an extreme current density of 100 A g^{-1}, the material showed excellent capacitance retention of 92%. When the entire device was considered, an energy density of 8–10 Wh kg^{-1} could be achieved at a charge time less than 6 s. On the other hand, heteroatom doping and compositing with functional nanoparticles are also effective ways to control their capacitive behaviors, which alter the electron distribution in the carbon layers to improve the wettability of the material and act as electron donors to introduce additional pseudocapacitance [65, 128]. With $ZnCl_2$ and $CaCl_2$ as activators for the carbonization of sugarcane bagasse, the obtained porous carbon materials all achieved excellent capacitance in a two-electrode system even under high current densities of 50 and 30 A g^{-1}, respectively, due to the high N content [78].

4.4.2 Capacitive Deionization Cells

Following the double-layer energy storage mechanism, researchers have developed CDI cells using a charged electrode to adsorb ions or charged particles in water onto the surface of the electrode to achieve water purification and desalination [129]. In recent years, the scarcity of fresh water resources has made the development of desalination technology an urgent task. Traditional desalination techniques including reverse osmosis and multistage flash distillation suffer from high cost and energy consumption, especially in the desalination of water with low or medium salinity (salt concentration < 3000 ppm). The energy consumption of the new CDI technology is 0.13–0.59 kWh m^{-3}, which is significantly lower than traditional techniques. Moreover, the electrochemical CDI system does not need complicated processes and has the advantage of low cost and environmental compatibility. The effective electrode material is the

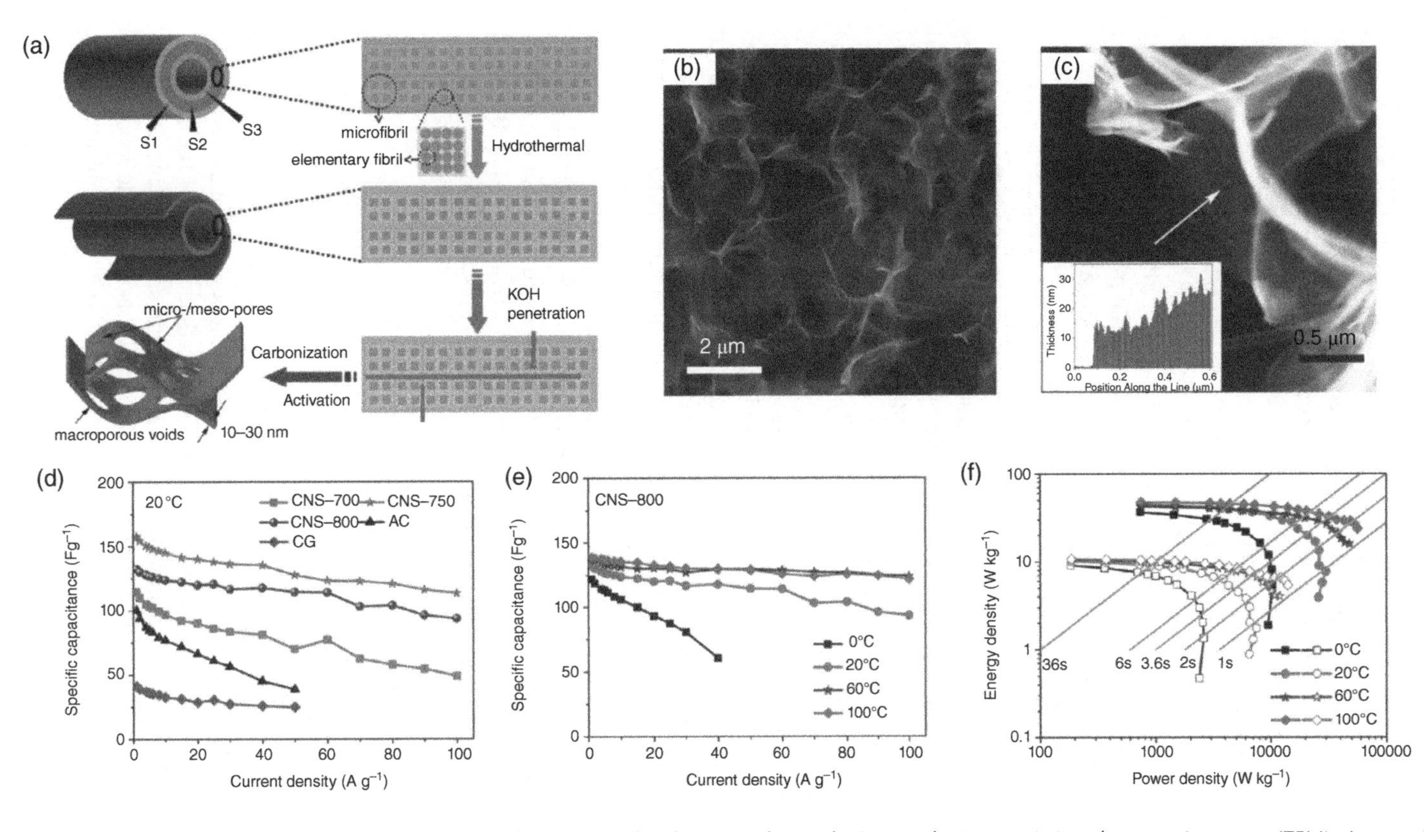

Figure 4.3 *(a) Schematic of the synthesis process for the hemp-derived carbon nanosheets, (b) SEM, and (c) transmission electron microscope (TEM)micrographs highlighting the interconnected 2D structure of carbon nanosheet. Specific capacitance, (d) versus current density for different samples, and (e) for the material (CNS-800) measured at different temperatures, (f) Ragone chart of CNS-800 evaluated at different temperatures. Source: Adapted from Wang et al. [86]. Reprinted with permission from American Chemical Society.*

key to developing CDI systems. In 1995, Farmer et al. [130] reported the use of a carbon aerogel as CDI electrodes. After that, the carbon material as CDI electrode has received extensive attention. Pure carbon materials, however, have a plateau of desalination capacity and there is a significant co-ion expulsion effect in high salinity water [131]. In 2012, Pasta et al. [132] proposed the concept of Faradaic CDI electrode and applied it to seawater desalination. This kind of seawater desalination cell can enhance the ion storage capacity, improve ion selection, and minimize the co-ion expulsion effect [133].

Biowaste-derived carbon materials have been used in CDI systems, owing to their unique microchannel structure and rich heteroatom doping [79]. Xu et al. [81] synthesized N and P co-doped meso-/microporous carbon derived from pomelo peel via a dual-activation strategy by using $NH_4H_2PO_4$ and $KHCO_3$ (Figure 4.4). The obtained material possessed a high SSA of 2726 $m^2\ g^{-1}$ with abundant mesopores and micropores, due to the synergistic activation of $NH_4H_2PO_4$ and $KHCO_3$. The obtained electrode showed high specific capacitance, low inner resistance, and good wettability for deionization capacitors, with an ultrahigh deionization capacity of 20.78 mg g^{-1} at 1.4 V in a 1000 mg l^{-1} NaCl solution. The regeneration performance is ascribed to the high SSA, hierarchical porous structure, and N, P co-doping. Barouds and Giannelis [134] synthesized HPCs via ice templation possessing a high porosity for CDI performance. The HPCs displayed a high salt capacity up to 13 mg g^{-1}. The results showed that hierarchical structures had higher salt removal ability, and the high mesoporosity enabled better utilization of the accessible surfaces of HPCs, while the introduction of micropores led to a more than 80% increase in the salt capacity.

In addition to the removal of NaCl or KCl, biowaste-derived carbon materials are also able to remove hard and heavy metal ions. Tsai and Doong [135] reported that after activation with HNO_3, hierarchically ordered mesoporous carbons derived from sugarcane bagasse showed excellent reversibility and ideal specific electrosorption capacity of 115.4 μmol g^{-1} at 1.2 V for Ca^{2+} removal. The excellent electrochemical performance of the resulting material is mainly attributed to the enhanced mesoporous structures and hydrophilic functional groups. Gaikwad and Balomajumder [91] reported a method to prepare microporous activated carbon from tea waste by chemical and thermal modification. The obtained material was used for simultaneous removal of hexavalent chromium [Cr(VI)] and fluoride (F) and showed excellent ion removal performance. The maximum electrosorption capacities of Cr(VI) and F were 0.77 and 0.74 mg g^{-1} for 10 mg l^{-1}, and 2.83 and 2.49 mg g^{-1} for 100 mg l^{-1}, respectively.

4.4.3 Hydrogen and Oxygen Evolution

Hydrogen is an attractive energy source due to its high energy density and wide utilization. The energy output of hydrogen combustion can reach 120 MJ kg^{-1}, which is much higher than gasoline (43 MJ kg^{-1}). Especially, the combustion product is water, which is of great significance for environmental protection. Electrolysis of water is an important way to produce hydrogen on an industrial scale [136]. For this purpose, electrocatalyst is the key to reducing the overpotential and realize high electrolytic efficiency. The cathode catalysts for hydrogen evolution reaction (HER) are normally transition metal alloys, rare earth element alloys, and nano-alloy materials. The main task in the design of electric catalysts is to construct appropriate metal centers and substrates with considerations on conductivity, electrochemical stability, and necessary catalytic activity of the electrodes [137]. Platinum group metals as well as some non-noble metal-based materials have significant catalytic activity and stability for hydrogen evolution but the cost is the main obstacle for their wide application [138].

The introduction of biowaste-derived carbon into HER catalysts can not only effectively reduce the synthesis cost and improve the distribution of the metal nanoparticles, but can also enhance the performance of HER electrocatalysts [92]. Zhou et al. [80] have proved that N and

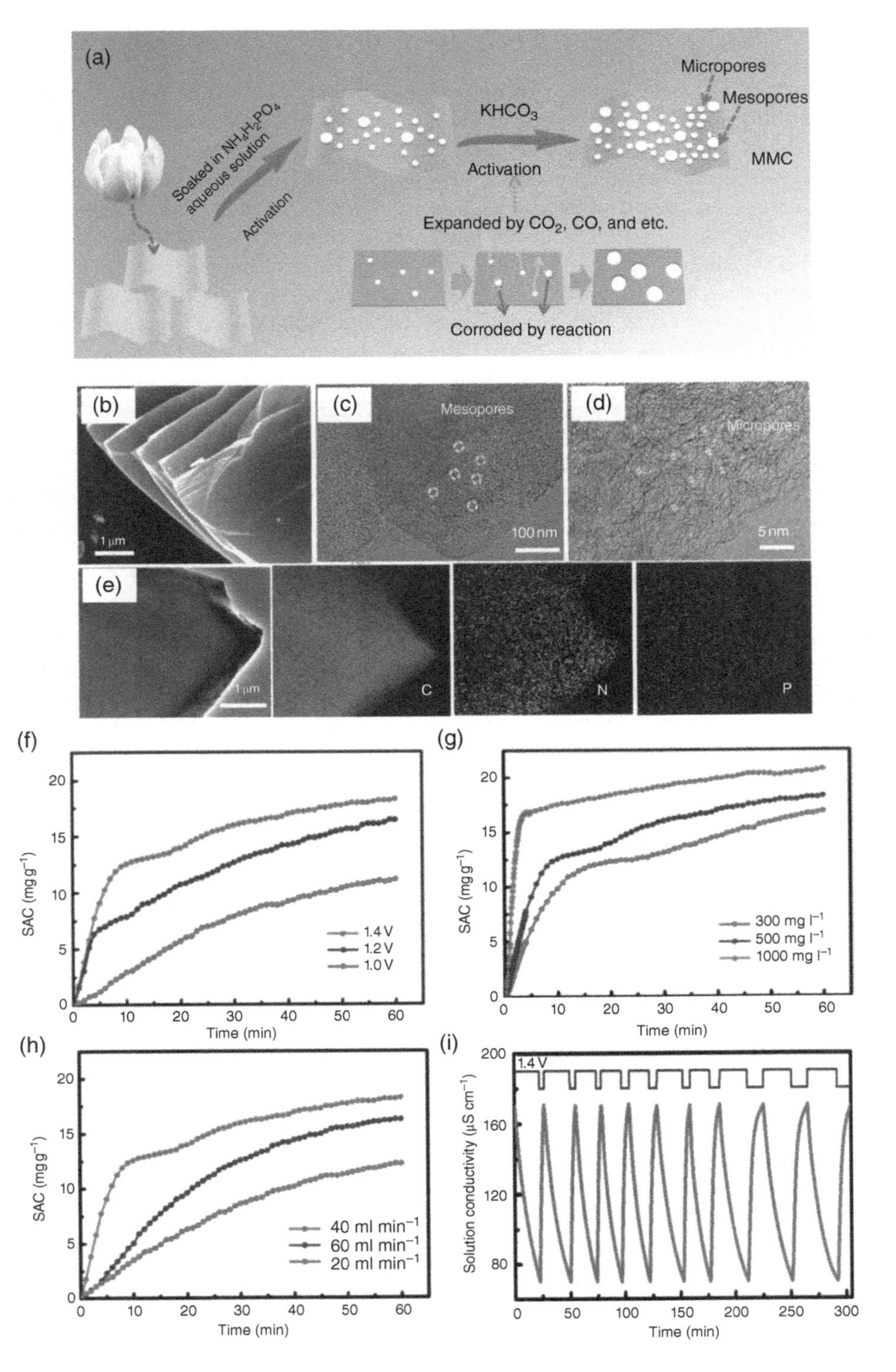

Figure 4.4 *(a) Schematic illustration of the preparation process of N, P-co-doped meso-/microporous carbon, (b) SEM images, and (c, d) TEM images of the carbon material, (e) energy-dispersive X-ray spectroscopy (EDX) elemental mapping images of C, N, and P. Plots of the material vs. deionization time, (f) for MMC electrodes at 1.0–1.4 V, (g) in a 300–1000 mg l^{-1} NaCl aqueous solution, and (h) at 20–60 ml min^{-1}. (i) Deionization–regeneration curves of MMC electrodes in 100 mg l^{-1} NaCl aqueous solution. Source: Adapted from Xu et al. [81]. Reprinted with permission from American Chemical Society.*

S atoms can change the electronic structure in the carbon layers and increase the electron density of surrounding carbon atoms. The lone pair electrons of N and S atoms can enhance the interaction between the catalyst and H^+, thus effectively improving the material's HER activity. Zhu et al. [66] demonstrated an effective strategy to prepare microporous heteroatom-doped carbon frameworks derived from pine needles with KOH activation. The resulting materials served as an HER electrocatalyst with a low onset potential of 4 mV, a small Tafel slope of 45.9 mV dec^{-1}, and remarkable stability. Further, the researchers matched the heteroatom-doped carbon materials with different active nanoparticles to improve the HER performance of the electrocatalyst. An et al. [93] prepared an N-doped carbon material from a sunflower seed shell integrated with Mo_2C nanoparticles. The Mo_2C particles with sizes from approximately 5 to 8 nm were evenly distributed in the S and N co-doped carbon matrix, and the composite formed with high SSA and a mesoporous character. The overpotential of the composite material was only 60 mV, with excellent durability and nearly 100% Faradic efficiency.

The oxygen evolution reaction (OER) is another half-reaction of water-splitting that involves a four-electron transfer process normally with slow kinetics. Therefore, highly active catalysts are needed to accelerate OER performance and reduce the overpotential [139]. Typical OER catalysts include rare metal oxides such as IrO_2 and RuO_2. These oxides have high catalytic activity and good stability, but their high cost and scarcity greatly limit their commercialization [140].

Biowaste-derived materials often show excellent performance in both the OER and ORR. Guan et al. [67] demonstrated a facile hydrothermal synthesis of a Co_3O_4 nanoparticles/N-doped hierarchical porous carbon (NHPC) material derived from cattle bone, which acted as an efficient ORR/OER bifunctional electrocatalyst (Figure 4.5). The material showed 3D hierarchically porous structure with interconnected meso-/macropores and well-dispersed Co_3O_4 nanoparticles on the carbon networks. Owing to the high SSA of 1070 $m^2 g^{-1}$, the well-defined hierarchically porous structure, the high content of N doping, and the homogeneous dispersion of Co_3O_4 nanoparticles, the Co_3O_4/NHPC electrocatalyst exhibited a superior OER activity with a lower evolution potential than that of commercial Pt/C and RuO_2. Furthermore, Co_3O_4/NHPC demonstrated superior electrochemical stability with approximately 87.6% of relative current retained after 12000 s of the test. In addition to their use as substrates for supporting metal-active catalysts, biowaste-derived carbon materials themselves show good OER electrocatalytic capability that is attributed to heteroatom doping. Amiinu et al. [94] synthesized a 3D porous graphene catalyst from sheep horn rich in both N and S via pyrolysis without adding any metal species. The obtained materials presented excellent ORR performance with the half-wave potential ($E_{1/2} = -0.23$ V) and diffusion-limited current density ($j_L = -5.41$ mA cm^{-2} at -1.0 V) closely approaching that of Pt/C. The material also achieved good OER performance with the operating potential of 0.69 V (at 10 mA cm^{-2}), which was comparable to that of IrO_2 (0.64 V vs. SCE).

These works have shown the potential of biowaste-derived carbon materials as electrocatalysts for hydrogen and oxygen evolution, but there are still some challenges. The relationship between the structure and active sites on carbon materials and their composites is still unclear. Determining the active sites and their composition and optimizing their amount and distribution is the main task to fully explore the potential of carbon materials as electrocatalysts.

4.4.4 Fuel Cells

Fuel cells are efficient devices to convert fuels into electric energy, which represent a clean use of fuels and lower greenhouse gas emission. Proton exchange membrane fuel cell (PEMFC) is one of the most important types of fuel cells. The carbon-based active materials are mainly used as fuel cell cathodes to catalyze the ORR [141].

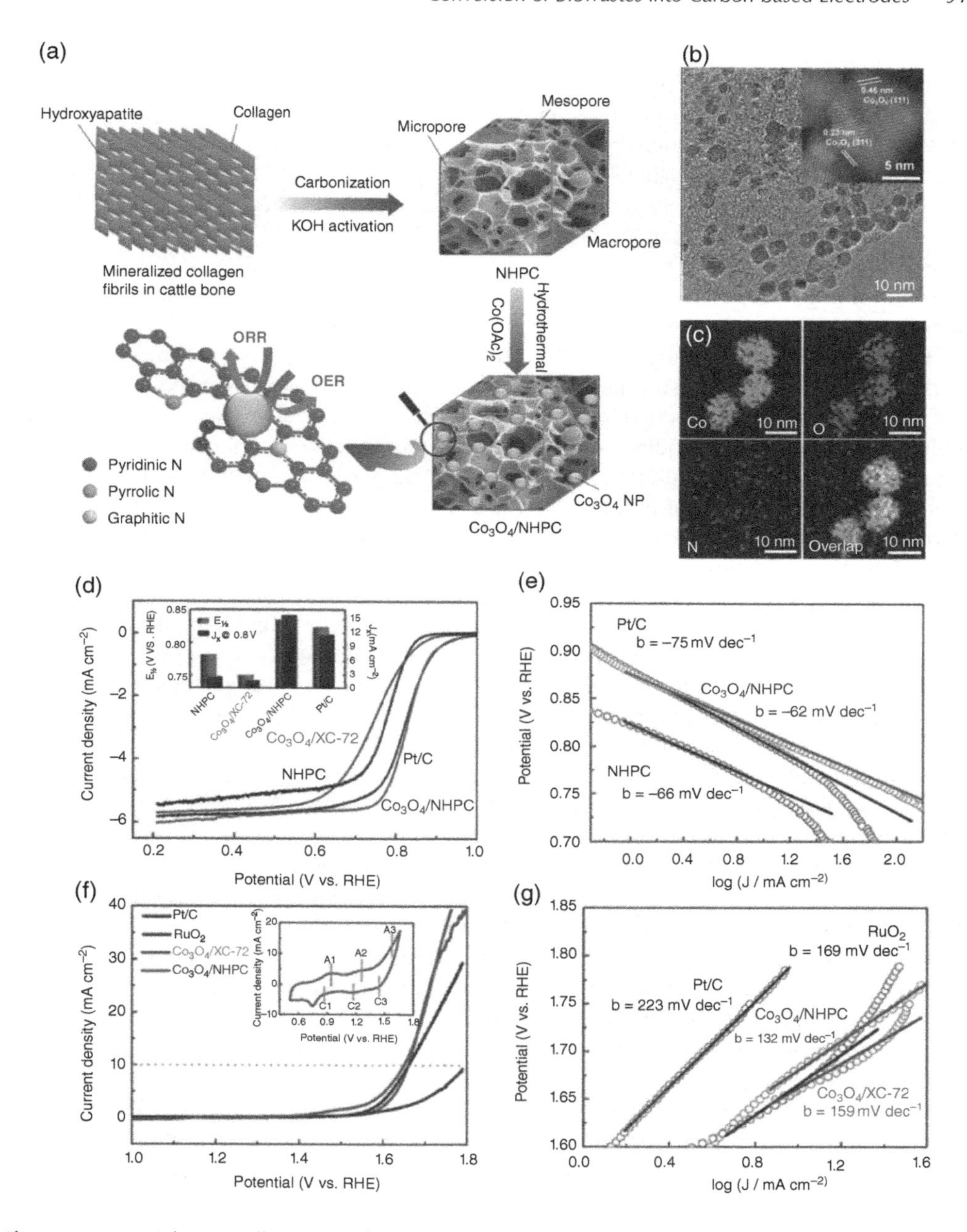

Figure 4.5 *(a) Schematic illustration of the synthetic process of the Co_3O_4/NHPC ORR/OER bifunctional electrocatalyst, (b) bright-field HR-TEM image (inset: the high-magnification dark-field image) of the material, (c) dark-field high resolution-scanning transmission electron microscope (HR-STEM) elemental mapping images of Co, O, N, and overlap, (d) linear sweep voltammetry (LSV) curves, and (e) Tafel plots of resulting materials and the commercial Pt/C for ORR (5 mV s^{-1}, 1600 rpm), (f) LSV curves, and (g) Tafel plots of resulting materials, Pt/C and RuO_2 for OER (20 mV s^{-1}). Source: Adapted from Guan et al. [67]. Reprinted with permission from American Chemical Society.*

The ORR is a key process and a major limiting factor of the energy conversion efficiency of fuel cells. The reaction rate of ORR is significantly limited by the slow kinetics. An efficient catalyst is mainly used for overcoming the energy barrier [142]. Traditional ORR electrocatalysts are mostly precious metals (such as Pt and Pd) and their oxides, but their poor stability, high cost, and scarcity limit their large-scale commercial application. Pt/C catalysts with 20–40% of Pt nanoparticles loading on the activated carbon have greatly reduced the cost, but still suffer from the problems of poor long-term stability and low methanol and CO tolerances.

Biowaste-derived carbon materials have been used as metal-free ORR catalysts [75, 143], which exhibit good catalytic performance with a high content of heteroatom doping, excellent long-term stability, and satisfactory methanol tolerance. For example, lignocellulose is rich in N, P, S, O, and B heteroatoms, which may produce synergetic effects to improve the catalytic performance of ORR. Transformation of livestock waste such as chicken bones, cow dung, and fish scales has been reported to produce N-doped carbon catalysts applicable for oxygen reduction [73, 95, 144]. Pyridine N and graphite N are recognized as the active sites in most researches [70]. Recently, studies on highly directional pyrolyzed graphite with precise control of the doping content of pyridine N or graphite N have shown that the Lewis base-type carbon atoms adjacent to pyridine N can be ORR active sites in acidic solutions, facilitating the initial oxygen adsorption step [145]. The role of different forms of N doping needs further investigation. The N content in the products can be increased by ammonia-assisted activation and additives such as melamine, or by using biowastes with high N content [102]. Pan et al. [96] synthesized metal-free N-doped porous carbon nanosheets from ginkgo leaves for ORR catalysis. Ammonia post-treatment was used instead of supporting templates and activation for ensuring a large proportion of active nitrogen species (N-graphitic and N-pyridinic). The resulting material showed a slightly negative ORR half-wave potential in O_2-saturated 0.1 M KOH solution and exhibited a high kinetic-limiting current density of 13.57 mA cm^{-2} at −0.25 V (vs. Ag/AgCl), which is similar to that of the commercial Pt/C catalyst (14 mA cm^{-2}). The material also showed a better durability than Pt/C, only a slight decrease (15%) in the current density after a continuous 14 hours' test, while the decrease was more than 27% for Pt/C. The excellent electrocatalytic properties originate from the hierarchical porous structure, SSA, and, particularly, the optimal N doping of the material.

The introduction of transition metals, such as Fe, Ni, and Co, into heteroatom-doped porous carbon is another way to obtain highly efficient ORR catalysts [146]. The catalytic activity may be attributed to the multivalent nature of transition metals and the unique structures of their nanoparticles. Some metal nanoparticles (like nano-Fe) may facilitate the formation of planar pyridinic-N, pyrrolic-N and defective edges, which could act as active sites with enhanced ORR performance through the four-electron transfer process [147]. Wu et al. [97] reported the synthesis of mesoporous carbon microspheres produced from eggs without the introduction of extrinsic dopants through a high-throughput spray-drying process. The final material acted as a bifunctional electrocatalyst for both OER and ORR (Figure 4.6). The sol-gel chemistry of egg content and silica was adopted to endow the final product (egg-CMS) with a large surface area, high pore volume, and intrinsic atom doping of N, P, and Fe. The obtained catalyst showed a good electrocatalytic activity with a high current density of 74.6 mA cm^{-2} at a potential of 1.6 V versus RHE, low onset potential of 59 mV dec^{-1}, which was lower than that of meso-C (122 mV dec^{-1}, synthesized by evaporation-induced self-assembling method), and excellent stability (95% current retention at an overpotential of 300 mV). Because of the similar role of ORR in metal–air batteries, the obtained bifunctional catalyst was also used as electrode material in rechargeable Zn – air batteries, with an output

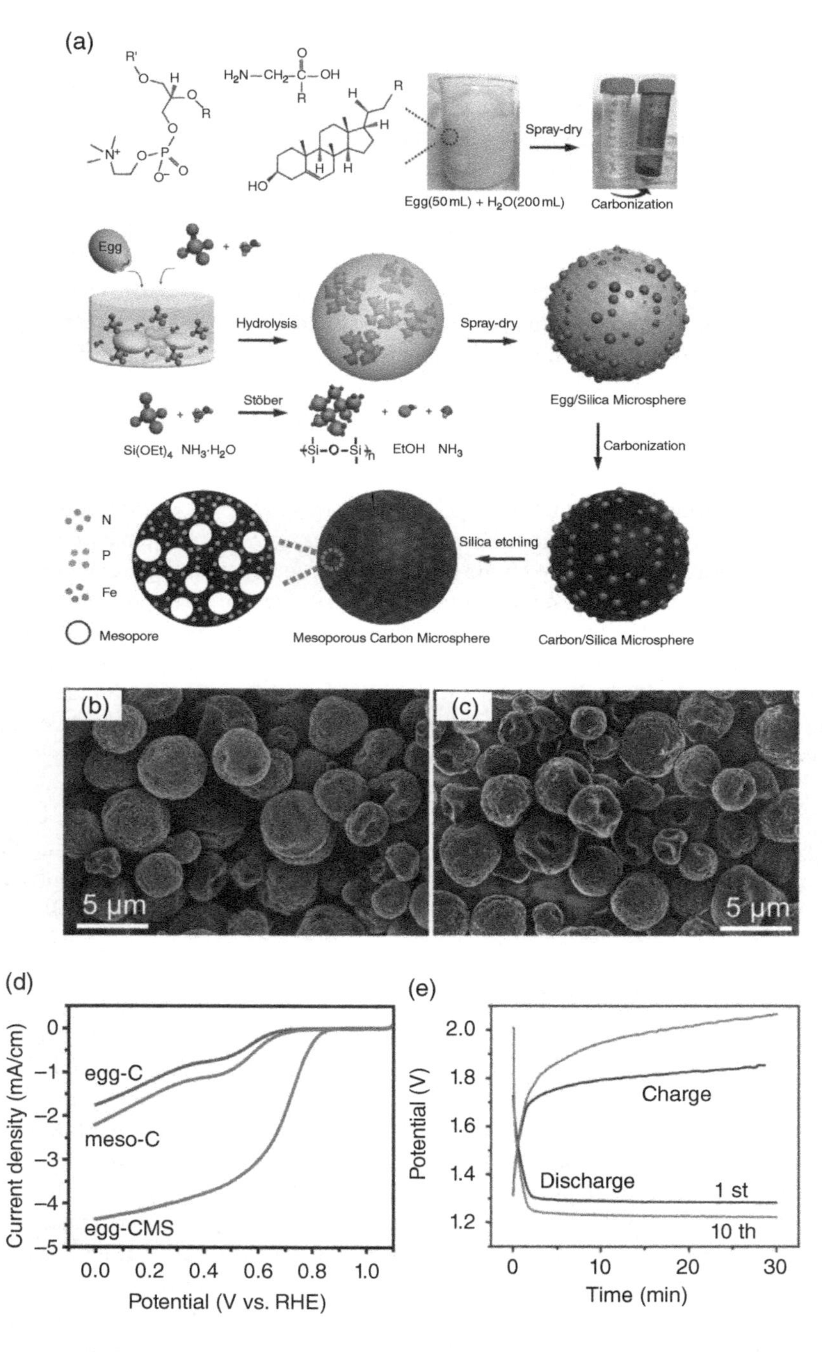

Figure 4.6 (a) Illustration of fabrication of co-doped egg-CMS. SEM images of egg-CMS (b) with and (c) without silica, (d) LSV curves of resulting materials at 1600 rpm in O_2 – saturated solution, (e) typical charge and discharge curves of rechargeable Zn–air battery. Source: Adapted from Wu et al. [97]. Reprinted with permission from Wiley-VCH.

potential of approximately 1 V for over 30 charge–discharge cycles. It should be mentioned that eggs cannot be classified as a biowaste, but because the main innate components of eggs are proteins, this work demonstrates the application potential of other protein-rich biowastes.

An "M–N" coordination structure of a transition metal atom with N atom(s) has been recognized as an active site to increase the catalytic activity of ORR [148]. Li et al. [98] reported the preparation of novel Fe and N co-doped CNT@hollow carbon fibers using catkin as the precursor, $FeCl_3$, and melamine as additives. The Fe catalyzed the growth of abundant CNTs along both the inner and outer walls of the product to form porous structures and ensured a high SSA. Furthermore, Fe element also facilitated the transformation of inactive oxidized N species to the highly active pyridinic-N, pyrrolic-N, and Fe-N clusters in resulting material, which contributed to the excellent ORR catalytic activity in both alkaline ($E_{1/2} = -194$ mV with $J_L = 4.08$ mA cm^{-2} at 0.4 V) and acidic media (E_{onset} was only 46 mV lower than Pt/C). The durability of the material was better than Pt/C as reflected by a relative current of 88.9% comparing with that of 73.5% for Pt/C after 20 000 s.

Biowaste-derived carbon materials have some additional functions in direct carbon fuel cells (DCFCs) and microbial fuel cells (MFCs). For example, in DCFCs, carbonaceous solid fuel is directly converted into electricity in molten carbonate systems. The molten carbonate is used as the electrolyte with good contact with the solid carbonaceous fuel at high temperature, enhancing the diffusivity of the carbon materials to the reaction site. Biowaste-derived carbon has the potential to replace coal, coke, and other petrochemical raw materials because of its unique physical properties including crystallinity, conductivity, structural morphology, and dopants. In particular, biowaste-derived carbon with a well-defined porous structure and high SSA can promote the transport of salt ions, thus improving the energy density [149]. However, the role of micropores in DCFCs is not clear, because the electrolyte has difficulty in entering the micropores. In MFCs, microorganisms convert organic and inorganic pollutants and nutrients in wastewater into electric energy through metabolism. Biowaste-derived carbon can be used as a promising cathodic catalyst for MFCs with high activity and durability, improving the enrichment, internal resistance, and electron transfer rate from electroactive bacteria to the anode and boosting the power-generation efficiency and conversion capacity of the cell [150].

4.4.5 Lithium-Ion Batteries and Others

LIBs are widely used high-performance energy storage devices with higher energy density and higher voltage than other secondary batteries such as lead-acid, Ni—Cd, and NiMH [151]. Carbon (graphite) is a typical anode material of LIBs and porous carbon has been used to enhance their performance. Biowaste-derived carbons may possess a well-defined porous structure and active sites, making them very suitable as electrodes of LIBs.

A variety of biowastes such as rice husk, peanut husk, CNS, and human hair have been successfully used in preparing high-performance lithium-ion battery electrode materials [53, 68]. Zhou et al. [152] prepared ultrathin graphitic carbon nanosheets (2–10 atomic layers) frameworks from wheat straw with mesoporous structure, which not only provides more Li ion-storage sites, but facilitates the rapid transfer of Li ions compared with traditional graphite electrodes. Lotfabad et al. [49] prepared graphene-like materials from banana peel for LIBs. The resulting carbon was promising for LIBs' application with a gravimetric capacity of 1090 mAh g^{-1} at 50 mA g^{-1}, although the SSA of the material was only 217 m^2 g^{-1}. The good performance is attributed to highly defective graphene sheets for the adsorption of Li ions and near-surface nanopores for Li ions' deposition. The pore morphology of materials has a great influence on the performance of LIBs. Hwang et al. [69] found that CNS carbonized at 900 °C by KOH activation tended to form the carbide material with smaller particle size and abundant pores. The obtained material showed a higher first lithium insertion capacity of 1714 mAh g^{-1} and lower irreversible capacity loss.

Transition metal oxides such as Co_3O_4, Fe_3O_4, and MnO_2 are often used as anode materials for LIBs, but the large volume expansion and contraction of these metal oxides and particle aggregation may lead to electrode pulverization, capacity loss, or poor cyclic stability, which limit their practical application [153]. Carbon/transition metal oxides' composite materials used in LIBs can effectively reduce the problems of pure metal oxides. Jiang et al. [100] reported a synthetic method to form carbon fibers using bamboo chopsticks' waste as a source and coat MnO_2 onto each fiber thread. The MnO_2 shells fully covered the entire carbon fiber core. The composite material as anode for LIBs showed a reversible capacity of 710 mAh g^{-1} up to 300 cycles at 0.2 A g^{-1} with good rate performance.

Sodium-ion batteries (NIBs), which are similar to LIBs, have also received attention due to their lower cost [154]. Hong et al. [76] synthesized a kind of hard porous carbon material by simple pyrolysis of pomelo peels with H_3PO_4 treatment. The obtained carbon showed a 3D-connected structure rich in O and P doping. When used as NIB anode, the material exhibited good cycling stability and attained a capacity of 181 mAh g^{-1} at 200 mA g^{-1}. Graphene-like materials from banana peel prepared by Lotfabad et al. [49] were also used in NIB with a gravimetric capacity of 355 mAh g^{-1} after 10 cycles at 50 mA g^{-1} achieved. The multiple components in the banana peel including pectin, lignin, free sugars, and cellulose were responsible for the NIB performance. During pyrolysis, pectin and lignin prevented further crystallization of free sugars, providing sufficient space for Na ion intercalation. However, because the radius of Na ion is larger than that of Li ion, sometimes NIBs are plagued by low capacity and poor cyclic stability. Therefore, it is still necessary to continuously explore ways to improve the performance of NIBs.

A lithium-sulfur battery (LSB) consists of a sulfur-containing cathode and a lithium-containing anode, and can achieve a higher energy density up to 2600 mAh g^{-1}, more than twice as efficient as LIBs. However, there are still some challenges of LSBs, such as low conductivity of sulfur products, large volume expansion during cycling, and severe shuttle effect caused by polysulfide dissolution [155]. Carbon-based materials can encapsulate sulfur and trap polysulfide during the lithiation-delithiation process, thus inhibiting the volume expansion of active substances and improving the stability of LSBs [156]. Fibrous biowastes such as cotton or bacterial cellulose could be converted to carriers with advantages of high sulfur loading and rapid electron transfer in a sulfur cathode [99].

4.5 Conclusion and Outlook

The research on the preparation of carbon electrodes using biowastes has achieved considerable progress over the past years. The primary driving force comes from their low cost and environmental compatibility, but the advantages of using biowastes as starting materials are far more than those. After billions of years of natural selection, biomass has unique internal morphology and element composition, which provide almost indefinite possibilities of improving the electrochemical performance of the resulting products. Biowaste-derived carbon-based electrode materials have been used in almost every area of electrochemistry and it is still in a rapid development regarding the synthesis procedure, biowaste selection, and performance of the products. Some issues that should be considered in the future include:

1. In the process of converting biowastes into porous carbon materials, it is frequently necessary to add a large amount of activators to obtain a higher SSA or special porous structure. However, these activators may cause a significant reduction in carbon production and form other byproducts, leading to the collapse of porous structures. Therefore, how to use the original structures of biowastes and decrease the amount of activators or other additives deserves more investigation. New conversion techniques, such as low-temperature plasma, should also be considered.
2. Critical steps that influence reaction kinetics in the pyrolysis process of biowastes include the breaking and redistribution of chemical bonds, changes in material geometry, and interface diffusion between reactants and products. Based on the thermogravimetric analysis, researchers have studied the quantitative control of the morphologies through pyrolysis reaction conditions, but due to the complexity of biowastes both in their origin and the way they are collected, more robust models should be established. More research on the mechanisms of pore formation and the ways for accurate control of surface properties should be conducted.
3. To ensure the quality of the final products, standard procedures for collecting biowastes are indispensable for large-scale production. Biowastes are generally "unwanted" and contamination may occur that can significantly affect the performance of the carbonized products or batch-to-batch reproducibility. If a complicated pre-purification process is required, it may largely compromise the advantage of low cost. On the other hand, the biowastes from agriculture may be contaminated with dust. In this circumstance, the carbonization and activation processes should be tuned to mitigate the negative effect of the contaminants or better to convert them to dopants for positive effects.

In any case, it can be expected that when biowastes are widely used in industry as raw materials, it is improper to call them "waste"; they deserve a more exciting and prestigious name.

Acknowledgments

The financial support from NSFC (No. 21527808) is gratefully acknowledged.

References

1. Fera, M., Macchiaroli, R., Iannone, R. et al. (2016). Economic evaluation model for the energy demand response. *Energy* 112: 457–468.
2. Liu, C., Li, F., Ma, L.P. et al. (2010). Advanced materials for energy storage. *Advanced Materials* 22: 28–62.
3. Zhang, Q., Uchaker, E., Candelaria, S.L. et al. (2013). Nanomaterials for energy conversion and storage. *Chemical Society Reviews* 42: 3127–3171.
4. Titirici, M.M., White, R.J., Brun, N. et al. (2015). Sustainable carbon materials. *Chemical Society Reviews* 44: 250–290.
5. Wyk, J.P.H.V. (2001). Biotechnology and the utilization of biowaste as a resource for bioproduct development. *Trends in Biotechnology* 19: 172–177.
6. Woolf, D., Amonette, J.E., Street-Perrott, F.A. et al. (2010). Sustainable biochar to mitigate global climate change. *Nature Communications* 1: 56.

7. Mensah-Darkwa, K., Zequine, C., Kahol, P. et al. (2019). Supercapacitor energy storage device using biowastes: a sustainable approach to green energy. *Sustainability* 11: 414–436.
8. Nabais, J.M.V., Teixeira, J.G., and Almeida, I. (2011). Development of easy made low cost bindless monolithic electrodes from biomass with controlled properties to be used as electrochemical capacitors. *Bioresource Technology* 102: 2781–2787.
9. Wang, H., Li, Z., and Mitlin, D. (2014). Tailoring biomass-derived carbon nanoarchitectures for high-performance supercapacitors. *ChemElectroChem* 1: 332–337.
10. Liu, W., Jiang, H., and Yu, H. (2015). Development of biochar-based functional materials: toward a sustainable platform carbon material. *Chemical Reviews* 115: 12251–12285.
11. Gu, J., Zhang, W., Su, H. et al. (2015). Morphology genetic materials templated from natural species. *Advanced Materials* 27: 464–478.
12. Thomas, P., Lai, C.W., and Johan, M.R.B. (2019). Recent developments in biomass-derived carbon as a potential sustainable material for super-capacitor-based energy storage and environmental applications. *Journal of Analytical and Applied Pyrolysis* 140: 54–85.
13. Zhang, Y., You, Y., Xin, S. et al. (2016). Rice husk-derived hierarchical silicon/nitrogen-doped carbon/carbon nanotube spheres as low-cost and high-capacity anodes for lithium-ion batteries. *Nano Energy* 25: 120–127.
14. Liu, J., Kopold, P., Aken, P.A.V. et al. (2015). Energy storage materials from nature through nanotechnology: a sustainable route from reed plants to a silicon anode for lithium-ion batteries. *Angewandte Chemie International Edition* 54: 9632–9636.
15. Siipola, V., Tamminen, T., Källi, A. et al. (2018). Effects of biomass type, carbonization process, and activation method on the properties of bio-based activated carbons. *Bioresources* 13: 5976–6002.
16. Dupont, C., Chen, L., Cances, J. et al. (2009). Biomass pyrolysis: kinetic modelling and experimental validation under high temperature and flash heating rate conditions. *Journal of Analytical and Applied Pyrolysis* 85: 260–267.
17. Chew, J.J. and Doshi, V. (2011). Recent advances in biomass pretreatment – torrefaction fundamentals and technology. *Renewable and Sustainable Energy Reviews* 15: 4212–4222.
18. Arvelakis, S., Crocker, C., Folkedahl, B. et al. (2010). Activated carbon from biomass for mercury capture: effect of the leaching pretreatment on the capture efficiency. *Energy & Fuels* 24: 4445–4453.
19. Huang, X., Liu, X., Shang, J. et al. (2012). Pretreatment methods for aquatic plant biomass as carbon sources for potential use in treating eutrophic water in subsurface-flow constructed wetlands. *Water Science and Technology* 66: 2328–2335.
20. Tao, Z., Zuo, S., Ling, X. et al. (2017). Study on pretreatment condition optimization for maize straw biomass carbon source. *Technology & Development of Chemical Industry* 8: 1–4,12.
21. Brown, T.R., Wright, M.M., and Brown, R.C. (2011). Estimating profitability of two biochar production scenarios: slow pyrolysis vs fast pyrolysis. *Biofuels, Bioproducts and Biorefining* 5: 54–68.
22. Puthusseri, D., Aravindan, V., Madhavi, S. et al. (2014). 3D micro – porous conducting carbon beehive by single step polymer carbonization for high performance supercapacitors: the magic of in situ porogen formation. *Energy & Environmental Science* 7: 728–735.
23. Calvo, E.G., Ferrera-Lorenzo, N., Menéndez, J.A. et al. (2013). Microwave synthesis of micro-mesoporous activated carbon xerogels for high performance supercapacitors. *Microporous and Mesoporous Materials* 168: 206–212.
24. Shang, H., Ran-Ran, L.U., and Sun, X. (2011). Research on microwave pyrolysis of biomass waste. *Renewable Energy Resources* 3: 25–29.
25. Shi, K., Yan, J., Lester, E. et al. (2014). Catalyst-free synthesis of multiwalled carbon nanotubes via microwave-induced processing of biomass. *Industrial & Engineering Chemistry Research* 53: 15012–15019.
26. Zhu, H., Wang, X., Yang, F. et al. (2011). Promising carbons for supercapacitors derived from fungi. *Advanced Materials* 23: 2745–2748.
27. Xie, Z., White, R.J., Weber, J. et al. (2011). Hierarchical porous carbonaceous materials via ionothermal carbonization of carbohydrates. *Journal of Materials Chemistry* 21: 7434–7442.
28. Yin, H., Lu, B., Xu, Y. et al. (2014). Harvesting capacitive carbon by carbonization of waste biomass in molten salts. *Environmental Science & Technology* 48: 8101–8108.

29. Deng, X., Zhao, B., Zhu, L. et al. (2015). Molten salt synthesis of nitrogen-doped carbon with hierarchical pore structures for use as high-performance electrodes in supercapacitors. *Carbon* 93: 48–58.
30. Sevilla, M. and Mokaya, R. (2014). Energy storage applications of activated carbons: supercapacitors and hydrogen storage. *Energy & Environmental Science* 7: 1250–1280.
31. Demiral, H., Demiral, L., Karabacakoğlu, B. et al. (2011). Production of activated carbon from olive bagasse by physical activation. *Chemical Engineering Research and Design* 89: 206–213.
32. Ould-Idriss, A., Stitou, M., Cuerda-Correa, E.M. et al. (2011). Preparation of activated carbons from olive-tree wood revisited. II. Physical activation with air. *Fuel Processing Technology* 92: 266–270.
33. Qu, W., Xu, Y., Lu, A. et al. (2015). Converting biowaste corncob residue into high value added porous carbon for supercapacitor electrodes. *Bioresource Technology* 189: 285–291.
34. Miguel, G., Fowler, G.D., and Sollars, C.J. (2003). A study of the characteristics of activated carbons produced by steam and carbon dioxide activation of waste tyre rubber. *Carbon* 41: 1009–1016.
35. Zhang, X., Zhang, S., Yang, H. et al. (2014). Nitrogen enriched biochar modified by high temperature CO_2-ammonia treatment: characterization and adsorption of CO_2. *Chemical Engineering Journal* 257: 20–27.
36. Molina-Sabio, M. and Rodríguez-Reinoso, F. (2004). Role of chemical activation in the development of carbon porosity. *Colloids and Surfaces A: Physicochemical and Engineering Aspects* 241: 15–25.
37. Wang, R., Wang, P., Yan, X. et al. (2012). Promising porous carbon derived from celtuce leaves with outstanding supercapacitance and CO_2 capture performance. *ACS Applied Materials & Interfaces* 4: 5800–5806.
38. Lillo-Ródenas, M.A., Juan-Juan, J., Cazorla-Amorós, D. et al. (2004). About reactions occurring during chemical activation with hydroxides. *Carbon* 42: 1371–1375.
39. Farma, R., Deraman, M., Awitdrus, A. et al. (2013). Preparation of highly porous binderless activated carbon electrodes from fibres of oil palm empty fruit bunches for application in supercapacitors. *Bioresource Technology* 132: 254–261.
40. Singh, G., Lakhi, K.S., Kim, I.Y. et al. (2017). Highly efficient method for the synthesis of activated mesoporous biocarbons with extremely high surface area for high-pressure CO_2 adsorption. *ACS Applied Materials & Interfaces* 9: 29782–29793.
41. Subha, R. and Namasivayam, C. (2009). Zinc chloride activated coir pith carbon as low cost adsorbent for removal of 2,4-dichlorophenol: equilibrium and kinetic studies. *Indian Journal of Chemical Technology* 16: 471–479.
42. Donald, J., Ohtsuka, Y., and Xu, C. (2011). Effects of activation agents and intrinsic minerals on pore development in activated carbons derived from a Canadian peat. *Materials Letters* 65: 744–747.
43. Senthilkumar, S.T., Selvan, A.R.K., Lee, Y.S. et al. (2013). Electric double layer capacitor and its improved specific capacitance using redox additive electrolyte. *Journal of Materials Chemistry A* 1: 1086–1095.
44. Rajendra, S., Shin-Ichiro, F., Shuhei, O. et al. (2009). Alkylation of aromatic compounds with multi-component Lewis acid catalysts of $ZnCl_2$ and ionic liquids with different organic cations. *Reaction Kinetics and Catalysis Letters* 96: 55–64.
45. Hsu, L.Y. and Teng, H. (2000). Influence of different chemical reagents on the preparation of activated carbons from bituminous coal. *Fuel Processing Technology* 64: 155–166.
46. Williams, P.T. and Reed, A.R. (2006). Development of activated carbon pore structure via physical and chemical activation of biomass fibre waste. *Biomass and Bioenergy* 30: 144–152.
47. Fu, Y., Shen, Y., Zhang, Z. et al. (2019). Activated bio-chars derived from rice husk via one- and two-step KOH-catalyzed pyrolysis for phenol adsorption. *Science of the Total Environment* 646: 1567–1577.
48. Schlee, P., Hosseinaei, O., Baker, D. et al. (2019). From waste to wealth: from kraft lignin to free-standing supercapacitors. *Carbon* 145: 470–480.
49. Lotfabad, E.M., Ding, J., Cui, K. et al. (2014). High-density sodium and lithium Ion battery anodes from banana peels. *ACS Nano* 8: 7115–7129.
50. Jin, Z., Yan, X., Yu, Y. et al. (2014). Sustainable activated carbon fibers from liquefied wood with controllable porosity for high-performance supercapacitors. *Journal of Materials Chemistry A* 2: 11706–11715.
51. Zhao, G., Chen, C., Yu, D. et al. (2018). One-step production of O-N-S co-doped three-dimensional hierarchical porous carbons for high-performance supercapacitors. *Nano Energy* 47: 547–555.
52. Qian, W., Sun, F., Xu, Y. et al. (2014). Human hair-derived carbon flakes for electrochemical supercapacitors. *Energy & Environmental Science* 7: 379–386.
53. Saravanan, K.R. and Kalaiselvi, N. (2015). Nitrogen containing bio-carbon as a potential anode for lithium batteries. *Carbon* 81: 43–53.

54. Li, Y., Wang, G., Wei, T. et al. (2016). Nitrogen and sulfur co-doped porous carbon nanosheets derived from willow catkin for supercapacitors. *Nano Energy* 19: 165–175.
55. Zhang, W., Lin, H., Lin, Z. et al. (2015). 3D hierarchical porous carbon for supercapacitors prepared from lignin through a facile template-free method. *ChemSusChem* 8: 2114–2122.
56. Song, S., Ma, F., Wu, G. et al. (2015). Facile self-templating large scale preparation of biomass-derived 3D hierarchical porous carbon for advanced supercapacitors. *Journal of Materials Chemistry A* 3: 18154–18162.
57. Liang, Q., Ye, L., Huang, Z. et al. (2014). A honeycomb-like porous carbon derived from pomelo peel for use in high-performance supercapacitors. *Nanoscale* 6: 13831–13837.
58. Wang, B., Wang, Y., Peng, Y. et al. (2018). 3-dimensional interconnected framework of N-doped porous carbon based on sugarcane bagasse for application in supercapacitors and lithium ion batteries. *Journal of Power Sources* 390: 186–196.
59. Qiu, Z., Wang, Y., Bi, X. et al. (2018). Biochar-based carbons with hierarchical micro-meso-macro porosity for high rate and long cycle life supercapacitors. *Journal of Power Sources* 376: 82–90.
60. Zhang, Q., Han, K., Li, S. et al. (2018). Synthesis of garlic skin-derived 3D hierarchical porous carbon for high-performance supercapacitors. *Nanoscale* 10: 2427–2437.
61. Hao, P., Zhao, Z., Tian, J. et al. (2014). Hierarchical porous carbon aerogel derived from bagasse for high performance supercapacitor electrode. *Nanoscale* 6: 12120–12129.
62. Hao, P., Zhao, Z., Leng, Y. et al. (2015). Graphene-based nitrogen self-doped hierarchical porous carbon aerogels derived from chitosan for high performance supercapacitors. *Nano Energy* 15: 9–23.
63. Chen, L., Zhang, Y., Lin, C. et al. (2014). Hierarchically porous nitrogen-rich carbon derived from wheat straw as an ultra-high-rate anode for lithium ion batteries. *Journal of Materials Chemistry A* 2: 9684–9690.
64. Chen, C., Yu, D., Zhao, G. et al. (2016). Three-dimensional scaffolding framework of porous carbon nanosheets derived from plant wastes for high-performance supercapacitors. *Nano Energy* 27: 377–389.
65. Liu, W., Tian, K., Ling, L. et al. (2016). Use of nutrient rich hydrophytes to create N, P-dually doped porous carbon with robust energy storage performance. *Environmental Science & Technology* 50: 12421–12428.
66. Zhu, G., Ma, L., Lu, H. et al. (2016). Pine needle-derived microporous nitrogen-doped carbon frameworks with high performances for electrocatalytic hydrogen evolution and supercapacitors. *Nanoscale* 9: 1237–1243.
67. Guan, J., Zhang, Z., Ji, J. et al. (2017). Hydrothermal synthesis of highly dispersed Co_3O_4 nanoparticles on biomass-derived nitrogen-doped hierarchically porous carbon networks as an efficient bifunctional electrocatalyst for oxygen reduction and evolution reactions. *ACS Applied Materials & Interfaces* 9: 30662–30669.
68. Gu, X., Wang, Y., Lai, C. et al. (2015). Microporous bamboo biochar for lithium-sulfur batteries. *Nano Research* 8: 129–139.
69. Hwang, Y.J., Jeong, S.K., Shin, J.S. et al. (2008). High capacity disordered carbons obtained from coconut shells as anode materials for lithium batteries. *Journal of Alloys and Compounds* 448: 141–147.
70. Liu, X., Zhou, Y., Zhou, W. et al. (2015). Biomass-derived nitrogen self-doped porous carbon as effective metal-free catalysts for oxygen reduction reaction. *Nanoscale* 7: 6136–6142.
71. Sun, L., Tian, C., Li, M. et al. (2013). From coconut shell to porous graphene-like nanosheets for high-power supercapacitors. *Journal of Materials Chemistry A* 1: 6462–6470.
72. Cheng, L., Guo, P., Wang, R. et al. (2014). Electrocapacitive properties of supercapacitors based on hierarchical porous carbons from chestnut shell. *Colloids and Surfaces: Physicochemical and Engineering Aspects* 446: 127–133.
73. Guo, C., Hu, R., Liao, W. et al. (2017). Protein-enriched fish "biowaste" converted to three-dimensional porous carbon nano-network for advanced oxygen reduction electrocatalysis. *Electrochimica Acta* 236: 228–238.
74. Huang, C., Puziy, A.M., Sun, T. et al. (2014). Capacitive behaviours of phosphorus-rich carbons derived from lignocelluloses. *Electrochimica Acta* 137: 219–227.
75. Borghei, M., Laocharoen, N., Kibena-Põldsepp, E. et al. (2017). Porous N, P-doped carbon from coconut shells with high electrocatalytic activity for oxygen reduction: alternative to Pt-C for alkaline fuel cells. *Applied Catalysis B: Environmental* 204: 394–402.
76. Hong, K., Qie, L., Zeng, R. et al. (2014). Biomass derived hard carbon used as a high performance anode material for sodium ion batteries. *Journal of Materials Chemistry A* 2: 12733–12738.
77. Cao, Y., Xie, L., Sun, G. et al. (2018). Hollow carbon microtubes from kapok fiber: structural evolution and energy storage performance. *Sustainable Energy & Fuels* 2: 455–465.

78. Liu, J., Deng, Y., Li, X. et al. (2016). Promising nitrogen-rich porous carbons derived from one-step calcium chloride activation of biomass-based waste for high performance supercapacitors. *ACS Sustainable Chemistry & Engineering* 4: 177–187.
79. Zhao, C., Liu, G., Sun, N. et al. (2018). Biomass-derived N-doped porous carbon as electrode materials for Zn-air battery powered capacitive deionization. *Chemical Engineering Journal*. 334: 1270–1280.
80. Zhou, Y., Leng, Y., Zhou, W. et al. (2015). Sulfur and nitrogen self-doped carbon nanosheets derived from peanut root nodules as high-efficiency non-metal electrocatalyst for hydrogen evolution reaction. *Nano Energy* 16: 357–366.
81. Xu, D., Tong, Y., Yan, T. et al. (2017). N, P-codoped meso-/microporous carbon derived from biomass materials via a dual-activation strategy as high-performance electrodes for deionization capacitors. *ACS Sustainable Chemistry & Engineering* 5: 5810–5819.
82. Zhao, Y., Lu, M., Tao, P. et al. (2016). Hierarchically porous and heteroatom doped carbon derived from tobacco rods for supercapacitors. *Journal of Power Sources* 307: 391–400.
83. Ferrero, G.A., Fuertes, A.B., and Sevilla, M. (2015). From soybean residue to advanced supercapacitors. *Scientific Reports* 5: 16618–16621.
84. Hao, E., Liu, W., Liu, S. et al. (2017). Rich sulfur doped porous carbon materials derived from ginkgo leaves for multiple electrochemical energy storage devices. *Journal of Materials Chemistry A* 5: 2204–2214.
85. Wei, T., Wei, X., Gao, Y. et al. (2015). Large scale production of biomass-derived nitrogendoped porous carbon materials for supercapacitors. *Electrochimica Acta* 169: 186–194.
86. Wang, H., Xu, Z., Kohandehghan, A. et al. (2013). Interconnected carbon nanosheets derived from hemp for ultrafast supercapacitors with high energy. *ACS Nano* 7: 5131–5141.
87. Wang, C., Xiong, H., Wang, H. et al. (2017). Naturally three-dimensional laminated porous carbon network structured short nano-chains bridging nanospheres for energy storage. *Journal of Materials Chemistry A* 5: 15759–15770.
88. Li, Q., Lu, C., Xiao, D. et al. (2018). β-$Ni(OH)_2$ nanosheet arrays grown on biomass-derived hollow carbon microtubes for high-performance asymmetric supercapacitors. *ChemElectroChem* 5: 1279–1287.
89. Li, Y., Liu, Y., Chen, S. et al. (2018). Self-templating synthesis nitrogen and sulfur co-doped hierarchical porous carbons derived from crab shells as a high-performance metal-free oxygen electroreduction catalyst. *Materials Today Energy* 10: 388–395.
90. Xiong, W., Gao, Y., Wu, X. et al. (2014). Composite of macroporous carbon with honeycomb-like structure from mollusc shell and $NiCo_2O_4$ nanowires for high-performance supercapacitor. *ACS Applied Materials & Interfaces* 6: 19416–19423.
91. Gaikwad, M.S. and Balomajumder, C. (2017). Tea waste biomass activated carbon electrode for simultaneous removal of Cr(VI) and fluoride by capacitive deionization. *Chemosphere* 184: 1141–1149.
92. Lai, F., Miao, Y., Huang, Y. et al. (2016). Nitrogen-doped carbon nanofiber/molybdenum disulfide nanocomposites derived from bacterial cellulose for high-efficiency electrocatalytic hydrogen evolution reaction. *ACS Applied Materials & Interfaces* 8: 3558–3566.
93. An, K., Xu, X., and Liu, X. (2017). Mo_2C-based electrocatalyst with biomass-derived sulfur and nitrogen co-doped carbon as a matrix for hydrogen evolution and organic pollutant removal. *ACS Sustainable Chemistry & Engineering* 6: 1446–1455.
94. Amiinu, I.S., Zhang, J., Kou, Z. et al. (2016). Self-organized 3D porous graphene dual-doped with biomass-sponsored nitrogen and sulfur for oxygen reduction and evolution. *ACS Applied Materials & Interfaces* 8: 29408–29418.
95. Zhang, Z., Li, H., Yang, Y. et al. (2015). Cow dung-derived nitrogen-doped carbon as a cost effective, high activity, oxygen reduction electrocatalyst. *RSC Advances* 5: 27112–27119.
96. Pan, F., Cao, Z., Zhao, Q. et al. (2014). Nitrogen-doped porous carbon nanosheets made from biomass as highly active electrocatalyst for oxygen reduction reaction. *Journal of Power Sources* 272: 8–15.
97. Wu, H., Geng, J., Ge, H. et al. (2016). Egg-derived mesoporous carbon microspheres as bifunctional oxygen evolution and oxygen reduction electrocatalysts. *Advanced Energy Materials* 6: 1600794.
98. Li, M., Xiong, Y., Liu, X. et al. (2015). Iron and nitrogen co-doped carbon nanotube@hollow carbon fibers derived from plant biomass as efficient catalysts for the oxygen reduction reaction. *Journal of Materials Chemistry A* 3: 9658–9667.
99. Zheng, B., Li, N., Yang, J. et al. (2019). Waste cotton cloth derived carbon microtube textile: a robust and scalable interlayer for lithium-sulfur batteries. *Chemical Communications* 55: 2289–2292.

100. Jiang, J., Zhu, J., Ai, W. et al. (2014). Evolution of disposable bamboo chopsticks into uniform carbon fibers: a smart strategy to fabricate sustainable anodes for Li-ion batteries. *Energy & Environmental Science* 7: 2670–2679.
101. Wang, H., Yi, H., Zhu, C. et al. (2015). Functionalized highly porous graphitic carbon fibers for high-rate supercapacitive electrodes. *Nano Energy* 13: 658–669.
102. Graglia, M., Pampel, J., Hantke, T. et al. (2016). Lignin-derived nitrogen-doped carbon as an efficient and sustainable electrocatalyst for oxygen reduction. *ACS Nano* 10: 4364–4371.
103. Cagnon, B., Py, X., Guillot, A. et al. (2009). Contributions of hemicellulose, cellulose and lignin to the mass and the porous properties of chars and steam activated carbons from various lignocellulosic precursors. *Bioresource Technology* 100: 292–298.
104. Teng, H. and Yeh, T.S. (1998). Preparation of activated carbons from bituminous coals with zinc chloride activation. *Industrial & Engineering Chemistry Research* 37: 58–65.
105. Khalil, H.P.S.A., Jawaid, M., Firoozian, P. et al. (2013). Activated carbon from various agricultural wastes by chemical activation with KOH: preparation and characterization. *Journal of Biobased Materials and Bioenergy* 7: 1–7.
106. Ramakrishnan, K. and Namasivayam, C. (2009). Development and characteristics of activated carbons from Jatropha husk, an agro industrial solid waste, by chemical activation methods. *Journal of Environmental Engineering and Landscape Management* 19: 173–178.
107. Simon, P. and Gogotsi, Y. (2008). Materials for electrochemical capacitors. *Nature Materials* 7: 845–854.
108. Wei, Q., Xiong, F., Tan, S. et al. (2017). Porous one-dimensional nanomaterials: design, fabrication and applications in electrochemical energy storage. *Advanced Materials* 29: 1602300.
109. Bi, Z., Kong, Q., Cao, Y. et al. (2019). Biomass-derived porous carbon materials with different dimensions for supercapacitor electrodes: A review. *J. Mater. Chem. A* 7: 16028–16045.
110. Tan, C., Cao, X., Wu, X. et al. (2017). Recent advances in ultrathin two-dimensional nanomaterials. *Chemical Reviews* 117: 6225–6331.
111. Yao, L., Wu, Q., Zhang, P. et al. (2018). Scalable 2D hierarchical porous carbon nanosheets for flexible supercapacitors with ultrahigh energy density. *Advanced Materials* 30: 1706054.
112. Jin, H., Li, J., Yuan, Y. et al. (2018). Recent progress in biomass-derived electrode materials for high volumetric performance supercapacitors. *Advanced Energy Materials* 8: 1801007.
113. Feng, X., Bian, L., Ma, J. et al. (2019). Distinct correlation between $(CN_2)_x$ units and pores: a low-cost method for predesigned wide range control of micropore size of porous carbon. *Chemical Communications* 55: 3963–3966.
114. Liang, C., Li, Z., and Dai, S. (2008). Mesoporous carbon materials: synthesis and modification. *Angewandte Chemie International Edition* 47: 3696–3717.
115. Guo, Z., Zhou, D., Dong, X. et al. (2013). Ordered hierarchical mesoporous/macroporous carbon: a high-performance catalyst for rechargeable Li-O_2 batteries. *Advanced Materials* 25: 5668–5672.
116. Yang, X., Chen, L., Li, Y. et al. (2017). Hierarchically porous materials: synthesis strategies and structure design. *Chemical Society Reviews* 46: 481–558.
117. Wang, J., Angnes, L., Tobias, H. et al. (1993). Carbon aerogel composite electrodes. *Analytical Chemistry* 65: 2300–2303.
118. Inagaki, M., Toyoda, M., Soneda, Y. et al. (2018). Nitrogen-doped carbon materials. *Carbon* 132: 104–140.
119. Cao, H., Peng, X., Zhao, M. et al. (2018). Oxygen functional groups improve the energy storage performances of graphene electrochemical supercapacitors. *RSC Advances* 8: 2858–2865.
120. Hou, J., Cao, C., Idrees, F. et al. (2015). Hierarchical porous nitrogen-doped carbon nanosheets derived from silk for ultrahigh-capacity battery anodes and supercapacitors. *ACS Nano* 9: 2556–2564.
121. Hulicova-Jurcakova, D., Seredych, M., Lu, G. et al. (2009). Combined effect of nitrogen- and oxygen-containing functional groups of microporous activated carbon on its electrochemical performance in supercapacitors. *Advanced Functional Materials* 19: 438–447.
122. Hulicova-Jurcakova, D., Puziy, A.M., Poddubnaya, O.I. et al. (2009). Highly stable performance of supercapacitors from phosphorus-enriched carbons. *Journal of the American Chemical Society* 131: 5026–5027.
123. Yu, P., Zhang, Z., Zheng, L. et al. (2016). A novel sustainable flour derived hierarchical nitrogen-doped porous carbon/polyaniline electrode for advanced asymmetric supercapacitors. *Advanced Energy Materials* 6: 1601111.

124. Liu, J., Wang, J., Xu, C. et al. (2017). Advanced energy storage devices: basic principles, analytical methods, and rational materials design. *Advanced Science* 5: 1700322.
125. González, A., Goikolea, E., Barrena, J.A. et al. (2016). Review on supercapacitors: technologies and materials. *Renewable Sustainable Energy Review* 58: 1189–1206.
126. Paraknowitsch, J.P. and Thomas, A. (2013). Doping carbons beyond nitrogen: an overview of advanced heteroatom doped carbons with boron, sulphur and phosphorus for energy applications. *Energy & Environmental Science* 6: 2839–2855.
127. Chmiola, J., Yushin, G., Gogotsi, Y. et al. (2006). Anomalous increase in carbon capacitance at pore sizes less than 1 nanometer. *Science* 313: 1760–1763.
128. Zhang, N., Liu, F., Xu, S. et al. (2017). Nitrogen-phosphorus co-doped hollow carbon microspheres with hierarchical micro-meso-macroporous shells as efficient electrodes for supercapacitors. *Journal of Materials Chemistry A* 5: 22631–22640.
129. Anderson, M.A., Cudero, A.L., and Palma, J. (2010). Capacitive deionization as an electrochemical means of saving energy and delivering clean water. Comparison to present desalination practices: will it compete? *Electrochimica Acta* 55: 3845–3856.
130. Farmer, J.C., Fix, D.V., Mack, G.V. et al. (1995). The use of capacitive deionization with carbon aerogel electrodes to remove inorganic contaminants from water. *Office of Science and Technology Integration* 15: 595–599.
131. Suss, M.E. and Presser, V. (2018). Water desalination with energy storage electrode materials. *Joule* 2: 10–15.
132. Pasta, M., Wessells, C.D., Cui, Y. et al. (2012). A desalination battery. *Nano Letters* 12: 839–843.
133. Han, B., Cheng, G., Wang, Y. et al. (2019). Structure and functionality design of novel carbon and faradaic electrode materials for high-performance capacitive deionization. *Chemical Engineering Journal* 360: 364–384.
134. Baroud, T.N. and Giannelis, E.P. (2018). High salt capacity and high removal rate capacitive deionization enabled by hierarchical porous carbons. *Carbon* 139: 614–625.
135. Tsai, Y.C. and Doong, R.A. (2015). Activation of hierarchically ordered mesoporous carbons for enhanced capacitive deionization application. *Synthetic Metals* 205: 48–57.
136. Ströbel, R., Garche, J., Moseley, P. et al. (2006). Hydrogen storage by carbon materials. *Journal of Power Sources* 159: 781–801.
137. Zhang, L., Xiao, J., Wang, H. et al. (2017). Carbon-based electrocatalysts for hydrogen and oxygen evolution reactions. *ACS Catalysis* 7: 7855–7865.
138. Cui, W., Liu, Q., Xing, Z. et al. (2015). MoP nanosheets supported on biomass-deliverd carbon flake: one-step facile preparation and application as a novel high-active electrocatalyst toward hydrogen evolution reaction. *Applied Catalysis B: Environmental* 164: 144–150.
139. Wang, H., Chen, R., Feng, J. et al. (2018). Freestanding non-precious metal electrocatalysts for oxygen evolution and reduction reactions. *ChemElectroChem* 5: 1786–1804.
140. Mccrory, C.C.L., Jung, S., Peters, J.C. et al. (2013). Benchmarking heterogeneous electrocatalysts for the oxygen evolution reaction. *Journal of the American Chemical Society* 135: 16977–16987.
141. Zhu, Y., Wang, X., Jia, C. et al. (2016). Redox-mediated ORR and OER reactions: redox flow lithium oxygen batteries enabled with a pair of soluble redox catalysts. *ACS Catalysis* 6: 6191–6197.
142. Zhou, M., Wang, H.L., and Guo, S. (2016). Towards high-efficiency nanoelectrocatalysts for oxygen reduction through engineering advanced carbon nanomaterials. *Chemical Society Reviews* 45: 1273–1307.
143. Borghei, M., Lehtonen, J., Liu, L. et al. (2018). Advanced biomass-derived electrocatalysts for the oxygen reduction reaction. *Advanced Materials* 30: 1703691.
144. Song, H., Li, H., Wang, H. et al. (2014). Chicken bone-derived N-doped porous carbon materials as an oxygen reduction electrocatalyst. *Electrochimica Acta* 147: 520–526.
145. Gao, K., Wang, B., Tao, L. et al. (2019). Efficient metal-free electrocatalysts from N-doped carbon nanomaterials: mono-doping and co-doping. *Advanced Materials* 31: 1805121.
146. Zhu, C., Li, H., Fu, S. et al. (2016). Highly efficient nonprecious metal catalysts towards oxygen reduction reaction based on three-dimensional porous carbon nanostructures. *Chemical Society Reviews* 45: 517–531.

147. Osgood, H., Devaguptapu, S.V., Xu, H. et al. (2016). Transition metal (Fe, Co, Ni, and Mn) oxides for oxygen reduction and evolution bifunctional catalysts in alkaline media. *Nano Today* 11: 601–625.
148. Wang, Y., Li, J., and Wei, Z. (2018). Transition-metal-oxide-based catalysts for the oxygen reduction reaction. *Journal of Materials Chemistry A* 6: 8194–8209.
149. Gür, T.M. (2013). Critical review of carbon conversion in "carbon fuel cells". *Chemical Reviews* 113: 6179–6206.
150. Wang, H., Park, J.D., and Ren, Z.J. (2015). Practical energy harvesting for microbial fuel cells: a review. *Environmental Science & Technology* 49: 3267–3277.
151. Ji, L., Lin, Z., Alcoutlabi, M. et al. (2011). Recent developments in nanostructured anode materials for rechargeable lithium-ion batteries. *Energy & Environmental Science* 4: 2682–2699.
152. Zhou, X., Chen, F., Bai, T. et al. (2016). Interconnected highly graphitic carbon nanosheets derived from wheat stalk as high performance anode materials for lithium ion batteries. *Green Chemistry* 18: 2078–2088.
153. Reddy, A.L.M., Shaijumon, M.M., Gowda, S.R. et al. (2009). Coaxial MnO_2/carbon nanotube array electrodes for high-performance lithium batteries. *Nano Letters* 9: 1002–1006.
154. Xia, X., Chao, D., Zhang, Y. et al. (2016). Generic synthesis of carbon nanotube branches on metal oxide arrays exhibiting stable high-rate and long-cycle sodium-ion storage. *Small* 12: 3048–3058.
155. Seh, Z.W., Sun, Y., Zhang, Q. et al. (2016). Designing high-energy lithium-sulfur batteries. *Chemical Society Reviews* 45: 5605–5634.
156. Yuan, H., Liu, T., Liu, Y. et al. (2019). A review of biomass materials for advanced lithium-sulfur batteries. *Chemical Science* 10: 7484–7495.

5

Bio-based Materials in Electrochemical Applications

Itziar Iraola-Arregui[1], Mohammed Aqil[2], Vera Trabadelo[1], Ismael Saadoune[3,4] and Hicham Ben Youcef[1]

[1]*High Throughput Multidisciplinary Research Laboratory (HTMR-Lab), Mohammed VI Polytechnic University (UM6P), Benguerir, Morocco*
[2]*Materials Science, Energy and Nanoengineering Department (MSN), Mohammed VI Polytechnic University (UM6P), Benguerir, Morocco*
[3]*Technology Development Cell (TechCell), Mohammed VI Polytechnic University (UM6P), Benguerir, Morocco*
[4]*IMED-Lab, Faculty of Science and Technology, Cadi Ayyad University, Marrakesh, Morocco*

5.1 Introduction

The world electrification and the technological revolution in the twentieth and the beginning of the twenty-first centuries played crucial roles in the world's increasing prosperity and human wellbeing [1]. This development was unfortunately accompanied by an increase in the exploitation of the earth's fossil sources and the carbon emissions that have created significant debate on the causes-effects of the climate-change risks. Nowadays, the world's fast-growing demand for energy makes it become the first preoccupation even ahead of the food security as their productivities are now closely connected [2]. With the fast-growing population and the modern society lifestyle, power consumption will continue to be a tremendous challenge worldwide. However, to secure sustainable use of energy, alternative renewable sources have been explored and developed (e.g. solar, wind, and water-based energy generation).

On the way to achieving sustainable and clean energy sources, the development of recyclable, performant, and cost-effective energy storage and conversion systems is a must.

High-Performance Materials from Bio-based Feedstocks, First Edition. Edited by Andrew J. Hunt, Nontipa Supanchaiyamat, Kaewta Jetsrisuparb and Jesper T.N. Knijnenburg.

Particularly, intermittent electricity generation based on different renewable sources is a key sector needing sustainable and scalable solutions based on electrochemical energy storage systems. Moreover, the roadmap drawn by numerous countries in reducing the CO_2 emission based on the electrification of the transportation sector will certainly increase the diversity of provided solutions on electrochemical energy storage and conversion.

This perspective puts the interest in developing bio-based materials as a sustainable, versatile, cost-effective, and eco-friendly source for several applications within electrochemical energy systems. Batteries, supercapacitors, and fuel cells are key examples where bio-based materials showed their applicability in numerous components, especially when recycling becomes a fundamental and crucial concept of sustainability.

5.2 Fundamentals of Bio-based Materials

Bio-based materials are produced or extracted from substances derived from renewable biological resources. Biomass, for example, can be directly used for energy production through combustion or gasification [3] but it can also give infinite possibilities to produce different kinds of 'safe' materials with various applications due to the diverse molecular structures available in nature. The composition of biomass varies significantly from sample to sample due to the different sources and origins but, in most cases, the main constituents are cellulose, hemicelluloses, and lignin (intricate polymers) [4, 5]. Most of the complex biological molecules consist of carbon, its different hybridization states give it the capability to form different bonds with many elements resulting in a vast number of compounds. Carbon is the main chemical element of the majority of polymers but it also has many applications on its own like adsorbent material [6] or as a precursor for high-performance engineering materials such as graphene [7].

Although there are infinite sources for obtaining carbon, the most popular today is the one that uses biomass (wood and woody materials, herbaceous and agricultural waste, aquatic biomass, and animal and human waste) to be part of a circular economy in which the agro-byproducts are valorised while avoiding the use of fossil fuel-based products.

Bio-based polymers and carbonaceous materials derived from those biomass sources will be briefly presented next (schematic diagram in Figure 5.1).

5.2.1 Bio-based Polymers

In the developed society, plastics have become part of almost all daily used materials, from the clothes we wear, to the food containers for take-away. Although they facilitate our lives enormously, they have great disadvantages such as the consumption of fossil fuels for their production (and its consequent CO_2 emission) and the fact that most of them are not biodegradable so they generate a great waste problem when reaching the end of their useful life. In order to solve these problems that significantly impact the environment, a very effective approach is the use of bio-based polymers. Usually, lignocellulosic biomass is employed as it does not compete with other materials that are used as food supplies.

Bio-based polymers are sustainable polymers synthesised from renewable resources such as biomass instead of conventional fossil resources including petroleum oil and natural gas. As the concept is quite new, there are several notions that are worth mentioning here. Biopolymers are biodegradable and natural polymers that are formed by plants, microorganisms, algae, and animals. They should not be confused with biodegradable polymers which can be either natural or synthetic. A way for classifying these materials is depending on their origin [8]. Agro-polymers

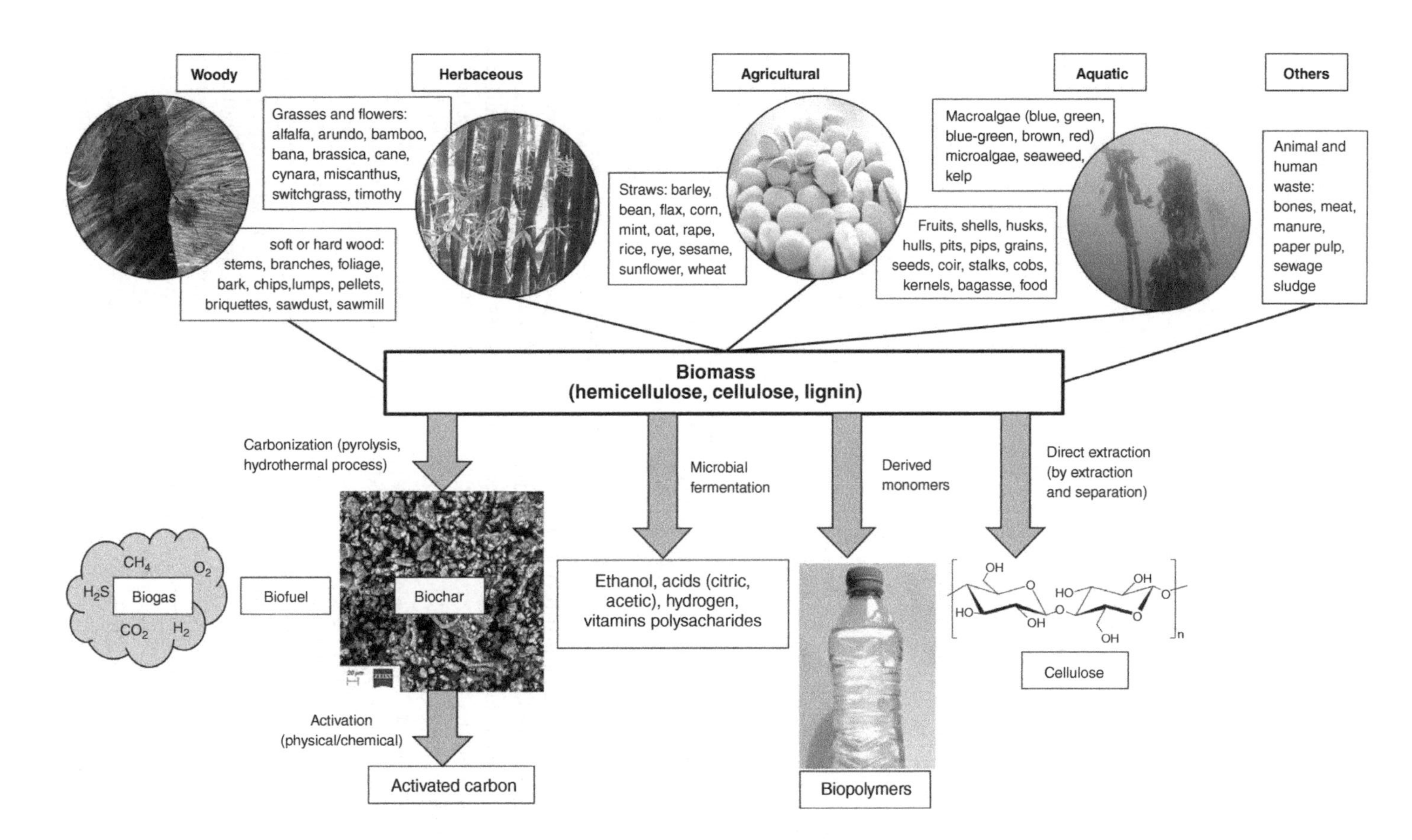

Figure 5.1 Schematic diagram showing the different products obtained from several biomasses by various processes.

(e.g. proteins or polysaccharides) are considered natural polymers and are directly obtained from biomass either by extraction and separation, or from microorganisms or by fermentation (e.g. microbial polyesters) [9], while the polymers that are produced following conventional synthetic reactions using monomers derived either from natural molecules or resulting from the chemical and biochemical breakdown processes of natural macromolecules are considered as synthetic [10].

The polymers extracted from biomass such as starch or cellulose and their derivatives have been widely used [11]. For instance, cellulose recovered from cardboard was cross-linked using natural agents such as citric acid to avoid the problems associated with the aldehydes classically used to manufacture superabsorbent polymers [12], creating a 100% sustainable material.

The materials directly obtained from living organisms also open the door to new structures that exist in nature and cannot be replicated at the laboratory or industrial level. For example, Yao et al. used the structure of the crab exoskeleton as a pattern in lithium–sulphur batteries [13].

Some of the most commonly used polymers, polyethylene (PE), or PE terephthalate (PET), can be substituted by greener alternatives. In the case of PE, this can be produced from monomers derived from bioethanol (obtained from the fermentation of sugar), such as bioethylene, which allows bio-PE to exhibit similar properties to synthetic PE. In the case of PET, large companies such as 'Coca-Cola Company' are using ethylene glycol from bio-based sources to produce more eco-friendly PET bottles. PET can as well be replaced for packaging applications [14] by poly(ethylene 2,5-furandicarboxylate) (PEF), a bio-based polymer that uses 2,5-furan dicarboxylic acid (FDCA) derived from fructose as a precursor for its production.

Considering the very complex structures of the precursor materials such as lignocellulosic wastes, their breakdown will allow monomers, exhibiting varied chemical structures, to form polymers with versatile configurations and thus interesting applications.

5.2.2 Carbonaceous Materials from Biological Feedstocks

Carbonization (obtained either by pyrolysis or hydrothermal routes) is the process that transforms the big macromolecules of the biomaterials into carbon and other byproducts.

Pyrolysis is an energy-recovering process where biomass is thermo-chemically decomposed into a solid (bio-char, BC), a liquid (bio-fuel, [15]), and a bio-gas product under high temperatures in the presence of an inert atmosphere [16]. The final product distribution and properties depend on the reactor, the feedstock, the heating rate (slow or fast), etc. BC is a carbon-rich product with a highly developed internal porosity, enhanced during the thermal decomposition of organic polymers, and a large surface area. The graphene-like structure of BC contains different chemical functional groups such as hydroxyl, carboxylic, phenolic, ketone, etc. [17], making it very suitable for different applications. Its structure allows the insertion and rejection of functional groups that can be beneficial for several purposes. The produced BC [18] is considered a low-grade carbon black due to the broad difference in particle size and higher ash content (inherited from the chemical composition of the precursor), whereas, the commercial carbon black exhibits a narrow particle distribution and contains mostly carbon with only very small amounts of other elements.

In order to transform the BC into activated carbon, an activation step (either physical or chemical) has to be carried out. The physical process consists of partial gasification of the carbon matrix by means of gas (water steam, air, or CO_2) at a high temperature [19]. The porosity is developed by increasing the surface area and the reactivity can be modified as well, but the yield of the process is relatively low. On the other hand, in the chemical activation, the carbonaceous material is impregnated with chemicals (e.g. H_3PO_4, H_2SO_4, H_2O_2, K_2CO_3, KOH, NaOH, $ZnCl_2$) and treated at elevated temperatures. Some chemicals like KOH or K_2CO_3 act as oxidants and dehydrating agents compared to others like $ZnCl_2$ that only catalyses the dihydroxylation and dehydration reaction [20]. At high temperatures, the chemical agent dehydrates the char while the formation of tar and the evolution of volatiles is inhibited, thereby increasing the yield of the carbonization reaction [21]. The activated carbon produced is washed with water to re-establish a neutral pH and to remove the residual chemicals (both the residual activation agent and the chemicals created during the reaction), revealing the newly developed porosity. Activated carbons have high surface areas, a well-developed porosity (macro-, meso-, or microporous depending on the parent material and process conditions) and some functional groups that can enhance certain properties, such as the adsorption capacity [22].

The different electrochemical applications of the materials described earlier will be detailed in the following sections.

5.3 Application of Bio-based Materials in Batteries

The science and technology of energy storage systems have received significant attention as an advanced power, particularly in the areas of portable electronic devices, electric engines, and stationary energy storage [23]. The exponential growth in electronic technology during the past decade has created an enormous interest in more reliable energy storage systems. The lithium-ion batteries (LiBs) are expected to be one of the most used energy storage devices in the near future and have already replaced many other batteries in the market (e.g. lead acid, nickel-cadmium, and nickel–metal hydride batteries). However, the requirements are far from being completely fulfilled and should possess high performance including superior storage capacity, fast charging–discharging process, long life cycle, adaptability to multiple scales, good processability, light weight, flexibility, and, most importantly, cost-effectiveness and non-toxicity [24, 25]. This circumstance has stimulated the interest in a new energy economy based on sustainable, inexpensive, metal-free active material, and high-performance LiBs through applying renewable resources (e.g. soy protein, chitosan (CS), cellulose, etc.) [26–30]. The next part will discuss the current status and recent research achievements and give an outlook on the future research challenges using such materials by means of making or modifying critical battery components (e.g. electrode, electrolyte, and separator).

5.3.1 General Concept of Metal-Ion Batteries

Since their introduction on the market in 1991, LiBs have been intensively investigated and developed. LiBs consist of two lithium insertion electrode materials. The first commercialised LiBs consisted of a graphite anode and a lithium metal oxide cathode, commonly $LiCoO_2$. With the use of an electrolyte containing a lithium salt, the lithium ions shuttle between the positive and negative electrodes. The schematic structures for the charged and discharged anode and cathode, along with the half-cell and full-cell reactions for this type of battery are presented in Figure 5.2 [31]. The factor x in these reactions varies usually between 0.5 and 1, depending on the selected transition metal oxide.

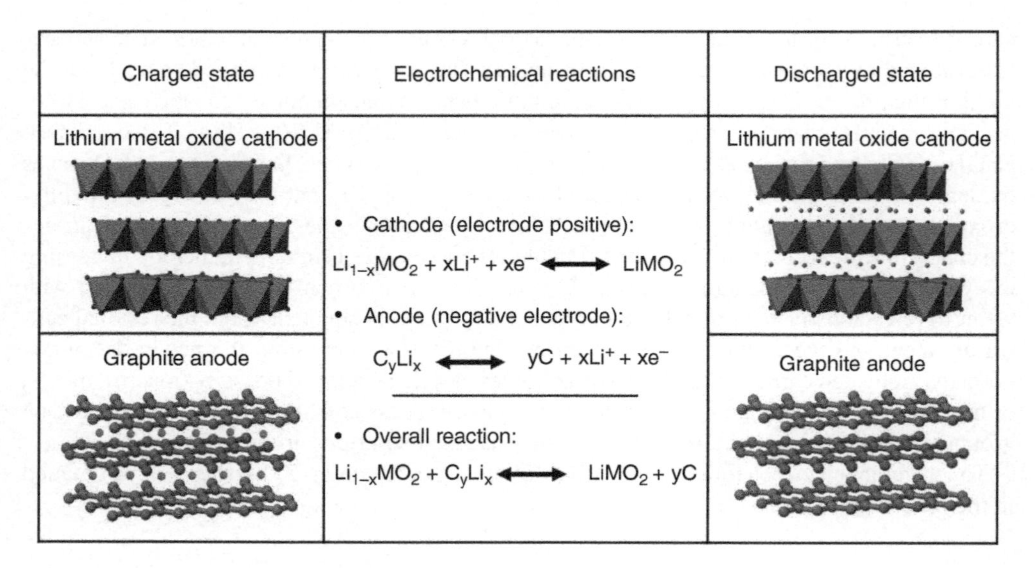

Figure 5.2 *Schematic presentation of the charged and discharged structure of a graphite anode and a layered oxide cathode, in addition to the half-cell reactions and full-cell reactions. Source: Figure adapted from International Atomic Energy Agency (graphite anode) and public domain (lithium metal oxide cathode).*

5.3.1.1 Electrode Materials

The most common objective of electrode development in LiBs or NiBs is to replace the actually used materials with inexpensive and abundant ones that can be scaled-up to match the need coming from the growing use of intermittent electricity sources (wind, turbine, and solar). Moreover, high electrochemical performances, stability, and safety are also targeted [32].

5.3.1.1.1 Anode Materials

Carbon-based electrode materials have played an intriguing and promising role in the development of high-performance LiB and NiB technologies due to their high capacity, cycling property, and stability/durability. The thermal treatment of biomass produces a typically non-graphitizable amorphous carbon. Indeed, depending on the biomass composition, different chemical and structural properties can be obtained [33].

Moreover, preliminary chemical treatment of the carbon precursor with, e.g. Na_2CO_3, K_2CO_3, NaOH, KOH, H_3PO_4, $AlCl_3$, $MgCl_2$, LiCl, or $ZnCl_2$, can produce a porous structure and particles with high specific surface area [34–36]. Stephan and co-authors treated banana peels with $ZnCl_2$ as a pore-forming substance. After the pyrolysis, the obtained carbon showed an improved surface area from 36 to 1285 $m^2\ g^{-1}$ and an increased pore diameter from 19 to 57 Å [37].

Besides the textural properties that can affect the cation intercalation process as well as the capacity, the introduction of functional groups, heteroatoms N, P, O, and/or S, in the carbon materials can produce larger asymmetrical spin and charge density. This results in enhanced defects, an increase of available active sites, and effective modulation of the electronic and chemical characteristics to reach satisfactory electrochemical performances in LiB and NiB applications [33].

Hou and co-authors studied biomass-derived natural silk using simultaneous activation and graphitization to obtain 4.7% N-doped porous carbon nanosheets. The obtained hierarchically structural carbon had a high microporous and mesoporous volume (~2.28 $cm^3\ g^{-1}$) and a high specific surface area of ~2494 $m^2 g^{-1}$ with stacking nanosheets' thickness ranging from 15 to 30 nm, thus having a large electrode/electrolyte interface and a short ion diffusion pathway. Hence, these materials showed high cyclic capacities of 1865 $mAh g^{-1}$, five times higher than the theoretical capacity of graphite [38].

Sodium, a much cheaper and more abundant element than lithium, becomes a reasonable and sustainable alternative for the development of cost-effective batteries for large-scale storage applications [39–41]. Unfortunately, graphite, the most used material in LiBs, shows poor electrochemical performance in NiBs. One of the main reasons is the larger radius of sodium ions in comparison with lithium ions (0.102 vs. 0.076 nm). Recently, a disordered non-graphitizable carbon, generally called 'hard carbon' derived from bio-waste has become one of the most promising anode materials in rechargeable NiBs [20]. Hard carbons are basically amorphous, highly irregular, and disordered single-layered carbon, and usually derived from pyrolysis of industrial products (e.g. sucrose, glucose, polyvinyl chloride (PVC), etc.) or bio-waste from the food industry. Passerini and coworkers successfully used a hard carbon obtained from apple bio-waste material as an anode in Na-ion cells. The electrode showed a very stable reversible capacity around 245 $mAh g^{-1}$ at 0.1 C rate. The prepared full cell based on hard carbon revealed stable cycling with specific capacities higher than 250 $mAh g^{-1}$ and a very promising rate capability of 220 and 183 $mAh g^{-1}$ at 1 C-rate and 2 C-rate, respectively [42].

Another approach to use bio-based materials in energy storage applications consists of fabricating hollow carbon nanofibres (HCNFs) using various sustainable templates from diverse biomaterials having naturally hierarchical structures. These nanofibres can encapsulate sulphur and silicon to form cathodes and anodes for LiBs (Figure 5.3a) [44]. Yao et al. reported a bio-template created from crab shells having twisted plywood structure and consisting of highly mineralised chitin–protein fibres. The silicon nanostructured electrode was obtained in two steps. First, through a simple thermal decomposition of the chitin–protein nanofibres in the stone crab shell, a thin layer of carbon was coated on the whole surface of the $CaCO_3$ framework. In the second step, silicon was loaded into the nanochannels by chemical vapour deposition (CVD). After dissolving the $CaCO_3$ framework using a diluted HCl solution, HCNF arrays encapsulating silicon were obtained. The resulting nanostructured electrode showed a high specific capacity of 3060 $mAh g^{-1}$ and excellent cycling performance until 200 cycles with 95% of capacity retention. The observed excellent performance is attributed to the rapid lithium-ion and electron transport and the sufficient available space for the active material volume expansion [13].

5.3.1.1.2 *Cathode Materials*

In the past few years, research in the area of anode materials has made great progress. In fact, new anode materials with excellent electrochemistry properties have been highlighted. However, the development of cathode materials is slower when targeting the improvement of capacity, which is vital in the determination of life cycle, energy density, and safety in LIBs. Currently, organic-based or polymer-based electrodes are potential candidates due to their unique characteristics (e.g. lightweight, environmentally benign, mechanical flexibility, and processing compatibility) [43, 45, 46]. Some of these cathode materials and more specific ones will be reported in the upcoming sections.

5.3.1.1.2.1 Organic Battery

Organic cathodes produced from biomass are environmentally friendly and originate from sustainable resources. For example, polydopamine (PDA) can be easily obtained from biomolecules

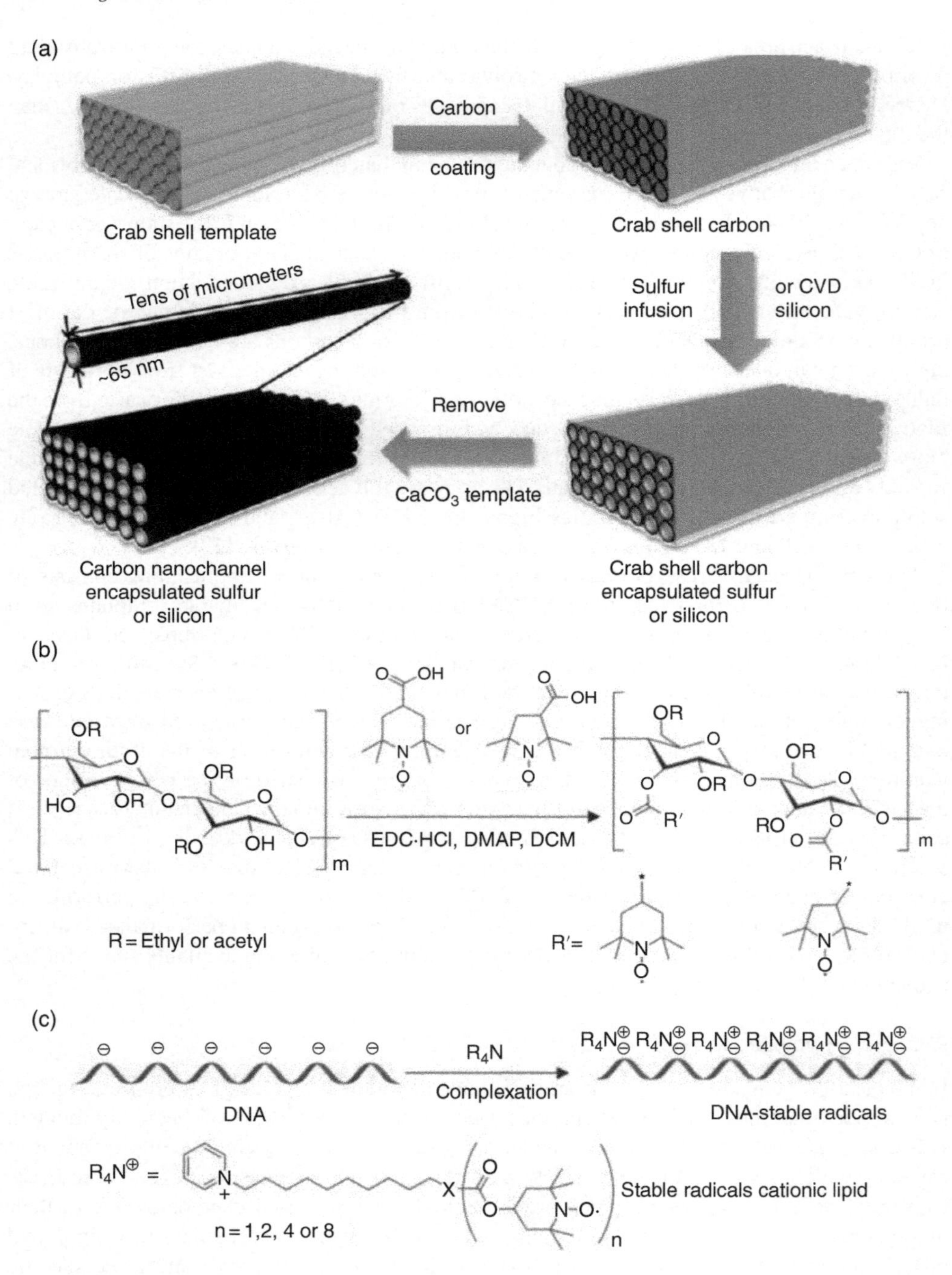

Figure 5.3 *(a) Schematic illustration of the fabrication procedure for hollow carbon nanofibre (HCNF) arrays encapsulating sulphur or silicon electrode materials based on stone Crab shell template; (b) cellulose through esterification between the hydroxyl groups of cellulose and carboxylic acid-functionalised nitroxide; (c) DNA polyplexes through electrostatic interactions between DNA and positively charged nitroxide molecule. Source: Yao et al. [13]. Reproduced with permission of American Chemical Society. Zhang et al. [43]. Reproduced with permission of Royal Society of Chemistry.*

such as dopamine with quinone or hydroquinone motifs. However, PDA has a low electronic conductivity, thus a combination with a conductive additive is crucial for charge storage applications. Recently, it has been reported as an organic electrode material upon mixing with [47] or polymerization onto [48] carbon nanotubes (CNTs). The CNTs not only provide the necessary conductivity to the system but also show capacitive charge storage behaviour. It is also possible to use conductive polymers to increase the conductivity and improve the adhesive properties. In this regard, dopamine has been incorporated into the polypyrrole (PPy) chain while hydrogen bonding and π-stacking of aromatic units of dopamine and pyrrole promote interchain charge transport. The electrical conductivity of a 1 : 1 mixture of dopamine and pyrrole was $1.5\,mS\,cm^{-1}$. The obtained pyrrole-dopamine copolymers behaved as high-performance cathodes in rechargeable lithium batteries with a charge storage capacity of 160 and $90\,mAh\,g^{-1}$ at current density of 100 and $800\,mA\,g^{-1}$, respectively [49]. Chitosan–vanillin secondary amines were obtained using the grafting onto method. Here, 62% of saccharide units were functionalised with vanillin-derived guaiacyl groups onto the CS backbone. To investigate their potential as cathode materials for electrochemical energy storage, the composite materials were directly mixed with carbon black and used as electrode materials without any binder additive. The electrodes showed a combination of Faradaic and non-Faradaic charge storage in the range of up to $80\,mAh\,g^{-1}$, resulting from the quinone–hydroquinone redox couple in the bio-based polymer and carbon black [50].

Ethyl cellulose (EC) derivatives and cellulose acetate (CA) derivatives carrying TEMPO (2, 2,6,6-tetramethylpiperidine-1-oxyl) were synthesised by the condensation reaction of 4-carboxy-TEMPO with the residual hydroxyl group of cellulosic polymers (Figure 5.3b). All the free radical-containing cellulose derivatives demonstrated a total degree of substitution (DS) [51]. It is also possible to obtain deoxyribonucleic acid (DNA)–lipid complexes that contain TEMPO radicals as an electroactive cathode material for an organic radical battery. The functionalization was obtained by simple cation exchange of the sodium counter-ions of the DNA with cationic amphiphilic lipids that carry TEMPO moieties (Figure 5.3c) [52]. Both cellulose and DNA electrodes demonstrated a reversible charge/discharge reaction, with a two-stage discharge process, and discharge capacities up to 190% of the theoretical value for a one-electron redox reaction. The two-stage redox reaction of TEMPO-based cathode materials corresponds to a stepwise reduction of the oxammonium cation via the free radical to the aminoxyl anion. The diversity of the synthesis procedures of polymers matrices functionalised with redox-active groups stabilise the anionic species which increase the battery capacity.

5.3.1.1.2.2 Bio-battery

Recently, cellulose has been (re)discovered as a smart material that can be used in electronics and energy storage applications [53]. One of the main interests regarding cellulose-based active materials relies on lightweight and flexible bio-batteries technology for supplying power to implantable medical devices for in vivo usage. Baptista and coworkers proposed a combination of the electrospinning technique [54, 55] with a nanocoating procedure of a conjugated polymer to improve the electrical conductivity of cellulose-based fibres from $(7.1 \pm 0.8)\cdot 10^{-11}$ to 10^{-2} and $10^{-1}\,S\,cm^{-1}$ after adding PPy and polyaniline (PANI), respectively. The obtained lightweight, non-toxic, and conductive cellulose-based electrospun fibres functionalised with PPy and PANI were used as the negative and positive electrodes, respectively, in the presence of simulated biological fluids (0.9%wt/v NaCl) as electrolytes. The fully polymeric biobattery based on CA, PPy and PANI materials generated a power density of $1.7\,mW\,g^{-1}$ ($0.8\,mW\,cm^{-3}$) at room temperature [53].

5.3.1.2 Battery Separators

The separator is an indispensable component in a battery that prevents electronic contact between the anode and the cathode and enables ionic transport. Continuous research is ongoing to improve the wetting properties of separators and electrolytes and to make them safer, flexible, electrochemically, thermally, and mechanically more stable and environmentally friendly [56]. Commonly used separators are typically made of PE, polypropylene (PP), polytetrafluoroethylene (PTFE), PVC, fibres of cotton, nylons, or polyesters. For high-capacity and longer life metal-ion batteries, many attempts in developing bio-based separators have been made and reported in the literature [57, 58]. Direct use of cellulose as battery separators can be problematic because of their high water content and the difficulty of tuning their porosities. In contrast, cellulosic derivative materials are studied to address all the needed properties for the Li-ion battery separators. A cellulose/PVDF-HFP, poly(vinylidene fluoride-co-hexafluoropropylene), composite membrane obtained by electrospinning had a superior electrolyte wettability, which may be attributed to its hydrophobic nature and low surface energy [59]. In addition, these membranes showed better thermal tolerance and thermal stability compared to a non-woven cellulose/PVDF-HFP composite and PP separator when treated at 200 °C for 0.5 hour [60].

A simple and facile process was developed by Xu and coworkers to fabricate cellulose/polysulfonamide (cellulose/PSA) composite membranes as Li-ion battery separators. The developed membranes showed enhanced mechanical and thermal stability, low interfacial resistance, better rate capability, and good cell cycle performances. Also, in this case, the cellulose/PSA composite membrane does not undergo any thermal damage at 200 °C. Furthermore, adding a liquid electrolyte (1 M $LiPF_6$ in EC/DMC) to the cellulose/PSA composite membrane revealed a superior ionic conductivity of $1.2{\cdot}10^{-3}$ S cm^{-1} due to the synergic effect between the electrolyte and the cellulose/PSA material [61].

In another study, pristine nanocellulose derived from green algae (Cladophora) was used directly as a separator because of its low water sorption, abundant mesopores, better electrolyte wettability, and thermal stability, which can suppress the growth of Li metal dendrites and improve the cycling stability of the batteries. Furthermore, the combination of traditional PE separators laminated between two layers of nanocellulose yielded a trilayer separator with high cycling stability and safety [62].

In addition to the developed materials mentioned earlier, this list can be extended with other polysaccharides used as matrices in the gel polymer electrolytes (GPEs) and composite electrolytes for different applications [63–66].

5.3.1.3 Solid Electrolytes

The battery research community is focusing efforts on the development of safe batteries based on solid electrolytes. Thus, the all-solid-state battery concept is extensively investigated based on the use of ceramic, polymeric or hybrid solid electrolytes. Among all studied systems, the solid polymer electrolyte is a potential candidate after the success of the electric car-sharing service ‘Autolib’ [67]. So far, cellulose derivatives were mainly reported as reinforcing agents, matrices for GPEs or as separators [68, 69]. Recently, a quasi-single ion-conducting polymer electrolyte based on chemically modified ethylcellulose was presented and the preparation method is patented [70]. The quasi-single Li/Na ion-conducting polymer electrolyte was prepared by covalently anchoring an anionic organic salt onto the cellulose chain. Thus, the use of this electrolyte avoids any issues related to concentration polarization and improves the mechanical properties and the material resistance to dendritic growth.

5.4 Application of Bio-based Polymers in Capacitors

5.4.1 General Concept of Electrochemical Capacitors

Capacitors consist of two conducting metallic plates or electrodes (positive, called cathode and negative, called anode) separated by an insulating dielectric material. When a voltage is applied to a capacitor, the surfaces of the electrodes accumulate charges from the opposite sign (anions are gathered on the cathode electrode while cations are collected on the anode) [71]. The charge separation by the dielectric creates an electric field, which influences the electric potential energy and voltage and allows energy storage.

Compared to electrochemical batteries or fuel cells, conventional capacitors have lower energy densities but higher power densities. This means that a capacitor can store less energy that can be delivered quickly. Supercapacitors are ruled by the same principle as normal capacitors but their higher surface area and the shorter distance between electrodes allowed by thinner dielectrics increase both their capacitance and energy. One of the problems of supercapacitors is that the materials used for their manufacturing come from fossil fuel sources.

Supercapacitors (schematic diagram given in Figure 5.4) are categorised depending on their charge storage mechanisms into electrochemical double-layer capacitors (EDLCs) and pseudocapacitors. EDLCs store energy non-Faradaically due to reversible ion adsorption using a high surface area carbon material to construct an electric double-layer at the interface between the electrode material and electrolyte solution. Pseudocapacitors store energy by Faradaic processes, such as reversible oxidation-reduction reactions at the surface of electroactive materials [72]. Hybrid capacitors combine the two categories defined earlier.

The type of electrolytes used impacts the performance of the electrode materials. Therefore, their choice will depend on the application of the supercapacitor. There are three types of electrolytes: aqueous (salts dissolved in acids or bases), organic (generally, tetraethylammonium tetrafluoroborate [TEATFB] salt dissolved in acetonitrile [AN] or propylene carbonate

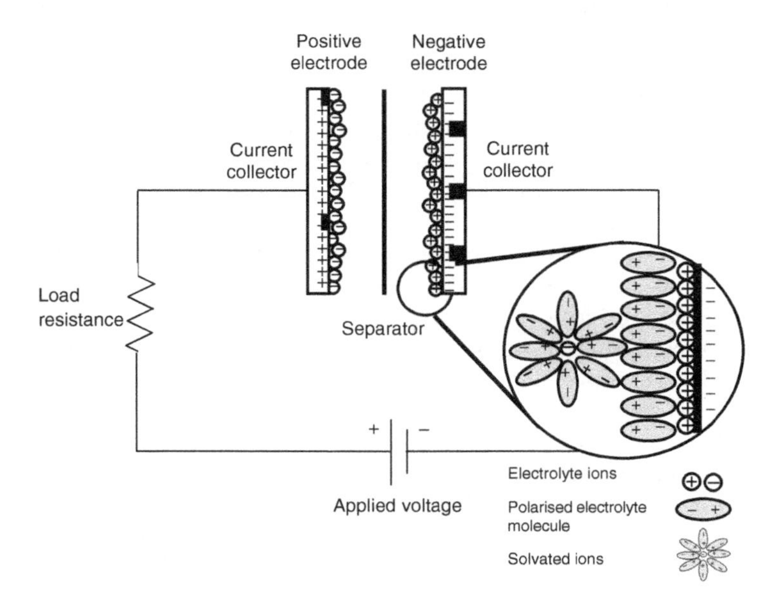

Figure 5.4 Schematic diagram of a supercapacitor.

[PC]) and ionic liquid (IL). Aqueous electrolytes are cheap, have high ionic conductivity, and need a smaller pore size for diffusion. Nevertheless, they cause corrosion when used at high temperature and voltage, therefore reducing the lifetime of the EDLC. Organic electrolytes allow high cycle life (as much as 500 000 cycles) and can operate at high voltage but they are flammable and have risk of explosion, thus, creating safety issues. ILs are non-flammable, non-volatile, and non-toxic, but extremely expensive [73].

5.4.2 Electrode Materials

Although the electrolyte used is highly important for the performance and design, supercapacitors are classified based on the energy storage mechanism (EDLC and pseudocapacitors) that is linked directly to the material used for the electrode [74].

5.4.2.1 Electrochemical Double-Layer Capacitor

The active material in both anode and cathode is usually a highly porous carbon due to the fact that the stored energy is directly related to the surface area [33]. The different carbon materials used are presented in detail hereafter.

5.4.2.1.1 Activated Carbon

Activated carbons are carbonaceous materials with a highly developed structure consisting of micropores (<20 Å wide), mesopores (20–500 Å), and macropores (>500 Å). Macro- and mesopores enable a fast ion transport, allowing a short ion-transport pathway, whereas micropores are electroactive sites to store energy. The pore size distribution is very important regarding its applicability in electrochemical devices as the diffusion of some electrolytes could be hampered due to the small dimensions of the micropores. This characteristic is linked to the production route of the AC, which can consequently be modified during the manufacturing process to fulfil the requirements. In this regard, the use of renewable, inexpensive, and eco-friendly precursors for AC production opens a new promising alternative.

Wei and Yushin [73] reviewed the use of several bio-based activated carbons as electrode materials in supercapacitors. They found that wheat straw-derived AC showed the highest capacitance in (methyltriethylammoniumtetrafluoroborat/AN) organic electrolyte, most probably due to the adequate pore size distribution. In this case, the material showed not only micro- but also mesopores that allowed a good impregnation of the electrolyte.

Regarding the use of aqueous electrolytes, the oxygen groups may contribute to the capacitance not only due to the better wettability of the electrode but also because of the pseudocapacitance influence linked to the Faradic reactions. These surface groups are obtained during the activation step either by physical or chemical activation, although the chemical route is preferred as the temperatures used are lower leading to a higher carbon yield.

Ghosh et al. [75] compared four different carbonaceous materials synthesised from acid (H_3PO_4)-treated banana stem, base (KOH)-activated banana stem, corn cob, and starch resulting from potato. They found that the best-performing carbon as supercapacitor electrode material was the KOH activated one whereas the starch-derived AC showed no porosity and, therefore, exhibited a poor performance for electrochemical energy storage devices.

In order to reduce the steps for producing AC, Chen et al. [76] used microwave heating under humidified nitrogen atmosphere to convert lignin. The porosity was highly developed, and the surface was rich in oxygen groups that allowed high energy density and rate capability.

In most of the works cited, the precursor material is lignin because of the high availability, low cost (considered as a waste product from the papermaking industry), and non-toxicity. Nevertheless, other waste biomasses derived from food such as egg white [77] have also been used to produce carbons with interesting electrode properties (high N content resulting from the precursor protein and hierarchical mesoporosity). Waste biomass is also used to produce other electrode materials mentioned in the next section.

5.4.2.1.2 Carbon Nanomaterials

The group of carbon nanomaterials comprises CNTs, carbon-derived carbons (CDCs), graphene, and CNFs. These materials have good electrical properties that together with the porosity and high surface area turn them into interesting alternatives to AC.

Zhao et al. [78] used biomass (*Typha orientalis*) as a precursor for CNTs. The treatment with hydrothermal and calcination techniques resulted in nitrogen-doped carbon nanotubes (N-doped CNTs) that performed better than commercial CNTs, most probably due to the O- and N-rich functional groups present on the surface that originate from the parent structure of the raw biomass.

Carbon nanomaterials usually have a lower surface area compared to AC but, in this case, due to the interconnected mesoporosity and the entanglement of the nanorods, they allow efficient transport of electrolytes through their structure. The surface area is used more efficiently as the charge is continuously distributed on the surface.

Eucalyptus kraft lignin was extracted by the electrospinning method and produced into a mat [79]. The resulting fibre mats consisted of randomly entangled fibres with a cylindrical shape. The material showed high porosity (mainly (ultra-)micro- and seldom meso-) and oxygen functional groups on the surface that improved the EDLC performance. The oxygen groups imparted pseudocapacitance, contributing to the charge storage. At high C-rates, the small pores limited the capacitance due to poor accessibility of oxygen.

Graphene is a monolayer of sp^2 carbon atoms densely packed into a two-dimensional (2D) honeycomb lattice [74]. Due to its unique structure, it possesses outstanding properties such as good chemical stability, structural flexibility, exceptional mechanical properties, good thermal and electrical conductivity, and a high surface-to-volume ratio. Graphene is usually produced by expensive techniques that use high amounts of organic solvents or energy. Furthermore, the production process is non-ecological since raw material derived from fossil fuel sources such as coal is used. Recently, Luong et al. [80] proposed a low-energy bottom-up synthesis method that allowed the use of inexpensive carbonaceous materials (derived from renewable resources and mixed-waste products) while getting a low defect material that does not need purification steps and that renders the procedure economically appealing.

Graphene is also considered as the building block for all the carbon nanostructures (0D fullerene, 1D nanotubes, 3D graphite, and CNT). Due to the Van der Waals forces between the adjacent sheets, it agglomerates and stacks itself, thus reducing the initial coulombic efficiency and the energy area as a result of available surface loss. Therefore, graphene is a good candidate to be combined either with other bio-based materials such as CNT, CNF, and porous carbon or with metal oxides.

For instance, Farma [81] used palm empty fruit bunch (EFB) fibres and CNT to obtain activated carbon monoliths (ACMs) for supercapacitor electrode material, while Raymundo-Piñero et al. [82] used oxygen-rich seaweed carbonised with CNTs to take advantage of the nanotexture of CNTs. However, Taberna et al. [83] found that the use of CNT on coconut shell carbon/CNT composites reduced the capacitance of the material, giving worse performance than both the coconut shell-derived AC or the CNT alone. An increase of CNT loading to 50% lowered the performance than when 15% CNT was used.

5.4.2.1.3 Carbon Aerogels

Carbon aerogels are relatively cheap ultralight materials with low mass densities, high surface areas, and excellent chemical stabilities. They consist of a 3D interconnected porous structure that allows the electrolyte molecules to easily access the material and offers low resistance for ion transport. Their main drawback is the brittleness derived from their skeleton morphology that hinders their use in applications where good mechanical performances are required.

The porosity results from a combination of interconnected colloidal particles. This continuous structure together with the ability to chemically react with the current collector allows this electrode material to be used without an additional binder [71].

Carbon aerogels are produced by pyrolysis of organic aerogels. For instance, Cheng et al. [84] produced a cotton base aerogel by calcination and chemical activation of cotton, which mainly consists of millimetre-scale inter-entangled cellulose fibres. Hao et al. [85] also used a bio-based raw material. In this case, sugar bagasse (a waste material produced during the extraction of sugar) was used to produce the aerogel for supercapacitor electrodes. The combination of macro-, meso-, and micro-pores together with a large surface area provided the material with high capacitance retention and high energy and power density that translate into a well-performing electrode for supercapacitors.

5.4.2.2 Pseudocapacitors

Contrary to EDLC that stores charge electrostatically, the pseudocapacitance emerges from the Faradaic process between the electrolyte and the electrode. This takes place by electrosorption, reversible redox reactions, and intercalation processes [71]. There are two kinds of pseudocapacitive materials, namely electronically conducting polymers and transition metal oxides.

5.4.2.2.1 Conductive Polymers

Redox-active group-embedded polymers or π-conjugated polymers possess an electrochemically active backbone, consisting of redox-active monomers with a closed-shell electronic valence structure and with a high degree of π-conjugation. The n/p-type polymer configuration is the most interesting one for getting the highest potential energy and power densities, consisting of one negatively charged (n-doped) and one positively charged (p-doped) conducting polymer electrode. Therefore, the doping level will influence electrochemical performance. The most common polymers used in this application are usually PANI, PPy, poly(3-4-ethylenedioxythiphene) (PEDOT), and polythiophene. To show high conductivity and electrochemical capacitance, they must have favourable charge transfer kinetics, charge mobility, charge carriers, and readily available solvated counterions [86].

A PANI-activated carbon composite electrode material was synthesised by the deposition of aniline in the carbonaceous material obtained by the carbonisation of natural bamboo [87]. The structure arising from the activated carbon in combination with the pseudocapacitance of PANI rendered the electrode with an extraordinary electrochemical performance. The main drawback for these materials is their poor cycling stability during long charge/discharge cycles. Lignin can be used to improve their capacity.

Milczarek and Inganäs [88] studied the interpenetrating polymer network formed by PPy and bio-based lignin. They observed that the quinone group (derived from the phenol by oxidation reactions) present in the lignin and lignosulphonates (another type of lignin produced by a different separation method) can be used as proton and electron storage and exchange during redox cycling. Phosphomolybdic acid (PMA), a renewable biopolymer, lignin, and PPy were used to synthesise a ternary composite in a one-step electrochemical deposition process to enhance the capacitance of the PPy–lignin composite [89]. This work also highlighted the role that bio-based quinone played in the pseudocapacitance.

Nowadays, alkali lignin (AL) accounts for most of the paper-processing-derived lignin. However, the use of AL has not been studied in detail for electrochemical storage due to the low solubility in inorganic acids. Leguizamon et al. [90] used an organic acid (acetic acid) to produce a conductive polymer using PPy that gave better performance than the common sodium lignosulphonate (SLS) derived one. The amount of phenol was the key to improve the performance.

5.4.2.2.2 Metal Oxides

Ruthenium oxide (RuO_2), manganese (IV) oxide (MnO_2), cobalt (II, III) oxides (Co_3O_4), nickel oxide (NiO), molybdenum trioxide (MoO_3), molybdenum (IV) oxide (MoO_2), and vanadium nitride (VN) have been generally used as electrochemically active materials [91]. They present high capacitance values, but because of their scarcity, the high price and processing variability restrict their use in large-scale systems [92]. In order to overcome the problems related to these materials, they are principally used in combination with carbon materials. For example, HCNFs were produced with poly(styrene-co-acrylonitrile) in the presence of iron (III) acetylacetonate as the core and acetic acid lignin as the shell. The high surface of the HCNF and the reversible redox reactions of the iron oxide particle improved the capacitance and therefore enhanced the electrochemical properties of the material [93]. Sponge-like carbonaceous hydrogels and aerogels were prepared using watermelon as the carbon source. Fe_3O_4 nanoparticles were then incorporated followed by calcination in order to obtain magnetite carbon aerogels (MCAs) [94]. The porous structure derived from the carbon material allowed optimal transport of the electrolyte ions and electrons to the porous carbon surface showing excellent capacitance. In both examples, there was a synergistic effect due to the combination of both organic and inorganic materials.

5.5 Alternative Binders for Sustainable Electrochemical Energy Storage

The main function of binders is to preserve the physical structure of the electrode. Although the binders represent less than 5% of the entire electrode weight, they are of vital importance. Electrodes consist of different materials that have different properties, such as the active material (fixes the energy density of the electrode), the conductive agent (improves the electron transport on the active material and current collector), and the binder (keeps the active material and the conductive material close to the current collector). Normally, binders are viscous polymers (made up of one or more polymers). In order to be suitable, the materials used as binders have to fulfil some general criteria: to warrant appropriate cohesion between components, to withstand dimensional changes during charging–discharging cycles, high chemical, thermal, and electrochemical stability, insolubility in the electrolyte and minimal swelling, to form a homogeneous slurry at room temperature, strong ion/electron conductivity and low cost.

PTFE or PVDF are commonly used as binders for both anodes and cathodes due to their good electrochemical stability and excellent binding strength with electrode materials.

However, they present several drawbacks. They require organic solvents such as N-methyl-2-pyrrolidone (NMP) which is volatile, flammable and explosive [95] that can react with lithium metal and form LiF-stable compounds that will deteriorate the battery performance. On top of that, the reaction between Li and PVDF is exothermic and can cause self-heating and thermal run-away reactions [96].

From a green perspective, the fluorinated binders cannot be separated from the rest of the components and require special treatments for disposal after they reach their end of life. This represents an extra environmental issue as most of the decomposition products of fluoropolymers in N_2 (e.g. cycloperfluorobutane, hexafluoropropene, perfluoroisobutene, fluorophosgene, and other fluorocarbons) are extremely toxic [97]. There are other alternatives that use polymers such as polyacrylates (polyacrylic acid [PAA]), polyolefins (such as PE), or aromatic synthetic polymers (such as polystyrene [PS]) [97] but these are still based on non-renewable materials. Binders derived from renewable green sources are therefore mandatory to have more eco-friendly energy storage alternatives.

5.5.1 Polysaccharides and Cellulose-based Binders

Cellulose is a linear polysaccharide consisting of β (1→4)-linked D-glucose units made of fibrils and microfibrils. This structure leads to hydrophobic nature and it is insoluble in most organic solvents. In order to produce soluble products, the hydroxyl groups of the cellulose can be substituted to obtain, e.g. carboxymethyl cellulose (CMC), CA, and EC. These nanofibre composites can be used as binders in electrochemical applications.

5.5.1.1 Cellulose-based Materials

In CMC, some of the –OH groups have been substituted by sodium carboxymethyl (–CH_2COONa) groups (see structure in Figure 5.5) making it water-soluble. Its physicochemical properties depend on the molecular weight, type of substituent, particle size, and DS. CMC is used as a thickener and it acts by stabilising the slurry and adjusting the viscosity. It is the most used organic binder to replace PVDF due to its greener nature and significantly lower cost. Several authors used it to overcome different problems found in electrochemistry. Hochgatterer et al. [106] analysed the effect of the DS in a Na–CMC binder system. A high level of substitution and high molecular weight of the base biopolymer were desirable properties to improve the cycling stability. Moreover, the covalent chemical bond between the binder and the active material is more important than the physical steadiness.

In Li-ion batteries, the most used carbon material is still natural graphite that is hydrophobic. However, when CMC is added to the system, the Na–CMC carboxylic group adsorbs on the graphite surface stabilising the suspension [97]. Drofenik et al. [107] compared four different cellulose-based materials (CMC, EC, methylcellulose [MC], and hydroxyethylcellulose [HEC]) as binders in Li-ion battery graphitic anode and studied the minimum amount necessary to obtain a good performance. They concluded that CMC was the best binding agent and that about 2 wt% was enough to obtain a good electrochemical functioning. Below that quantity, due to the dilatation of the material, the anode was detached from the electrode resulting in a lower reversibility. At a sufficiently high concentration, the cellulose creates continuous networks that hold together all the electrode constituents.

5.5.1.2 Natural Cellulose and Nanocellulose

Natural cellulose (structure given in Figure 5.5) is very abundant and finding a way to use it without modification will reduce the cost of the fabrication process. Cellulose has been used

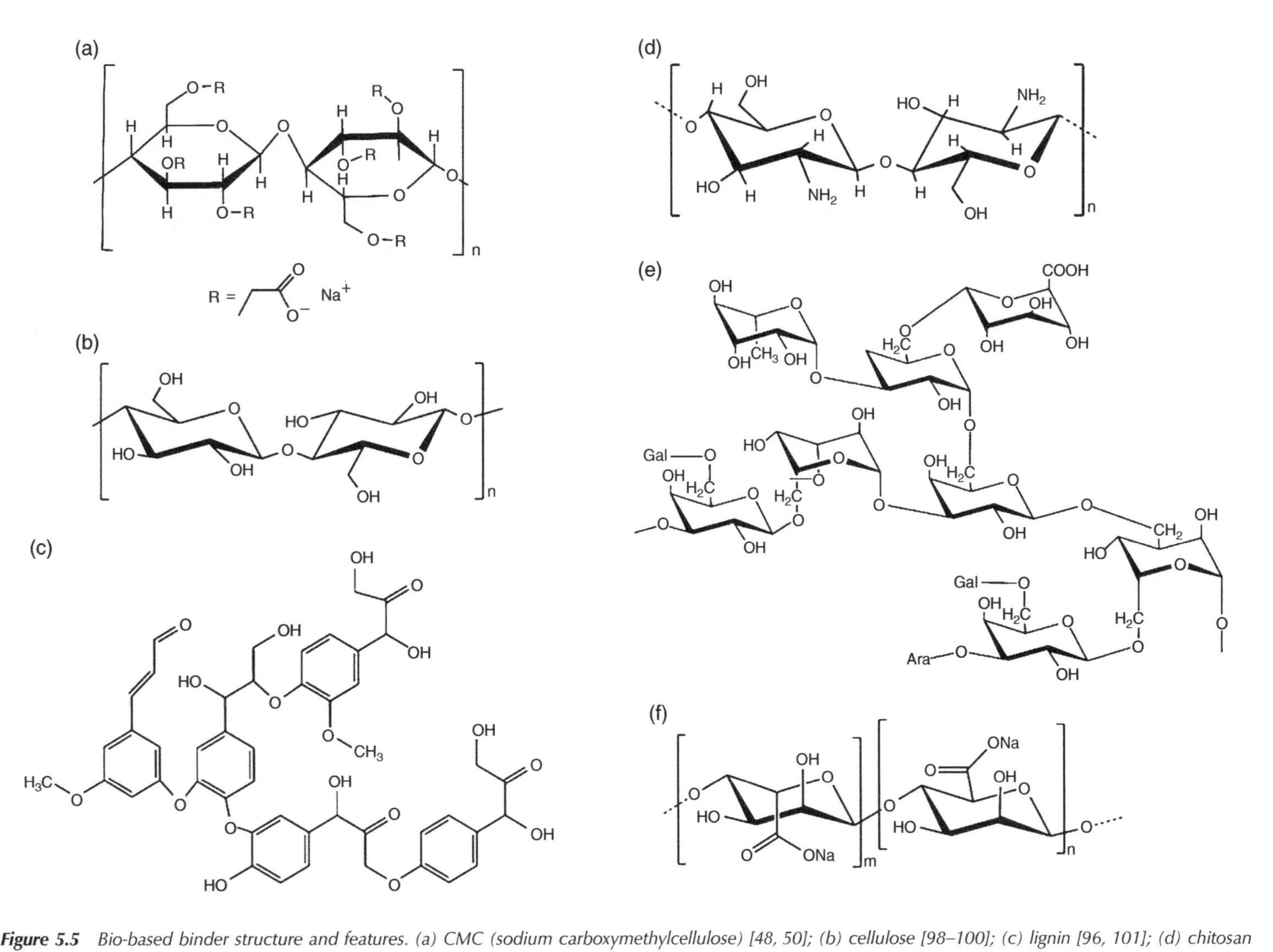

Figure 5.5 *Bio-based binder structure and features. (a) CMC (sodium carboxymethylcellulose) [48, 50]; (b) cellulose [98–100]; (c) lignin [96, 101]; (d) chitosan (CS) [102]; (e) Gums: arabic gum [103]; (f) sodium alginate (SA) [95, 104, 105].*

as a binder in EDLC with activated carbon as active material [98]. The composite electrodes were prepared by IL-based slurries or by aqueous suspension containing a polymeric stabiliser and a surfactant. The performance of the electrodes was good as they showed uniform morphology and stabilities similar to those of classical electrodes.

Nyström et al. [100] used wood-based cellulose as a binder to produce an electronically conductive high surface area composite. The chemical polymerisation of pyrrole on wood-derived nanofibres was done in situ, avoiding the use of laborious techniques. The created structure did not break after drying and showed a good conductivity and surface area that are necessary for electrochemical applications. Microfibrillated cellulose nanoparticles were also analysed as binders in Li-ion batteries by Jabbour et al. [99]. The cellulose allowed a highly porous structure based on a web-like network around graphite platelets. A water-based solution of synthetic graphite and microfibrillated cellulose (MFC) was dried and the obtained anodes exhibited outstanding flexibility, highly porous structure, and good cycling performance.

5.5.1.3 Other Polysaccharides

Due to their polar functional groups (structures are summarised in Figure 5.5), stability at temperatures of up to 200 °C (compatible with electrode drying temperature) and easy availability, polysaccharides such as CS, alginate, or natural gums (arabic, xanthan, guar, etc.) are potential materials for binder application. They hold functional groups on their structure that make them water-soluble and, therefore, the use of organic solvents can be avoided.

For example, Chen et al. [102] cross-linked CS with glutaraldehyde (GA) to form an interconnected 3D network. Once the film was formed, it showed a yellowish colour resulting from the formation of new bonds and became insoluble in acetic acid. The high density of the network limits the swelling of the material when in contact with the electrolyte, thereby limiting the movement of active anode material particles. The material showed a high reversible capacity attributed to the strong chemical bonds between the CS and active material particles that enhanced the electrochemical performance.

Among the naturally derived gums, Arabic gum shows a dual functionality derived from its chemical composition consisting not only of polysaccharides but also glycoproteins. The hydroxyl groups present on the surface warrant a strong binding with electrode materials such as silicon while the long protein chains improve the tolerance toward the huge volume change. The anodes show exceptional long-term stability and an excellent capacity [103].

Another well-known polysaccharide that has been exploited extensively in different applications is alginate. Alginates, present on the cell walls of brown algae, are the salts of alginic acid. These polysaccharides containing surface-active carboxylic groups form chemical bonds and create stable solid electrolyte interfaces (SEIs) while offering a high elastic modulus and accommodating the volume expansion suffered by active materials in electrochemical applications. Even if it does not swell, alginate provides good access to the Li by hopping between the carboxylic acids present on its chain [105]. Shi et al. [105] reviewed the use of alginate as a water-soluble binder in high-performance electrochemical applications. Alginate has been used combined with several different products such as $CaCl_2$, CS, and PAA. Ryou et al. [104] studied a mixture of PAA and alginate treated with dopamine (catechol) as an anode in a Li-ion battery. The hydrogen bonds and catecholic interaction with Si nanoparticles, derived in an electrically efficient conducting pathway (with decreased dead portion) resulting in high performance and superior capacity. Tran et al. [95] compared the performance of different water-based binders (PAA, sodium polyacrylate, and sodium alginate [SA]) with Li-salt containing electrolytes.

The nature of the binder affected the behaviour of the capacitor. Alginate was found to be the best alternative to conventional fluorinated polymers used as binders. It showed an accessible porosity (for extended charge–discharge cycles) and did not break the polymer network during stress relaxation tests.

5.5.2 Lignin

The aromatic nature of lignin (chemical structure shown in Figure 5.5) together with its low density, high hardness, resistance to heat, chemicals, friction, and humidity makes it interesting to be used as a binder in LIBs [47]. Taking advantage of the high carbon content and long and resisting chains of lignin, Chen et al. [101] used a low-temperature carbonised lignin to replace both the conventional binder (polymer) and the conductive additive (carbon black). The low pyrolysis temperature allowed to keep the polymeric flexibility (that accommodated the volume expansion) while maintaining the conductivity and high performance. The fact that only one renewable material was used to replace two non-renewable ones reduced the cost of the entire battery in an environmentally friendly manner.

5.6 Application of Bio-based Polymers in Fuel Cells

A fuel cell is an environmentally friendly electrochemical device that converts the chemical energy obtained from a redox reaction into electrical energy. The fuel oxidation at the anode and the counter-reaction at the cathode generates a voltage and, consequently, an electron flux takes place. The best performance in terms of power output is achieved when pure hydrogen reacts with oxygen, but other fuels such as methanol and ethanol are also commonly used.

Fuel cells basically consist of an electrolyte material packed between two porous electrodes. The fuel and the oxidant are not integral parts of the cell but are supplied externally, and they are dissociated catalytically into ions and electrons when passing over the anode and the cathode. The electrolyte membrane allows the flow of ions between the anode and the cathode and it must be a good electronic insulator.

There are different types of fuel cells that are generally classified according to their electrolyte. The main types are: alkaline fuel cell (AFC); molten carbonate fuel cell (MCFC); phosphoric acid fuel cell (PAFC); solid oxide fuel cell (SOFC); biofuel cell; and proton exchange membrane fuel cell (PEMFC), also known as polymer electrolyte membrane fuel cell.

There are two types of PEMFCs, namely hydrogen PEMFCs, that use hydrogen as fuel and direct methanol fuel cells (DMFCs) in which methanol is the fuel. Both types utilise a proton exchange membrane for the transfer of protons. The polymer membrane in a PEMFC has a two-fold function. It acts as an electrolyte providing ionic communication between the electrodes and, at the same time, is a separator for the reactants. A PEMFC is schematically represented in Figure 5.6. The main characteristics of a high-performant proton-conducting polymer electrolyte membrane are low cost, high proton conductivity, good fuel barrier properties, high mechanical strength, thermal and chemical stability, and they must be electronically non-conducting.

A polymer commercially called Nafion® that was developed by DuPont de Nemours, USA, in the 1960s fulfills almost all the necessary requirements for a good fuel cell membrane. However, Nafion membranes are expensive to produce, their dimensional stability is limited and, in addition, methanol crossover through the Nafion membrane is a key issue that has not been resolved yet.

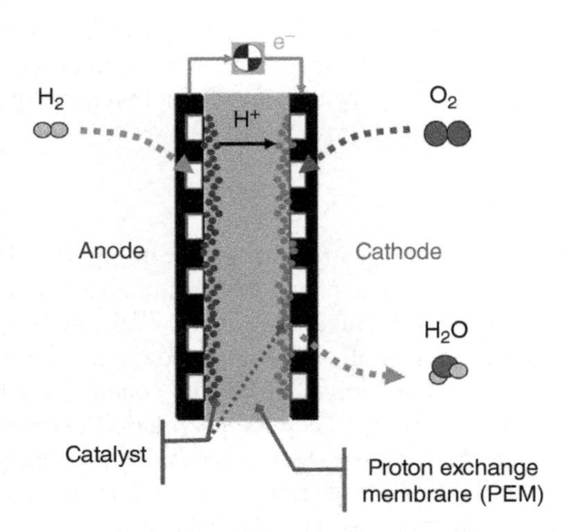

Figure 5.6 *Schematic representation of the polymer electrolyte membrane fuel cell.*

In that sense, cost-effective and eco-friendly polymer electrolytes from biosources can become a promising substitute for synthetic polymers. In particular, CS offers several advantages for its use as a polymer electrolyte membrane.

5.6.1 Chitosan

CS is a biopolymer resulting from the deacetylation of chitin, which is largely found in the exoskeletons of crustaceans such as crabs, lobsters, krill, and crayfish; insects; molluscan organs; and fungi. The advantages of CS as a polymer electrolyte membrane include: (i) CS is low-cost and eco-friendly; (ii) its hydrophilicity opens the possibility to use it in a high-temperature and low-relative humidity environment; (iii) it has low methanol permeability; (iv) the backbone of CS has functional groups that can be chemically modified to tailor its properties.

There are various studies in the literature dealing with the use of CS to produce high-performance polymer electrolytes for fuel cells. The importance of CS in the development of novel cost-effective proton exchange membranes is supported by the considerable quantity of reviews and book chapters referring to that topic [108–114].

Mukoma et al. [115] synthesised CS membranes cross-linked in sulphuric acid as alternative proton exchange membrane material for application in fuel cells. The developed membranes had better water-uptake properties, but poorer thermal stability and proton conductivity than Nafion 117. Sulphuric acid was also used as the cross-linking agent by Ma et al. [116] who reported a CS membrane used as the polymer electrolyte and separator in a direct borohydride fuel cell (DBFC). Additionally, a chitosan chemical hydrogel (CCH) was prepared and used as a binder for anode catalysts. The fabricated membrane exhibited higher ionic conductivity in an alkaline medium and a higher borohydride crossover rate than Nafion 212. When using CCH as an anode binder, the performance of the DBFC was further improved.

The exoskeleton of Cape rock lobsters from South Africa was the raw material to extract chitin for the preparation of CS flakes that were, in turn, the raw material for H_2SO_4 cross-linked membranes [117]. Although the performance of the prepared membranes was very low compared to Nafion 117, they were still suitable for other fuel cell applications (e.g. PEMFC) due to their good proton transport properties. This is supported by the fact that the water uptake of the reported CS membranes was three times higher than that of Nafion 117.

On the other hand, mixing CS and different heteropolyacids, i.e. phosphomolybdic acid (PMA), phosphotungstic acid (PWA) and silicotungstic acid (SiWA), was a successful approach to prepare novel proton-conducting membranes for DMFCs. The resulting CS/PMA membrane was ideal for use in DMFC as it exhibited a low methanol permeability (2.7×10^{-7} cm^2 s^{-1}) and comparatively high proton conductivity (0.015 S cm^{-1} at 25 °C) [118].

Shakeri et al. [119] prepared a polyelectrolyte complex of CS and PWA as a proton-conducting membrane for DMFC applications. Different loading weights of montmorillonite (MMT) nanoclays were introduced to the system to reduce the methanol permeability. The optimum developed membrane showed a higher power density (49.7 mW cm^{-2}) than Nafion 117. The interest of MMT as an inorganic filler in biopolymers is reflected by its use in several recent papers [120–122].

Another approach concerning membranes for DMFC is presented in the work of Mohanapriya et al. [123]. CS was combined with plant hormones belonging to the auxin group to make proton-conducting bio-composite membranes for DMFC. It was found that the addition of those hormones improved the ionic conductivity and also the tensile strength of the CS membranes.

GA was used as a cross-linking agent by Wan et al. [124] to prepare CS-based composite membranes in a two-step pathway. These new alkaline composite membranes were prepared by incorporating KOH as the functional ionic source to promote the ionic migration rate through the membrane. Feketeföldi et al. [125] synthesised novel cross-linked highly quaternised CS and quaternised poly(vinyl alcohol) membranes to be applied in alkaline direct ethanol fuel cells. Different amounts of GA and ethylene glycol diglycidyl ether were used as cross-linkers. The membranes showed excellent transport and ionic properties for anions and low ethanol permeability.

Some recent and innovative research about the use of CS for fuel cell applications deals with the preparation of self-humidifying proton exchange hybrid membranes from CS wherein heptanedioic acid (HA) is used as cross-linker and geopolymer (GP), an inorganic aluminosilicate material, as the filler [126]. The presence of hydrophilic voids in the GP that hold the water molecules are considered responsible for the optimum water uptake and self-humidification of the hybrid membranes. Furthermore, the incorporation of GP in the polymer matrix resulted in superior thermal, mechanical, and oxidation stability, making these novel membranes promising candidates for PEMFC applications. Ryu et al. [127] recently synthesised a novel CS-based anion exchange membrane (AEM) modified with pyridine and a vinylimidazole derivative that exhibits high ionic conductivity and structural stability, even under high pH conditions.

5.6.2 Other Biopolymers

Besides CS, there are other biomaterials used in fuel cell applications, although there are less numerous studies. Cellulose, which is the most abundant biopolymer, and, in particular, cellulose nanocrystals and cellulose nanofibrils are also valid materials to be used in fuel cells. Vilela et al. [128] reviewed the most recent efforts in the domain of bio-based PEFC components, meaning the ion-exchange membrane and the electrodes, by focusing on the use of nanoscale forms of cellulose as a renewable and easily foldable or stackable nanomaterial.

Bacterial cellulose is a type of pure cellulose generated by the genus *Gluconacetobacter*. Compared to plant cellulose, it has high crystallinity, high tensile strength, high water-binding capacity, low gas permeation, and good biocompatibility. In fact, novel nanocomposite membranes based on blending bacterial nanocellulose pulp and Nafion were developed, exhibiting high strength, good proton conductivity, and low methanol permeability [129]. The annealed membrane having a nanocellulose/Nafion mass ratio of 1/7 showed maximum power densities for PEMFC and DMFC of 106 and 20.4 $mW\,cm^{-2}$, respectively, which are much higher than those of other reported bacterial cellulose-based membranes. Gadim et al. [130] studied how the preferential orientation of supporting bacterial cellulose nanofibrils affected the conductivity of composite proton-conducting electrolytes with poly(4-styrene sulphonic acid). The fuel cell tests at room temperature yield 40 $mW\,cm^{-2}$ at 125 $mA\,cm^{-2}$, which is one of the highest values reported for a biopolymer-based electrolyte membrane.

Agar-agar is a polysaccharide derived from red seaweeds and is characterised by its chemical repeating units of 3–6, anhydro-L-galactose. Boopathi et al. [131] reported a proton-conducting polymer electrolyte based on agar-agar and ammonium nitrate (NH_4NO_3). The highest proton conductivity polymer membrane, based on agar-agar/NH_4NO_3: 40/60 mol%, was tested in a single fuel cell finding a value for the open-circuit voltage of 558 mV. On the other hand, Selvalakshmi et al. [132] assembled a single PEM fuel cell with a selected proton-conducting membrane based on a mixture of agar-agar (50 mol%) and NH_4I (50 mol%) and it exhibited an output voltage of 408 mV.

Another natural polymer is carrageenan extracted from red algae, which has been used for the preparation of fuel cell membranes [133]. I-Carrageenan-based biopolymer membranes doped with ammonium bromide (NH_4Br) were prepared successfully [133]. The conductivity value was improved from 1.46×10^{-5} up to 1.08×10^{-3} $S\,cm^{-1}$ by simple doping with 20 mol% of NH_4Br salt.

Alginate is a heteropolymer containing mannuronic acid and guluronic acid groups, which is commonly found in seaweeds and its use for the preparation of membranes is reviewed by several scholars [110, 112]. A simple route for the preparation of proton exchange membranes using sulphonated side-chain graphite oxides and cross-linked SA was recently reported [134]. According to the authors, the performances of some of the developed membranes are superior to the commercially available Nafion 117 membrane. SA was also used to prepare a polymer electrolyte membrane with low methanol permeability and high proton conductivity [135]. The membrane was based on the SA polymer as the matrix and sulphonated graphene oxide (SGO) as an inorganic filler. The SGO improved the properties of the SA/SGO proton conductivity, methanol permeability, and mechanical properties to a high level comparable to commercial membranes.

In a different approach, calcined crab shell (CCS) forming nanocomposites with $La_{0.6}Sr_{0.4}Co_{0.2}Fe_{0.8}O_3$ (LSCF) perovskite were highlighted as functional electrolytes for low-temperature SOFCs (LTSOFCs) [136]. The CCS-LSCF composite fuel cell showed a six-fold enhancement of power output compared to the CCS electrolyte. This fact is attributed to the ion channel construction and interface effect built in the CCS-LSCF composite.

5.7 Conclusion and Outlook

With the rise of new concepts based on sustainability and circular economy, bio-based materials are a key piece of the electrochemical energy conversion and storage ecosystem. Among the existing bio-based materials, no clear candidate is competitive with commercially available ones in terms of energy density. However, the potential growth of bio-based materials has not been reached and their full capabilities have not been completely exploited yet. The present-day renaissance of organic and polymer-based batteries and supercapacitors is encouraging but has not yet moved much beyond the performance that was prevalent in earlier years. Recent developments on solid-state batteries using polysaccharides showed high performances and enormous options to fine-tune several parameters by targeted functional groups.

In fact, the largest potential is predicted in the development within the batteries field since intense research in nanoscale bio-based materials was started just a few years ago. The efforts on green extraction processes, chemical modifications, device fabrication, and materials processing must consider challenges that are directly related to the engineering/design part, which have a significant impact on the overall end user application, overall performance, and materials' durability.

References

1. Chu, S. and Majumdar, A. (2012). Opportunities and challenges for a sustainable energy future. *Nature* 488 (7411): 294–303.
2. Ringler, C., Willenbockel, D., Perez, N. et al. (2016). Global linkages among energy, food and water: an economic assessment. *Journal of Environmental Studies and Sciences* 6 (1): 161–171.
3. Farzad, S., Mandegari, M.A., and Görgens, J.F. (2016). A critical review on biomass gasification, co-gasification, and their environmental assessments. *Biofuel Research Journal* 3 (4): 483–495.
4. Collard, F.-X. and Blin, J. (2014). A review on pyrolysis of biomass constituents: mechanisms and composition of the products obtained from the conversion of cellulose, hemicelluloses and lignin. *Renewable and Sustainable Energy Reviews* 38: 594–608.
5. Vassilev, S.V., Baxter, D., Andersen, L.K., and Vassileva, C.G. (2010). An overview of the chemical composition of biomass. *Fuel* 89 (5): 913–933.
6. Ahmad, M., Rajapaksha, A.U., Lim, J.E. et al. (2014). Biochar as a sorbent for contaminant management in soil and water: a review. *Chemosphere* 99: 19–33.
7. Choi, W., Lahiri, I., Seelaboyina, R., and Kang, Y.S. (2010). Synthesis of graphene and its applications: a review. *Critical Reviews in Solid State and Materials Sciences* 35 (1): 52–71.
8. Sharma, B., Malik, P., and Jain, P. (2018). Biopolymer reinforced nanocomposites: a comprehensive review. *Materials Today Communications* 16: 353–363.
9. Vroman, I. and Tighzert, L. (2009). Biodegradable polymers. *Materials* 2 (2): 307–344.
10. Nakajima, H., Dijkstra, P., and Loos, K. (2017). The recent developments in biobased polymers toward general and engineering applications: polymers that are upgraded from biodegradable polymers, analogous to petroleum-derived polymers, and newly developed. *Polymers* 9 (10): 523.
11. Shaghaleh, H., Xu, X., and Wang, S. (2018). Current progress in production of biopolymeric materials based on cellulose, cellulose nanofibers, and cellulose derivatives. *RSC Advances* 8 (2): 825–842.
12. Lacoste, C., Lopez-Cuesta, J.-M., and Bergeret, A. (2019). Development of a biobased superabsorbent polymer from recycled cellulose for diapers applications. *European Polymer Journal* 116: 38–44.
13. Yao, H., Zheng, G., Li, W. et al. (2013). Crab shells as sustainable templates from nature for nanostructured battery electrodes. *Nano Letters* 13 (7): 3385–3390.
14. Bajpai, P. (2019). Chapter 2 – Description of biobased polymers. In: *Biobased Polymers* (ed. P. Bajpai), 13–23. Amsterdam, Netherlands: Elsevier.
15. Jahirul, M.I., Rasul, M.G., Chowdhury, A.A., and Ashwath, N. (2012). Biofuels production through biomass pyrolysis – a technological review. *Energies* 5 (12): 4952–5001.
16. Carpenter, D., Westover, T.L., Czernik, S., and Jablonski, W. (2014). Biomass feedstocks for renewable fuel production: a review of the impacts of feedstock and pretreatment on the yield and product distribution of fast pyrolysis bio-oils and vapors. *Green Chemistry* 16 (2): 384–406.
17. Qu, X., Fu, H., Mao, J. et al. (2016). Chemical and structural properties of dissolved black carbon released from biochars. *Carbon* 96: 759–767.
18. Yargicoglu, E.N., Sadasivam, B.Y., Reddy, K.R., and Spokas, K. (2015). Physical and chemical characterization of waste wood derived biochars. *Waste Management* 36: 256–268.

19. Ioannidou, O. and Zabaniotou, A. (2007). Agricultural residues as precursors for activated carbon production – a review. *Renewable and Sustainable Energy Reviews* 11 (9): 1966–2005.
20. Zhu, J., Yan, C., Zhang, X. et al. (2020). A sustainable platform of lignin: from bioresources to materials and their applications in rechargeable batteries and supercapacitors. *Progress in Energy and Combustion Science* 76: 100788.
21. Sajjadi, B., Chen, W.-Y., and Egiebor Nosa, O. (2019). A comprehensive review on physical activation of biochar for energy and environmental applications. *Reviews in Chemical Engineering* 35 (6): 735–776.
22. Ummartyotin, S. and Pechyen, C. (2016). Strategies for development and implementation of bio-based materials as effective renewable resources of energy: a comprehensive review on adsorbent technology. *Renewable and Sustainable Energy Reviews* 62: 654–664.
23. Chen, H., Cong, T.N., Yang, W. et al. (2009). Progress in electrical energy storage system: a critical review. *Progress in Natural Science* 19 (3): 291–312.
24. Larcher, D. and Tarascon, J.M. (2015). Towards greener and more sustainable batteries for electrical energy storage. *Nature Chemistry* 7 (1): 19–29.
25. Thackeray, M.M., Wolverton, C., and Isaacs, E.D. (2012). Electrical energy storage for transportation – approaching the limits of, and going beyond, lithium-ion batteries. *Energy & Environmental Science* 5 (7): 7854–7863.
26. Fu, X. and Zhong, W.-H. (2019). Biomaterials for high-energy lithium-based batteries: strategies, challenges, and perspectives. *Advanced Energy Materials* 9 (40): 1901774.
27. Liedel, C., Wang, X., and Antonietti, M. (2018). Biobased polymer cathodes with enhanced charge storage. *Nano Energy* 53: 536–543.
28. Mohanty, A.K., Misra, M., and Drzal, L.T. (2002). Sustainable bio-composites from renewable resources: opportunities and challenges in the green materials world. *Journal of Polymers and the Environment* 10 (1): 19–26.
29. Poizot, P., Gaubicher, J., Renault, S. et al. (2020). Opportunities and challenges for organic electrodes in electrochemical energy storage. *Chemical Reviews* 120 (14): 6490–6557.
30. Zhu, Y., Romain, C., and Williams, C.K. (2016). Sustainable polymers from renewable resources. *Nature* 540 (7633): 354–362.
31. Nitta, N., Wu, F., Lee, J.T., and Yushin, G. (2015). Li-ion battery materials: present and future. *Materials Today* 18 (5): 252–264.
32. Etacheri, V., Marom, R., Elazari, R. et al. (2011). Challenges in the development of advanced Li-ion batteries: a review. *Energy & Environmental Science* 4 (9): 3243–3262.
33. Correa, C.R. and Kruse, A. (2018). Biobased functional carbon materials: production, characterization, and applications – a review. *Materials* 11 (9): 1568.
34. Caballero, A., Hernán, L., and Morales, J. (2011). Limitations of disordered carbons obtained from biomass as anodes for real lithium-ion batteries. *ChemSusChem* 4 (5): 658–663.
35. Long, W., Fang, B., Ignaszak, A. et al. (2017). Biomass-derived nanostructured carbons and their composites as anode materials for lithium ion batteries. *Chemical Society Reviews* 46 (23): 7176–7190.
36. Wang, Y.-Y., Hou, B.-H., Lü, H.-Y. et al. (2015). Porous N-doped carbon material derived from prolific chitosan biomass as a high-performance electrode for energy storage. *RSC Advances* 5 (118): 97427–97434.
37. Stephan, A.M., Kumar, T.P., Ramesh, R. et al. (2006). Pyrolitic carbon from biomass precursors as anode materials for lithium batteries. *Materials Science and Engineering: A* 430 (1): 132–137.
38. Hou, J., Cao, C., Idrees, F., and Ma, X. (2015). Hierarchical porous nitrogen-doped carbon nanosheets derived from silk for ultrahigh-capacity battery anodes and supercapacitors. *ACS Nano* 9 (3): 2556–2564.
39. Balogun, M.-S., Luo, Y., Qiu, W. et al. (2016). A review of carbon materials and their composites with alloy metals for sodium ion battery anodes. *Carbon* 98: 162–178.
40. Kim, S.-W., Seo, D.-H., Ma, X. et al. (2012). Electrode materials for rechargeable sodium-ion batteries: potential alternatives to current lithium-ion batteries. *Advanced Energy Materials* 2 (7): 710–721.
41. Liang, Y., Lai, W.-H., Miao, Z., and Chou, S.-L. (2018). Nanocomposite materials for the sodium–ion battery: a review. *Small* 14 (5): 1702514.
42. Wu, L., Buchholz, D., Vaalma, C. et al. (2016). Apple-biowaste-derived hard carbon as a powerful anode material for Na-ion batteries. *ChemElectroChem* 3 (2): 292–298.

43. Zhang, K., Monteiro, M.J., and Jia, Z. (2016). Stable organic radical polymers: synthesis and applications. *Polymer Chemistry* 7 (36): 5589–5614.
44. Yao, Y. and Wu, F. (2015). Naturally derived nanostructured materials from biomass for rechargeable lithium/sodium batteries. *Nano Energy* 17: 91–103.
45. Aqil, A., Vlad, A., Piedboeuf, M.-L. et al. (2015). A new design of organic radical batteries (ORBs): carbon nanotube buckypaper electrode functionalized by electrografting. *Chemical Communications* 51 (45): 9301–9304.
46. Aqil, M., Aqil, A., Ouhib, F. et al. (2015). RAFT polymerization of an alkoxyamine bearing acrylate, towards a well-defined redox active polyacrylate. *RSC Advances* 5 (103): 85035–85038.
47. Patil, N., Aqil, M., Aqil, A. et al. (2018). Integration of redox-active catechol pendants into poly(ionic liquid) for the design of high-performance lithium-ion battery cathodes. *Chemistry of Materials* 30 (17): 5831–5835.
48. Liu, T., Kim, K.C., Lee, B. et al. (2017). Self-polymerized dopamine as an organic cathode for Li- and Na-ion batteries. *Energy & Environmental Science* 10 (1): 205–215.
49. Zhang, W., Pan, Z., Yang, F.K., and Zhao, B. (2015). A facile in situ approach to polypyrrole functionalization through bioinspired catechols. *Advanced Functional Materials* 25 (10): 1588–1597.
50. Ilic, I.K., Meurer, M., Chaleawlert-umpon, S. et al. (2019). Vanillin decorated chitosan as electrode material for sustainable energy storage. *RSC Advances* 9 (8): 4591–4598.
51. Qu, J., Khan, F.Z., Satoh, M. et al. (2008). Synthesis and charge/discharge properties of cellulose derivatives carrying free radicals. *Polymer* 49 (6): 1490–1496.
52. Qu, J., Morita, R., Satoh, M. et al. (2008). Synthesis and properties of DNA complexes containing 2,2,6,6 -tetramethyl-1-piperidinoxy (TEMPO) moieties as organic radical battery materials. *Chemistry – A European Journal* 14 (11): 3250–3259.
53. Baptista, A.C., Ropio, I., Romba, B. et al. (2018). Cellulose-based electrospun fibers functionalized with polypyrrole and polyaniline for fully organic batteries. *Journal of Materials Chemistry A* 6 (1): 256–265.
54. Ketpang, K. and Park, J.S. (2010). Electrospinning PVDF/PPy/MWCNTs conducting composites. *Synthetic Metals* 160 (15): 1603–1608.
55. Zhang, Y. and Rutledge, G.C. (2012). Electrical conductivity of electrospun polyaniline and polyaniline-blend fibers and mats. *Macromolecules* 45 (10): 4238–4246.
56. Jabbour, L., Bongiovanni, R., Chaussy, D. et al. (2013). Cellulose-based Li-ion batteries: a review. *Cellulose* 20 (4): 1523–1545.
57. Sheng, J., Chen, T., Wang, R. et al. (2020). Ultra-light cellulose nanofibril membrane for lithium-ion batteries. *Journal of Membrane Science* 595: 117550.
58. Wang, S., Zhao, C., Han, W. et al. (2019). A polyimide/cellulose lithium battery separator paper. E3S Web of Conferences, Wuhan, China (01 December 01 2019). 03004.
59. Spirk, S. (2018). *Polysaccharides as Battery Components*. Cham, Switzerland: Springer International Publishing.
60. Zhang, J., Liu, Z., Kong, Q. et al. (2013). Renewable and superior thermal-resistant cellulose-based composite nonwoven as lithium-ion battery separator. *ACS Applied Materials & Interfaces* 5 (1): 128–134.
61. Xu, Q., Kong, Q., Liu, Z. et al. (2014). Cellulose/polysulfonamide composite membrane as a high performance lithium-ion battery separator. *ACS Sustainable Chemistry & Engineering* 2 (2): 194–199.
62. Zhou, S., Nyholm, L., Strømme, M., and Wang, Z. (2019). Cladophora cellulose: unique biopolymer nanofibrils for emerging energy, environmental, and life science applications. *Accounts of Chemical Research* 52 (8): 2232–2243.
63. Du, Z., Su, Y., Qu, Y. et al. (2019). A mechanically robust, biodegradable and high performance cellulose gel membrane as gel polymer electrolyte of lithium-ion battery. *Electrochimica Acta* 299: 19–26.
64. Leng, L., Zeng, X., Chen, P. et al. (2015). A novel stability-enhanced lithium-oxygen battery with cellulose-based composite polymer gel as the electrolyte. *Electrochimica Acta* 176: 1108–1115.
65. Wang, J., Song, S., Gao, S. et al. (2017). Mg-ion conducting gel polymer electrolyte membranes containing biodegradable chitosan: preparation, structural, electrical and electrochemical properties. *Polymer Testing* 62: 278–286.
66. Zhao, L., Fu, J., Du, Z. et al. (2020). High-strength and flexible cellulose/PEG based gel polymer electrolyte with high performance for lithium ion batteries. *Journal of Membrane Science* 593: 117428.

67. Ma, Q., Zhang, H., Zhou, C. et al. (2016). Single lithium-ion conducting polymer electrolytes based on a super-delocalized polyanion. *Angewandte Chemie International Edition* 55 (7): 2521–2525.
68. Kuribayashi, I. (1996). Characterization of composite cellulosic separators for rechargeable lithium-ion batteries. *Journal of Power Sources* 63 (1): 87–91.
69. Nair, J.R., Chiappone, A., Gerbaldi, C. et al. (2011). Novel cellulose reinforcement for polymer electrolyte membranes with outstanding mechanical properties. *Electrochimica Acta* 57: 104–111.
70. Ben Youcef, H., Orayech, B., Del Amo, J.M.L. et al. (2020). Functionalized cellulose as quasi single-ion conductors in polymer electrolyte for all-solid–state Li/Na and LiS batteries. *Solid State Ionics* 345: 115168.
71. Halper, M.S. and Ellenbogen, J.C. (2006). *Supercapacitors: A Brief Overview*. McLean, Virginia, USA: The MITRE Corporation.
72. Choi, D. and Kim, K. (2015). Reduced graphene oxide-coated polyvinyl alcohol sponge as a high-performance supercapacitor. *Bulletin of the Korean Chemical Society* 36 (2): 692–695.
73. Wei, L. and Yushin, G. (2012). Nanostructured activated carbons from natural precursors for electrical double layer capacitors. *Nano Energy* 1 (4): 552–565.
74. Dubey, R. and Guruviah, V. (2019). Review of carbon-based electrode materials for supercapacitor energy storage. *Ionics* 25 (4): 1419–1445.
75. Ghosh, S., Santhosh, R., Jeniffer, S. et al. (2019). Natural biomass derived hard carbon and activated carbons as electrochemical supercapacitor electrodes. *Scientific Reports* 9 (1): 16315.
76. Chen, W., Wang, X., Feizbakhshan, M. et al. (2019). Preparation of lignin-based porous carbon with hierarchical oxygen-enriched structure for high-performance supercapacitors. *Journal of Colloid and Interface Science* 540: 524–534.
77. Li, Z., Xu, Z., Tan, X. et al. (2013). Mesoporous nitrogen-rich carbons derived from protein for ultra-high capacity battery anodes and supercapacitors. *Energy & Environmental Science* 6 (3): 871–878.
78. Zhao, J.-R., Hu, J., Li, J.-F., and Chen, P. (2020). N-doped carbon nanotubes derived from waste biomass and its electrochemical performance. *Materials Letters* 261: 127146.
79. Schlee, P., Hosseinaei, O., Baker, D. et al. (2019). From waste to wealth: from kraft lignin to free-standing supercapacitors. *Carbon* 145: 470–480.
80. Luong, D.X., Bets, K.V., Algozeeb, W.A. et al. (2020). Gram-scale bottom-up flash graphene synthesis. *Nature* 577 (7792): 647–651.
81. Farma, R., Deraman, M., Awitdrus, A. et al. (2013). Physical and electrochemical properties of supercapacitor electrodes derived from carbon nanotube and biomass carbon. *International Journal of Electrochemical Science* 8 (1): 257–273.
82. Raymundo-Piñero, E., Cadek, M., Wachtler, M., and Béguin, F. (2011). Carbon nanotubes as nanotexturing agents for high power supercapacitors based on seaweed carbons. *ChemSusChem* 4 (7): 943–949.
83. Taberna, P.-L., Chevallier, G., Simon, P. et al. (2006). Activated carbon–carbon nanotube composite porous film for supercapacitor applications. *Materials Research Bulletin* 41 (3): 478–484.
84. Cheng, P., Li, T., Yu, H. et al. (2016). Biomass-derived carbon fiber aerogel as a binder-free electrode for high-rate supercapacitors. *The Journal of Physical Chemistry C* 120 (4): 2079–2086.
85. Hao, P., Zhao, Z., Tian, J. et al. (2014). Hierarchical porous carbon aerogel derived from bagasse for high performance supercapacitor electrode. *Nanoscale* 6 (20): 12120–12129.
86. Bryan, A.M., Santino, L.M., Lu, Y. et al. (2016). Conducting polymers for pseudocapacitive energy storage. *Chemistry of Materials* 28 (17): 5989–5998.
87. Zhou, X., Li, L., Dong, S. et al. (2012). A renewable bamboo carbon/polyaniline composite for a high-performance supercapacitor electrode material. *Journal of Solid State Electrochemistry* 16 (3): 877–882.
88. Milczarek, G. and Inganäs, O. (2012). Renewable cathode materials from biopolymer/conjugated polymer interpenetrating networks. *Science* 335 (6075): 1468–1471.
89. Admassie, S., Elfwing, A., Jager, E.W.H. et al. (2014). A renewable biopolymer cathode with multivalent metal ions for enhanced charge storage. *Journal of Materials Chemistry A* 2 (6): 1974–1979.
90. Leguizamon, S., Díaz-Orellana, K.P., Velez, J. et al. (2015). High charge-capacity polymer electrodes comprising alkali lignin from the Kraft process. *Journal of Materials Chemistry A* 3 (21): 11330–11339.
91. Enock, T.K., King'ondu, C.K., Pogrebnoi, A., and Jande, Y.A.C. (2017). Status of biomass derived carbon materials for supercapacitor application. *International Journal of Electrochemistry* 2017: 1–14.

92. Espinoza-Acosta, J.L., Torres-Chávez, P.I., Olmedo-Martínez, J.L. et al. (2018). Lignin in storage and renewable energy applications: a review. *Journal of Energy Chemistry* 27 (5): 1422–1438.
93. Yu, B., Gele, A., and Wang, L. (2018). Iron oxide/lignin-based hollow carbon nanofibers nanocomposite as an application electrode materials for supercapacitors. *International Journal of Biological Macromolecules* 118: 478–484.
94. Wu, X.-L., Wen, T., Guo, H.-L. et al. (2013). Biomass-derived sponge-like carbonaceous hydrogels and aerogels for supercapacitors. *ACS Nano* 7 (4): 3589–3597.
95. Tran, H.Y., Wohlfahrt-Mehrens, M., and Dsoke, S. (2017). Influence of the binder nature on the performance and cycle life of activated carbon electrodes in electrolytes containing Li-salt. *Journal of Power Sources* 342: 301–312.
96. Nirmale, T.C., Kale, B.B., and Varma, A.J. (2017). A review on cellulose and lignin based binders and electrodes: small steps towards a sustainable lithium ion battery. *International Journal of Biological Macromolecules* 103: 1032–1043.
97. Bresser, D., Buchholz, D., Moretti, A. et al. (2018). Alternative binders for sustainable electrochemical energy storage – the transition to aqueous electrode processing and bio-derived polymers. *Energy & Environmental Science* 11 (11): 3096–3127.
98. Böckenfeld, N., Jeong, S.S., Winter, M. et al. (2013). Natural, cheap and environmentally friendly binder for supercapacitors. *Journal of Power Sources* 221: 14–20.
99. Jabbour, L., Gerbaldi, C., Chaussy, D. et al. (2010). Microfibrillated cellulose–graphite nanocomposites for highly flexible paper-like Li-ion battery electrodes. *Journal of Materials Chemistry* 20 (35): 7344–7347.
100. Nyström, G., Mihranyan, A., Razaq, A. et al. (2010). A nanocellulose polypyrrole composite based on microfibrillated cellulose from wood. *The Journal of Physical Chemistry B* 114 (12): 4178–4182.
101. Chen, T., Zhang, Q., Pan, J. et al. (2016). Low-temperature treated lignin as both binder and conductive additive for silicon nanoparticle composite electrodes in lithium-ion batteries. *ACS Applied Materials & Interfaces* 8 (47): 32341–32348.
102. Chen, C., Lee, S.H., Cho, M. et al. (2016). Cross-linked chitosan as an efficient binder for Si anode of Li-ion batteries. *ACS Applied Materials & Interfaces* 8 (4): 2658–2665.
103. Ling, M., Xu, Y., Zhao, H. et al. (2015). Dual-functional gum arabic binder for silicon anodes in lithium ion batteries. *Nano Energy* 12: 178–185.
104. Ryou, M.-H., Kim, J., Lee, I. et al. (2013). Mussel-inspired adhesive binders for high-performance silicon nanoparticle anodes in lithium-ion batteries. *Advanced Materials* 25 (11): 1571–1576.
105. Shi, Y., Zhou, X., and Yu, G. (2017). Material and structural design of novel binder systems for high-energy, high-power lithium-Ion batteries. *Accounts of Chemical Research* 50 (11): 2642–2652.
106. Hochgatterer, N.S., Schweiger, M.R., Koller, S. et al. (2008). Silicon/graphite composite electrodes for high-capacity anodes: influence of binder chemistry on cycling stability. *Electrochemical and Solid-State Letters* 11 (5): A76.
107. Drofenik, J., Gaberscek, M., Dominko, R. et al. (2003). Cellulose as a binding material in graphitic anodes for Li ion batteries: a performance and degradation study. *Electrochimica Acta* 48 (7): 883–889.
108. Kraytsberg, A. and Ein-Eli, Y. (2014). Review of advanced materials for proton exchange membrane fuel cells. *Energy & Fuels* 28 (12): 7303–7330.
109. Ma, J. and Sahai, Y. (2013). Chitosan biopolymer for fuel cell applications. *Carbohydrate Polymers* 92 (2): 955–975.
110. Muthumeenal, A., Pethaiah, S.S., and Nagendran, A. (2017). Biopolymer composites in fuel cells. In: *Biopolymer Composites in Electronics* (ed. K.K. Sadasivuni, J.J. Cabibihan, D. Ponnamma, et al.), 185–217. Amsterdam: Elsevier.
111. Rosli, N.A.H., Loh, K.S., Wong, W.Y. et al. (2020). Review of chitosan-based polymers as proton exchange membranes and roles of chitosan-supported ionic liquids. *International Journal of Molecular Sciences* 21 (2): 632.
112. Shaari, N. and Kamarudin, S.K. (2015). Chitosan and alginate types of bio-membrane in fuel cell application: an overview. *Journal of Power Sources* 289: 71–80.
113. Tripathi, B.P. and Shahi, V.K. (2011). Organic–inorganic nanocomposite polymer electrolyte membranes for fuel cell applications. *Progress in Polymer Science* 36 (7): 945–979.

114. Vaghari, H., Jafarizadeh-Malmiri, H., Berenjian, A., and Anarjan, N. (2013). Recent advances in application of chitosan in fuel cells. *Sustainable Chemical Processes* 1 (1): 16.
115. Mukoma, P., Jooste, B.R., and Vosloo, H.C.M. (2004). Synthesis and characterization of cross-linked chitosan membranes for application as alternative proton exchange membrane materials in fuel cells. *Journal of Power Sources* 136 (1): 16–23.
116. Ma, J., Choudhury, N.A., Sahai, Y., and Buchheit, R.G. (2011). A high performance direct borohydride fuel cell employing cross-linked chitosan membrane. *Journal of Power Sources* 196 (20): 8257–8264.
117. Osifo, P.O. and Masala, A. (2010). Characterization of direct methanol fuel cell (DMFC) applications with H_2SO_4 modified chitosan membrane. *Journal of Power Sources* 195 (15): 4915–4922.
118. Cui, Z., Xing, W., Liu, C. et al. (2009). Chitosan/heteropolyacid composite membranes for direct methanol fuel cell. *Journal of Power Sources* 188 (1): 24–29.
119. Shakeri, S.E., Ghaffarian, S.R., Tohidian, M. et al. (2013). Polyelectrolyte nanocomposite membranes, based on chitosan-phosphotungstic acid complex and montmorillonite for fuel cells applications. *Journal of Macromolecular Science, Part B* 52 (9): 1226–1241.
120. Kalaiselvimary, J., Selvakumar, K., Rajendran, S. et al. (2017). Effect of surface-modified montmorillonite incorporated biopolymer membranes for PEM fuel cell applications. *Polymer Composites* 40 (S1): E301–E311.
121. Nataraj, S.K., Wang, C.-H., Huang, H.-C. et al. (2015). Functionalizing biomaterials to be an efficient proton-exchange membrane and methanol barrier for DMFCs. *ACS Sustainable Chemistry & Engineering* 3 (2): 302–308.
122. Purwanto, M., Atmaja, L., Mohamed, M.A. et al. (2016). Biopolymer-based electrolyte membranes from chitosan incorporated with montmorillonite-crosslinked GPTMS for direct methanol fuel cells. *RSC Advances* 6 (3): 2314–2322.
123. Mohanapriya, S., Sahu, A.K., Bhat, S.D. et al. (2011). Bio-composite membrane electrolytes for direct methanol fuel cells. *Journal of the Electrochemical Society* 158 (11): B1319.
124. Wan, Y., Creber, K.A.M., Peppley, B., and Tam Bui, V. (2006). Chitosan-based solid electrolyte composite membranes: I. Preparation and characterization. *Journal of Membrane Science* 280 (1): 666–674.
125. Feketeföldi, B., Cermenek, B., Spirk, C. et al. (2016). Chitosan-based anion exchange membranes for direct ethanol fuel cells. *Journal of Membrane Science & Technology* 6 (1): 1000145.
126. Al, A.K.S. and Cindrella, L. (2019). Self-humidifying novel chitosan-geopolymer hybrid membrane for fuel cell applications. *Carbohydrate Polymers* 223: 115073.
127. Ryu, J., Seo, J.Y., Choi, B.N. et al. (2019). Quaternized chitosan-based anion exchange membrane for alkaline direct methanol fuel cells. *Journal of Industrial and Engineering Chemistry* 73: 254–259.
128. Vilela, C., Silvestre, A.J.D., Figueiredo, F.M.L., and Freire, C.S.R. (2019). Nanocellulose-based materials as components of polymer electrolyte fuel cells. *Journal of Materials Chemistry A* 7 (35): 20045–20074.
129. Jiang, G.-P., Zhang, J., Qiao, J.-L. et al. (2015). Bacterial nanocellulose/Nafion composite membranes for low temperature polymer electrolyte fuel cells. *Journal of Power Sources* 273: 697–706.
130. Gadim, T.D.O., Loureiro, F.J.A., Vilela, C. et al. (2017). Protonic conductivity and fuel cell tests of nanocomposite membranes based on bacterial cellulose. *Electrochimica Acta* 233: 52–61.
131. Boopathi, G., Pugalendhi, S., Selvasekarapandian, S. et al. (2017). Development of proton conducting biopolymer membrane based on agar–agar for fuel cell. *Ionics* 23 (10): 2781–2790.
132. Selvalakshmi, S., Mathavan, T., Selvasekarapandian, S., and Premalatha, M. (2018). A study of electrochemical devices based on Agar-Agar-NH4I biopolymer electrolytes. *AIP Conference Proceedings* 1942 (1): 140019.
133. Karthikeyan, S., Selvasekarapandian, S., Premalatha, M. et al. (2017). Proton-conducting I-carrageenan-based biopolymer electrolyte for fuel cell application. *Ionics* 23 (10): 2775–2780.
134. Munavalli, B., Torvi, A., and Kariduraganavar, M. (2018). A facile route for the preparation of proton exchange membranes using sulfonated side chain graphite oxides and crosslinked sodium alginate for fuel cell. *Polymer* 142: 293–309.
135. Shaari, N., Kamarudin, S.K., Basri, S. et al. (2018). Enhanced proton conductivity and methanol permeability reduction via sodium alginate electrolyte-sulfonated graphene oxide bio-membrane. *Nanoscale Research Letters* 13 (1): 82.
136. Liu, L., Li, L., Yang, Y. et al. (2019). Fast ion channels for crab shell-based electrolyte fuel cells. *International Journal of Hydrogen Energy* 44 (29): 15370–15376.

6

Bio-based Materials Using Deep Eutectic Solvent Modifiers

Wanwan Qu, Sarah Key and Andrew P. Abbott

School of Chemistry, University of Leicester, Leicester, UK

6.1 Introduction

All living things are made up of a variety of complex biomaterials and humans have been using and modifying these throughout their existence. It is only relatively recently that high-performance materials have been developed through biopolymer modification and it is useful to reflect initially on the reasons for biopolymer modification. These include:

- Up-cycling material, particularly from waste streams.
- Producing biocompatible materials.
- Enabling biodegradation pathways.
- Creating novel frameworks/structures.
- Producing new chemical activities.

The aim of this article is to create a forward-looking preview of potential directions that this fast-moving topic could progress with the aid of a relatively new processing medium and to explain how this medium can aid in; chemical modification, bring about morphological or mechanical modification, enabling the composite formation and gel liquids to allow delivery systems, and forming membranes.

High-Performance Materials from Bio-based Feedstocks, First Edition. Edited by Andrew J. Hunt, Nontipa Supanchaiyamat, Kaewta Jetsrisuparb and Jesper T.N. Knijnenburg.

Biopolymers can be divided into two categories, one naturally produced by living organisms and the other synthesised from naturally occurring monomers. Polynucleotides (such as the nucleic acids – DNA and RNA), polypeptides (proteins), and polysaccharides (polymeric carbohydrates) are the main three classes of biopolymers in which the monomeric units are nucleotides, amino acids, and sugars, respectively [1]. One of the main differences between biopolymers and synthetic polymers is their structures. Many biopolymers spontaneously fold into characteristic compact shapes, which determine their biological functions and depend in a complicated way on their primary structures, while most synthetic polymers have much simpler and more random structures made up of one type of monomer [2]. The synthesis of most natural polymers is controlled by a template-directed process. So, all biopolymers are alike with similar sequences and the polymers are generally monodispersed [2]. In contrast, most synthetic biopolymers are polydispersed [3]. While this may seem obvious, the distinction is important as most natural polymers use secondary and tertiary structures to impart properties such as elasticity and toughness.

Nature has designed degradation pathways for biopolymers through enzymes, which can hydrolyse peptide and saccharide links. The use of polyolefins such as polystyrene (PS) or polyethylene (PE) for short-term applications is clearly causing issues with environmental build-up [4]. Some biopolymers such as polylactic acid and semi-synthetic polymers such as cellulose acetate are not only biodegradable but also compostable, as they can be put through an industrial composting process with 90% being broken down within six months [5]. However, it should be noted that specific moisture, oxygen, light, and temperature conditions are required for biodegradability and these are often not met in landfills [6]. Biopolymers offer important contributions by decreasing the dependence on fossil fuels and the related environmental impacts such as carbon dioxide emissions. As a result, there is a worldwide interest in replacing petroleum-based polymers with renewable resource-based raw materials. However, in spite of all the advancements, there are still some shortcomings stopping the wider commercialisation of bio-based polymers in many applications. This is mainly due to the performance and commercial reasons when comparing with their conventional counterparts, which remains a significant challenge for bio-based polymers.

6.2 Bio-based Materials

A bio-based material is a material made from biopolymers derived from natural resources [7]. The distinction should be drawn at this point between bio-based materials and bioplastics, the latter of which are synthetic plastics made from renewable resources. Some of these can be the same as polymers derived from oil, e.g. PE can be derived from bioethanol, but is not biodegradable. Bio-based materials have strictly undergone significant physical and chemical modification to change their properties from those found in nature. Unprocessed materials such as wood, cotton, and silk can be called biotic materials. Bio-based materials or biomaterials fall under the broader category of bio-products or bio-based products that include materials, chemicals, and energy derived from renewable biological resources.

While attention has focused on issues with polyolefins used primarily for packaging and, in particular, food packaging, this is probably the sector in which bio-based materials have the least chance of being effective due to the need for sterility and stability, albeit for a relatively short shelf life. It is more probable that the main markets for bio-based materials will be in a short lifetime, smaller volume, and single-use applications. This may be in areas such as scaffold materials, short-lived construction materials, transdermal drug delivery, or horticultural applications [7].

One of the fundamental issues with bio-based materials is the variability of the starting material and the generally protracted methods required to obtain pure materials. Although bio-based materials may appear to be green as they come from a renewable resource and are readily biodegradable, the inherent energy requirement needed to grow and process them combined with the associated waste produced in their manufacture makes the metrics for production poor from a viewpoint of energy efficiency, carbon footprint, and carbon efficiency. Producing PE by cracking hydrocarbons and then gas phase polymerisation produces a material with a much lower carbon footprint, a much higher energy efficiency, and a lower E-factor (mass ratio of waste to product) than any biopolymer. The main issue with synthetic polymers is their environmental legacy caused by their incorrect disposal.

Biomaterials can be divided into those based on carbohydrates and those derived from proteins. By far the most abundant of these are carbohydrates which are predominantly polymers of D-glucose. Starch and cellulose are two of the most abundant polysaccharides, and both are homoglycan polymers. These two materials have very different mechanical and chemical properties due to the stereochemistry of the etheric linker between the chains. Subsequently, the alkyl alcohol branch causes interaction between cellulose chains by strong hydrogen bonding, leading to a more crystalline material.

Carbohydrates in nature have different mechanical properties in plants depending on composition and placement in the plant, e.g. stem, trunk leaf, and root. The properties are moderated by the presence of water and modifiers such as sugars, polar molecules, and salts. Wood, for example, contains 40–55% cellulose and the strength and durability depend on the cellular structure and material density.

The same is true to some extent for protein-based materials although, unlike carbohydrates, which are made up of one type of monomer, the primary structure of proteins is governed by the order and type of amino acids in the polymer chain. The primary sequence causes the protein chains to coil and fold giving rise to complex secondary and tertiary structures including coils and sheets.

Compared to synthetic polymers, bio-based materials have the advantage of being hydrophilic, biocompatible, biodegradable, and are generally non-toxic. Furthermore, many natural animal proteins such as keratin, collagen, elastin, and silk are relatively cost-effective and sustainable with many originating as waste from the meat industry. They are easily derived from their natural sources and can be simple to process under mild conditions. There have been many attempts on the synthesis of multifunctional protein-engineered materials with a wide range of customisable properties and activities, including protein-based composite from elastin, collagens, silks, keratins, and resilins [8]. These have been used for a variety of applications including films, foams, composites, and gels, in particular, for biomedical applications such as biosensors, scaffolds for tissue regeneration, [9, 10] and drug delivery systems [11].

In the same way that carbohydrates maintain their mechanical properties using complex aqueous solutions, the same is true for proteinaceous materials. The problem with natural materials is that their properties are controlled by the water content, which means that they can be used over a limited ambient temperature range and are susceptible to atmospheric humidity.

To circumvent this water sensitivity, numerous semi-synthetic materials have been made using polar modifiers to substitute for water. Alcohols and amides have predominantly been used to modify materials such as starch [12]. These compounds break up the intra- and intermolecular interactions decreasing the crystalline structure and lowering phase transition temperatures as they act as plasticisers. Thermomechanical processes such as extrusion can have a significant effect on the structure of the material [13]. More in-depth reviews of molecular modifiers have recently been published [14, 15].

What follows is a discussion of processing bio-based materials using ionic solvents and this needs to be prefaced with an explanation of the need for alternative processing media. Most biomaterials have a high molecular weight, significant crystallinity, and relatively low solubility in molecular solvents. Complex biomaterials such as wood, silk, hair, and skin have multifaceted cellular and layered structures with different hydrophobicities/hydrophilicities making uniform solvent penetration difficult. Ionic solvents can show enhanced solubilities for biopolymers and high penetration into biomaterials due to tuneable surface tensions combined with their amphiphilic components.

6.2.1 Ionic Liquids

Ionic liquids (ILs) are pure compounds that consist only of ions and should melt at or below 100 °C [16]. Many ILs are liquid at room temperature and have been applied in areas such as catalysis, electrochemistry, and material chemistry. The properties of these liquids can be tuned through judicious choice of the cation and/or anion. It has been suggested that as many as 10^{18} different ILs could theoretically be produced, and to date, less than 10^3 have been prepared and characterised [16]. ILs have extremely low vapour pressures and high boiling points, which has led to certain applications, although toxicity, poor biodegradability, and high cost of most ILs have meant that they have not been widely applied as bulk solvents. The complexity of their synthesis can result in low yields, E-factors, and carbon efficiency. Nevertheless, a few processes have been commercialised using ILs albeit on a relatively small scale. Plechkova and Seddon carried out a comprehensive review of applications, although these are now somewhat outdated and have not been comprehensively revised [17]. Yet one of the issues that remain is that with each new combination of anion and cation, a novel material is made and, as such, needs new toxicological testing and registration [18].

6.2.2 Deep Eutectic Solvents

A deep eutectic solvent (DES) is a liquid formed from mixtures of a quaternary ammonium salt (QAS) and either a metal salt or a hydrogen bond donor (HBD). Carboxylic acids, amides, and alcohols are examples of appropriate HBDs [19]. The first of these systems to be described was the choline chloride (ChCl)/urea system mixed in a 1 : 2 mole ratio [20]. ChCl is one of the most commonly used QASs as it is inexpensive, biodegradable, and has low toxicity. It is classified as a pro-vitamin and is produced on the Mtonne p.a. scale as an additive to animal feed and a common additive for household products. The HBD interacts with the chloride anion decreasing the lattice energy of both components and the melting point of the mixture is lower than that of individual constituents. Many DESs have melting points at or below ambient temperature. The advantages of these liquids are that they are easier to make than ILs, they can be less toxic, and many can be made from naturally occurring compounds such as creatine, urea, oxalic, citric and succinic acids, glycerol, and glucose. The fact that DESs are just mixtures of well-characterised compounds simplifies their registration and, in most cases, their toxicity is simply the additive of their constituents. DESs are now gaining interest in many fields of research including metal deposition, electropolishing, metal oxide processing, and some of these applications have been scaled up to >1 ton [21].

DESs can be designed to have low toxicity and biocompatibility making them suitable for biomaterial modification. It is suspected that some living cells may have a composition close to a DES when dehydrated and some natural fluids such as maple syrup could be classed as DESs [22]. The topic of DESs has been reviewed in depth by Smith et al. [21].

In the following section, the different ways of modifying biopolymers/biomaterials will be discussed with emphasis on the unique properties that ionic processing media can have on the finished material. As ILs pre-date DESs, considerably more work has been carried out on the former, but primarily due to cost and toxicity issues, DESs may be more applicable to bulk material modification, particularly where the solvent remains in the material after synthesis.

6.2.3 Morphological/Mechanical Modification

The main interest in biopolymer processing stems from the desire to process cellulose into new materials. One of the largest markets for bio-based materials are cellulose derivatives such as viscose and cellulose acetate. Very few solvents are able to solubilise cellulose but some imidazolium acetates and chlorides demonstrate moderate solubility enabling cellulose to be cast into films and fibres [23]. This solubility for cellulose is only exhibited for a small number of ionic fluids. It has been suggested that the reason that biopolymers are soluble in ionic solvents is partly not only due to the enthalpy of solvation but also due to the entropy of the solution [24]. ILs, for example, are good solvents for cellulose because they are highly ordered and the solvation of macromolecules decreases the entropy of the liquids. DESs are strong HBDs and acceptors and are therefore good at dissolving hydrogen-bonding solutes. For some solutes, the solubility can be related to the enthalpy of solvation [24]. This can be determined using calorimetry and Hess cycles [25]. Cellulose is only soluble in very ordered ILs, such as imidazolium chlorides and acetates, where a large entropy change of dissolution is sufficient to thermodynamically overcome the lattice energy for cellulose [23].

DESs show a more limited solubility for cellulose but they can be used for lignocellulosic biomass fractionation and this area has been recently reviewed [26–28]. DESs can be used as pre-treatments to clean biomass of lignins and lower molecular weight oligomers. This has been practically applied for biomass pre-treatment to a variety of materials including wheat straw, poplar wood, corn stover, and palm oil trunk [29–32].

ILs have been extensively used to modify the structure of biopolymers, mostly by dissolving them and reformulating them into different structures. Dissolving biopolymers in ILs enables changes to the polymer morphology to form sponges, films, and particles of different sizes and aerogels. These new materials have been investigated for drug delivery systems due to their ability to control the rate of drug release [33]. Cellulose, chitin, poly-caprolactone, starch, DNA, cholesterol, and gelatin are amongst the numerous materials that have been modified, many as blends. Microspheres have been prepared through a sol-gel approach [34] whereas nanofibers have been electrospun [35]. Drying using supercritical fluids has enabled the formation of aerogels [36] and chemical modification can lead to biopolymeric ILs [37]. This topic has recently been thoroughly reviewed [33].

In comparison to ILs, DESs have not been extensively studied for morphological changes but they could be used in an analogous method, and it is likely that the same antisolvents (typically water and alcohols) will be appropriate as they solubilise the DES but not the biopolymers. This is an area that could receive considerable attention due to the ability to create biocompatible DESs where the toxicity of IL residues can be negated. Materials such as starch, lignin, and gelatin show high solubility in hydrophilic DESs but hydrophobic proteins such as soy and zein show much lower solubility. A few examples demonstrate the applicability of this approach. A ChCl-thiourea DES was used to dissolve and reprecipitate chitin without significant damage to the primary structure but the process resulted in significant morphology changes [38]. Other studies on chitin extraction and regeneration found that the yield and protein content extracted from shrimp shells varied significantly depending on the HBD of the DES [39]. Jiang et al. dissolved and regenerated wool keratin using a DES formed from ChCl and urea [40]. A similar process was carried out with feathers [41] and with rabbit hair [42], the former using aqueous DESs.

Only a limited number of pH-neutral DESs have been tested to date and a greater change in physical and mechanical properties can be achieved through different DES designs. It has been shown that some of the DESs, such as those based on oxalic acid, can have pH values in the range 1–1.5, whereas those using urea can have values above 7 [43]. DESs, like ILs, can be modified to change their hydrophilicity. The hydrophobicity of the DES can be modified using QAS of longer chain length or long-chain fatty acids [44]. This has not been attempted but could provide some significantly different materials.

The biopolymer does not necessarily need to be soluble in order to be modified. By far, the easiest way in which DESs can be used to modify biopolymers is by acting as lubricants or plasticisers, breaking up the hydrogen bond structure of the biopolymers, and decreasing the degree of crystallinity. This has been demonstrated for materials such as starch, cellulose, and gelatin [45–47].

Starch is one of the most widely studied biomaterials as it is produced in large quantities by most plants. The glass transition temperature of starch could be reduced using DESs and ILs. The tensile strength increased, and the material was found to retain its properties for more than one year if kept in anhydrous conditions. The water sensitivity of thermoplastic starch (TPS) can be regulated such that it can be reprocessed and recycled [48, 49]. Since starch is omnipresent in nature, the materials are rapidly biodegraded or composted under the appropriate conditions. What has still to be tested is the extent to which starch-based materials can be mechanically modified. It has been shown that they can be vacuum formed from thin sheets and injection moulded, as shown in Figure 6.1, but calendaring and blowing have not yet been demonstrated. The effect of using more hydrophobic DESs has also not been tested. In this respect, hydrophobic DESs [44] are likely to form more ductile starch-based plastics that may have less water sensitivity, in the same way that waxed papers are currently produced.

Carbohydrates and proteins differ in that the former are homopolymers made up predominantly of glucose monomers, whereas the latter has a random sequence of different monomers and an

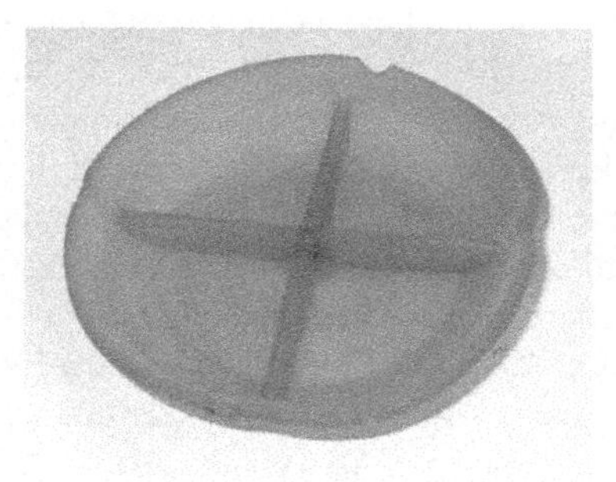

Figure 6.1 *Sample of starch plasticised with glycerol and choline chloride (ChCl) injection moulded at 200 °C and 60 bar. Sample diameter 7 cm. Source: Abbott et al. [45].*

ordered secondary structure. Most carbohydrates are very hydrophilic, whereas proteins differ depending on their primary and secondary structure. Starch, for example, is soluble up to 40 wt% in some ChCl-based DESs [50, 51], whereas zein, a globular plant protein, is almost insoluble. Starches modified with DESs tend to be elastomeric due to the coiled structure of amylopectin but modified proteins behave differently depending on the secondary and tertiary structure of the proteins. The secondary structure of some proteins is generally preserved, but the tertiary structure can be disrupted by some DESs [52, 53]. DES does not tend to improve the thermostability of the proteins which mostly denature above 80 °C. While active proteins may decrease their activity in DESs, much of this loss is recovered when they are diluted into aqueous solutions [9].

There are few studies on DES-modified bioplastics from proteins. Leroy et al. plasticised starch/zein blends (90 : 10) using 1-butyl-3-methyl imidazolium chloride which gave materials with improved mechanical properties compared with pure TPS [46]. Qu investigated both globular and fibrous proteins modified with DESs and found that homogeneous materials could be produced [47, 54]. Zein-based material modified with Glyceline was weak and brittle, supporting the study by Leroy et al., whereas soy protein was slightly stronger and more ductile, particularly when 20 wt% Glyceline was added. Both of these materials were, however, weaker and less ductile than the starch-based materials previously reported. When gelatin was plasticised by DES, a very different material was obtained. Qin et al. [51] showed that 22 wt% gelatin mixed with ethylene glycol/ChCl DES resulted in a highly stretchable, conductive gel suitable for ionic skin applications. Qu found that the properties of gelatin-DES mixtures were dependent on the mixing time prior to pressing [47, 54]. If the mixtures were immediately pressed into a sheet, then a very strong material was produced which displayed a strength similar to high-density PE (HDPE) or polypropylene. If the DES and gelatin were allowed to mix for 24 hours prior to compression moulding, a very flexible, transparent material was obtained. The longer the contact time, the more amorphous and more flexible the material became. NMR relaxation measurements combined with conductivity measurements showed that the DES phase was continuous and fluid in the gel. Pressure-moulded samples did not change their properties depending on how long they were stored. This shows that penetration of the viscous liquid into the crystalline regions is relatively slow. It offers advantages due to tunability but difficulty in terms of reproducibility [47, 54].

Lignin is often viewed as a problematic biopolymer as it decreases the optical appearance of cellulose. ILs and DESs have been used to extract lignin from biomass but relatively few have focused on the unique properties of this useful biomaterial. The anti-microbial and UV-shielding properties of lignin have also been demonstrated using IL-stabilised gelatin–lignin films [55]. It was proposed that these materials could find use in food packaging and would eventually be compostable. A bio-derived IL, choline citrate, was used for the process. The polyphenolic polymer is naturally acidic, an effective antioxidant, and has a very high extinction coefficient due to its high degree of conjugation. On its own, lignin has often been ignored as a material but mixed with a DES, it does have interesting adhesive properties which were demonstrated by Abolibda [56]. When lignin was mixed with starch and a DES made from glycerol and ChCl, an adhesive with a strength of >1.5 MPa could be produced.

6.2.4 Chemical Modification

ILs and DES have been extensively studied as media in which to carry out chemical reactions [57]. There are some inherent advantages to carrying out reactions in these media such as the ability to solubilise both hydrophobic and hydrophilic molecules. ILs can be regarded as co-catalysts because of the coordinative properties of the anion, e.g. chloroaluminate melts can act as strong Lewis acid catalysts. The properties of an IL can be tuned to create biphasic reaction media.

It should, however, be noted that there are also some drawbacks to using ionic fluids for chemical reactions. The most obvious of these is that solute separation from ILs is difficult because ILs have negligible vapour pressure making distillation impossible. The cost of ILs is often prohibitive given that for most reactions, there is generally a 10 : 1 solvent : solute ratio. The final issue is that most ILs are relatively toxic. So, residues may cause difficulties with some applications. Accordingly, most modification studies of bio-based materials have focused on insoluble polymers that can be regenerated using an anti-solvent to make product separation easier.

As in the previous section, the fact that ILs and DESs are good at dissolving hydrogen-bonding solutes makes them good media for extraction and subsequent reactions. It is useful to consider the DES formulation as an opportunity. The acidity and reactivity of the HBD can be viewed as a limitation as it interferes with potential reactions, or it could be viewed as an opportunity to formulate the DES such that the HBD or QAS is the reagent required to modify the biopolymer. This has been demonstrated with cellulose where choline was grafted onto the carbohydrate backbone through an etherification reaction. For this purpose, the cellulose did not need to be soluble as the DES penetrated well into the structure achieving a high degree of quaternisation [58]. A similar approach was used for the acylation of cellulose using a Lewis-acidic DES based on $ZnCl_2$ [59]. Modification of proteins has been demonstrated using ILs using the benzoylation of zein [50] and the sulfation of silk [60].

Protein extraction using DESs has been carried out to a much lower extent than using ILs. The first attempts were made using a betaine-urea eutectic to extract bovine serum albumin, trypsin, and ovalbumin [61]. In all cases, high extraction efficiencies were noted with little loss of protein activity. A recent paper showed that proteins could be extracted from the porcine pancreas using DESs based on molecules such as L-proline, lysine, sorbitol, and xylitol [62]. The use of liquids as reactive media has not been tested to date but may be an interesting avenue of research.

DESs have been applied to natural silk fibres and it was shown that silk nanofibrils could be extracted by using a DES of urea/guanidine hydrochloride which acted as a protein denaturant [63]. This is a good example of designing a DES which has a specific property. The recovered silk nanofibrils were used in a solid-state supercapacitor.

The literature has tended to focus on making pure solubilised proteins but some of the largest potentials may come in the modification of commodity biomaterials such as leather, wool, and silk. These materials are already mechanically and chemically processed but the advantages of using a DES strategy may originate from the capability to formulate the DES into a liquid-active reagent. We have previously shown that leather can be tanned, lubricated, and dyed using DESs [64]. The tanning of leather is either carried out using aqueous chromium sulfate solutions or an organic cross-linking agent. It was shown that eutectic mixtures of 1 ChCl: 2 $CrCl_3{\cdot}6H_2O$ and 2 urea: 1 $CrCl_3{\cdot}6H_2O$ could be used to tan leather. The advantage of a DES approach is that the liquid can be applied as a cream which is fully absorbed in the tanning process resulting in negligible waste. The DES acts as a very concentrated reagent with effectively no solvent (e.g. 1 ChCl: 2 $CrCl_3{\cdot}6H_2O$ has 4.75 mol dm^{-3} chromium salt). DESs can also show high solubility for polar compounds, such as vegetable-tanning agents speeding up the tanning process. It has been shown that DESs are rapidly absorbed into the leather and can act as lubricants known as fat-liquors. These are absorbed into the

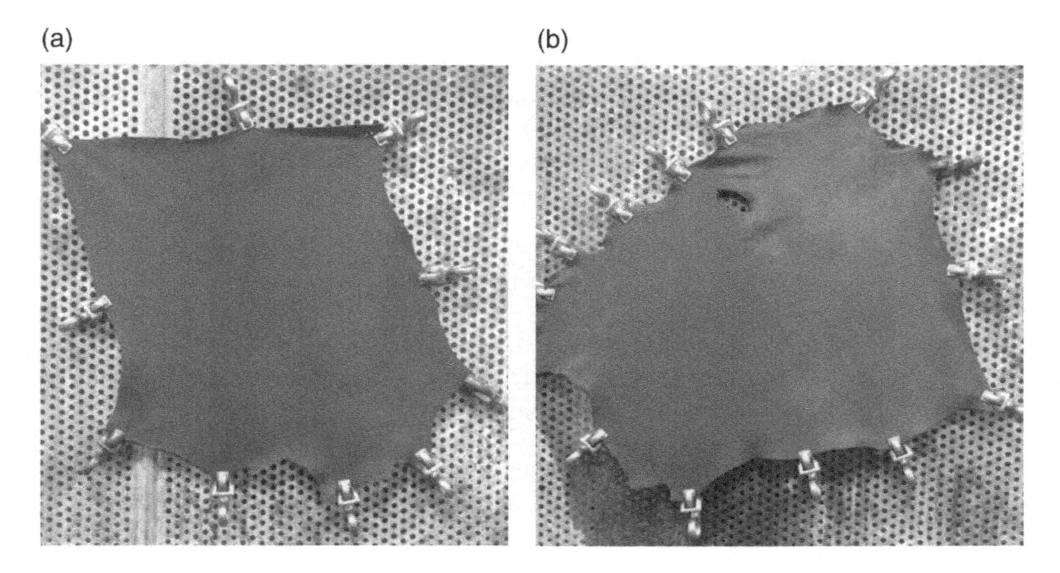

Figure 6.2 Dry leather samples aqueous post-tanned in (a) and DES post-tanned in (b).

large void volume and do not denature the collagen to any measurable extent. The other area of interest for natural polymer modification is in dyeing. Removing dye from aqueous solutions is a ubiquitous problem but many dyes are also QAS and/or HBD; so, they could be formulated into DESs. We have shown that this works in our laboratory but have never applied it to polymer dyeing. It was demonstrated that very hydrophobic dyes such as Sudan Black B could be used to dye leather and none of the dyes was leached when the sample was washed in water [64]. Figure 6.2 shows halves of sheep leather, which were retanned, dyed, and fat-liquored using both conventional aqueous solutions (a) and a single DES liquid (b) [65]. The samples had similar appearances and almost identical mechanical properties. Similar approaches could be made with other biopolymers such as keratin and silks, which could be used for a less aggressive hair dye and which might lead to fabrics that fade less easily. This is a large and untapped field of research.

6.2.5 Composite Formation

The synthesis of biopolymer composites is rapidly expanding due to the desire to reduce the dependency on fossil-fuel-based polymers and the need to create biocompatible materials. The main materials of interest include polylactic acid, chitin, chitosan, cellulose, starch, and PE glycol due to their availability and cost on a large scale. These materials find applications in areas such as packaging, bio-implants, horticulture, and constructional materials [66]. DESs can be used in composite formation due to their amphiphilic properties. The ability to bring together hydrophobic and hydrophilic materials is the complex characteristic achieved by natural biomaterials. Hydrophobic DESs can be made using fatty acids, fatty alcohols, or long-chain QAS. While not as numerous in the literature as hydrophilic DESs, hydrophobic DESs are starting to receive increased interest and recent reviews detail some of their applications [67, 68].

Biocomposites have been prepared using ILs (Park et al. 2018) [69]. Several studies have shown that carbohydrates and proteins can be blended using ILs. These include keratin and cellulose and cellulose acetate with wool. Keratins and silks have been extensively studied and are subject to comprehensive review [70]. The issues with processing these materials come in the high temperatures required to solubilise some of the polymers (particularly silk) and there is a tendency for some proteins to denature when they are precipitated, often using an anti-solvent [71, 72].

There is significant potential to use ILs and DESs as wetting agents [17]. The ability to tune the hydrophobicity and hydrophilicity of these liquids could enable them to act as agents to enable the homogeneous mixing of contrasting material types. This has been applied to paint additives where ILs have brought together hydrophilic pigments with hydrophobic binders and solvents [73, 74]. This simple concept could be brought into a variety of applications for mixing hydrophobic materials such as graphite, graphene, or polyolefins with biomaterials. It could be used to increase the conductivity of resins, e.g. for battery applications.

Two studies have shown that hydrophobic and hydrophilic polymers can be homogenised using simple DESs. The first mixed HDPE and starch [48]. Normally these two materials cannot be blended as the former is hydrophobic while the latter is hydrophilic but the addition of a small amount of DES to both polymers enables good miscibility. DESs have also been used to aid the blending of hydrophobic and hydrophilic polymers. One strategy to accelerate the mechanical degradation of polyolefin plastics in the environment is to blend them with carbohydrate-based polymers. Unfortunately, polyolefins are hydrophobic, whereas carbohydrates tend to be hydrophilic. So, the two do not blend without chemical modification of the carbohydrate. In this study [75], HDPE and TPS are used as the polymers with DESs as the modifiers. Both TPS and DESs are biodegradable and the DESs are water-miscible and biocompatible ensuring that the composite plastic contains a biodegradable flaw, which should enable mechanical and chemical degradation. It is shown that DESs enable the facile mixing of the two polymers. The composite has strength similar to TPS but a ductility greater than either of the two components. The glass transition temperature of the composite plastic shows that HDPE and TPS are homogeneously mixed and data suggests that the DESs act as lubricants rather than plasticisers. The same approach was applied to agar-HDPE-DES mixtures but here surfactants resulted in a more homogeneous blend [76]. Polysaccharides are usually chemically modified (e.g. carboxymethyl cellulose) so that they become miscible with hydrophobic polyolefins. The extrusion of polyolefins with biopolymers produces materials with faster environmental degradation. However, it still leads to micro-plastics that can bio-accumulate.

It is not only as a solid plastic that thermoplastic biopolymers can find use. Glues and adhesives are commonly derived from biopolymer, for example, starch is commonly used in cardboard and wallpaper paste, and proteins form the basis of many types of glues. DES-modified biopolymers could therefore be used as resins for composite materials. So far, this has only been demonstrated as a binder for wood particles. Wood composites are usually held together using non-biodegradable thermoset resins such as urea-formaldehyde (UF) and phenol-formaldehyde (PF). TPS has been used to bind wood fibres and wood particles, and $1\,m^{-2}$ boards have been constructed with similar mechanical properties to conventional UF boards [49]. The biopolymer has the advantage of being biodegradable but also recyclable. Figure 6.3 shows samples of boards made from TPS which have simply been reground and repressed. While these boards may have short-lifetime applications, their higher costs compared with UF make their use less technically viable. Other DESs based on group 1 inorganic salts have been developed [77] and these have been applied to composite board manufac-

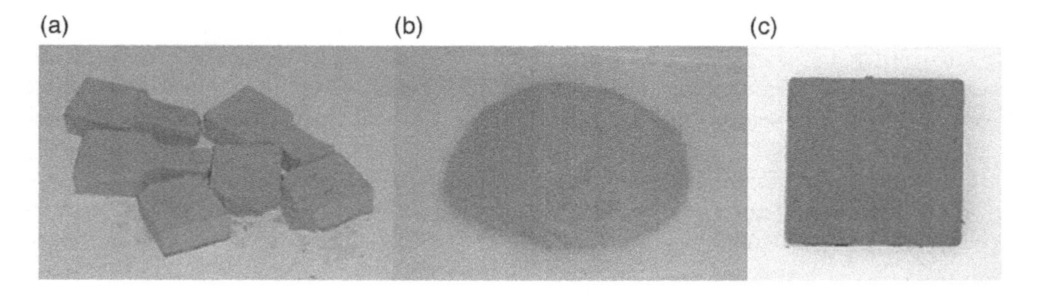

Figure 6.3 Starch-based thermoplastic wood after forming (a), after grinding (b), and then after reforming (c). This shows that thermoplastic wood can be effectively recycled.

Figure 6.4 From the top going clockwise: starch composites made with orange peel, banana peel, bagasse (sugar cane) and straw, and homogenised using a ChCl-glycerol DES.

ture [78]. These resins are comparable in cost to UF boards. Nature uses water, starch, and salts to adhere cells together in wood, which is similar to the approach used here except ChCl replaces inorganic salts and glycerol replaces water.

There are many forms of cellulose-based agricultural waste that go to animal feed (where appropriate), to anaerobic digestion, or to simple composting. Many of these wastes could be upcycled into constructional materials using starch-DES binders. Figure 6.4 shows examples of DES-modified starch composites with straw, orange peel, banana peel, and bagasse (sugar cane) but samples using hemp, eggshell, and paper have also been prepared. While these samples are water sensitive, they are no different to cardboard which also uses starch as a binder and there are numerous sealants, which have been used.

6.2.6 Gelation

The final area where biopolymers have an application with DESs is to produce high-performance materials for gel formation. In the area of morphological and mechanical modification of biopolymers described above, the DES was the minor component, but for gels, the DES is the major component. To this end, the biopolymers act as structural elements decreasing the flow of the ionic fluid. Clearly, the less biopolymer, the more mobile the DES, and the less rigid the material. This is really being bio-inspired and the same process is carried out in nature where biomaterials are modified with aqueous solutions of salts and polar molecules. This process can build materials as diverse as tissue and teeth, wood, and leaf stems. Gelation occurs in nature in cell membranes and structural proteins.

Numerous types of gelling agents have been used, primarily with ILs, and these are covered in detail in a recent review [79]. These gelling agents are either relatively low-molecular-mass components, which interact strongly with the ionic phase pulling the combined phase together, or they are macromolecular species, which physically restrict the movement of the liquid phase.

ILs and DESs are of interest in electrochemical applications where they can act as electrolytes. They are of particular interest due to their wide potential window and their wide temperature range. The negligible vapour pressure of ILs has led to their use in dye-sensitised solar cells [80]. Gelatin has been used to gel imidazolium ILs to produce solid electrolytes with a wide temperature window [81]. These were originally tested with electrochromic devices containing metal ion salts with the aim of making smart windows, but they could also be used for membranes and separation devices.

Gels have been formed from natural biomacromolecules. Kadokawa et al. gelled chitin [82] and chitin/cellulose mixtures [83] with a variety of imidazolium halides. The gels were relatively slow to form due to the limited solubility of the polymers in the IL. Other studies have investigated chitin and agarose mixtures [84]. Gelatin is the most studied macromolecular gelling agent due to its high solubility in most ionic fluids [81, 85, 86]. From a practical perspective, gelatin is one of the easiest gelling agents to produce from a range of collagen-containing materials and its properties have been well studied with a range of fluids.

Gelation using ILs and DESs can also be used in formulation science. Many active ingredients can be formulated into a liquid by introducing an ionic or hydrogen-bonding functional group and these can then be combined with an inert or active ionic component. The area that has received the most attention in this respect is the delivery of active pharmaceutical ingredients (APIs). Several studies have shown that numerous pharmaceuticals and other bioactive groups can be formulated into ILs with a variety of counter-ions. One example of this approach was demonstrated using propantheline acesulfamate. The cation (propantheline) is an anti-muscarinic API used to treat excessive sweating, cramps, and spasms of the stomach, whereas the anion (acesulfamate) is an artificial sweetener [87]. The same approach can be used for a variety of active materials and the properties of the resultant liquid can be tuned depending on the desired hydrophobicity/hydrophilicity, acidity/basicity, and/or viscosity. Some solid drugs have issues such as low bioavailability, polymorphic conversion, low solubility, and a tendency of amorphous forms to spontaneously crystallise which can be overcome using liquid APIs [88]. Gelled ILs and DESs have also been considered as a way of delivering pharmaceutical ingredients into the human body. More than 50% of medicines on the market are now sold as organic salts [89]. As a crucial step in drug development, salt formation using a drug substance can have a huge impact on the final product's solubility, dissolution rate, hygroscopicity, stability, impurity profile, and particle characteristics.

The potential of pharmaceutical-based eutectic mixtures has been explored to increase the pharmacological growth of the desired API. The high thermodynamic functions of eutectics, such as free energy, enthalpy and entropy, can enhance solubility and dissolution of poor solubility drugs, similar to the more popular amorphous solids and solid dispersions [90]. Solid dispersions sometimes also exhibit a eutectic-like behaviour with improved solubility, e.g. fenofibrate–PE glycol [91]. On the other hand, the low melting point of organic eutectics can also pose stability issues. As have often been confused with solid dispersions, eutectics share the problems associated with solid dispersions as well. The absence of a clear distinction between phases of a given drug can, in principle, be separately patented. Though it is accepted by regulatory bodies in the US that different solid forms of a drug are effectively different drugs (i.e. have different pharmacokinetics and pharmacodynamics), European regulators still view them to be the same.

Recent work by Abbott et al. showed that highly concentrated liquid formulations could be produced from APIs in a DES formulation [92]. Many APIs are either HBD or QAS, or drugs with amine functionalities that can be converted to the QAS with HCl aiding its solubility. It was shown that compounds with poor solubility in water can be made miscible with water through a DES formulation. The study also showed that drug molecules that exhibit polymorphism such as adiphenine and ranitidine can be made into liquids and the HBD can also be an API such as aspirin.

A study showed that pharmaceutical DESs could be used to modify gelatin and these materials were tested as transdermal drug delivery systems. It was found that having the APIs in a DES formulation enabled four times faster delivery of the API across the skin barrier than using a solid API [47].

6.3 Conclusion

This review has shown that ionic fluids are useful modifiers for a range of biomaterials. Many of the principles of plasticisers, chemical modification, composite formation, and gelation have been established using ILs, and some have been demonstrated using DESs. The ability to produce biocompatible DESs suggests that they may have advantages over conventional ILs from the ease of preparation, toxicity, and cost perspective. This will become particularly relevant where these materials are used for applications such as the replacement for oil-based plastics, drug delivery, and natural material modification. The potential to create biomimetic systems using inorganic salts and complex organic HBDs is only in its infancy but is surely a major direction in which this topic can progress.

DESs encompass a wide range of synthetic and natural HBDs and QASs which are used as active ingredients such as APIs, dyes, reagents, anti-oxidants, etc. The future of this research area depends on making use of the DES as a liquid-active ingredient. This will aid formulation and negate the use of solvents for material modification. The other important aspect of using DESs is the ability to modify the hydrophobicity/hydrophilicity of the liquids by changing the properties of the components. This can have a significant effect on phase behaviour and solubility. This chapter shows that the directions of biomaterial valorisation have been signposted by work carried out using ILs and practical materials can now be exploited using DESs.

References

1. Mohanty, A.K., Misra, M., and Drzal, L.T. (2005). *Natural Fibers, Biopolymers, and Biocomposites*. Boca Raton, FL: CRC Press.
2. Sionkowska, A. (2011). Current research on the blends of natural and synthetic polymers as new biomaterials. *Progress in Polymer Science* 36: 1254–1276.
3. Stupp, S.I. and Braun, P.V. (1997). Molecular manipulation of microstructures: biomaterials, ceramics, and semiconductors. *Science* 277: 1242–1248.
4. Babu, R.P., O'Connor, K., and Seeram, R. (2013). Current progress on bio-based polymers and their future trends. *Progress in Biomaterials* 2: 8.
5. Degli Innocenti, F. (2003). *Biodegradability & Compostability. Biodegradable Polymers and Plastics*, 33–45. Boston, MA: Springer.
6. Qi, X., Ren, Y., and Wang, X. (2017). New advances in the biodegradation of poly(lactic) acid. *International Biodeterioration and Biodegradation* 117: 215–223.

7. Oledzka, E. and Sobczak, M. (2012). Polymers in the pharmaceutical applications-natural and bioactive initiators and catalysts in the synthesis of biodegradable and bioresorbable polyesters and polycarbonates. In: *Innovations in Biotechnology*, 139–160. Rijeka, Croatia: Intech Publishing.
8. Hu, X., Cebe, P., Weiss, A.S. et al. (2012). Protein-based composite materials. *Materials Today* 15: 208–215.
9. Gomes, S., Leonor, I.B., Mano, J.F. et al. (2012). Natural and genetically engineered proteins for tissue engineering. *Progress in Polymer Science* 37: 1–17.
10. Jonker, A.M., Löwik, D.W., and van Hest, J.C. (2012). Peptide-and protein-based hydrogels. *Chemistry of Materials* 24: 759–773.
11. Silva, N.H.C.S., Vilela, C., Marrucho, I.M. et al. (2014). Protein-based materials: from sources to innovative sustainable materials for biomedical applications. *Journal of Materials Chemistry B* 2: 3715–3740.
12. Janssen, L.P. and Moscicki, L. (2009). *Thermoplastic Starch: A Green Material for Various Industries*. Weinheim, Germany: Wiley-VCH.
13. Lourdin, D., Coignard, L., Bizot, H., and Colonna, P. (1997). Influence of equilibrium relative humidity and plasticizer concentration on the water content and glass transition of starch materials. *Polymer* 38: 5401–5406.
14. Shanks, R. and Kong, I. (2012). *Thermoplastic Elastomers* (ed. A.Z. El-Sonbati), 95–116. Rijeka, Croatia: Intech Publishing.
15. Zhang, Y., Rempel, C., and Liu, Q. (2014). Thermoplastic starch processing and characteristics – a review. *Critical Reviews in Food Science and Nutrition* 54: 1353–1370.
16. Hallett, J.P. and Welton, T. (2011). Room-temperature ionic liquids: solvents for synthesis and catalysis 2. *Chemical Reviews* 111: 3508–3576.
17. Plechkova, N.V. and Seddon, K.R. (2008). Applications of ionic liquids in the chemical industry. *Chemical Society Reviews* 37: 123–150.
18. Zhao, Y., Zhao, J., Huang, Y. et al. (2014). Toxicity of ionic liquids: database and prediction via quantitative structure–activity relationship method. *Journal of Hazardous Materials* 278: 320–329.
19. Abbott, A.P., Boothby, D., Capper, G. et al. (2004). Deep eutectic solvents formed between choline chloride and carboxylic acids. *Journal of the American Chemical Society* 126: 9142–9147.
20. Abbott, A.P., Capper, G., Davies, D.L. et al. (2003). Novel solvent properties of choline chloride/urea mixtures. *Chemical Communications* 70–71.
21. Smith, E.L., Abbott, A.P., and Ryder, K.S. (2014). Deep eutectic solvents (DESs) and their applications. *Chemical Reviews* 114: 11060–11082.
22. Choi, Y.H., van Spronsen, J., Dai, Y. et al. (2011). Are natural deep eutectic solvents the missing link in understanding cellular metabolism and physiology? *Plant Physiology* 156: 1701–1705.
23. Wang, H., Gurau, G., and Rogers, R.D. (2012). Ionic liquid processing of cellulose. *Chemical Society Reviews* 41: 1519–1537.
24. Häkkinen, R. and Abbott, A.P. (2019). Solvation of carbohydrates in five choline chloride-based deep eutectic solvents and the implication for cellulose solubility. *Green Chemistry* 21: 4673–4682.
25. Abbott, A.P., Al-Murshedi, A.Y.M., Alshammari, O.A.O. et al. (2017). Thermodynamics of phase transfer with deep eutectic solvents. *Fluid Phase Equilibria* 448: 99–104.
26. Loow, Y.L., New, E.K., Yang, G.H. et al. (2017). Potential use of deep eutectic solvents to facilitate lignocellulosic biomass utilization and conversion. *Cellulose* 24: 3591–3618.
27. Tang, X., Zuo, M., Li, Z. et al. (2017). Green processing of lignocellulosic biomass and its derivatives in deep eutectic solvents. *ChemSusChem* 10: 2696–2706.
28. van Osch, D.J.G.P., Kollau, L.J.B.M., van den Bruinhorst, A. et al. (2017). Ionic liquids and deep eutectic solvents for lignocellulosic biomass fractionation. *Physical Chemistry Chemical Physics* 19: 2636–2665.
29. Abdulmalek, E., Zulkefli, S., and Rahman, M.B.A. (2017). Deep eutectic solvent as a media in swelling and dissolution of oil palm trunk. *Malaysian Journal of Analytical Sciences* 21: 20–26.
30. Chen, Z. and Wan, C. (2018). Ultrafast fractionation of lignocellulosic biomass by microwave-assisted deep eutectic solvent pretreatment. *Bioresource Technology* 250: 532–537.
31. Jablonský, M., Škulcová, A., Kamenská, L. et al. (2015). Deep eutectic solvents: fractionation of wheat straw. *Bioresources* 10: 8039–8047.
32. Liu, Y., Chen, W., Xia, Q. et al. (2017). Efficient cleavage of lignin-carbohydrate complexes and ultrafast extraction of lignin oligomers from wood biomass by microwave-assisted treatment with deep eutectic solvent. *ChemSusChem* 10: 1692–1700.

33. Chen, J., Xie, F., Li, X., and Chen, L. (2018). Ionic liquids present huge potential in the fabrication of biopolymer-based pharmaceutical materials for accurately controlled drug/gene delivery. *Green Chemistry* 20: 4169–4200.
34. Silva, S.S., Duarte, A.R.C., Mano, J.F., and Reis, R.L. (2013). Design and functionalization of chitin-based microsphere scaffolds. *Green Chemistry* 15: 3252–3258.
35. Hou, L., Udangawa, W.M.R.N., Pochiraju, A. et al. (2016). Synthesis of heparin-immobilized, magnetically addressable cellulose nanofibers for biomedical applications. *ACS Biomaterials Science & Engineering* 2: 1905–1913.
36. Lopes, J.M., Mustapa, A.N., Pantić, M. et al. (2017). Preparation of cellulose aerogels from ionic liquid solutions for supercritical impregnation of phytol. *The Journal of Supercritical Fluids* 130: 17–22.
37. Sahiner, N. and Sagbas, S. (2018). Polymeric ionic liquid materials derived from natural source for adsorption purpose. *Separation and Purification Technology* 196: 208–216.
38. Sharma, M., Mukesh, C., Mondala, D., and Prasad, K. (2013). Dissolution of α-chitin in deep eutectic solvents. *RSC Advances* 3: 18149–18155.
39. Sivagnanam Saravana, P., Ho, T.C., Chae, S.-J. et al. (2018). Deep eutectic solvent-based extraction and fabrication of chitin films from crustacean waste. *Carbohydrate Polymers* 195: 622–630.
40. Jiang, Z., Yuan, J., Wang, P. et al. (2018). Dissolution and regeneration of wool keratin in the deep eutectic solvent of choline chloride-urea. *International Journal of Biological Macromolecules* 119: 423–430.
41. Nuutinen, E.-M., Willberg-Keyriläinen, P., Virtanen, T. et al. (2019). Green process to regenerate keratin from feathers with an aqueous deep eutectic solvent. *RSC Advances* 9: 19720–19728.
42. Wang, D., Yang, X.-H., Tang, R.-C., and Yao, F. (2018). Extraction of keratin from rabbit hair by a deep eutectic solvent and its characterization. *Polymers* 10: 993.
43. Abbott, A.P., Alabdullah, S.S.M., Al-Murshedi, A.Y.M., and Ryder, K.S. (2018). Brønsted acidity in deep eutectic solvents and ionic liquids. *Faraday Discussions* 206: 365–377.
44. van Osch, D.J.G.P., Zubeir, L.F., van den Bruinhorst, A. et al. (2015). Hydrophobic deep eutectic solvents as water-immiscible extractants. *Green Chemistry* 17: 4518–4521.
45. Abbott, A.P., Ballantyne, A.D., Palenzuela Conde, J. et al. (2012). Salt modified starch: sustainable, recyclable plastics. *Green Chemistry* 14: 1302–1307.
46. Leroy, E., Decaen, P., Jacquet, P. et al. (2012). Deep eutectic solvents as functional additives for starch based plastics. *Green Chemistry* 14: 3063–3069.
47. Qu, W., Häkkinen, R., Allen, J. et al. (2019). Globular and fibrous proteins modified with deep eutectic solvents: materials for drug delivery. *Molecules* 24: 3583.
48. Abbott, A.P., Harris, R.C., and Ryder, K.S. (2011). Glycerol eutectics as sustainable solvent systems. *Green Chemistry* 13: 82–90.
49. Abbott, A.P., Palenzuela Conde, J., Davis, S., and Wise, W.R. (2012). Starch as a replacement for urea-formaldehyde in medium density fibreboard. *Green Chemistry* 14: 3067–3070.
50. Biswas, A., Shogren, R.L., Stevenson, D.G. et al. (2006). Ionic liquids as solvents for biopolymers: acylation of starch and zein protein. *Carbohydrate Polymers* 66: 546–550.
51. Qin, H., Owyeung, R.E., Sonkusale, S.R., and Panzer, M.J. (2019). Highly stretchable and nonvolatile gelatin-supported deep eutectic solvent gel electrolyte-based ionic skins for strain and pressure sensing. *Journal of Materials Chemistry C* 7: 601–608.
52. Esquembre, R., Sanz, J.M., Wall, J.G. et al. (2013). Thermal unfolding and refolding of lysozyme in deep eutectic solvents and their aqueous dilutions. *Physical Chemistry Chemical Physics* 15: 11248–11256.
53. Sanchez-Fernandez, A., Edler, K.J., and Arnold, T. (2017). Protein conformation in pure and hydrated deep eutectic solvents. *Physical Chemistry Chemical Physics* 19: 8667–8670.
54. Qu, W. (2019). DES modified protein-based materials and their use in pharmaceutical applications. PhD thesis. University of Leicester.
55. Mehta, M.J. and Kumar, A. (2019). Ionic liquid assisted gelatin films: green, UV shielding, antioxidant, and antibacterial food packaging materials. *ACS Susustainable Chemistry & Engineering* 7: 8631–8636.
56. Abolibda, T. (2015). Starch based bio-plastics. PhD thesis. University of Leicester.
57. Wasserscheid, P. and Welton, T. (2006). *Ionic Liquid in Synthesis*. Weinheim, Germany: Wiley-VCH.
58. Abbott, A.P., Bell, T.J., Handa, S., and Stoddart, B. (2006). Cationic functionalisation of cellulose using ionic liquids. *Green Chemistry* 8: 784–786.

59. Abbott, A.P., Bell, T.J., Handa, S., and Stoddart, B. (2005). O-acetylation of cellulose and monosaccharides using zinc based ionic liquids. *Green Chemistry* 7: 705–707.
60. Liu, X., Xu, W., Zhang, C. et al. (2015). Homogeneous sulfation of silk fibroin in an ionic liquid. *Materials Letters* 143: 302–304.
61. Li, N., Wang, Y., Xu, K. et al. (2016). Development of green betaine-based deep eutectic solvent aqueous two-phase system for the extraction of protein. *Talanta* 52: 23–32.
62. Meng, J., Wang, Y., Zhou, Y. et al. (2019). Development of different deep eutectic solvent aqueous biphasic systems for the separation of proteins. *RSC Advances* 9: 14116–14125.
63. Tan, X., Zhao, W., and Mu, T. (2018). Controllable exfoliation of natural silk fibers into nanofibrils by protein denaturant deep eutectic solvent: nanofibrous strategy for multifunctional membranes. *Green Chemistry* 20: 3625–3633.
64. Abbott, A.P., Alaysuy, O., Antunes, A.P.M. et al. (2015). Processing of leather using deep eutectic solvents. *ACS Susustainable Chemistry & Engineering* 3: 1241–1247.
65. Alaysuy, O. (2019). Processing leather using deep eutectic solvents. PhD thesis. University of Leicester.
66. Chen, G. and Patel, M. (2012). Plastics derived from biological sources: present and future: a technical and environmental review. *Chemical Reviews* 112: 2082–2099.
67. Lee, J., Jung, D., and Park, K. (2019). Hydrophobic deep eutectic solvents for the extraction of organic and inorganic analytes from aqueous environments. *Trends in Analytical Chemistry* 118: 853–868.
68. Leroy, E., Jacquet, P., Coativy, G. et al. (2012). Compatibilization of starch–zein melt processed blends by an ionic liquid used as plasticizer. *Carbohydrate Polymers* 89: 955–963.
69. Park, S., Oh, K.K., and Lee, S.H. (2018). Biopolymer-based composite materials prepared using ionic liquids. In: *Application of Ionic Liquids in Biotechnology. Advances in Biochemical Engineering/Biotechnology*, vol. 168 (ed. T. Itoh and Y.M. Koo). Cham, Switzerland: Springer.
70. Díez-Pascual, A.M. (2019). Synthesis and applications of biopolymer composites. *International Journal of Molecular Sciences* 20: 2321.
71. Goujon, N., Wang, X., Rajkowa, R., and Byrne, N. (2012). Regenerated silk fibroin using protic ionic liquids solvents: towards an all-ionic-liquid process for producing silk with tunable properties. *Chemical Communications* 48: 1278–1280.
72. Zhang, Z., Nie, Y., Zhang, Q. et al. (2017). Quantitative change in disulfide bonds and microstructure variation of regenerated wool keratin from various ionic liquids. *ACS Sustainable Chemistry & Engineering* 5: 2614–2622.
73. Lehmann, K., Silber, S., and Weyershausen, B. (2016). Process for the preparation of homogeneous storage-stable pastes, paints, laquers by using ionic liquids as dispersing aids. EP 1566413 B1.
74. Weyershausen, B. and Lehmann, K. (2005). Industrial application of ionic liquids as performance additives. *Green Chemistry* 7: 15–19.
75. Abbott, A.P., Abolibda, T.Z., Qu, W. et al. (2017). Thermoplastic starch–polyethylene blends homogenised using deep eutectic solvents. *RSC Advances* 7: 7268–7273.
76. Shamsuri, A.A., Daik, R., Zainudin, E.S., and Tahir, P.M. (2014). Compatibilization of HDPE/agar biocomposites with eutectic ionic liquid containing surfactant. *Journal of Reinforced Plastics and Composites* 33: 440–453.
77. Abbott, A.P., D'Agostino, C., Davis, S.J. et al. (2016). Do group 1 metal salts form deep eutectic solvents? *Physical Chemistry Chemical Physics* 18: 25528–25537.
78. Davis, S.J. (2016). Deep eutectic solvents derived from inorganic salts. PhD thesis. University of Leicester.
79. Marr, P.C. and Marr, A.C. (2016). Ionic liquid gel materials: applications in green and sustainable chemistry. *Green Chemistry* 18: 105–128.
80. Kubo, W., Kambe, S., Nakade, S. et al. (2003). Photocurrent-determining processes in quasi-solid-state dye-sensitized solar cells using ionic gel electrolytes. *Journal of Physical Chemistry B* 107: 4374–4381.
81. Leonesa, R., Sentanin, F., Rodrigues, L.C. et al. (2012). Novel polymer electrolytes based on gelatin and ionic liquids. *Optical Materials* 35: 187–195.
82. Prasad, K., Murakami, M., Kaneko, Y. et al. (2009). Weak gel of chitin with ionic liquid, 1-allyl-3-methylimidazolium bromide. *International Journal of Biological Macromolecules* 45: 221–225.
83. Takegawa, A., Murakami, M., Kaneko, Y., and Kadokawa, J. (2010). Preparation of chitin/cellulose composite gels and films with ionic liquids. *Carbohydrate Polymers* 79: 85–90.

84. Singh, T., Trivedi, T.J., and Kumar, A. (2010). Dissolution, regeneration and ion-gel formation of agarose in room-temperature ionic liquids. *Green Chemistry* 12: 1029–1035.
85. Lourenço, N.M.T., Nunes, A.V.M., Duarte, C.M.M., and Vidinha, P. (2011). *Ionic Liquids Gelation with Polymeric Materials: The Ion Jelly Approach in Applications of Ionic Liquids in Science and Technology* (ed. S. Handy). Rijeka, Croatia: InTech.
86. Vidinha, P., Lourenco, N.M.T., Pinheiro, C. et al. (2008). Ion jelly: a tailor-made conducting material for smart electrochemical devices. *Chemical Communications* 44: 5842–5844.
87. Stoimenovski, J., MacFarlane, D.R., Bica, K., and Rogers, R.D. (2010). Crystalline vs. ionic liquid salt forms of active pharmaceutical ingredients: a position paper. *Pharmaceutical Research* 27: 521–526.
88. Paulekuhn, G.S., Dressman, J.B., and Saal, C. (2007). Trends in active pharmaceutical ingredient salt selection based on analysis of the Orange Book database. *Journal of Medicinal Chemistry* 50: 6665–6672.
89. Hillery, A.M. and Park, K. (2016). *Drug Delivery: Fundamentals and Applications*. Boca Raton, FL: CRC Press.
90. Cherukuvada, S. and Nangia, A. (2014). Eutectics as improved pharmaceutical materials: design, properties and characterization. *Chemical Communications* 50: 906–923.
91. Ghosh, M.K., Wahed, M.I.I., Ali, M.A., and Barman, R.K. (2019). Formulation and characterization of fenofibrate loaded solid dispersion with enhanced dissolution profile. *Pharmacology & Pharmacy* 10: 343–355.
92. Abbott, A.P., Ahmed, E.I., Prasad, K. et al. (2017). Liquid pharmaceuticals formulation by eutectic formation. *Fluid Phase Equilibria* 448: 2–8.

7

Biopolymer Composites for Recovery of Precious and Rare Earth Metals

Jesper T.N. Knijnenburg[1] **and Kaewta Jetsrisuparb**[2]

[1]*International College, Khon Kaen University, Khon Kaen, Thailand*
[2]*Department of Chemical Engineering, Khon Kaen University, Khon Kaen, Thailand*

7.1 Introduction

Nowadays, a significantly higher number of elements is being incorporated into our daily lives. In fact, practically every stable element in the periodic table currently has an application [1]. Unfortunately, a large number of these elements are now considered "critical" due to their risk of supply shortage and economic importance [2]. Among those critical elements are precious metals and rare earth metals. Precious metals include silver (Ag), gold (Au), and the platinum group metals (ruthenium [Ru], rhodium [Rh], palladium [Pd], osmium [Os], iridium [Ir], and platinum [Pt]), which find applications in, e.g. electronics, jewelry, catalysis, and medicine [3]. Their natural abundance in the Earth's crust is among the lowest of elements, ranging from 0.001 to 0.1 ppm (Figure 7.1). The rare earth elements (REEs) consist of a group of 15 elements called "lanthanides" ranging from lanthanum (La, atomic number 57) to lutetium (Lu, atomic number 71). Yttrium (Y, atomic number 39), and often also scandium (Sc, atomic number 21) are included in the list of REEs due to their similar chemical properties and occurrence in the same ores. These REEs have a wide range of applications including permanent magnets, phosphors, battery alloys, catalysts, glass additives, and ceramics [5, 6]. Most of the REEs, however, are not as rare as the name suggests; their natural abundance (except promethium) is higher than that of Ag (Figure 7.1).

High-Performance Materials from Bio-based Feedstocks, First Edition. Edited by Andrew J. Hunt, Nontipa Supanchaiyamat, Kaewta Jetsrisuparb and Jesper T.N. Knijnenburg.

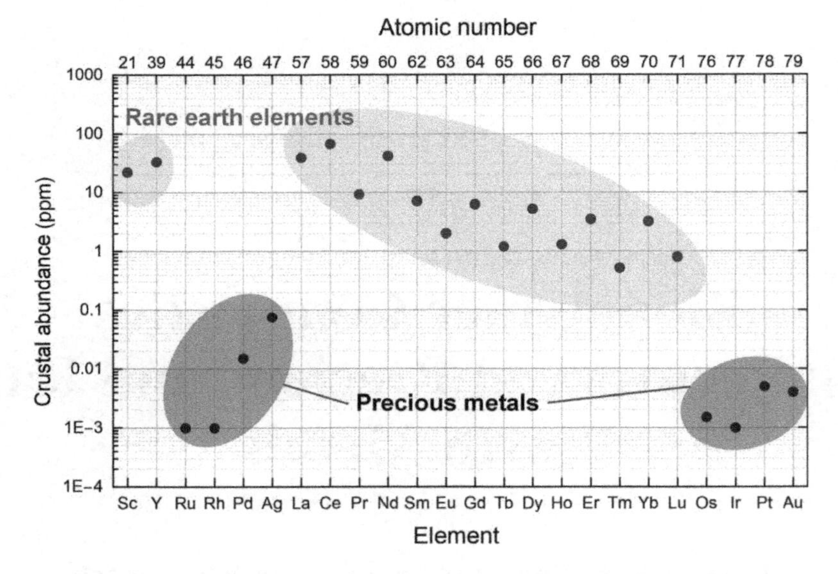

Figure 7.1 *Natural abundance of precious metals and rare earth elements (REEs) in the Earth's crust in order of increasing atomic number. Source: Data taken from Haynes et al. [4].*

The applications of precious metals and REEs in consumer products are increasing and become widely diverse. More than half of the precious elements and REEs can be found in typical consumer electronics like hard drives or mobile phones, albeit at low concentrations (<1% per element) [7]. It is often not technologically and economically attractive to extract the low concentrations of these metals from end-of-life products, meaning that only very little (or none at all) is being recycled [1, 8].

The natural reserves of REEs are estimated to be exhausted in 100–500 years, whereas the precious metal reserves, with current rates of extractions, may already be exhausted within the next 5–50 years [9]. To prevent their depletion, the substitution of precious metals and REEs with less critical elements remains challenging, since these elements have very few (if any) sustainable substitutes with similar performance [8]. Combined with the low recycling rates, this means that these elements will end up being dispersed in electronic wastes and contaminated water sources at relatively low concentrations. There is thus a strong need for selective elemental recovery from such waste sources.

Among techniques available for the recovery of dissolved REEs and precious metals (such as chemical precipitation, solvent extraction, or electroextraction), adsorption is one of the most effective methods for dissolved metals at low concentrations due to its versatility and simplicity [10, 11]. A wide range of bio-based sorbents have been developed for the extraction of REEs [12] and precious metals [13–15]. To increase their stability, adsorption performance, and/or ease of separation, the bio-based materials are often physically and/or chemically modified or combined with one or more phases to form composite sorbents.

This chapter presents the use of biopolymer composite sorbents for the recovery of precious and rare earth metals. The composites of primarily four biopolymers (cellulose, chitosan, alginate, and lignin) and their adsorption performance are presented. We present an overview of the developments in this field since 2009. Several review articles have been published on sorbents based on cellulose [16, 17], chitosan [18, 19], alginate [20, 21], and lignin [22, 23] for adsorption of heavy metals, dyes and/or other water pollutants, which are not in the scope of this chapter. Apart from the four materials discussed in this chapter, other promising bio-based sorbents include tannins [24], pectin-based hydrogels [25, 26], gums [27, 28], humic acids [29], and clays [30], to name a few.

7.2 Mechanisms of Metal Adsorption

7.2.1 Silver

Silver is usually present in aqueous solutions in cationic form [13]. Adsorption of Ag(I) generally starts with the electrostatic attraction of the positively charged Ag(I) toward negatively charged functional groups on the sorbent surface [31], where the Ag(I) can bind to the sorbent via complexation, coordination, or chelation with electron-rich oxygen-, nitrogen- and/or sulfur-containing functional groups [31–35]. In acidic conditions, the adsorption is typically higher at higher (i.e. near-neutral) pH [31, 35–37] due to increased surface protonation and competition with protons at lower pH. For example, Ag(I) adsorption on nanostructured cellulose and chitin was highest at near-neutral pH, since at lower pH, there is competition with H^+ adsorption on the surface functional groups ($—SO^{3-}$, $—COO^-$, —NH—, and $—NH_2$) [31]. Similar results were observed for Ag(I) adsorption onto chitosan [36] or chitosan/bamboo charcoal beads [37]. Sericin and alginate-based sorbents had a poor affinity for Ag(I) at pH 2.5–3.0 due to electrostatic repulsion by protonated surface groups [38]. Apart from electrostatic interaction, due to its large standard electrode potential, Ag(I) can be easily reduced to metallic Ag^0 in the presence of reducing agents such as peptides [39] and nitrogen-containing groups [40], or by a combination of chelation (by oxygen/sulfur groups) and reduction (by nitrogen-containing groups) [41–43] resulting in the formation of Ag^0 nanoparticles.

7.2.2 Gold and Platinum Group Metals

After leaching from secondary wastes using acids such as aqua regia or concentrated HCl, precious metals like Au, Pt, Pd, and Rh are usually present as anionic chloro-metal complexes (e.g. $[AuCl_4]^-$, $[PtCl_4]^{2-}$, $[PtCl_6]^{2-}$, $[PdCl_4]^{2-}$, $[PdCl_6]^{2-}$, or $[RhCl_6]^{3-}$). Other common ligands instead of Cl^- are other halides, cyanide (CN^-) or thiosulfate ($S_2O_3^{2-}$) [14, 15]. Adsorption generally starts with the electrostatic attraction of these negatively charged complexes toward positively charged surface groups such as protonated amine groups [44–46] or sulfur-containing groups [47]. Adsorption then takes place through a combination of coordination, electrostatic attraction, and/or ion exchange [44, 46]. Electrostatic attraction between negatively charged metal ion complexes and positively charged amine groups has been observed for, e.g. polyethyleneimine (PEI) [48, 49] and chitosan [50], whereas ion exchange between the chloro-metal complexes with anions like Cl^- was proposed for the adsorption of Au(I) [51] and Rh(III) [52].

The anionic chloro-metal complexes are typically best adsorbed at pH 1–4 [13] at which amine sites are positively charged [15, 44]. At higher pH, the capacity is reduced due to electrostatic repulsion [45], whereas at lower pH, adsorption decreases due to competition with Cl^- or other anions [46, 53]. Adsorption of Pd(II) and Pt(IV) onto ethylenediamine-modified magnetic cross-linked chitosan nanoparticles was optimal at pH 2 [46] and pH 2.5 was optimal for Pd(II) adsorption onto chitosan-coated wool fabrics [54]. Maximum uptake of Au(III), Pd(II), and Pt(IV) was obtained at pH 2 for glycine-modified cross-linked chitosan resin [45]. Apart from adsorption, there is evidence for the reductive recovery of precious metals. For example, the hydroxyl groups of persimmon tannin could reduce palladium to Pd^0 [55] and reduction of Au(III) toward Au^0 was observed for sericin-alginate cross-linked by proanthocyanidins (SAPAs) promoted by oxidation of hydroxyl groups [38].

Recovery of Rh(III) from acidic solutions is more challenging than Au(III), Pt(IV), and Pd(II); at high Cl^- concentrations, anionic complexes such as $(RhCl_6)^{3-}$ and $(RhCl_5)^{2-}$ are dominant, whereas in systems with low Cl^- levels, positively charged complexes such as Rh^{3+}, $RhCl^{2+}$, and $RhCl_2^+$ are formed [56, 57]. The octahedral $(RhCl_6)^{3-}$ complex has a weaker binding than comparable complexes of Pt, Pd, or Au [58] which may be due to steric effects [56], making it more difficult to recover by adsorption. Instead, adsorption studies have been carried out to adsorb the positively charged Rh(III) complexes at near-neutral conditions. Labib et al. [52] found a maximum Rh(III) adsorption capacity at pH 6.9 onto amine- and amine-/thio-modified cellulose/silica hybrids. The decreased adsorption at lower pH was ascribed to competitive adsorption of $(RhCl_6)^{3-}$ with Cl^-. Similarly, the optimal Rh(III) adsorption was at pH 7 onto cellulose/graphite oxide [59] and at pH 6.5 for imidodiacetate-modified cellulose [60].

7.2.3 Rare Earth Metals

The REEs are predominantly in their trivalent (3+) state in aqueous solutions, whereas some ions may exist as divalent (Sm^{2+}, Eu^{2+}, or Yb^{2+}) or tetravalent (Ce^{4+}) ions [61]. Adsorption typically takes place through mechanisms such as cation exchange, electrostatic interaction, coordination, complexation, or chelation with functional groups such as amine [62, 63], carboxylic acid [64, 65], thiol [66], and sulfonate [67]. Adsorption is generally optimal at pH 5–7 [12] at which surface groups are negatively or neutrally charged with minimal electrostatic repulsion. In more acidic media, the electrostatic interactions are not optimal, while at higher pH, precipitation occurs. Adsorption of Gd(III) onto magnetically retrievable ion-imprinted chitosan/carbon nanotube (CNT) nanocomposites was best at pH 7, which took place via coordination of Gd(III) with amino groups of chitosan [62]. Similarly, Nd(III), Dy(III), and Eu(III) onto diethylenetriamine (DETA)-modified magnetic chitosan nanoparticles were best adsorbed at pH 7 by coordination with $-NH_2$ groups in DETA [63].

7.3 Composite Materials and Their Adsorption

Bio-based resources have been widely studied for the adsorption of numerous metals, dyes, and other compounds. Whereas the pure biomaterials intrinsically work, they frequently have some drawbacks, thus the adsorption performance is often enhanced by a chemical or physical modification to improve, for instance, the sorption capacity; chemical, thermal, or mechanical stability; and/or ease of separation. The physical modification of bio-based sorbents can be roughly divided into the incorporation of (combinations of) three material types, namely polymers (such as PEI or polyaniline, PANI), magnetic materials (often Fe_3O_4), and/or other compounds (such as clays or nanocarbons). In this section, the adsorption performance of selected composites prepared from four different biopolymers, namely cellulose (Section 7.3.1), chitosan (Section 7.3.2), alginate (Section 7.3.3), and lignin (Section 7.3.4) is discussed.

7.3.1 Cellulose-based Composite Adsorbents

7.3.1.1 Structure and Chemistry of Cellulose

Cellulose is a polysaccharide consisting of linear chains of D-glucose units connected by β-(1–4)-glycosidic bonds. As a building block of cell walls of green plants, it is the most abundant naturally occurring organic polymer. Pure, unmodified cellulose makes a poor sorbent due to its low affinity for metals [16]. However, the numerous hydroxyl groups on the cellulose backbone can be easily modified to improve sorption properties. Specific functional groups (such as carboxyl, ether, ester, sulfur, or amine) can be attached to enhance adsorption capacity [16, 68] or modify specific physicochemical properties (such as hydrophilicity, thermal stability, or water adsorption). Physical modification by mixing with polymers or inorganic compounds has also been studied to improve cellulose properties, as will be described next. An overview of the adsorption performance of various cellulose composites is given in Table 7.1.

Table 7.1 *Adsorption performance of precious and rare earth metals of selected cellulose-based composites.*

Material	Element	Maximum adsorption capacity ($mg g^{-1}$)	pH	T (°C)	Time until equilibrium (min)	Reference
Nano-magnetic cellulose	Ag(I)	140.2	6.3	25	3–5	[69]
Magnetic cellulose xanthate (MCX1, MCX2)	Ag(I)	166^a, 15.6^b	0.1 M HNO_3	30	5	[32]
Avicel-bound AgBP-CBD	Ag(I)	332	—	RT	—	[39]
Cellulose filter paper-g-PHCTMA (Cell-g-PHCTMA)	Ag(I)	78.7	3	25	120	[34]
Cellulose filter paper-g-PHCTMA (CFP-g-PHCTMA)	Ag(I)	69.5	1	25	90	[70]
PANI-coated CA	Ag(I)	0.97	—	25	180	[71]
PPy-coated CA	Ag(I)	0.95	—	25	180	[71]
CMC/CMCt/SSS hydrogels	Ag(I)	0.45	5	25	600	[35]
CA-PANI membranes	Au(I)	1.2	—	25	180	[72]
PANI-coated CA	Au(I)	3.9	—	25	360	[51]
PPy-coated CA	Au(I)	3.0	—	25	360	[51]
PANI-coated CA	Au(III)	0.72	—	25	900	[71]
PPy-coated CA	Au(III)	0.51	—	25	900	[71]
Cellulose filter paper-g-PHCTMA (Cell-g-PHCTMA)	Au(III)	153.6	3	25	60	[34]
Cellulose filter paper-g-PHCTMA (CFP-g-PHCTMA)	Au(III)	137.6	1	25	150	[70]
Cellulose filter paper-g-PHCTMA (Cell-g-PHCTMA)	Pd(II)	92.6	3	25	60	[34]
Cellulose filter paper-g-PHCTMA (CFP-g-PHCTMA)	Pd(II)	84.5	1	25	90	[70]
Magnetic cellulose with amine/thiol (GMC-N/S)	Pt(IV)	40.5	2	45	60	[73]
Magnetic cellulose with quaternary amine	Pt(IV)	178	2	25	55	[53]
PEI-modified nanocellulose	Pt(IV)	625^c, 123.5^d, 47.4^e	1	25	240	[74]
Cellulose/graphite oxide	Rh(III)	371	7	—	1	[59]
Cellulose-silica-N (R—N)	Rh(III)	51, 76.5^f	1.1	25	5	[52]

(Continued)

Table 7.1 *(Continued)*

Material	Element	Maximum adsorption capacity ($mg\,g^{-1}$)	pH	T (°C)	Time until equilibrium (min)	Reference
Cellulose-silica-NS (R-NS)	Rh(III)	60, 90[f]	1.1	25	15	[52]
Carboxymethylcellulose/ poly(acrylic acid) hydrogel (CMC-g-PAA)	La(III)	384.6	5.58	30	20	[75]
Zn/Al LDH intercalated cellulose (CL-Zn/Al) nanocomposite	La(III)	92.5	7	25	1	[76]
Cellulose-based silica (CLx/SiO_2) nanocomposite	La(III)	29.5	6	25	50	[77]
PEI-cross-linked cellulose nanocrystals	La(III)	84.9	5.4	30	120	[78]
Carboxymethylcellulose/ poly(acrylic acid) hydrogel (CMC-g-PAA)	Ce(III)	333.33	5.83	30	30	[75]
Zn/Al LDH intercalated cellulose (CL-Zn/Al) nanocomposite	Ce(III)	96.3	7	25	10	[76]
Graphene oxide/ cellulose composite films (CGC)	Ce(III)	109.1	6	25	40	[79]
Cellulose-based silica (CLx/SiO_2) nanocomposite	Sc(III)	23.8	6	25	50	[77]
Cellulose-based silica (CLx/SiO_2) nanocomposite	Eu(III)	24.3	6	25	50	[77]
PEI-cross-linked cellulose nanocrystals	Eu(III)	101.8	5.4	30	120	[78]
Yeast embedded in cellulose matrix	Eu(III)	25.9	6	25	180	[80]
PEI-cross-linked cellulose nanocrystals (PEI-CNCs)	Er(III)	120.3	5.4	30	120	[78]
o-CNCs/o-MWCNTs films	Dy(III)	41.2[g], 42.8[h]	7	—	240	[81]
o-CNCs/GO films	Dy(III)	42.2[g], 44.4[h]	7	—	240	[81]

[a] MCX1.
[b] MCX2.
[c] T-CNF.
[d] P-CNC.
[e] P-CNF.
[f] Column studies, 10 ppm Rh(III), pH 1.1, 15 $ml\,min^{-1}$.
[g] Non-ion-imprinted.
[h] Ion-imprinted.
RT, room temperature.

7.3.1.2 Cellulose Composites for Precious Metal Adsorption

7.3.1.2.1 Magnetic Cellulose Composites

Magnetic sorbents based on cellulose or other materials are widely popular due to their ease of separation from solution. To further improve the sorbent performance, functional groups such as amine [53, 69], amino/thiol [73], and xanthate [32] have been grafted onto the magnetic cellulose composite. Nano-magnetic cellulose with amine-functional groups for adsorption of Ag(I) and other metals were prepared by precipitation of cellulose in the presence of Fe(II) and Fe(III) followed by modification with glycidylmethacrylate and tetraethylenepentamine (TEP) [69]. This microporous material had high Ag(I) adsorption capacity of $140\,mg\,g^{-1}$ ($1.3\,mmol\,g^{-1}$) and adsorption equilibrium was reached in 3–5 minutes. Thermodynamic calculations showed that adsorption was spontaneous and less favorable at higher temperatures. Regeneration using acidified thiourea was successful with comparable uptake capacity over eight adsorption/desorption cycles [69]. Beyki et al. [32] prepared magnetic cellulose xanthate (MCX) composites for the recovery of Ag(I). Two preparation methods were employed, namely one-step preparation of Fe_3O_4/cellulose xanthate from solution (MCX1) and two-step preparation by synthesis of Fe_3O_4/cellulose followed by xanthate formation (MCX2). The magnetic properties of the two composites were similar. However, MCX1 had a superior Ag(I) adsorption capacity over MCX2, likely due to the higher surface area and sulfur functional group loading of MCX1. Rapid adsorption was achieved (<5 minutes), and both composites showed high selectivity toward Ag(I) over various divalent and trivalent metal ions. Silver desorption was effectively done using 0.4 M $Na_2S_2O_3$ with little leaching of the magnetic core; after three desorption cycles, only 3–4% of the Fe had been leached, indicating good stability and reusability [32].

To recover Pt(IV) from solution, a magnetic cellulose has been functionalized with amino/thiol functional groups (GMC-N/S) through grafting with glycidyl methacrylate, thiourea, and ethylenediamine [73]. The authors used a central composite design to optimize adsorption conditions for maximum Pt(IV) recovery. A maximum capacity of $40.48\,mg\,g^{-1}$ at pH 2 and 45 °C was obtained. After four regeneration cycles with 0.5 M H_2SO_4/1 M thiourea, the sorbent retained 93% of its initial capacity. A magnetic cellulose composite with quaternary amine functional groups for recovery of Pt(IV) was developed by Yousif et al. [53]. The chlorinated cellulose was coprecipitated with Fe^{2+}/Fe^{3+} followed by a reaction with TEP and immobilization of glycidyl trimethylammonium chloride (GTA). The resulting magnetic cellulose composite with quaternary amine functional groups had a specific surface area (SSA) of $117.1\,m^2\,g^{-1}$. Maximum Pt(IV) adsorption ($178\,mg\,g^{-1}$) took place within 55 minutes at pH 2 according to electrostatic attraction and ion exchange. Regeneration of up to 90% after four cycles was obtained using a mixture of 2 M HCl and 0.5 M thiourea.

7.3.1.2.2 Polymer-Cellulose Composites

Nitrogen-containing polymers such as PANI, polypyrrole (PPy), or polyethylenimine (PEI) are popular sorbent materials for their high abundance of primary, secondary, and/or tertiary amine groups that can coordinate with various metal ions. Researchers at the University of Sonora in Mexico have studied the use of cellulose acetate (CA) membranes coated with PANI or PPy for adsorption of Ag(I) (as $AgBr_2^-$), Au(I) (as AuI_2^-), and Au(III) (as $AuBr_4^-$) [51, 71, 72]. Adsorption of all ions was higher onto CA/PANI compared to CA/PPy. In mixed

Ag(I)/Au(III) systems, a higher selectivity was found for Ag(I) over Au(III) for short times (3–9 hours), but the selectivity was approximately equal for adsorption times of 15 hours and longer. This favorable adsorption of Ag(I) was explained by its lower charge density and the steric hindrance of the larger $AuBr_4^-$ complex [71]. Partial metal desorption (15–55%) was observed after exposure to 3 M NH_4OH. The CA/PANI membranes could potentially be used for six cycles, whereas the CA/PPy membranes showed degradation after two cycles, limiting their reusability [71].

A PEI-modified nanocellulose for recovery of Pt from chloroplatinic acid has been prepared using cellulose nanofibrils (CNFs) and cellulose nanocrystals (CNCs) from hardwood pulp (P-CNF and P-CNC, respectively), and CNF from tunicates (T-CNF) [74]. The Pt adsorption capacity of PEI-modified composites was in the order T-CNF (600 mg g^{-1}) > P-CNC (100 mg g^{-1}) > P-CNF (50 mg g^{-1}). The superior performance of the T-CNF-containing composite was ascribed to its high surface area, open structure and negative charge of T-CNF that resulted in a higher PEI loading (Figure 7.2).

Polymers with sulfur-based functional groups have also been incorporated into cellulose for their high affinity for soft metal ions such as Au(III), Pd(II), and Ag(I). A thiocarbamate-containing polymer (PHCTMA) was grafted onto cellulose filter paper (CFP) for adsorption of Ag(I), Au(III), and Pd(II) from highly acidic media (0.1–6 M HCl) [34, 70]. Adsorption equilibrium was reached within 60–150 minutes and high selectivity toward Ag(I), Au(III), and Pd(II) was achieved from mixtures containing various metal ions. These precious metals could be selectively recovered from waste printed circuit boards leachate in aqua regia, demonstrating the viability of this cellulose-based composite adsorbent [34, 70].

Hybrid hydrogels containing carboxymethyl cellulose (CMC), carboxymethyl chitosan (CMCt), and sodium sulfonate styrene (SSS) for selective adsorption of Ag(I) were prepared by Tran et al. [35]. The CMC/CMCt/SSS hydrogels had high porosity and contained various functional (e.g. carboxyl, hydroxyl, amino, and sulfonate) groups for adsorption. While the sorbent showed good selectivity toward Ag(I) over various cations, the adsorption capacity was low (0.451 mg g^{-1} at pH 5) and equilibrium was reached only after 10 hours. Interaction between Ag(I) and the hydrogel was weak; physical adsorption through ion exchange was proposed to be the main mechanism.

7.3.1.2.3 Other Cellulose Composites

Cellulose/graphite oxide composites for recovery of Rh(III) were prepared by Yavuz et al. [59]. Graphite oxide was produced by oxidation of graphite powder, which was ultrasonically mixed with cellulose in the presence of NaOH and thiourea in aqueous media. The Rh(III) adsorption was highest at pH 6–9. So, pH 7 was selected as optimal, where the adsorption capacity was 37 mg g^{-1}. At lower pH, competitive adsorption between Rh(III) and protons reduced the capacity. Adsorption was very rapid (one minute) and selective toward Rh(III) with high tolerance to interference caused by various cations and anions. The Rh(III) could be fully recovered using 2.5 M H_2SO_4, and the composite could be used for as many as 21 cycles without reduction in Rh(III) adsorption. Excellent recovery (94–103%) of Rh(III) was obtained from a number of samples including water sources, street dust, and catalytic converters.

Labib et al. [52] prepared mesoporous cellulose/silica hybrid particles for Rh(III) sorption by precipitation from aqueous solution, which were subsequently functionalized with amine (R—N) and amine/thio (R—NS) moieties. For both hybrids, the adsorption increased with pH and was highest at pH 6.9. However, research focused on adsorption at pH 1.1 for

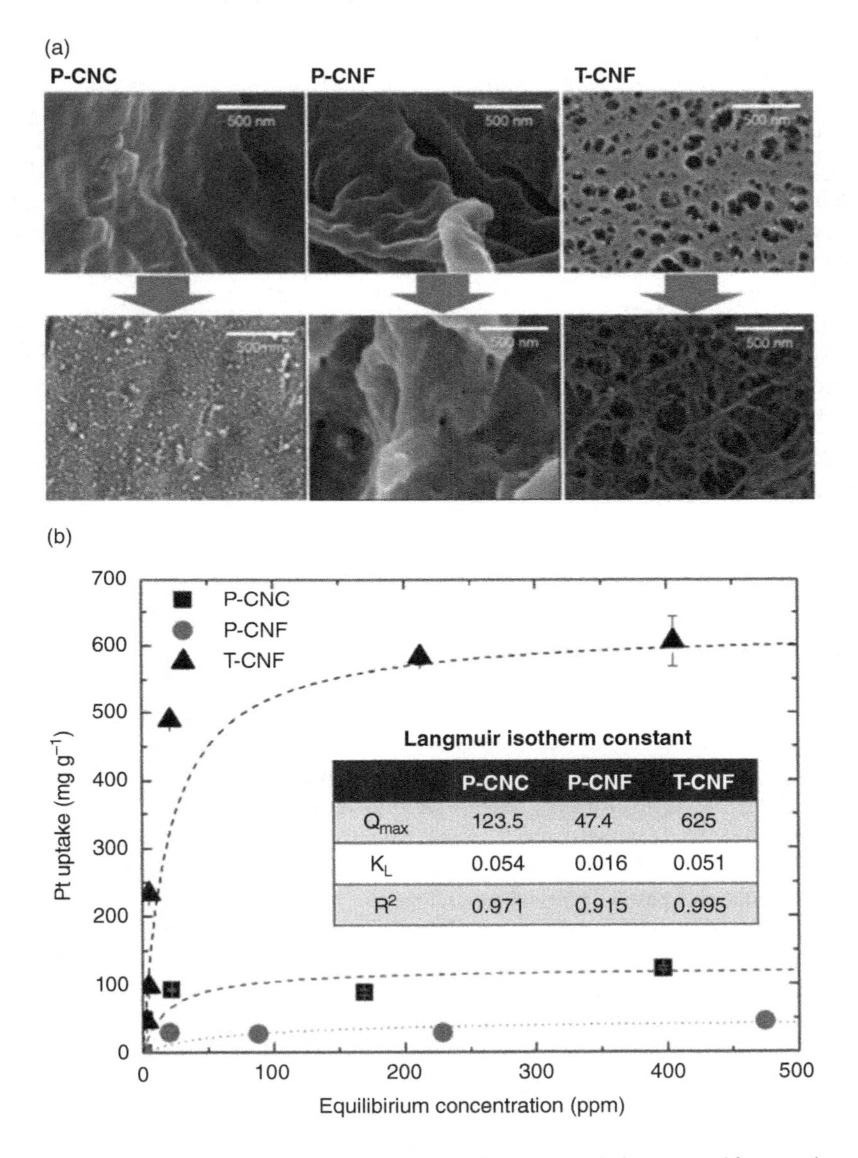

Figure 7.2 (a) SEM images of P-CNC, P-CNF, and T-CNF before (top) and after PEI modification (bottom), and (b) Pt adsorption isotherm of PEI-modified materials. Source: Hong et al. [74]. Reproduced with permission from Elsevier.

industrial applications. At this acidic pH, a maximum adsorption capacity of 51 mg g^{-1} (0.51 mmol g^{-1}) and 60 mg g^{-1} (0.59 mmol g^{-1}) was found for R—N and R—NS, respectively. The authors highlight that this relatively low adsorption capacity compared to functional group loading (5 and 7.6 mmol g^{-1} for R—N and R—NS, respectively) indicated that not all active sites were available for adsorption. Despite its lower SSA and pore volume, Rh(III) adsorption was higher on R—NS compared to R—N. Adsorption on R—N was dominated by intraparticle diffusion, whereas in the case of R—NS, the chemical reaction between active sites and Rh(III) ions was the rate-limiting step. Continuous experiments employing packed columns yielded higher Rh(III) adsorption capacities of 75 mg g^{-1} (R—N) and 90 mg g^{-1} (R—NS). Regeneration using a mixture of HCl/HNO_3 was more difficult for R—NS compared to R—N due to the stronger interaction between Rh(III) and the sulfur groups.

A genetically engineered designer biomolecule consisting of a silver-binding peptide (AgBP) and cellulose-binding domain (CBD) could effectively recover Ag(I) in the form of antibacterial Ag^0 nanoparticles [39]. In this work, the biomolecule was attached to commercial microcrystalline cellulose (Avicel) to form Avicel-bound AgBP-CBD complexes. A high Ag(I) adsorption capacity of 332 mg g^{-1} was measured, with high selectivity (≥85%) toward Ag(I) over K^+ and Cu^+. The adsorbed Ag(I) was reduced by the AgBP to form 20 nm Ag^0 nanoparticles on the cellulose surface. Adsorption was not significantly changed for pH 3–5, but a decrease in adsorption at pH 2 was observed due to the denaturation of the biomolecules. This green bionanocomposite showed desirable bactericidal activity against *Staphylococcus aureus* and *Escherichia coli*.

7.3.1.3 Cellulose Composites for Rare Earth Metal Adsorption

7.3.1.3.1 Polymer-Cellulose Composites

As was described in Section 7.3.1.2.2, PEI is a popular material for sorbents due to its ability to form complexes with various metal ions. Zhao et al. [78] prepared PEI-cross-linked CNCs (PEI-CNCs) for the recovery of REEs from aqueous solution (pH 1.1–6.9). The PEI-CNC was prepared by carbodiimide-mediated peptide coupling between carboxylate groups on TEMPO-oxidized CNC and amino groups of PEI. The resulting PEI-CNC contain primary (3–5 mmol g^{-1}) and secondary amines (3–5 mmol g^{-1}) and some residual carboxylic acid groups (1–1.3 mmol g^{-1}) as determined by titration, with isoelectric point 9.1–9.9. The REEs' adsorption increased with pH due to reduced protonation of amine groups resulting in a stronger chelating ability. At pH 5.4, the maximum molar adsorption capacity was Er(III) (0.719 mmol g^{-1}) > Eu(III) (0.670 mmol g^{-1}) > La(III) (0.611 mmol g^{-1}), in the order of increasing ionic radius (i.e. Er(III) has the smallest radius). In ternary systems, there was strong competitive adsorption, with preferential adsorption of Er(III) over the other REEs, indicating that the PEI-CNC could be used to separate Er(III) from other REEs such as Eu(III) and La(III). The adsorption capacity was decreased in seawater due to the presence of other cations. The X-ray photoelectron spectroscopy (XPS) and Fourier-transform infrared (FTIR) data and elemental mapping after adsorption demonstrated that the adsorption mechanism was through a coordination bond formation between rare earth ions and the N atoms of primary and secondary amine groups. Regeneration efficiency for La(III) was above 95% after three adsorption/desorption cycles using 0.1 M and 1 M HNO_3, indicating good regeneration and reusability.

A hydrogel adsorbent (CMC-g-PAA) based on carboxymethylcellulose (CMC) and poly(acrylic acid) (PAA) was prepared by Zhu et al. [75] using high internal phase emulsions (HIPEs) for adsorption of La(III) and Ce(III) (Figure 7.3a). Emulsion formation using p-xylene resulted in the formation of a porous monolith adsorbent with an average pore size of 1.18 μm (Figure 7.3b). Adsorption equilibrium was reached rapidly within 20–30 minutes and the sorbent had a very high capacity for La(III) (384.62 mg g^{-1}, pH 5.58) and Ce(III) (333.33 mg g^{-1}, pH 5.83) (Figure 7.3c,d). This high capacity was explained by the hierarchical pore structure of the monolithic open-cellular hydrogel sorbent. Adsorption was suggested to take place via interaction of the carboxyl functional groups of PAA with the REEs. Regenerability using 0.5 M HCl was excellent with only a slight loss after five regeneration cycles, demonstrating very promising use of the sorbent.

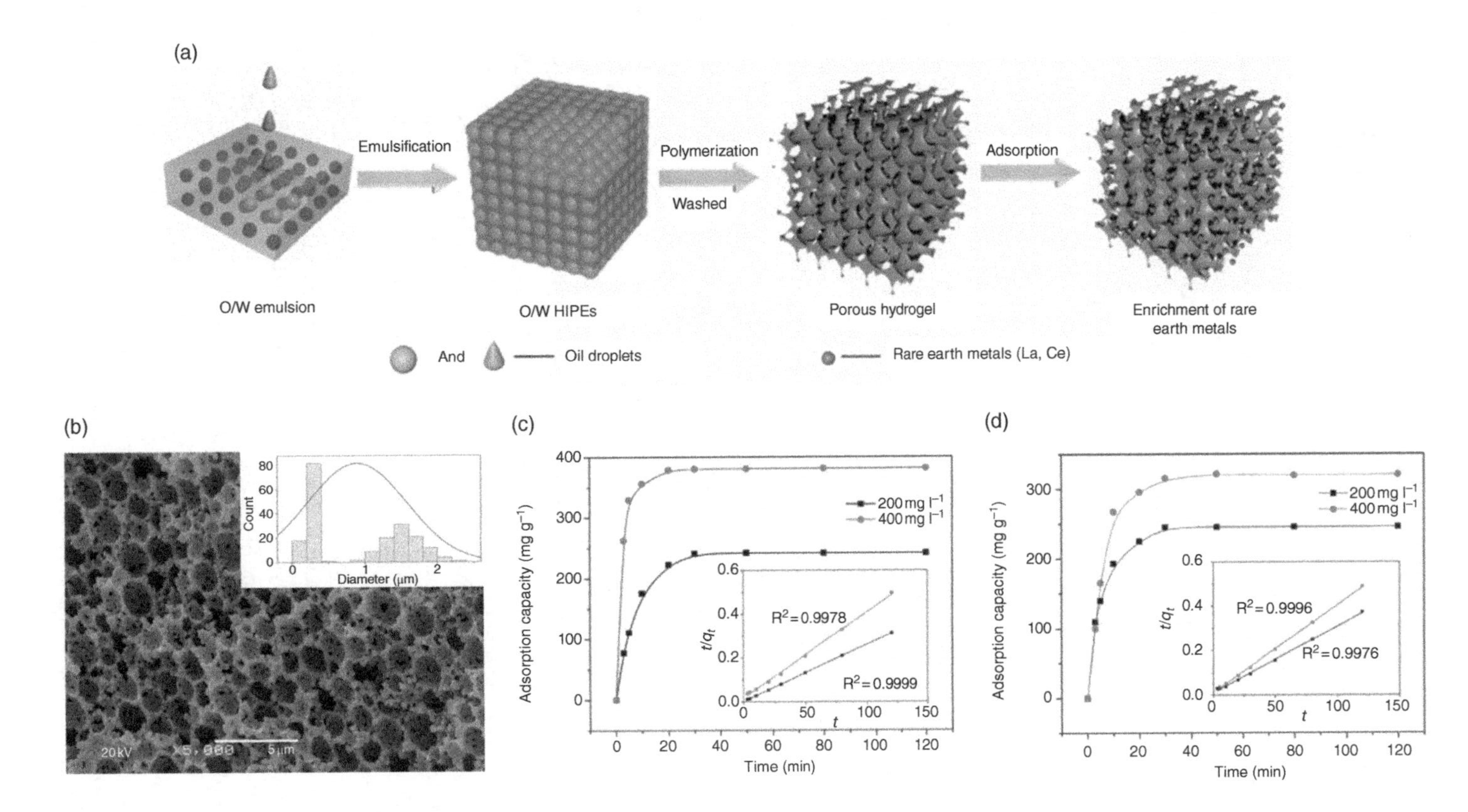

Figure 7.3 *(a) Synthetic route for the preparation of the CMC-g-PAA hydrogel adsorbent for the enrichment of La and Ce, (b) SEM and pore size distribution, (c) adsorption kinetics of La(III), and (d) adsorption kinetics of Ce(III). Source: Zhu et al. [75]. Reproduced with permission from Elsevier.*

7.3.1.3.2 Other Cellulose Composites

Iftekhar et al. [76] used Zn—Al-layered double hydroxide (LDH) intercalated cellulose prepared by co-precipitation for uptake of Y(III), La(III), and Ce(III). Higher adsorption was found for a higher Zn/Al ratio. Maximum adsorption was at pH 7, with a capacity of 102.25 mg g^{-1} for Y(III), 92.51 mg g^{-1} for La(III), and 96.25 mg g^{-1} for Ce(III). Adsorption was spontaneous and endothermic, and took place primarily at hydroxyl groups of Zn(II) as well as the inner layer anion exchange site. Regeneration using 0.1 M HCl demonstrated a slight loss in adsorption capacity over five cycles. Selectivity toward the REEs was good; uptake only slightly decreased in the presence of high (10-fold) concentrations of Na(I), K(I), Ca(II), Mg(II), and Al(III). The same researchers have developed cellulose-based silica nanocomposites for the adsorption of rare earth metals [77]. Two modified celluloses were tested, namely H_2SO_4-modified cellulose (CLN) and citric acid-modified cellulose (CLCA). Adsorption of Eu(III), La(III), and Sc(III) was higher onto CLN/SiO_2 than on CLCA/SiO_2, likely due to its higher SSA (169.74 m^2 g^{-1} compared to 110.29 m^2 g^{-1}). The maximum adsorption capacity at pH 6 on CLN/SiO_2 was similar for all three ions (24.27, 29.48, and 23.76 mg g^{-1} for Eu(III), La(III), and Sc(III), respectively). Adsorption of Eu(III) and La(III) took place by chemisorption which could best be described by the Langmuir isotherm. Adsorption of Sc(III), on the other hand, was primarily through physisorption, and followed the Freundlich isotherm. Regeneration using 0.5 M HCl reduced the adsorption capacity to 50–60% after three cycles.

Hao et al. [79] prepared graphene oxide (GO)/cellulose composites in ionic liquid for adsorption of Ce(III). GO is difficult to use. So, incorporating the GO into the cellulose matrix would improve its ease of use and adsorption capacity. The GO/cellulose composite reached adsorption equilibrium in 40 minutes and a maximum Ce(III) adsorption capacity of 109.1 mg g^{-1} was obtained. Adsorption was high over a wide pH range of 3–8, where the optimal adsorption was obtained at pH 6. At low pH, there was competitive binding between H^+ and Ce(III), whereas, at higher pH, Ce(III) hydrolysis took place. The main adsorption mechanism was through ion exchange.

Ion imprinting of polymers can greatly enhance the selectivity of sorbents toward specific metal ions due to the "recognition" of the ion by the sorbent, similar to enzyme-substrate interaction. Zheng et al. [81] prepared oxidized CNCs (o-CNCs) bound to acidified multi-walled CNT (o-MWCNT or GO. Nanomaterials like o-MWCNT and GO are promising adsorbents but are difficult to disperse in water. Thus, the authors embedded these materials in cellulose to improve dispersibility and reduce agglomeration. To further improve selectivity, the o-CNCs/o-MWCNTs and o-CNCs/GO composites were ion-imprinted (IIPs) with Dy(III) or non-imprinted (NIIPs). These composites showed significantly higher adsorption over o-CNTs-NIIPs and o-CNTs-IIPs. Maximum Dy(III) adsorption was found at pH 7 (capacity 41.24–44.36 mg g^{-1}), with higher capacity for o-CNCs/GO over o-CNCs/o-MWCNTs composites. At this pH, the ion-imprinting only slightly improved adsorption, but this effect was more pronounced at pH 4, where high selective adsorption of Dy(III) over other trivalent cations was achieved (in order Dy(III) > Gd(III) > Nd(III) > Pr(III) > Fe(III)). The adsorption capacity decreased to 75–89% after five adsorption/desorption cycles, mainly due to incomplete removal of Dy(III) and loss of binding sites.

The cell wall of yeast (*Saccharomyces cerevisiae*) contains polysaccharides like chitin, mannan, and glucans that have various functional groups capable of adsorbing cations. In the work of Arunraj et al. [80], yeast was embedded in cellulose for recovery of Eu(III) from an aqueous solution. At pH 6, the capacity of the yeast-immobilized cellulose was 25.9 mg g^{-1}, which was higher than the separate cellulose (18.5 mg g^{-1}) or yeast (14.2 mg g^{-1}). Adsorption was spontaneous and exothermic. Regeneration of the sorbent using EDTA resulted in percentage adsorption of 97.5% after one cycle, which decreased to 83.2% after four adsorption/desorption cycles. The authors also demonstrated the recovery of Eu(III) from a fluorescent lamp phosphor in the presence of competing cations.

7.3.2 Chitosan-based Composite Adsorbents

7.3.2.1 Structure and Chemistry of Chitosan

Chitosan is a polyaminosaccharide derived from naturally occurring chitin, the second most abundant biopolymer after cellulose. The high value of chitosan as a sorbent comes from the high abundance of hydroxyl and amine functional groups. The amine groups are easily protonated in acidic solutions, which can cause electrostatic attraction of anionic compounds (such as dyes and anionic metal ion complexes), Adsorption of cations, on the other hand, usually occurs at near to neutral pH by interaction with the electron lone pair of nitrogen [44]. In addition, the hydroxyl and amine groups provide platforms for chemical or physical modification to further enhance adsorption capacity or sorbent stability [44, 82]. Major drawbacks of chitosan, however, are its poor chemical, mechanical, and thermal stability [19, 83]. For example, chitosan is soluble in weak acidic solution, limiting its applicability at low pH [84]. Therefore, chitosan is often physically or chemically modified to improve its stability and/or adsorption performance [19, 82, 85]. Table 7.2 gives an overview of the adsorption performance of selected chitosan composites.

7.3.2.2 Chitosan Composites for Precious Metal Adsorption

7.3.2.2.1 Magnetic Chitosan Composites

To improve the ease of separation, magnetic (nano)particles have been incorporated into chitosan for the recovery of precious metals. Such magnetic composites can either be used without chemical modification [91, 93] or by grafting with various functional groups such as amine [46, 89] or thiourea [33] to further improve adsorption performance.

Ion-imprinted thiourea-chitosan was coated on Fe_3O_4 for the removal of Ag(I) [33]. Adsorption equilibrium was reached within 50 minutes with a maximum Ag(I) adsorption capacity of 4.93 mmol g^{-1} at pH 5, 30 °C. The material was stable and could be easily recovered; adsorption capacity was 90% of its initial capacity after five cycles. The authors showed that there was a tradeoff between adsorption capacity and saturation magnetization; a higher adsorption capacity was achieved at higher chitosan content, but at the cost of lower magnetization and thus slower recovery. To this end, Fan et al. [91] optimized the composition of magnetic chitosan beads. A chitosan: Fe_3O_4 mass ratio of 0.3 showed the best absorption performance for Ag(I) and other metal ions. Composites prepared by either embedding Fe_3O_4 particles into chitosan beads or by co-precipitation of Fe_3O_4 and chitosan in the same solution gave similar absorption performance. The high-saturation magnetization (46.97 and 29.18 emu g^{-1}, respectively) resulted in rapid recovery of the magnetic material within several minutes.

A magnetic cross-linked chitosan resin modified with pentaamine moieties for adsorption was developed by Donia et al. [89] for recovery of Ag(I) and Au(III). In batch experiments, the resin with 6.4 mmol g^{-1} active amine groups had an adsorption capacity of 183 mg g^{-1} (1.7 mmol g^{-1}) for Ag(I) at pH 6.9 and 453 mg g^{-1} (2.3 mmol g^{-1}) for Au(III) at pH 0.5. Adsorption was also tested in continuous flow experiments using packed beds.

Table 7.2 *Adsorption performance of precious and rare earth metals of selected chitosan composites.*

Material	Element	Maximum adsorption capacity (mg g^{-1})	pH	T (°C)	Time until equilibrium (min)	Reference
Ion-imprinted chitosan/PVA cross-linked membrane	Ag(I)	125	6	25	40	[86]
Ion-imprinted biosorbent	Ag(I)	199.2	7	25–27	90	[87]
Ag(I)-imprinted magnetic thiourea-chitosan beads	Ag(I)	531.8	5	30	50	[33]
Fe_3O_4/ATP@(CS/cys-β-CD)$_8$	Ag(I)	—	4	25	60	[88]
Magnetic chitosan with pentaamine moieties	Ag(I)	183.4	6.9	28	60	[89]
Chitosan/bamboo charcoal	Ag(I)	52.9	6	30	180	[37]
Chitosan/montmorrilonite	Ag(I)	43.8	6	30	150	[90]
Magnetic chitosan beads	Ag(I)	107[a], 117.7[b]	4	30	—	[91]
PEI-modified chitosan (PCSF)	Au(I)	251.7	5.5	25	300	[48]
PEI-modified bacterial biosorbent fiber (PBBF)	Au(I)	421.1	5.5	25	300	[48]
Chitosan/graphene oxide (CSGO)	Au(III)	1076.7	4	RT	780	[50]
Magnetic chitosan with pentaamine moieties	Au(III)	453.0	0.5	28	60	[89]
PEI-modified bacterial biosorbent fiber (PBBF)	Ru(III)	110.5	Acetic acid waste solution	25	90	[92]
Ethylenediamine-modified magnetic chitosan nanoparticles (EMCN)	Pd(II)	138	2	25	40	[46]
Chitosan-coated wool fabrics	Pd(II)	35.0	2.5	60	30	[54]
Chitosan/graphene oxide (CSGO)	Pd(II)	216.9	3	RT	720	[50]
Chitosan grafted on persimmon tannin extract	Pd(II)	330	5	50	120	[55]
Magnetic chitosan nanoparticles	Pd(II)	192.3	6	20	10	[93]
Ethylenediamine-modified magnetic chitosan nanoparticles (EMCN)	Pt(IV)	171	2	25	20	[46]
PEI-loaded chitosan hollow beads (CHBs)	Pt(IV)	815.2	1	25	—	[49]
Magnetic alginate-chitosan beads	La(III)	97.1	2.8	25	600	[94]
Chitosan/TiO_2 nanocomposites (NTiO_2-Glu-NChit)	La(III)	167, 146, 160	1, 3, 6	RT	30	[95]
Chitosan/PVA/3-mercaptopropyltrimethoxysilane (CTS/PVA/TMPTMS)	La(III)	263.2, 460.9[c]	5	25	360	[66]
Ion-imprinted potassium tetratitanate whisker with chitosan	Ce(III)	42.7	6	25	—	[96]
Chitosan-g-(AA-co-SS)/ISC	Ce(III)	174.1	5	30	15	[67]

Chitosan/PVA/3-mercaptopropyltrimethoxysilane (CTS/PVA/TMPTMS)	Ce(III)	251.4, 374.9[c]	5	25	360	[66]
DETA-functionalized chitosan magnetic nanoparticles	Nd(III)	49–51	5	27–47	190–250	[97]
DETA-modified magnetic chitosan nanoparticles (FCCD)	Nd(III)	27.1	7	25	240–300	[63]
DETA-modified magnetic chitosan nanoparticles (FCCD)	Eu(III)	30.6	7	25	240–300	[63]
Amidoxime-functionalized magnetic chitosan microparticles	Eu(III)	375.3	5	20	60–90	[98]
Ion-imprinted chitosan/CNT composite	Gd(III)	68–104	7	20–40	120	[62]
Chitosan-g-(AA-co-SS)/ISC	Gd(III)	224.0	5	30	15	[67]
Magnetic-chitosan nano-based particles grafted with amino acids	Dy(III)	8–18	5	27	240–360	[99]
DETA-functionalized chitosan magnetic nanoparticles	Dy(III)	51–52	5	27	190–250	[97]
DETA-modified magnetic chitosan nanoparticles (FCCD)	Dy(III)	28.3	7	25	240–300	[63]
DETA-functionalized chitosan magnetic nanoparticles	Yb(III)	51–52	5	27	190–250	[97]

[a] Prepared by co-precipitation.
[b] Prepared by embedding.
[c] Continuous column system (4 ml min^{-1}, C_0 300 mg l^{-1}).
RT, room temperature.

Magnetic cross-linked chitosan nanoparticles (TEM [transmission electron microscopy] size 15–40 nm) modified with ethylenediamine for adsorption of Pt(IV) and Pd(II) were prepared by Zhou et al. [46]. The increase in amine functional groups by ethylenediamine modification resulted in a high capacity (171 mg g^{-1} for Pt(IV) and 138 mg g^{-1} for Pd(II) at pH 2.0) and rapid adsorption equilibrium (20–40 minutes). While both Pt(IV) and Pd(II) competed for the same adsorption sites, the sorbent had a greater affinity for Pt(IV) than for Pd(II). Separation of adsorbed Pd(II) and Pt(IV) could be achieved by first selective desorption of Pd(II) using 5 M ammonia followed by HNO_3/thiourea to desorb the remaining metal. After five successive adsorption/desorption cycles, the adsorption capacity was over 90%, but during each cycle, 0.8–1% of the Fe_3O_4 was dissolved. Omidinasab et al. [93] developed magnetic chitosan nanoparticles for adsorption of Pd(II). A monolayer adsorption capacity for Pd(II) was 192.3 mg g^{-1} at pH 6 (calculated from the Langmuir model). Adsorption of Pd(II) decreased in the presence of V(V). The sorbent could recover over 90% of Pd(II) from jewelry wastewater.

Mu et al. [88] have deposited chitosan and L-cysteine-modified β-cyclodextrin on Fe_3O_4-decorated attapulgite. Adsorption of Ag(I), Pt(IV), and Pd(II) was optimal at pH 4 but the maximum adsorption capacity was not determined. The composite showed preferential adsorption of Ag(I) and Pd(II) over Pt(IV). The XPS analysis before and after adsorption indicated that adsorption took place through electron donation by nitrogen, oxygen, and/or sulfur atoms to the metal ions.

7.3.2.2.2 *Polymer-Chitosan Composites*

Pure chitosan membranes have poor chemical resistance and mechanical strength, which can be overcome by the incorporation of PVA, a biodegradable polymer with good membrane-forming properties. Cross-linked ion-imprinted chitosan/polyvinyl alcohol (PVA) (2 : 1 weight ratio) membranes showed a higher adsorption capacity (125 mg g^{-1} at pH 6, 25 °C) and higher removal rate of Ag(I) compared to non-imprinted ones. Moreover, easy regeneration was possible with 0.01 M EDTA with 15% loss in capacity after five cycles [86].

Chitosan hollow beads (CHBs) loaded with PEI had an adsorption capacity for Pt(IV) as high as 815.2 mg g^{-1} at pH 1.0, which was 2.5 times higher than that of a commercial resin. At this pH, the dominant species were $(PtCl_4)^{2-}$ and $(PtCl_6)^{2-}$, which were attracted to the positively charged beads. Using 0.1 M thiourea for desorption, adsorption efficiency in the third adsorption/desorption cycle was only 45%, probably due to the destruction of the bead structure caused by the thiourea. The Pt(IV) recovery was over 97% after 10 cycles using a sequential fill-and-draw process [49].

7.3.2.2.3 *Other Chitosan Composites*

Chitosan/bamboo charcoal composite beads were prepared by Nitayaphat and Jintakosol [37]. The composite containing 50 wt% bamboo charcoal (SSA 34.34 m^2 g^{-1} with pore volume 0.052 cm^3 g^{-1}) had an Ag(I) adsorption capacity of 52.91 mg g^{-1} at pH 6, twice as high as that of bare chitosan. The higher capacity was attributed to the higher SSA and porosity of the composite. Regeneration of the composite was limited; only 31% of the adsorbed Ag(I) could be desorbed. The same researchers prepared chitosan/montmorillonite composite beads for adsorption of Ag(I) [90]. Adsorption equilibrium was reached after 150 minutes and the maximum adsorption capacity for Ag(I) was 43.48 mg g^{-1} at pH 6, slightly higher than that of bare chitosan beads (38.46 mg g^{-1}).

Liu et al. [50] have developed chitosan/ GO composites for the adsorption of Au(III) and Pd(II). The addition of GO extended the adsorption pH range to higher pH values, likely due to the coordination of metal ions with the oxygen atoms of GO. The composite containing 5 wt% GO had the best performance: the adsorption capacity for Au(III) was as high as

1076.6 $mg\,g^{-1}$ at pH 4.0 and 216.9 $mg\,g^{-1}$ for Pd(II) at pH 3.0. The increased capacity over GO-free sorbent was explained by the increased SSA (4.23 versus 1.56 $m^2\,g^{-1}$). Adsorption equilibrium was reached only after 13 and 12 hours for Au(III) and Pd(II), respectively. Adsorption was suggested to take place primarily by ion interaction between protonated amines and tetrachloroaurate and chloropalladate complexes, and coordination of Au(III) and Pd(II) with nitrogen and oxygen moieties. Metal desorption was carried out using thiourea-HCl and sorbent reusability after three cycles was higher than 95%.

Huo et al. [87] prepared Ag-imprinted biosorbents based on mycelium (waste biomass from *Penicillium chysogenum*) with chitosan (mass ratio 200 : 1). The maximum adsorption capacity for Ag(I) was 199.2 $mg\,g^{-1}$ at pH 7, which is six times higher than bare chitosan. The adsorption mechanism was chelation of Ag(I) with $—NH_2$ followed by reduction to metallic Ag: deposited Ag^0 nanoparticles of 10–100 nm were visible on the composite surface.

Wool fabrics were coated with chitosan for adsorption of Pd(II). The maximum adsorption capacity for Pd(II) was 35.014 $mg\,g^{-1}$ at pH 2.5 and 60 °C. The mechanism was mainly through physical adsorption [54]. A comparably higher Pd(II) adsorption capacity was obtained for chitosan grafted on persimmon tannin extract (PTCS, Figure 7.4a) [55]. Adsorption increased with pH up to pH 5 where the measured maximum adsorption capacity for Pd(II) was 310–330 $mg\,g^{-1}$ (30–50 °C). The Pd(II) was reduced to metallic Pd^0 (as evidenced by XRD) by the phenolic hydroxyl groups of tannin that were oxidized into quinine (Figure 7.4a). The sorbent showed good selectivity toward Pd(II) over various metal ions (Figure 7.4b).

7.3.2.3 Chitosan Composites for Rare Earth Metal Adsorption

7.3.2.3.1 Magnetic Chitosan Composites

Various research groups have developed magnetic chitosan composites for the adsorption of REEs. The adsorption performance of these composites could be improved by incorporating functional groups such as amine [63, 97], amidoxime [98], and amino acids [99]. Using a non-chemically modified chitosan, Wu et al. [94] prepared magnetic alginate-chitosan gel beads (average diameter 0.85 mm) for removal of La(III). Despite a higher adsorption capacity at higher pH (tested in the range of pH 1–6), a higher selectivity for La(III) over divalent cations was found at pH 2.8, where the maximum La(III) uptake was 97.1 $mg\,g^{-1}$ (27 °C). Oxygen atoms rather than nitrogen atoms were most likely the main adsorptive binding sites as supported by XPS and FTIR analysis. Increasing the ionic strength reduced the adsorption capacity due to the competitive adsorption of Na(I). Regeneration using 0.1 M HCl resulted in the dissolution of the magnetic core; as much as 8.7% Fe was dissolved after three adsorption/desorption cycles.

Magnetic chitosan nano-based particles grafted with amino acids (alanine, serine, and cysteine) were developed by Galhoum et al. [99] for recovery of Dy(III). Superparamagnetic FeO_4 was used as magnetic core and chitosan as coating support for adsorption (1 : 1 chitosan : Fe_3O_4). Alanine-, serine-, and cysteine-modified composites were nanosized (15–40 nm from TEM) with Dy(III) adsorption capacity of 14.8, 8.9, and 17.6 $mg\,g^{-1}$, respectively (pH 5, 27 °C). The highest capacity for the cysteine-modified material was explained by its thiol end group that has a strong affinity for metals, and reduced adsorption due to steric hindrance in serine and alanine. The primary adsorption mechanism was through chelation and anion-exchange mechanisms. Regeneration using acidified thiourea for four sorption/desorption cycles slightly reduced adsorption capacity. The same research group co-precipitated Fe_3O_4 with chitosan followed by grafting of DETA for adsorption of Nd(III), Dy(III), and Yb(III) [97]. The final product contained 47% magnetite and had a saturation magnetization of approximately 20 $emu\,g^{-1}$. The authors estimated that more than 90% of nitrogen from the chitosan was substituted with DETA. The pH did

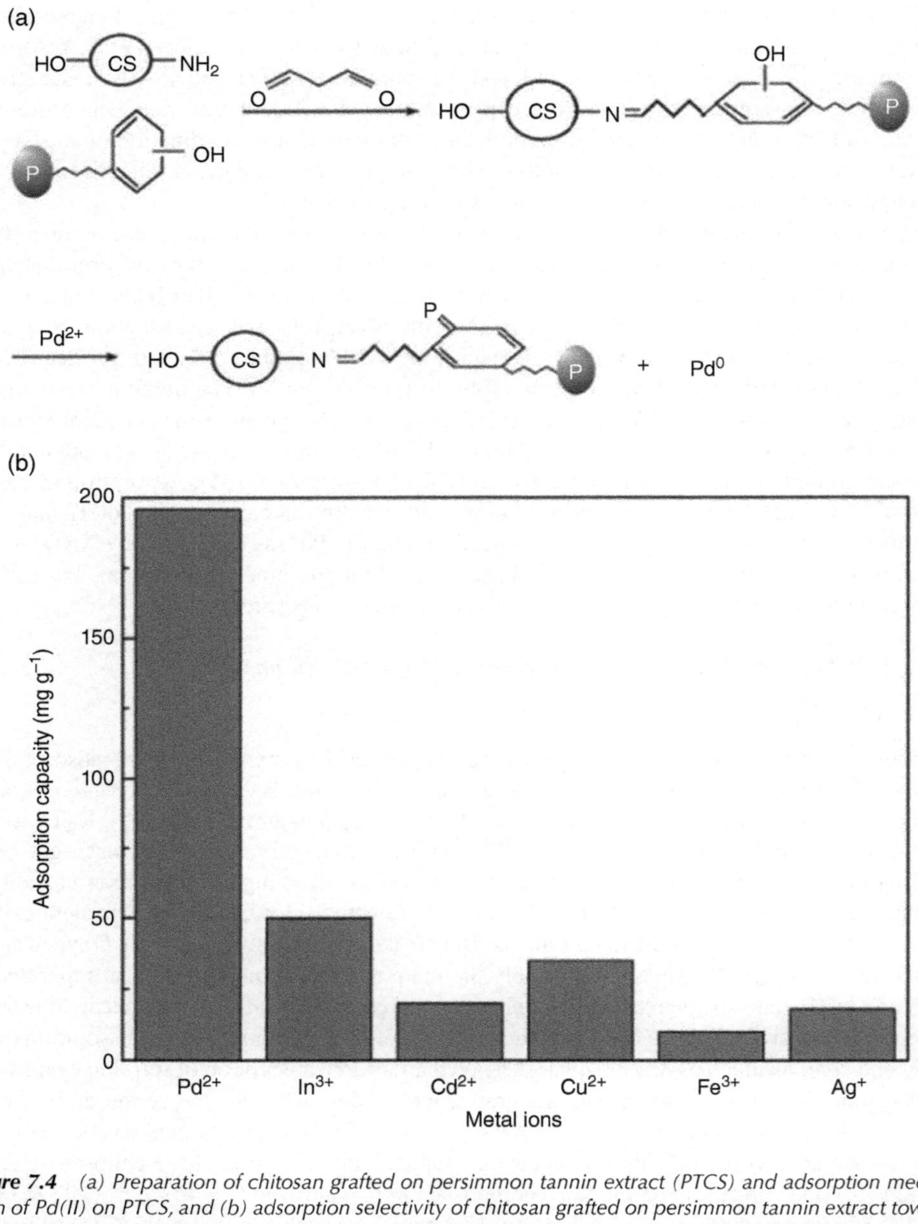

Figure 7.4 *(a) Preparation of chitosan grafted on persimmon tannin extract (PTCS) and adsorption mechanism of Pd(II) on PTCS, and (b) adsorption selectivity of chitosan grafted on persimmon tannin extract toward Pd(II) over various metal ions at pH 5.0. Source: Zhou et al. [55]. Reproduced with permission from Elsevier.*

not affect adsorption selectivity; all metal ions were adsorbed to similar content. The adsorption capacity was 47.4, 50.4, and 50.7 $mg\,g^{-1}$ for Nd(III), Dy(III), and Yb(III), respectively (pH 5, 27 °C). Sorbent regeneration using thiourea showed less than 6% activity loss after five cycles.

Using a different approach, Liu et al. [63] also prepared DETA-modified magnetic chitosan nanoparticles for adsorption of Dy(III), Nd(III), and Er(III). The Fe_3O_4 was first modified with octadecyltriethoxysilane and coated with chitosan/polyphosphate, which was subsequently modified with DETA. These Fe_3O_4-C_{18}-chitosan-DETA (FCCD) composites had SSA $20.86\,m^2\,g^{-1}$, pore volume $0.04\,cm^3\,g^{-1}$ and high saturation magnetization of $63\,emu\,g^{-1}$. The adsorption capacity at pH 7 was $28.3\,mg\,g^{-1}$ for Dy(III), $27.1\,mg\,g^{-1}$ for Nd(III), and $30.6\,mg\,g^{-1}$ for Eu(III). Desorption using 0.05 M HCl showed a slight decrease in adsorption capacity over time and leaching of the magnetic core, highlighting the need for a more hermetic coating.

Amidoxime-modified magnetic chitosan was prepared by Hamza et al. [98] for Eu(III) and U(VI) adsorption. The Fe_3O_4 was prepared by co-precipitation followed by mixing with chitosan and EPI for cross-linking. These magnetic composites were subsequently reacted with malononitrile and hydroxylamine to graft amidoxime functional groups (Figure 7.5). Titration revealed an —NH functional group content of $6.91\,mmol\,g^{-1}$ in the magnetic composite. Adsorption of Eu(III) increased almost linearly between pH 2.3 and 6; under acidic conditions, protonation of the amidoxime functional groups would repel Eu(III) ions. Chelation with the metal ion was thought to be simultaneous with amine and hydroxyl groups of amidoxime. The adsorption capacity for Eu(III) was $375\,mg\,g^{-1}$ ($2.47\,mmol\,g^{-1}$) at pH 5, the highest reported value until now. In a binary system with U(VI), preferential adsorption of U(VI) took place at pH 5, but selectivity for Eu(III) increased at a lower pH at the expense of adsorption capacity. Easy and fast desorption was realized in 0.5 M HCl.

7.3.2.3.2 Polymer-Chitosan Composites

Chitosan/PVA/3-mercaptopropyltrimethoxysilane (CTS/PVA/TMPTMS) beads for the adsorption of La(III) and Ce(III) in batch and continuous flow systems were developed by Najafi Lahiji et al. [66]. In this work, the chitosan was first blended with PVA followed by modification with thiol groups using TMPTMS. The CTS/PVA/TMPTMS beads had maximum adsorption capacities of $263.16\,mg\,g^{-1}$ for La(III) and $251.41\,mg\,g^{-1}$ for Ce(III) at pH 5 in the batch system, which was 1.5 times higher than beads without TMPTMS. In column adsorption tests, the capacities for La(III) and Ce(III) were as high as 460.94 and $374.83\,mg\,g^{-1}$, respectively.

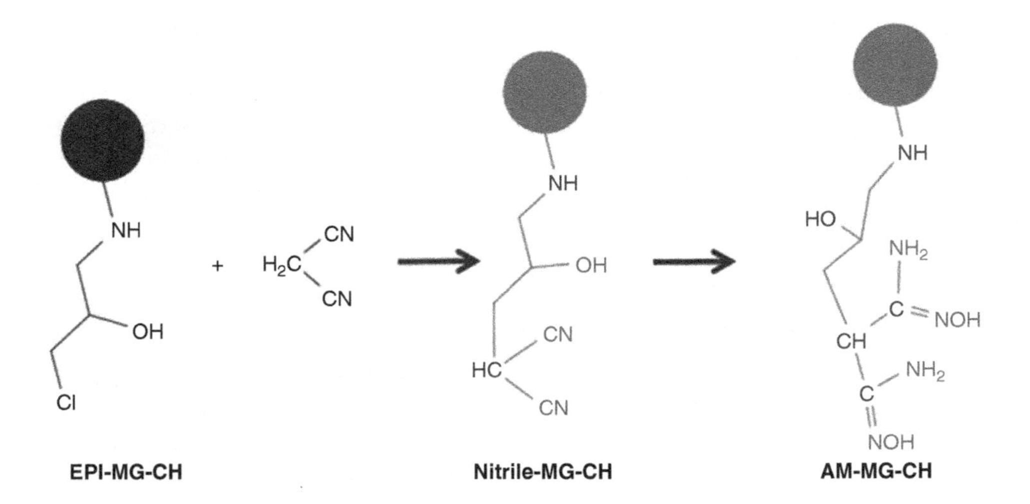

Figure 7.5 *Reaction pathway for the synthesis of amidoxime-modified magnetic chitosan (epichlorohydrin-activated magnetic chitosan [EPI-MG-CH] → malononitrile-grafted magnetic chitosan [Nitrile-MG-CH] → amidoxime-functionalized magnetic chitosan [AM-MG-CH]). Source: Hamza et al. [98]. Reproduced with permission from Elsevier.*

Wang et al. [67] grafted acrylic acid (AA) and sodium p-styrenesulfonate (SS) onto chitosan using illite/smectite clay (ISC) as an inorganic additive. The CTS-g-(AA-co-SS)/ISC hydrogel composites were used for the adsorption of Ce(III) and Gd(III). An AA/SS ratio of 19 : 1 was found optimal for adsorption. The adsorption capacity increased with pH and was highest for pH 4–6, with a maximum adsorption capacity at pH 5 of 174.05 mg g^{-1} for Ce(III) and 223.97 mg g^{-1} for Gd(III). The FTIR spectra indicated that adsorption was through electrostatic attraction and complexation of the metal ions with $-NH_2$, $-OH$, $-COOH$, and $-SO_3^-$ moieties.

7.3.2.3.3 Other Chitosan Composites

Ion-imprinted potassium tetratitanate whiskers with chitosan prepared by Zhang et al. [96] had a maximum Ce(III) adsorption capacity of 42.71 mg g^{-1}, twice as high as of non-imprinted material, with high selectivity over several cations. There was a slight decrease in adsorption capacity after five regeneration cycles. The sorbent could be used to detect trace amounts of Ce(III) in environmental samples.

Ion-imprinted chitosan/CNT composites have been developed for the removal of Gd(III) [62]. The chitosan was first imprinted with Gd(III) and deposited onto CNT, followed by cross-linking of chitosan and removal of imprinted Gd(III). These chitosan-coated CNT (SSA 44.8 m^2 g^{-1}, pore volume 0.20 cm^3 g^{-1}) could be easily magnetized by blending with silica-coated magnetite in an aqueous solution. The Gd(III) adsorption was optimal at pH 7 (maximum capacity 68, 88, and 104 mg g^{-1} at 20, 30, and 40 °C, respectively) and took place via coordination between Gd(III) and amino groups. The sorbent showed good selectivity for Gd(III) over La(III) and Ce(III), primarily due to ion imprinting. The adsorption capacity was relatively constant over five adsorption/desorption cycles using 1 M HCl.

Mahmoud et al. [95] prepared chitosan nanoparticles cross-linked with TiO_2 for adsorption of La(III). The spherical composite particles (SEM size 52–58 nm, SSA 9.869 m^2 g^{-1}, pore volume 0.02 cm^3 g^{-1}) had a maximum La(III) adsorption capacity of 167, 146, and 160 mg g^{-1} (3.5, 2.0, and 2.5 mmol g^{-1}, respectively) at pH 1.0, 3.0, and 6.0, respectively. Adsorption equilibrium was reached within 30 minutes. Interfering ions such as K^+, NH_4^+, and Ca^{2+} had minimal to no effect, but Na^+ and Mg^{2+} caused a significant decrease, which was thought to be mainly due to possible interference via binding with active sites.

Cellulose, chitin, and chitosan modified with Ti or Ni oxide nanoparticles were tested for removal of Eu(III) and other metals from radioactive wastewater [100]. All sorbents showed comparable performance: Eu(III) adsorption was negligible at pH 2, whereas 99% of the Eu(II) could be removed at pH 4–8. The authors suggested that adsorption was through inner-sphere surface complexation.

7.3.3 Alginate-based Adsorbents

7.3.3.1 Structure and Chemistry of Alginate

Alginic acid or alginate is a naturally occurring polysaccharide found in the cell walls of brown algae. Most commonly used as sodium alginate, it is a nontoxic and stable material with gel-forming and complexing abilities. In the presence of polyvalent cations like Ca(II), alginate can form a gel through ionotropic gelation [101]. The carboxyl and hydroxyl groups on the polymer chain can bind to cations. The disadvantages of alginate, however, are its high rigidity with poor mechanical strength, which can be overcome by physical and chemical modification. Alginate beads can serve as a matrix for fine particles like GO or biochar that are difficult to separate from water [21]. The adsorption performance of selected alginate-containing composites is given in Table 7.3.

Table 7.3 *Adsorption performance of precious and rare earth metals of selected alginate-based composites.*

Material	Element	Maximum adsorption capacity (mg g^{-1})	pH	T (°C)	Time until equilibrium (min)	Reference
Algal beads/PEI (AB/PEI)	Pd(II)	136.2	2.5	20	—	[102]
Laponite-calcium alginate (CA-Lap)	Pd(II)	69–170[a]	2–4[a]	25	—	[103]
Montmorrilonite-calcium alginate (CA-Mt)	Pd(II)	65–116[a]	2–4[a]	25	—	[103]
Algal beads/PEI (AB/PEI)	Pt(IV)	115.1	2.5	20	—	[102]
Iron oxide loaded calcium alginate beads	La(III)	123.5	5	25	1320	[65]
Magnetic alginate beads	La(III)	250	4	25	220	[104]
Alginate-clay-PNIPAm	La(III)	182	5	25	80	[105]
Sodium alginate-poly-γ-glutamate (SA-PGA) hydrogel	La(III)	163.9	5	25	90	[106]
Sodium alginate-poly-γ-glutamate (SA-PGA) hydrogel	Ce(III)	153.9	5	25	90	[106]
Alginate/SiO_2	Nd(III)	161.5	3.5	25	360	[107]
Alginate/P507/$CaSiO_3$	Nd(III)	193.3	3.6	25	540	[64]
Alginate/PGA	Nd(III)	238	3.6	25	360	[108]

[a] Some experiments were carried out in 0.01 M NaCl.

7.3.3.2 Alginate Composites for Precious Metal Adsorption

There are only a few reports on the use of alginate composites for the adsorption of precious metals. Santos et al. [38] prepared sericin-alginate cross-linked by SAPAs and PVA (SAPVA) to reduce the water solubility of alginate. At pH 2.5–3.0, both composites showed metal affinity in the order $AuCl_4^-$ > $PdCl_4^{2-}$ > $PtCl_6^{2-}$ > Ag^+ (Figure 7.6). The low affinity for Ag^+ was explained by electrostatic repulsion caused by the positively charged sericin and alginate under acidic conditions, while the other metals existed as their anionic chloro-metal complexes. The SAPAs had very low SSA (<0.009 m^2 g^{-1}) and porosity (1.964%). Metallic Au nanoparticles were visually observed after adsorption of Au(III), indicating reduction to be the dominant mechanism. Adsorption of Au(III) was mainly controlled by external diffusion, but the maximum capacity was not measured.

Glutaraldehyde-cross-linked PEI (GLA-PEI) was incorporated into algal biomass beads (AB/PEI) for adsorption of Pd(II) and Pt(IV) [102]. The authors showed that drying conditions had a critical impact on diffusion properties and therewith adsorption capacity; freeze-drying prevented structure collapse that commonly occurs with air-drying. The sorbent had a preference for Pd(II) over Pt(IV) in both single and binary Pd(II)/Pt(IV) solutions. At pH 2.5, the maximum adsorption capacity for Pd(II) was 136.2 mg g^{-1} (1.28 mmol g^{-1}), higher than for Pt(IV) (115.1 mg g^{-1}, 0.59 mmol g^{-1}). The resistance to intraparticle diffusion played a major role in the uptake kinetics.

Cataldo et al. [103] incorporated laponite (Lap) and montmorillonite (Mt) clays into calcium alginate gel beads for adsorption of Pd(II). In the presence of 0.01 M NaCl, the Pd(II) adsorption was optimal at pH 3 with adsorption capacities of 74, 170, and 94 mg g^{-1} dry beads for CA, CA-Lap, and CA-Mt, respectively. Adsorption was suggested to take place via ion exchange between Ca(II) and Pd(II) (through interaction with carboxyl groups of alginate) and binding of Pd(II) with silanol groups on the clay surface. At pH 2, in the absence of NaCl, however, there was little difference between the sorbents (119, 100, and 116 mg g^{-1} dry beads, respectively). The authors suggested that this was due to the protonation of surface silanol groups making them unavailable for Pd(II) binding.

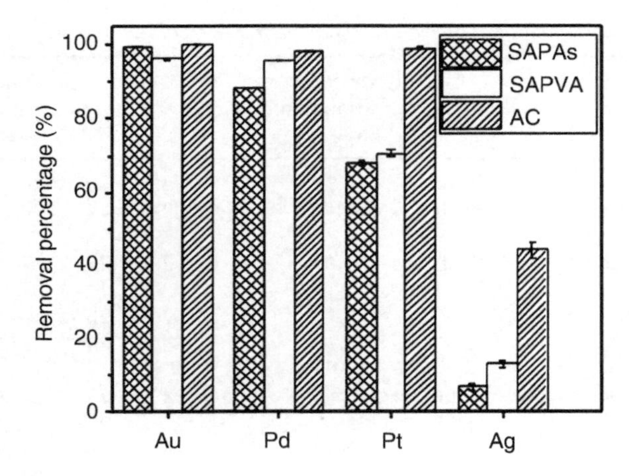

Figure 7.6 *Removal of Au, Pd, Pt, and Ag by SAPAs, SAPVA, and activated carbon (AC) at a fixed initial metal concentration (0.94 ± 0.08 mmol l⁻¹) at 25 °C and pH = 2.5–3.0. Source: Santos et al. [38]. Reproduced with permission from Springer Nature.*

7.3.3.3 Alginate Composites for Rare Earth Metal Adsorption

Alginate-based composites have been widely studied for the recovery of rare earth metals (Table 7.3). Wu et al. [65] have prepared iron oxide-loaded calcium alginate beads for adsorption of La(III). Adsorption capacity increased with pH with optimal adsorption at pH 5.0, where the maximum adsorption capacity of 123.5 mg g^{-1} was obtained (25 °C). Uptake, however, was slow: equilibrium was only reached after 22 hours. A drying step after preparation may have caused the porous beads to collapse. Magnetic alginate beads prepared by Elwakeel et al. [104] had a higher adsorption capacity of 250 mg g^{-1} (1.8 mmol g^{-1}) at pH 4 (25 °C) by using wet beads without drying. Moreover, equilibrium was reached more rapidly after 220 minutes.

Wu et al. [105] have prepared alginate-clay-poly(n-isopropylacrylamide) hydrogels for selective adsorption of La(III). The self-standing gels were prepared in a one-pot photoinitiated polymerization technique at subzero temperatures. Adsorption of La(III) proceeded via ion exchange and the maximum capacity for La(III) was 182 mg g^{-1} at pH 5. La(III) could be effectively separated from mixtures with several common cations and even over other trivalent rare earth metal ions like Yb(III), Eu(III), Ce(III), and Gd(III); the high selectivity was explained by the larger ionic radius of La(III) that had a high affinity for —COOH groups on the alginate surface. Moreover, the hydrogel demonstrated good mechanical properties and excellent stability in acidic, alkaline, and organic media. The authors further highlight the facile method of preparation and operation, tunable adsorption capacity and mechanical strength, reversible temperature-initiated swelling/deswelling behavior, and acceptable cost of gel preparation.

Alginate composites with nano-SiO_2 [107], $CaSiO_3$ [64], and γ-poly glutamic acid (PGA, Wang et al. [108]) have been developed for the adsorption of a large number of trivalent rare earth metals. In-depth studies were carried out using Nd(III) as a model ion. Each of the composites had a maximum Nd(III) adsorption capacity at pH close to 3.5 and showed high selectivity over a large number of divalent metal ions. Among materials tested, the calcium alginate/PGA gel had the highest capacity of 238 mg g^{-1} (1.65 mmol g^{-1}) at pH 3.6, higher than that of pure alginate (194.73 mg g^{-1}) [108]. Moreover, the incorporation of PGA increased the selectivity toward REEs over non-REEs. Adsorption took place according to a cation exchange mechanism. The alginate/$CaSiO_3$ composite suffered from loss of P507 ionophore during desorption resulting in a decreased adsorption activity [64]. In the case of alginate/SiO_2, the addition of SiO_2 (25 and 40 wt%) improved bead mechanical strength but reduced the maximum adsorption capacity over bare alginate due to a reduction in the alginate content. Recovery of these sorbents was excellent; the Nd(III) adsorption capacity slightly increased after the first desorption cycle due to removal of Ca(II) from alginate and remained constant over 12 cycles [107]. Xu et al. [106] also prepared alginate/PGA composites and obtained similar results as Wang et al. [108]: The adsorption capacity was 163.93 mg g^{-1} for La(III) and 153.85 mg g^{-1} for Ce(III), twice as high as on alginate alone. The optimum pH for adsorption was 4.0–5.0.

A different approach to prepare alginate/silica hybrid materials for the recovery of Nd(III) is by using vibrating nozzle technology [109]. In this work, two composite preparation methods were explored, namely the addition of commercial silica particles (alginate-SiO_2) and in situ silica polymerization using TMOS silica precursor (alginate-TMOS). Detailed structural characterization was carried out to compare the two composites. Despite its lower SSA (4.6 m^2 g^{-1}) the alginate-SiO_2 adsorbed more Nd(III) than alginate-TMOS (216.1 m^2 g^{-1}) due to its higher alginate content. Porosity and SSA, however, were still important for adsorption because dried spheres had a lower capacity than wet ones due to a collapse of the porous structure. Adsorption was optimal at pH 5–6 where 0.4–0.6 mmol g^{-1} of Nd(III) could be adsorbed, but the maximum capacity was not determined. Adsorption efficiency dropped almost 20% after the first desorption cycle (using 0.25 M HCl) due to conformational changes but remained over 80% after the seventh cycle.

7.3.4 Lignin-based Composite Adsorbents

7.3.4.1 Lignin Structure and Chemistry

Lignin is an amorphous biopolymer that is widely abundant in biomass including wood and agricultural wastes. Its structure is derived from monolignols, primarily coniferyl alcohol, sinapyl alcohol, and p-coumaryl alcohol, but the composition and structure depend on the biomass [110]. The largest proportion of lignin is not isolated directly from biomass but exists as a by-product of industrial paper manufacturing in the form of alkaline lignin, Kraft lignin, or lignosulfonates (LSs) that contain small amounts of sulfur [23, 111]. Non-modified lignins have been used as sorbents for heavy metals but have a relatively low adsorption capacity [22]. A large number of hydroxyl groups of lignin can be chemically modified with oxygen-, nitrogen-, or sulfur-containing functional groups to increase the affinity for metal ions [22, 23]. Physical modification of lignin has also been carried out to improve mechanical properties or adsorption performance. Modified lignin materials for the adsorption of heavy metals, dyes, and gases have been reviewed previously [22, 23]. Table 7.4 gives an overview of the adsorption capacity for precious metals of selected lignin-based composites.

7.3.4.2 Lignin Composites for Precious Metal Adsorption

A composite containing PANI and enzymatic hydrolysis lignin (EHL) for adsorption of Ag(I) was prepared by He et al. [40]. The EHL was dissolved in ammonia and mixed with aniline, which was polymerized through the addition of ammonium persulfate as an oxidizer. The PANI-EHL composite material consisted of hierarchical structures of approximately 480 nm diameter. The maximum Ag(I) adsorption capacity of the PANI-EHL was 662 mg g^{-1} with an equilibrium time of 22 hours. The capacity was higher than EHL and PANI alone, suggesting the presence of synergistic effects. The authors propose that the oxygen-containing

Table 7.4 Adsorption performance of precious metals of selected lignin-based composites.

Material	Element	Maximum adsorption capacity (mg g^{-1})	pH	T (°C)	Time until equilibrium (min)	Reference
Polyaniline-lignin (PANI-EHL)	Ag(I)	662	3.28–2.64[a]	25	1320	[40]
Poly(N-butylaniline)-lignosulfonate (PBA-LS)	Ag(I)	1148.6	4.96–1.94[a]	25	5760	[42]
Poly(N-methylaniline)-lignosulfonate (PNMA-LS)	Ag(I)	1978	4.96–2.02[a]	25	2880	[43]
Poly(N-methylaniline)-enzymatic hydrolysis lignin (lignin-PNMA)	Ag(I)	1801.1	5.65	25	4320	[41]

[a] pH at the start of the experiment – pH at the end of the experiment.

groups of EHL could chelate Ag(I) whereas the nitrogen-containing functional groups of PANI could both chelate and reduce Ag(I). Silver adsorption was primarily through the reduction of Ag(I) to Ag^0 threads and crystals.

The group of Lü et al. [42, 43] prepared electroconductive LS composites with amino-N-substituted PANI, namely poly(N-butylaniline)-lignosulfonate (PBA) [42] and poly(N-methylaniline)-lignosulfonate (PNMA) [43]. In the first process, the addition of LS to PBA reduced the particle size and improved morphology control through dispersion action. The optimal LS content was 10%, with good stability (no aggregation for at least one year), and a maximum Ag(I) adsorption capacity as high as 1148.6 mg g^{-1} at 25 °C. This high capacity could be achieved by the combination of imino and sulfonic groups on PBA-LS composite chains: the oxygen/sulfur-containing groups of LS can chelate Ag(I), whereas the nitrogen-containing groups of PBA can chelate and reduce Ag(I) at the same time. Indeed, Ag^0 nanoparticles (6.8–55 nm) were formed on the composite surface as evidenced by TEM analysis. Similar results were obtained for the PNMA-LS composite spheres (Figure 7.7a). Adsorption equilibrium was reached after 48 hours due to slow Ag(I) diffusion and adsorption caused by chelation and deposition of Ag^0 (Figure 7.7c). Despite their larger size, the maximum Ag(I) adsorption capacity onto PNMA-LS was even higher with 1978 mg g^{-1} at 25 °C. The formed nanosilver (2–18 nm, Figure 7.7b) showed good antimicrobial properties against *E. coli* and *S. aureus*. The same authors have also prepared composites of PNMA with EHL, which had a calculated maximum Ag(I) adsorption capacity of 1801.1 mg g^{-1} [41].

7.3.4.3 Lignin Composites for Rare Earth Metal Adsorption

Silica-supported pyrolyzed Kraft lignin (pLG@silica) for adsorption of REEs from natural (tap, surface, and sea) waters was prepared by adsorption of 5% lignin onto silica microparticles followed by vacuum pyrolysis at 600 °C [112]. These pyrolysis conditions were selected to obtain sp^2 carbon structures with oxygen-containing functional groups. Quantitative adsorption was found for all REEs at ng/L concentrations, without any evident effects from the presence of other dissolved ions or organic matter. The composite sorbent showed good batch-to-batch reproducibility and could be used for at least 20 adsorption/desorption cycles, even in highly saline conditions like seawater. Adsorption was likely primarily due to surface complexation between metal ions and carboxyl groups [112].

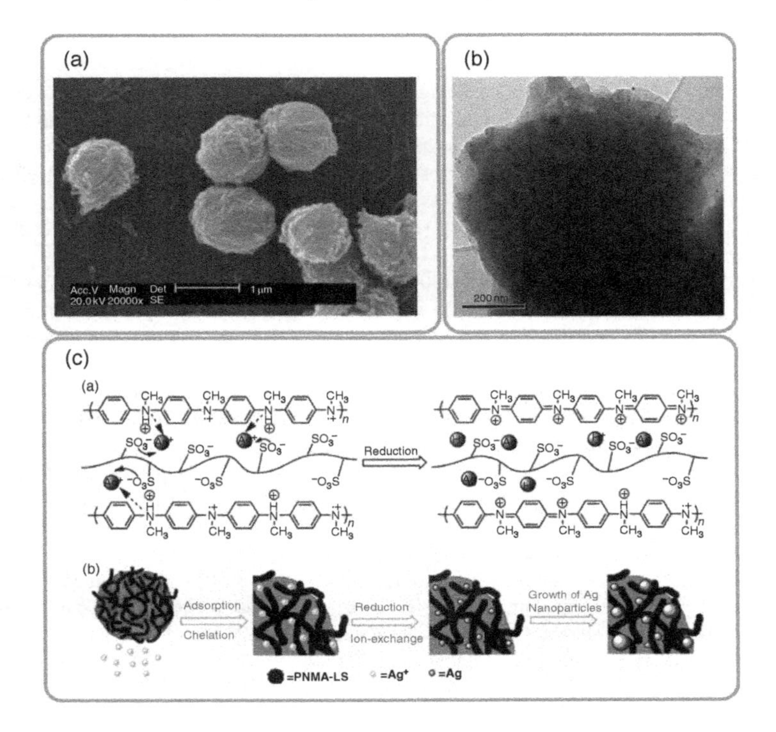

Figure 7.7 (a) SEM image of the PNMA-LS composite spheres, (b) TEM image of Ag^0 nanoparticles formed after Ag(I) adsorption on the PNMA-LS composite spheres, and (c) possible adsorption mechanisms of Ag(I) ions onto the PNMA-LS composite. Source: Lü et al. [43]. Reproduced with permission from Wiley-VCH.

7.4 Conclusion and Outlook

This chapter presents an overview of the recent advancements in biopolymer composite sorbents for the recovery of precious metals and REEs, with the focus on composites based on cellulose, chitosan, alginate, and lignin. Incorporation of one or more solid phases can significantly improve the physical properties (e.g. increase surface area, mechanical stability, and ease of separation) and/or adsorption capacity of the biomass. Additional chemical modification with electron-rich functional groups (such as sulfur, nitrogen, and oxygen) has also shown to improve ion uptake. In addition, the environment such as pH can also play an important role in the adsorption capacity as H^+ or Cl^- can compete with ion exchange of the precious metals and REEs. The composite sorbents have shown great promise. However, their application on an industrial scale needs to be demonstrated. Preparation of adsorbent, regeneration, and ease of sorbent recovery and metal recovery are the minimal criteria that should be considered. Extensive laborious work during these steps should be avoided or minimized for large-scale applications.

After adsorption, the recovered elements are most often desorbed using thiourea and/or acids, allowing the sorbent to be reused. A large opportunity exists in the direct valorization of the adsorbed metals by conversion of the metal-loaded composite into value-added products. For example, after reductive adsorption of Ag(I), the nanosilver-loaded sorbents can be used as antibacterial materials, of which some examples have been given in this chapter. Noble metal-loaded sorbents could be directly converted to catalysts and similar applications can be envisioned for REE-loaded sorbent materials.

References

1. Reck, B.K. and Graedel, T.E. (2012). Challenges in metal recycling. *Science* 337 (6095): 690–695.
2. European Commission. (2017). Communication from the Commission to the European Parliament, the Council, the European Economic and Social Committee and the Committee of the Regions on the 2017 list of Critical Raw Materials for the EU. https://ec.europa.eu/transparency/regdoc/rep/1/2017/EN/COM-2017-490-F1-EN-MAIN-PART-1.PDF (accessed 11 November 2021).
3. Wang, M., Tan, Q., Chiang, J.F., and Li, J. (2017). Recovery of rare and precious metals from urban mines – a review. *Frontiers of Environmental Science and Engineering* 11 (5): 1–17.
4. Haynes, W.M., Lide, D.R., and Bruno, T.J. (2014). Abundance of elements in the Earth's crust and in the sea. In: *CRC Handbook of Chemistry and Physics* (ed. W.M. Haynes), 14–19. Boca Raton, FL: CRC Press.
5. Jordens, A., Cheng, Y.P., and Waters, K.E. (2013). A review of the beneficiation of rare earth element bearing minerals. *Minerals Engineering* 41: 97–114.
6. Long, K.R., Van Gosen, B.S., Foley, N.K., and Cordier, D. (2010). *The Principal Rare Earth Elements Deposits of the United States – A Summary of Domestic Deposits and a Global Perspective*. Reston, VA: U.S. Geological Survey Scientific Investigations Report 2010–5220.
7. Buechler, D.T., Zyaykina, N.N., Spencer, C.A. et al. (2020). Comprehensive elemental analysis of consumer electronic devices: rare earth, precious, and critical elements. *Waste Management* 103: 67–75.
8. Supanchaiyamat, N. and Hunt, A.J. (2019). Conservation of critical elements of the periodic table. *ChemSusChem* 12 (2): 397–403.
9. Hunt, A.J. (2013). *Element Recovery and Sustainability*. Cambridge: RSC Publishing.
10. Lu, Y. and Xu, Z. (2016). Precious metals recovery from waste printed circuit boards: a review for current status and perspective. *Resources, Conservation and Recycling* 113: 28–39.
11. Sethurajan, M., van Hullebusch, E.D., Fontana, D. et al. (2019). Recent advances on hydrometallurgical recovery of critical and precious elements from end of life electronic wastes – a review. *Critical Reviews in Environmental Science and Technology* 49 (3): 212–275.
12. Anastopoulos, I., Bhatnagar, A., and Lima, E.C. (2016). Adsorption of rare earth metals: a review of recent literature. *Journal of Molecular Liquids* 221: 954–962.
13. Das, N. (2010). Recovery of precious metals through biosorption – A review. *Hydrometallurgy* 103 (1–4): 180–189.
14. Mack, C., Wilhelmi, B., Duncan, J.R., and Burgess, J.E. (2007). Biosorption of precious metals. *Biotechnology Advances* 25 (3): 264–271.
15. Won, S.W., Kotte, P., Wei, W. et al. (2014). Biosorbents for recovery of precious metals. *Bioresource Technology* 160: 203–212.
16. Hokkanen, S., Bhatnagar, A., and Sillanpää, M. (2016). A review on modification methods to cellulose-based adsorbents to improve adsorption capacity. *Water Research* 91: 156–173.
17. Suhas, D., Gupta, V.K., Carrott, P.J.M. et al. (2016). Cellulose: a review as natural, modified and activated carbon adsorbent. *Bioresource Technology* 216: 1066–1076.
18. Wang, J. and Zhuang, S. (2017). Removal of various pollutants from water and wastewater by modified chitosan adsorbents. *Critical Reviews in Environmental Science and Technology* 47 (23): 2331–2386.
19. Zhang, L., Zeng, Y., and Cheng, Z. (2016). Removal of heavy metal ions using chitosan and modified chitosan: a review. *Journal of Molecular Liquids* 214: 175–191.
20. Thakur, S., Sharma, B., Verma, A. et al. (2018). Recent progress in sodium alginate based sustainable hydrogels for environmental applications. *Journal of Cleaner Production* 198: 143–159.
21. Wang, B., Wan, Y., Zheng, Y. et al. (2019). Alginate-based composites for environmental applications: a critical review. *Critical Reviews in Environmental Science and Technology* 49 (4): 318–356.
22. Ge, Y. and Li, Z. (2018). Application of lignin and its derivatives in adsorption of heavy metal ions in water: a review. *ACS Sustainable Chemistry and Engineering* 6 (5): 7181–7192.
23. Supanchaiyamat, N., Jetsrisuparb, K., Knijnenburg, J.T.N. et al. (2019). Lignin materials for adsorption: current trend, perspectives and opportunities. *Bioresource Technology* 272: 570–581.
24. Santos, S.C.R., Bacelo, H.A.M., Boaventura, R.A.R., and Botelho, C.M.S. (2019). Tannin-adsorbents for water decontamination and for the recovery of critical metals: current state and future perspectives. *Biotechnology Journal* 14 (12): 1–14.
25. Bhuyan, M.M., Okabe, H., Hidaka, Y., and Hara, K. (2018). Pectin-[(3-acrylamidopropyl) trimethylammonium chloride-*co*-acrylic acid] hydrogel prepared by gamma radiation and selectively silver (Ag) metal adsorption. *Journal of Applied Polymer Science* 135 (8): 1–14.

26. Kusrini, E., Wicaksono, W., Gunawan, C. et al. (2018). Kinetics, mechanism, and thermodynamics of lanthanum adsorption on pectin extracted from durian rind. *Journal of Environmental Chemical Engineering* 6 (5): 6580–6588.
27. Iftekhar, S., Srivastava, V., Ramasamy, D.L. et al. (2018). A novel approach for synthesis of exfoliated biopolymeric-LDH hybrid nanocomposites via in-stiu coprecipitation with gum Arabic: application towards REEs recovery. *Chemical Engineering Journal* 347: 398–406.
28. Padil, V.V.T., Stuchlík, M., and Černík, M. (2015). Plasma modified nanofibres based on gum kondagogu and their use for collection of nanoparticulate silver, gold and platinum. *Carbohydrate Polymers* 121: 468–476.
29. Lee, Y.C., Rengaraj, A., Ryu, T. et al. (2016). Adsorption of rare earth metals (Sr^{2+} and La^{3+}) from aqueous solution by Mg-aminoclay-humic acid (MgAC-HA) complexes in batch mode. *RSC Advances* 6 (2): 1324–1332.
30. Mukhopadhyay, R., Bhaduri, D., Sarkar, B. et al. (2020). Clay–polymer nanocomposites: progress and challenges for use in sustainable water treatment. *Journal of Hazardous Materials* 383: 1–17.
31. Liu, P., Sehaqui, H., Tingaut, P. et al. (2014). Cellulose and chitin nanomaterials for capturing silver ions (Ag^+) from water via surface adsorption. *Cellulose* 21 (1): 449–461.
32. Beyki, M.H., Bayat, M., Miri, S. et al. (2014). Synthesis, characterization, and silver adsorption property of magnetic cellulose xanthate from acidic solution: prepared by one step and biogenic approach. *Industrial and Engineering Chemistry Research* 53 (39): 14904–14912.
33. Fan, L., Luo, C., Lv, Z. et al. (2011). Removal of Ag^+ from water environment using a novel magnetic thiourea-chitosan imprinted Ag^+. *Journal of Hazardous Materials* 194: 193–201.
34. Hyder, M.K.M.Z. and Ochiai, B. (2017). Synthesis of a selective scavenger for Ag(I), Pd(II), and Au(III) based on cellulose filter paper grafted with polymer chains bearing thiocarbamate moieties. *Chemistry Letters* 46 (4): 492–494.
35. Tran, T.H., Okabe, H., Hidaka, Y., and Hara, K. (2019). Equilibrium and kinetic studies for silver removal from aqueous solution by hybrid hydrogels. *Journal of Hazardous Materials* 365: 237–244.
36. Lasko, C.L. and Hurst, M.P. (1999). An investigation into the use of chitosan for the removal of soluble silver from industrial wastewater. *Environmental Science and Technology* 33 (20): 3622–3626.
37. Nitayaphat, W. and Jintakosol, T. (2015). Removal of silver(I) from aqueous solutions by chitosan/bamboo charcoal composite beads. *Journal of Cleaner Production* 87 (1): 850–855.
38. Santos, N.T.G., da Silva, M.G.C., and Vieira, M.G.A. (2018). Development of novel sericin and alginate-based biosorbents for precious metal removal from wastewater. *Environmental Science and Pollution Research* 26: 28455–28469.
39. Shih, T.Y. and Tsai, S.L. (2014). Simultaneous silver recovery and bactericidal bionanocomposite formation via engineered biomolecules. *RSC Advances* 4 (77): 40994–40998.
40. He, Z.W., Lü, Q.F., and Zhang, J.Y. (2012). Facile preparation of hierarchical polyaniline-lignin composite with a reactive silver-ion adsorbability. *ACS Applied Materials and Interfaces* 4 (1): 369–374.
41. Lü, Q.F., Luo, J.J., Lin, T.T., and Zhang, Y.Z. (2014). Novel lignin-poly(*N*-methylaniline) composite sorbent for silver ion removal and recovery. *ACS Sustainable Chemistry and Engineering* 2 (3): 465–471.
42. Lü, Q.F., Zhang, J.Y., and He, Z.W. (2012). Controlled preparation and reactive silver-ion sorption of electrically conductive poly(*N*-butylaniline)-lignosulfonate composite nanospheres. *Chemistry – A European Journal* 18 (51): 16571–16579.
43. Lü, Q.F., Zhang, J.Y., Yang, J. et al. (2013). Self-assembled poly(*N*-methylaniline)-lignosulfonate spheres: from silver-ion adsorbent to antimicrobial material. *Chemistry – A European Journal* 19 (33): 10935–10944.
44. Guibal, E. (2004). Interactions of metal ions with chitosan-based sorbents: a review. *Separation and Purification Technology* 38 (1): 43–74.
45. Ramesh, A., Hasegawa, H., Sugimoto, W. et al. (2008). Adsorption of gold(III), platinum(IV) and palladium(II) onto glycine modified crosslinked chitosan resin. *Bioresource Technology* 99 (9): 3801–3809.
46. Zhou, L., Xu, J., Liang, X., and Liu, Z. (2010). Adsorption of platinum(IV) and palladium(II) from aqueous solution by magnetic cross-linking chitosan nanoparticles modified with ethylenediamine. *Journal of Hazardous Materials* 182 (1–3): 518–524.
47. Donia, A.M., Atia, A.A., and Elwakeel, K.Z. (2007). Recovery of gold(III) and silver(I) on a chemically modified chitosan with magnetic properties. *Hydrometallurgy* 87 (3–4): 197–206.
48. Park, S.I., Kwak, I.S., Bae, M.A. et al. (2012). Recovery of gold as a type of porous fiber by using biosorption followed by incineration. *Bioresource Technology* 104: 208–214.

49. Song, M.H., Kim, S., Reddy, D.H.K. et al. (2017). Development of polyethyleneimine-loaded core-shell chitosan hollow beads and their application for platinum recovery in sequential metal scavenging fill-and-draw process. *Journal of Hazardous Materials* 324: 724–731.
50. Liu, L., Li, C., Bao, C. et al. (2012). Preparation and characterization of chitosan/graphene oxide composites for the adsorption of Au(III) and Pd(II). *Talanta* 93: 350–357.
51. Castillo-Ortega, M.M., Santos-Sauceda, I., Encinas, J.C. et al. (2011). Adsorption and desorption of a gold-iodide complex onto cellulose acetate membrane coated with polyaniline or polypyrrole: a comparative study. *Journal of Materials Science* 46 (23): 7466–7474.
52. Labib, S.A., Yousif, A.M., Ibrahim, I.A., and Atia, A.A. (2018). Adsorption of rhodium by modified mesoporous cellulose/silica sorbents: equilibrium, kinetic, and thermodynamic studies. *Journal of Porous Materials* 25 (2): 383–396.
53. Yousif, A.M., Labib, S.A., Ibrahim, I.A., and Atia, A.A. (2019). Recovery of Pt (IV) from aqueous solutions using magnetic functionalized cellulose with quaternary amine. *Separation Science and Technology (Philadelphia)* 54 (8): 1257–1268.
54. Yu, D., Wang, W., and Wu, J. (2011). Preparation of conductive wool fabrics and adsorption behaviour of Pd (II) ions on chitosan in the pre-treatment. *Synthetic Metals* 161 (1–2): 124–131.
55. Zhou, Z., Liu, F., Huang, Y. et al. (2015). Biosorption of palladium(II) from aqueous solution by grafting chitosan on persimmon tannin extract. *International Journal of Biological Macromolecules* 77: 336–343.
56. Benguerel, E., Demopoulos, G.P., and Harris, G.B. (1996). Speciation and separation of rhodium (III) from chloride solutions: a critical review. *Hydrometallurgy* 40 (1–2): 135–152.
57. Morisada, S., Rin, T., Ogata, T. et al. (2012). Adsorption recovery of rhodium(III) in acidic chloride solutions by amine-modified tannin gel. *Journal of Applied Polymer Science* 126 (SUPPL. 2): E34–E38.
58. Bernardis, F.L., Grant, R.A., and Sherrington, D.C. (2005). A review of methods of separation of the platinum-group metals through their chloro-complexes. *Reactive and Functional Polymers* 65 (3): 205–217.
59. Yavuz, E., Tokalıoğlu, Ş., Şahan, H. et al. (2017). Dispersive solid-phase extraction of rhodium from water, street dust, and catalytic converters using a cellulose–graphite oxide composite. *Analytical Letters* 50 (1): 63–79.
60. Yousif, A.M. (2011). Rapid and selective adsorption of Rh(III) from its solutions using cellulose-based biosorbent containing iminodiacetate functionality. *Separation Science and Technology* 46 (15): 2341–2347.
61. Wood, S.A. (1990). The aqueous geochemistry of the rare-earth elements and yttrium. 1. Review of available low-temperature data for inorganic complexes and the inorganic REE speciation of natural waters. *Chemical Geology* 82 (C): 159–186.
62. Li, K., Gao, Q., Yadavalli, G. et al. (2015). Selective adsorption of Gd^{3+} on a magnetically retrievable imprinted chitosan/carbon nanotube composite with high capacity. *ACS Applied Materials and Interfaces* 7 (38): 21047–21055.
63. Liu, E., Zheng, X., Xu, X. et al. (2017). Preparation of diethylenetriamine-modified magnetic chitosan nanoparticles for adsorption of rare-earth metal ions. *New Journal of Chemistry* 41 (15): 7739–7750.
64. Wang, F., Zhao, J., Li, W. et al. (2013). Preparation of several alginate matrix gel beads and their adsorption properties towards rare earths (III). *Waste and Biomass Valorization* 4 (3): 665–674.
65. Wu, D., Zhao, J., Zhang, L. et al. (2010). Lanthanum adsorption using iron oxide loaded calcium alginate beads. *Hydrometallurgy* 101 (1–2): 76–83.
66. Najafi Lahiji, M., Keshtkar, A.R., and Moosavian, M.A. (2018). Adsorption of cerium and lanthanum from aqueous solutions by chitosan/polyvinyl alcohol/3-mercaptopropyltrimethoxysilane beads in batch and fixed-bed systems. *Particulate Science and Technology* 36 (3): 340–350.
67. Wang, F., Wang, W., Zhu, Y., and Wang, A. (2017). Evaluation of Ce(III) and Gd(III) adsorption from aqueous solution using CTS-g-(AA-co-SS)/ISC hybrid hydrogel adsorbent. *Journal of Rare Earths* 35 (7): 697–708.
68. O'Connell, D.W., Birkinshaw, C., and O'Dwyer, T.F. (2008). Heavy metal adsorbents prepared from the modification of cellulose: a review. *Bioresource Technology* 99 (15): 6709–6724.
69. Donia, A.M., Atia, A.A., and Abouzayed, F.I. (2012). Preparation and characterization of nano-magnetic cellulose with fast kinetic properties towards the adsorption of some metal ions. *Chemical Engineering Journal* 191: 22–30.

70. Hyder, M.K.M.Z. and Ochiai, B. (2018). Selective recovery of Au(III), Pd(II), and Ag(I) from printed circuit boards using cellulose filter paper grafted with polymer chains bearing thiocarbamate moieties. *Microsystem Technologies* 24 (1): 683–690.
71. Rascón-Leon, S., Castillo-Ortega, M.M., Santos-Sauceda, I. et al. (2018). Selective adsorption of gold and silver in bromine solutions by acetate cellulose composite membranes coated with polyaniline or polypyrrole. *Polymer Bulletin* 75 (7): 3241–3265.
72. Rodríguez, F., Castillo-Ortega, M.M., Encinas, J.C. et al. (2009). Adsorption of a gold-iodide complex (AuI_2^-) onto cellulose acetate-polyaniline membranes: equilibrium experiments. *Journal of Applied Polymer Science* 113 (4): 2670–2674.
73. Anbia, M. and Rahimi, F. (2017). Adsorption of platinum(IV) from an aqueous solution with magnetic cellulose functionalized with thiol and amine as a nano-active adsorbent. *Journal of Applied Polymer Science* 134 (39): 1–8.
74. Hong, H.J., Yu, H., Park, M., and Jeong, H.S. (2019). Recovery of platinum from waste effluent using polyethyleneimine-modified nanocelluloses: effects of the cellulose source and type. *Carbohydrate Polymers* 210: 167–174.
75. Zhu, Y., Wang, W., Zheng, Y. et al. (2016). Rapid enrichment of rare-earth metals by carboxymethyl cellulose-based open-cellular hydrogel adsorbent from HIPEs template. *Carbohydrate Polymers* 140: 51–58.
76. Iftekhar, S., Srivastava, V., and Sillanpää, M. (2017). Synthesis and application of LDH intercalated cellulose nanocomposite for separation of rare earth elements (REEs). *Chemical Engineering Journal* 309: 130–139.
77. Iftekhar, S., Srivastava, V., and Sillanpää, M. (2017). Enrichment of lanthanides in aqueous system by cellulose based silica nanocomposite. *Chemical Engineering Journal* 320: 151–159.
78. Zhao, F., Repo, E., Song, Y. et al. (2017). Polyethylenimine-cross-linked cellulose nanocrystals for highly efficient recovery of rare earth elements from water and a mechanism study. *Green Chemistry* 19 (20): 4816–4828.
79. Hao, Y., Cui, Y., Peng, J. et al. (2019). Preparation of graphene oxide/cellulose composites in ionic liquid for Ce (III) removal. *Carbohydrate Polymers* 208: 269–275.
80. Arunraj, B., Sathvika, T., Rajesh, V., and Rajesh, N. (2019). Cellulose and *Saccharomyces cerevisiae* embark to recover europium from phosphor powder. *ACS Omega* 4 (1): 940–952.
81. Zheng, X., Zhang, Y., Bian, T. et al. (2020). Oxidized carbon materials cooperative construct ionic imprinted cellulose nanocrystals films for efficient adsorption of Dy(III). *Chemical Engineering Journal* 381: 1–10.
82. Wan Ngah, W.S., Teong, L.C., and Hanafiah, M.A.K.M. (2011). Adsorption of dyes and heavy metal ions by chitosan composites: a review. *Carbohydrate Polymers* 83 (4): 1446–1456.
83. Chang, M.Y. and Juang, R.S. (2004). Adsorption of tannic acid, humic acid, and dyes from water using the composite of chitosan and activated clay. *Journal of Colloid and Interface Science* 278 (1): 18–25.
84. Salehi, E., Daraei, P., and Arabi Shamsabadi, A. (2016). A review on chitosan-based adsorptive membranes. *Carbohydrate Polymers* 152: 419–432.
85. Wu, F.C., Tseng, R.L., and Juang, R.S. (2010). A review and experimental verification of using chitosan and its derivatives as adsorbents for selected heavy metals. *Journal of Environmental Management* 91 (4): 798–806.
86. Shawky, H.A. (2009). Synthesis of ion-imprinting chitosan/PVA crosslinked membrane for selective removal of Ag(I). *Journal of Applied Polymer Science* 114 (5): 2608–2615.
87. Huo, H., Su, H., and Tan, T. (2009). Adsorption of Ag^+ by a surface molecular-imprinted biosorbent. *Chemical Engineering Journal* 150 (1): 139–144.
88. Mu, B., Kang, Y., and Wang, A. (2013). Preparation of a polyelectrolyte-coated magnetic attapulgite composite for the adsorption of precious metals. *Journal of Materials Chemistry A* 1 (15): 4804–4811.
89. Donia, A.M., Yousif, A.M., Atia, A.A., and Elsamalehy, M.F. (2014). Efficient adsorption of Ag(I) and Au(III) on modified magnetic chitosan with amine functionalities. *Desalination and Water Treatment* 52 (13–15): 2537–2547.
90. Jintakosol, T. and Nitayaphat, W. (2016). Adsorption of silver (I) from aqueous solution using chitosan/montmorillonite composite beads. *Materials Research* 19 (5): 1114–1121.
91. Fan, C., Li, K., He, Y. et al. (2018). Evaluation of magnetic chitosan beads for adsorption of heavy metal ions. *Science of the Total Environment* 627: 1396–1403.

92. Kwak, I.S., Won, S.W., Chung, Y.S., and Yun, Y.S. (2013). Ruthenium recovery from acetic acid waste water through sorption with bacterial biosorbent fibers. *Bioresource Technology* 128: 30–35.
93. Omidinasab, M., Rahbar, N., Ahmadi, M. et al. (2018). Removal of vanadium and palladium ions by adsorption onto magnetic chitosan nanoparticles. *Environmental Science and Pollution Research* 25 (34): 34262–34276.
94. Wu, D., Zhang, L., Wang, L. et al. (2011). Adsorption of lanthanum by magnetic alginate-chitosan gel beads. *Journal of Chemical Technology and Biotechnology* 86 (3): 345–352.
95. Mahmoud, M.E., Ali, S.A.A.A., Nassar, A.M.G. et al. (2017). Immobilization of chitosan nanolayers on the surface of nano-titanium oxide as a novel nanocomposite for efficient removal of La(III) from water. *International Journal of Biological Macromolecules* 101: 230–240.
96. Zhang, X., Li, C., Yan, Y. et al. (2010). A Ce^{3+}-imprinted functionalized potassium tetratitanate whisker sorbent prepared by surface molecularly imprinting technique for selective separation and determination of Ce^{3+}. *Microchimica Acta* 169 (3): 289–296.
97. Galhoum, A.A., Mahfouz, M.G., Abdel-Rehem, S.T. et al. (2015). Diethylenetriamine-functionalized chitosan magnetic nano-based particles for the sorption of rare earth metal ions (Nd(III), Dy(III) and Yb(III)). *Cellulose* 22 (4): 2589–2605.
98. Hamza, M.F., Roux, J.C., and Guibal, E. (2018). Uranium and europium sorption on amidoxime-functionalized magnetic chitosan micro-particles. *Chemical Engineering Journal* 344: 124–137.
99. Galhoum, A.A., Atia, A.A., Mahfouz, M.G. et al. (2015). Dy(III) recovery from dilute solutions using magnetic-chitosan nano-based particles grafted with amino acids. *Journal of Materials Science* 50 (7): 2832–2848.
100. Pospěchová, J., Brynych, V., Štengl, V. et al. (2016). Polysaccharide biopolymers modified with titanium or nickel nanoparticles for removal of radionuclides from aqueous solutions. *Journal of Radioanalytical and Nuclear Chemistry* 307 (2): 1303–1314.
101. Patel, M.A., AbouGhaly, M.H.H., Schryer-Praga, J.V., and Chadwick, K. (2017). The effect of ionotropic gelation residence time on alginate cross-linking and properties. *Carbohydrate Polymers* 155: 362–371.
102. Wang, S., Vincent, T., Roux, J.C. et al. (2017). Pd(II) and Pt(IV) sorption using alginate and algal-based beads. *Chemical Engineering Journal* 313: 567–579.
103. Cataldo, S., Muratore, N., Orecchio, S., and Pettignano, A. (2015). Enhancement of adsorption ability of calcium alginate gel beads towards Pd(II) ion. A kinetic and equilibrium study on hybrid Laponite and Montmorillonite-alginate gel beads. *Applied Clay Science* 118: 162–170.
104. Elwakeel, K.Z., Daher, A.M., Abd El-Fatah, A.I.L. et al. (2017). Biosorption of lanthanum from aqueous solutions using magnetic alginate beads. *Journal of Dispersion Science and Technology* 38 (1): 145–151.
105. Wu, D., Gao, Y., Li, W. et al. (2016). Selective adsorption of La^{3+} using a tough alginate-clay-poly(*n*-isopropylacrylamide) hydrogel with hierarchical pores and reversible re-deswelling/swelling cycles. *ACS Sustainable Chemistry and Engineering* 4 (12): 6732–6743.
106. Xu, S., Wang, Z., Gao, Y. et al. (2015). Adsorption of rare earths(III) using an efficient sodium alginate hydrogel cross-linked with poly-γ-glutamate. *PLoS One* 10 (5): 1–12.
107. Wang, F., Zhao, J., Pan, F. et al. (2013). Adsorption properties toward trivalent rare earths by alginate beads doping with silica. *Industrial and Engineering Chemistry Research* 52 (9): 3453–3461.
108. Wang, F., Zhao, J., Wei, X. et al. (2014). Adsorption of rare earths (III) by calcium alginate-poly glutamic acid hybrid gels. *Journal of Chemical Technology and Biotechnology* 89 (7): 969–977.
109. Roosen, J., Pype, J., Binnemans, K., and Mullens, S. (2015). Shaping of alginate-silica hybrid materials into microspheres through vibrating-nozzle technology and their use for the recovery of neodymium from aqueous solutions. *Industrial and Engineering Chemistry Research* 54 (51): 12836–12846.
110. Li, M., Pu, Y., and Ragauskas, A.J. (2016). Current understanding of the correlation of lignin structure with biomass recalcitrance. *Frontiers in Chemistry* 4 (NOV): 1–8.
111. Thakur, V.K., Thakur, M.K., Raghavan, P., and Kessler, M.R. (2014). Progress in green polymer composites from lignin for multifunctional applications: a review. *ACS Sustainable Chemistry and Engineering* 2 (5): 1072–1092.
112. Maraschi, F., Speltini, A., Tavani, T. et al. (2018). Silica-supported pyrolyzed lignin for solid-phase extraction of rare earth elements from fresh and sea waters followed by ICP-MS detection. *Analytical and Bioanalytical Chemistry* 410 (29): 7635–7643.

8

Bio-Based Materials in Anti-HIV Drug Delivery

Oranat Chuchuen[1] and David F. Katz[2]

[1]*Department of Chemical Engineering, Khon Kaen University, Khon Kaen, Thailand*
[2]*Departments of Biomedical Engineering & Obstetrics and Gynecology, Duke University, Durham, North Carolina, USA*

8.1 Introduction

Since the human immunodeficiency virus (HIV) was recognized as the causative agent of the acquired immunodeficiency syndrome (AIDS) in the 1980s [1, 2], the HIV/AIDS pandemic has grown to be a major global health crisis. According to the Joint United Nations Programme on HIV/AIDS (UNAIDS) report (2020), since the beginning of the AIDS epidemic approximately 75.7 million people have been infected with HIV; of these, 32.7 million have died of AIDS-related complications. In 2019, 1.7 million new infections occurred leaving approximately 38 million HIV-infected people worldwide [3]. Through over three decades of research on HIV/AIDS, substantial knowledge has been gained in understanding the infection and ensuing disease, including the biological structure of HIV [4], the modes of HIV transmission [5], the cellular immune responses to HIV [6] and the pathogenesis of HIV/AIDS [7, 8]. These advances in the biomedical understanding of HIV/AIDS have contributed to the development of antiretroviral drugs (ARVs) that target various stages of the viral life cycle, along with antiretroviral therapy (ART) for the treatment of HIV infections. Even though ART has shown tremendous successes in reducing the HIV mortality rate among individuals with HIV/AIDS, it has limitations associated with patient adherence issues and drug adverse and resistance effects [9]. The cure for HIV has not yet been discovered, rendering HIV/AIDS a continuing major global health challenge.

High-Performance Materials from Bio-based Feedstocks, First Edition. Edited by Andrew J. Hunt, Nontipa Supanchaiyamat, Kaewta Jetsrisuparb and Jesper T.N. Knijnenburg.

However, with an advanced understanding of the biology and biophysics of HIV, there has been scientific and clinical interest in developing multiple biomedical prevention methods against HIV, to combat, control and, finally, end the HIV pandemic. Anti-HIV drug delivery systems contribute to a biomedical strategy for HIV prophylaxis, and are being developed for women and men to protect themselves against sexual transmission of HIV. Various anti-HIV drug delivery systems have been developed utilizing renewable, bio-based materials. This chapter starts with an overview of the current biomedical HIV preventive strategies, followed by a discussion of the development of anti-HIV drug delivery systems in terms of their properties, as well as rationale and stage of development. Then, it presents major types of bio-based materials that have been employed in the formulation of anti-HIV drug delivery systems. The chapter concludes with perspectives for the future development of anti-HIV drug delivery systems using bio-based materials.

8.2 Biomedical Strategies for HIV Prophylaxis

Strategies for HIV prophylaxis are needed and must be developed for long-term and effective control of the HIV/AIDS pandemic. Although male and female condoms have been available for many years, there are several barriers to condom use associated with personal and socio-economic limitations [10]. In some underdeveloped countries, the ability of women to negotiate condom use with their partners is limited [11, 12]. In addition, the effectiveness of condoms in preventing the transmission of HIV and other sexually transmitted infections (STIs) depends upon the correctness and consistency of condom use [13, 14].

Anti-HIV vaccine and pre-exposure prophylaxis (PrEP) products are two major biomedical prevention technologies being developed to provide protection against HIV infection. Despite many advances in our knowledge of the cellular immunology of HIV and the immunopathogenesis of AIDS, anti-HIV vaccine development has still confronted several difficulties and challenges, such as the antigenic diversity and hypervariability of HIV [15, 16]. PrEP strategies, an alternative to anti-HIV vaccine, can provide protection against HIV orally, topically, or systematically. Figure 8.1 summarizes biomedical strategies and drug delivery systems that have been developed for HIV-1 PrEP. An oral PrEP product (Truvada® pills) has now been approved by the Food and Drug Administration (FDA) for HIV protection in HIV-negative high-risk groups [28]. The pills combine two potent HIV-1 nucleoside reverse transcriptase inhibitors (NRTIs): emtricitabine and tenofovir disoproxil fumarate. The daily use of Truvada pills was found effective in reducing the risk of contracting HIV to varying degrees ranging from 44% in men who have sex with men [29] to 62.2 and 75% in seronegative [30] and sero-discordant [31] heterosexual couples, respectively. However, oral PrEP is still facing issues with user adherence [32], which is an important factor to ensure product effectiveness. In addition, oral PrEP may not provide protection against other STIs. In fact, oral PrEP use has recently been shown to be associated with the increased rates of certain STIs [33].

Anti-HIV drug delivery systems are intended to offer PrEP, user-controlled products to men and women to protect themselves against HIV sexual transmission. The approach to HIV prevention is not a one-size-fits-all approach; a variety of product types is needed so that users can choose within the constraints of their personal choices and sociocultural and economical factors. User compliance and willingness to use products have a direct impact on overall product efficacy in HIV prevention. Several vaginal and rectal platforms have been developed for the delivery of antiretroviral agents to prevent the transmission of HIV and other STIs. The earlier developments utilized polymeric products for topical drug delivery,

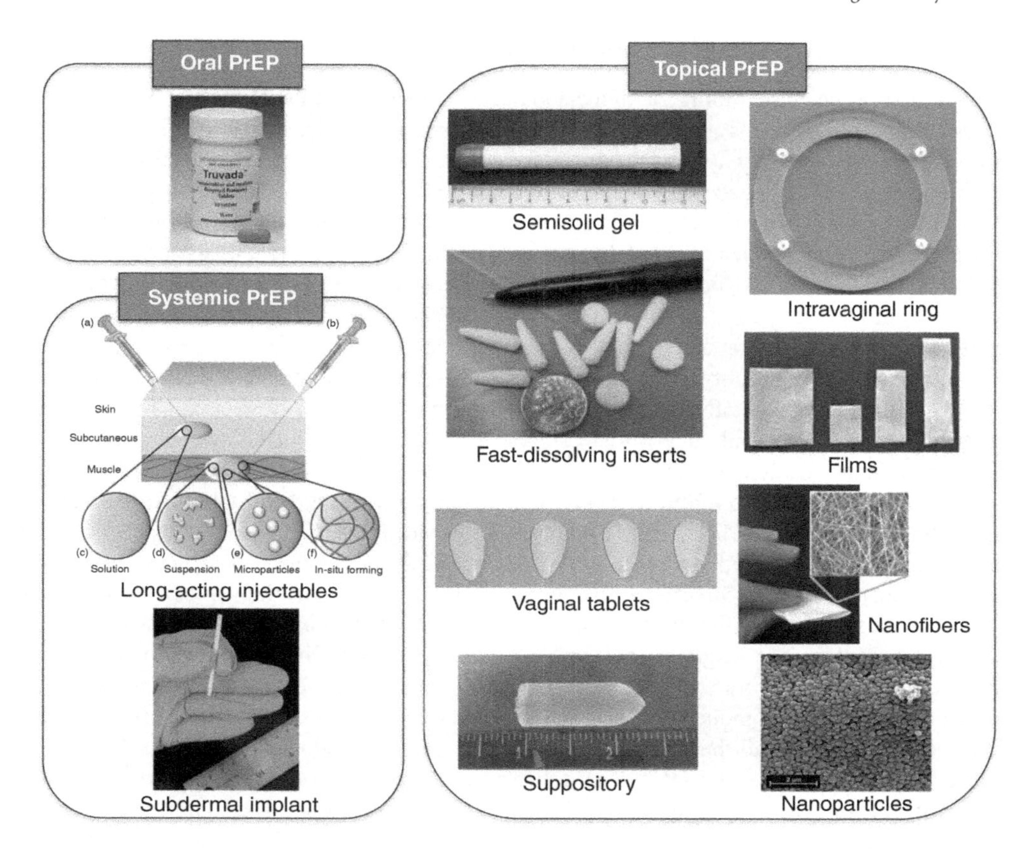

Figure 8.1 *Representative images of various drug delivery systems in development for HIV-1 pre-exposure prophylaxis (PrEP). The oral PrEP pills Truvada® have been approved by the FDA and are in global use (Source: Adapted from De Clercq [17]. Reproduced with permission of John Wiley & Sons). Topical PrEP strategies include semisolid gels (Source: Adapted from Rosenberg and Devlin [18]. Reproduced with permission of Elsevier), fast-dissolving inserts (Source: Adapted from Derby et al. [19]. Licensed by CC BY 4.0), vaginal tablets (Source: Garg et al. [20]. Reproduced with permission of Elsevier), suppositories (Source: Zaveri et al. [21]. Licensed under CC BY 3.0), intravaginal rings (Source: Adapted from Baum et al. [22]. Reproduced with permission of Elsevier), vaginal films (Source: Adapted from Fan et al. [23]. Reproduced with permission of Springer Nature), electrospun nanofibers (Source: Adapted from Blakney et al. [24]. Reproduced with permission of Springer Nature), and nanoparticles (Source: Adapted from Ham et al. [25]. Reproduced with permission of Springer Nature). Systemic PrEP strategies include long-acting injectables (Source: Owen and Rannard [26]. Reproduced with permission of Elsevier) and subdermal implants (Source: Adapted from Johnson et al. [27]. Licensed under CC BY 4.0).*

including semisolid gels and creams [34–42]. More recently, more fast-acting, on-demand products, such as fast-dissolving films [43–48], inserts [19, 49], and suppositories [50, 51], have been, and continue to be, developed for multiple microbicide drugs. Several micro/nanoparticle formulations have been explored to offer sustained drug release to mucosal tissues over several hours [52–56]. In addition, long-acting dosage forms, including intravaginal rings (IVRs) [57–62], long-acting injectables [63–65], and implants [27, 66–68], are being developed to provide extended release of antiretroviral agents without the need of daily product reinsertion/reapplication.

8.3 Properties of Anti-HIV Drug Delivery Systems

Clearly, there is an epidemiological niche and need for a variety of PrEP product types to meet the preferences of various users from different socioeconomic backgrounds. These include non-oral, as well as orally applied products, for on-demand as well as sustained protection.

The functioning of a successful anti-HIV drug delivery product depends upon several interacting factors, including the efficacy of the antiretroviral agents, safety, and biological and physicochemical properties of the delivery vehicle, the release kinetics of the drug from the vehicle, and the characteristics and adherence of the user. An effective anti-HIV drug delivery system should deliver and distribute antiretroviral agents to the mucosal tissue and/or luminal fluid prior to sexual intercourse, establishing prophylactic drug concentrations there to prevent infection by semen-borne HIV during exposure. This could involve direct neutralization of HIV virions or preventative action against viral infection in the host cells. A soft topical product (e.g. gel, film, and suppository) should also have good mucoadhesive properties to ensure prolonged residence time and drug delivery in the vagina or rectum and possibly providing a physical barrier to inhibit HIV or other sexually transmitted pathogens from reaching and initiating infection in host cells. The need for broad drug distribution along the vaginal canal extends to solid products such as fast-dissolving inserts. Long-acting, non-dissolving solid products, viz. IVRs, directly contact only a finite proportion of the vaginal mucosal surfaces. For these products, the distribution and flow of vaginal fluid can act to distribute their drug payloads along the length of the canal. Of course, any anti-HIV product should be safe, nonirritating, and noncytotoxic to the mucosal cells and tissues. Epithelial inflammation of cervicovaginal mucosa induced by vaginal products could increase mucosal HIV-1 infection [69]. Collectively, all these factors must be addressed in product design.

The earliest candidate anti-HIV drug delivery products were semisolid gels (vaginal and rectal applications) and rapid-dissolving films (vaginal application). The majority were gels, intended to deliver and distribute ARVs to target mucosal tissue while simultaneously creating a thin protective layer between infectious semen and mucosal tissue [70]. Examples of anti-HIV topical gels or creams that were formulated from bio-based polymeric materials are ones based on carrageenan [36, 71–73], dextrin sulfate [39], cellulose acetate phthalate (CAP) [41], cellulose sulfate [35, 74], methylcellulose [34], and hydroxyethyl cellulose [37, 38, 40, 42, 75, 76]. Limitations associated with gel use are the need for an applicator and possible messiness during product application and use. Behaviorally, different users may have different preferences over their sensory impressions of gels, for example, those that provide lubrication, vs. those that are undetectable by their partners [77, 78]. Consequently, acceptability and willingness to use have constrained gel success. In addition, gels provide relatively short-lasting protection; thus, they constitute a coitally dependent modality that requires administration before every act of intercourse.

Rapid dissolving films are semisolid dosage forms that are intended to provide topical protection against HIV sexual transmission while potentially improving user acceptability vs. other semisolid dosage forms [20, 43, 47, 79]. A rapid dissolving film quickly releases antiretroviral agents to the target tissue and luminal fluid. Polymeric films may not require an applicator (this is still being studied) and result in less messiness and product leakage compared to gels.

Although earlier development of anti-HIV drug delivery systems focused mainly on coitally dependent, on-demand products, the current focus has moved toward sustained-release long-acting dosage forms. IVRs were originally developed to provide contraception [80]. Thus, they became a coitally-independent PrEP modality that could offer controlled and sustained release of antiretroviral agents to mucosal tissues over extended times, from several weeks to potentially many months [81–83]. IVRs are torus-shaped and can be loaded with ARVs within the ring polymer matrix, a reservoir core at the ring's center, or in pods distributed around the ring [81–83]. The mechanisms of drug release and delivery for IVRs involve both diffusive drug transport from the ring surface into the contacting vaginal fluid and tissue and advective drug transport in vaginal fluid from the anterior to the posterior vagina [82]. Specifically, drug release rates from rings can be controlled by incorporating controlled-release inserts or pods within the IVR bodies [81, 82]. Since an IVR requires a low-dosage frequency, once a month or lower, it can overcome adherence issues associated with other coitally dependent topical PrEP products that require more frequent product reapplication.

Long-acting injectables represent a more recent, long-acting drug delivery modality that requires once-monthly or less frequent dosing schedules [84]. Thus, they have the potential to overcome adherence barriers posed by the daily oral PrEP regimens and traditional, coitally dependent topical PrEP dosage forms. A long-acting injectable formulation of cabotegravir, an integrase strand transfer inhibitor, has shown safety and promising pharmacokinetic data in phase I and IIa/b clinical trials and has now advanced to phase III studies to further evaluate its safety and efficacy for HIV PrEP [63, 65].

Long-acting subcutaneous implants offer another useful long-acting PrEP modality. Implants may remain in place for several years and present several advantages over the injectable formulations such as the ability to provide more consistent and predictable drug release kinetics [85]. For the purpose of HIV prophylaxis, long-acting implants are being developed for delivery of antiretroviral agents including tenofovir alafenamide [66, 68], islatravir [67], and nevirapine [86]. The implants formulated with alafenamide [66], and islatravir [67] have now advanced to clinical testing in humans. Recently, a reservoir-style subcutaneous implant made of biodegradable poly(ε-caprolactone) was developed for the long-term delivery of tenofovir alafenamide [27]. The implant provided sustained drug release for a period of 180 days in vitro. Thus, long-acting implants represent a promising HIV-1 PrEP strategy capable of providing long-term release of ARVs without the need for frequent product application. This could potentially overcome user adherence challenges of traditional topical semisolid gel dosage forms.

8.4 Bio-based Materials for Anti-HIV Drug Delivery Systems

Bio-based materials are sustainable materials produced fully or in part from biomass. Bio-based polymers may be divided into three categories according to their sources and manufacturing methods: those obtained from natural resources, chemical synthesis, or microbial synthesis [87]. The natural bio-based polymers include polysaccharide (e.g. cellulose, starch, chitosan, and alginate), protein (e.g. collagen, zein, and soy) and lipids (e.g. paraffin), while chemical synthesized bio-based polymers include, for example, polylactic acid (PLA) and poly (lactic-co-glycolic acid) (PLGA); and microbial synthesized bio-based polymers include poly(3-hydroxybutyrate) and polyhydroxyalkanoates (PHA) [87]. Bio-based polymeric materials have been widely used for biomedical applications, including drug delivery, tissue engineering, and regenerative medicine. In drug delivery, bio-based polymers have been used as a polymer matrix either alone, or in combination with, other bio-based or synthetic polymers to formulate various types of drug delivery vehicles, such as hydrogels [88, 89], transdermal patches [90, 91], electrospun fibers [92], and micro- and nanoparticles [93, 94]. The drug release kinetics of several drug delivery systems could be tunable to give either a rapid or sustained release profile by adjusting the compositions of the polymer components. For example, chitosan has been used to develop controlled-release hydrogels [88], transdermal microneedle patches for sustained topical delivery of protein drugs [95], and microspheres incorporated within poly-L-lactic acid electrospun fiber mats that provided dual-drug release profiles in a controlled manner [96]. Bacterial cellulose and gelatin were crosslinked by glutaraldehyde to form a hydrogel composite with a high swelling ratio for potential drug delivery applications [97]. Rapidly dissolving bio-based polymers such as alginate, chitosan, and bio-based polycarbonates are useful for triggering a quick burst release of loaded drugs from the drug delivery carriers. For example, tyrosine-derived polycarbonate terpolymers were used to form electrospun fiber mats for a complete and quick release of a hydrophilic peptide within nine hours [98]. Due to their biocompatibility and biodegradability, bio-based polymers have also been widely used in tissue engineering and regenerative medicine, in the fabrication of various tissue-engineered constructs and wound-healing products, such as scaffolds for articular cartilage tissue engineering [99] and bone regeneration [100], biodegradable sutures [101], tissue repair patches [102], and wound dressings [103].

In the field of HIV-prophylaxis, different types of bio-based materials have been widely explored for potential uses in anti-HIV drug-delivery systems. These include cellulose, chitosan, PLA, carrageenan, alginate, hyaluronic acid, and pectin. Table 8.1 summarizes bio-based materials that have been used in the development of anti-HIV drug delivery systems.

8.4.1 Cellulose

Cellulose is an abundant natural bio-based long-chain polymer made up of repeating units of glucose, which can be found in algae, fungi, some bacteria species, leaf and wood fibers, and cotton linters [128, 129]. Cellulose ethers and cellulose esters are the two cellulose derivatives used extensively in the pharmaceutical industry, for example, as gelling agents in semisolid dosage forms, thickeners and stabilizers in liquid dosage forms, and binders and fillers in tablets [129, 130]. These cellulose derivatives have also been used to form controlled release matrices in a variety of extended release dosage forms, and to promote bioadhesive and mucoadhesive properties of tablets and topical films [129, 130].

Hydrophilic cellulose ethers, such as hydroxyethyl cellulose and hydroxypropyl methylcellulose (HPMC), possess good swelling characteristics [131] and have been used widely in hydrophilic polymer matrices and gel-forming applications [129, 130]. Cellulose esters are typically not water-soluble, exhibit good film-forming ability, and have high water permeability and wet strength [132]. They have been generally used in preparing controlled-release formulations such as enteric-coated solid dosage forms [130, 132]. Examples of cellulose esters include CAP, HPMC phthalate, cellulose nitrate, and cellulose sulfate [130]. Moreover, microcrystalline cellulose (MCC), produced by treating powdered pure cellulose with hydrochloric acid, is commonly used as a diluent and binder in tablets, as a compressibility enhancer and free-flowing agent in solid dosage forms, and as a thickening agent in liquid dosage forms [129, 130].

In anti-HIV drug delivery, cellulose plays an important role in a variety of delivery systems for HIV prophylaxis. Several cellulose derivatives have been used to formulate both fast disintegrating and sustained-release tablets for vaginal ARV delivery. Unlike conventional tablets used for systemic purposes, anti-HIV vaginal tablets are intended for topical use. Their dissolution behavior within the vaginal canal is thus central to their drug delivery functioning. This involves interactions with ambient vaginal fluid and cervical mucus that may have leaked into the canal. Several vaginal tablets have been designed as topical PrEP products. A compressed tablet for vaginal delivery loaded with a non-NRTI (dapivirine) and a gp120 blocker (DS003) was developed by the International Partnership for Microbicides (IPM) [107]. MCC

Table 8.1 *Bio-based materials used in the development of anti-HIV drug delivery systems.*

Drug-delivery dosage form	Bio-based materials	Encapsulated antiretroviral agents	Drug release behavior	References
Semi-solid gels	Hydroxyethyl cellulose	Tenofovir, IQP-0528, dapivirine, maraviroc	Diffusion-controlled release	[37, 38, 40, 42, 75, 76]
	Methylcellulose	UC781	Diffusion-controlled release	[34]
	Carrageenan	MIV150	Diffusion-controlled release	[36, 71–73]
Vaginal films	Alginate, hydroxypropyl methylcellulose	Abacavir	Sustained release (over 4 h)	[104]
	Chitosan derivatives	Tenofovir	Sustained release (up to 5 days) with pH-sensitive behavior	[105]
	Hydroxylpropyl methyl cellulose	Dapivirine, IQP-0528, CSIC, EFdA, UAMC01398	Quick-dissolving properties	[43–48]
	Hydroxylpropyl methyl cellulose, zein	Tenofovir	Sustained release for 120 h	[106]
Vaginal tablets	Microcrystalline cellulose, lactose, hydroxyethyl cellulose	Dapivirine, DS003, tenofovir, emtricitabine	Quick-dissolving properties	[79, 107, 108]
	Hydroxypropyl methylcellulose, chitosan	Tenofoivir	Sustained release over 3–9 days depending upon formulations	[109–112]
	Chitosan, pectin, locust bean gum	Dapivirine, tenofovir	Sustained release (up to 4–5 days)	[113, 114]
Vaginal inserts	Carrageenan (API), griffithsin (API), dextran, hydroxyethyl cellulose	Carrageenan/griffithsin	Fast-dissolving in less than 60 s	[19, 49]
Suppositories	Carrageenan	Tenofovir	45–50% drug released by diffusion in 2 h	[51]
Nanoparticles	Cellulose acetate phthalate	Dolutegravir	pH-responsive release	[115]
	Chitosan	Tenofovir	Controlled release behavior following Higuchi and first order release models	[54, 55]
	Chitosan/PLGA	Tenofovir	pH-responsive, burst and sustained release	[52]
	Hyaluronic acid	Tenofovir	Semen (hyaluronidase enzyme)-triggered release	[116]
	PLGA	PSC-RANTES	Burst release (0–5 days) followed by a slow steady release (25 days)	[25]

(*Continued*)

Table 8.1 *(Continued)*

Drug-delivery dosage form	Bio-based materials	Encapsulated antiretroviral agents	Drug release behavior	References
	PLGA	Tenofovir, tenofovir disoproxil fumarate	pH-responsive release	[117]
	PLGA	Griffithsin, dapivirine	Burst release followed by a sustained release	[56]
Microparticles	Alginate	Silver complexes	pH-responsive release	[118]
	Alginate/chitosan	Tenofovir	Burst release followed by a sustained release	[53]
Electrospun nanofibers	Cellulose acetate phthalate	Etravirine or tenofovir disoproxil fumarate	pH-responsive release	[119]
	Ethyl cellulose	Maraviroc	Sustained release (up to several days) with tunable release rates	[120]
	Hyaluronic acid	Tenofovir	Semen (hyaluronidase enzyme)-triggered release	[121]
	PLLA, PDLLA	Maraviroc, AZT	Burst release or sustained release over several days	[122]
	PLGA	Griffithsin	pH-responsive release	[123]
Bigels	Pectin	Tenofovir	Sustained release (~72 h)	[124]
Intravaginal rings	Carrageenan (API)	MIV 150, carrageenan, zinc acetate, levonorgestrel	Sustained release over 94 days in vitro and 28 days in macaques.	[61]
	PLA (core coating polymer)	Tenofovir, tenofovir disoproxil fumarate, acyclovir	Sustained release in a controlled manner in vitro and in vivo over 28 days	[22, 57, 60, 125, 126]
Biodegradable implants	PLGA	MK-2048, rilpivirine, darunavir, ritonavir, atazanavir, dolutegravir	Sustained release over several weeks to up to a year in vivo	[127]

and compressible lactose were used as diluents with mannitol as a filler material and tartaric acid as an acidifier [79]. The MCC diluent serves as a drug dispersant, distributing the drug uniformly throughout the granulation. MCC and hydroxyethyl cellulose were also used as excipients to formulate fast-dissolving vaginal tablets (disintegration time of 60–75 seconds) loaded with two nucleoside-reverse transcriptase inhibitors, tenofovir and emtricitabine [108]. Moreover, HPMC has been used alone as a single sustained-release polymer [109] or in combination with chitosan [110] to formulate vaginal tablets loaded with tenofovir. The HPMC/chitosan tablets had a prolonged mucoadhesion residence time of four days and provided a controlled release of tenofovir for three days [110], offering promise as a sustained-release formulation for vaginal HIV PrEP.

Cellulose derivatives have also been widely used in semisolid dosage forms for topical PrEP. Various gels containing cellulose derivatives have been explored as microbicide products, including sodium carboxymethylcellulose [133], CAP [41], cellulose sulfate [35, 74], methylcellulose [34], and hydroxyethyl cellulose [37, 38, 40, 42, 75, 76]. These cellulose derivatives possess good mucoadhesive properties for intravaginal delivery by adhering to the vaginal mucosal surface and prolonging residence time with the mucous membrane [134]. Hydroxyethyl cellulose and methylcellulose gels have been demonstrated to create physical barriers to infectious HIV in vitro. The gels slowed down HIV transport, causing a reduction in viral diffusion coefficient by 10 000 times and a reduction in HIV transport through the mucosa in vitro by approximately 30% [135].

Hydroxyethyl cellulose has been used to formulate several gel formulations loaded with different antiretroviral agents, including dapivirine [37], tenofovir [38], IQP-0528 [40, 75], and maraviroc [76]. The tenofovir hydroxyethyl cellulose gel was found effective in reducing the transmission of HIV in women by 39% in a first phase-III study (CAPRISA 004) [38]. However, it failed to demonstrate efficacy in the second (VOICE, MTN003) [136] and third phase-III trials (FACTS 001) [137], mainly due to poor user compliance [138]. Interestingly, the gel was shown to be effective against HIV among the group of women who frequently used the gel [139, 140].

The failures of these microbicide gel products were in a large part due to issues with user compliance. Thus, a better dosage regimen design that can promote user acceptability of the gel products is clearly needed. A better understanding of the pharmacokinetics (PK) and pharmacodynamics (PD) of products is a key factor in designing those dosage regimens. The drug transport behaviors of the tenofovir gel [38] into vaginal and rectal mucosal tissues have been extensively explored by the confocal Raman spectroscopic technique [141–145]. The methodology can be extended to diverse drugs delivered in multiple vaginal and rectal topical dosage forms. Clearly, the knowledge of mucosal transport is important for all topical delivery modalities. The mucosal transport properties, including the diffusion and partition coefficients, of antiretroviral microbicides are important parameters in governing their transport behaviors through the vaginal and rectal mucosa. Thus, knowing them, together with clearance rates, is central to learning critical PK that can inform about the initial time to protection after product insertion, the degree of protection, and the duration of protection after product removal.

Besides tablets and semisolid gels, cellulose derivatives (e.g. ethyl cellulose and CAP) have been used to produce electrospun nanofibers as a drug delivery system for topical HIV PrEP [119, 120]. Ball et al. [120] formulated core-shell fibers made from ethyl cellulose (shell) and polyvinylpyrrolidone (core) using coaxial electrospinning for the delivery of maraviroc. The fibers were found to be noncytotoxic and provided tunable and sustained release of maraviroc over several days; the release rates were tunable by modulating the core-shell structure in the coaxial fibers [120]. CAP has been found to exhibit anti-HIV activity by binding to HIV and blocking the binding site on the HIV envelope glycoprotein gp120 [146]. CAP also exhibits a pH-responsive behavior, a desirable property in the design of semen-sensitive intravaginal drug delivery systems. Huang et al. [119] utilized CAP to fabricate pH-sensitive electrospun nanofibers loaded with an anti-HIV reverse transcriptase inhibitor (etravirine or tenofovir disoproxil fumarate). The CAP nanofibers remained stable at the acidic pH of simulated vaginal fluid (SVF), but rapidly disintegrated and released the incorporated drugs upon exposure to small amounts of human semen (pH 7.4–8.4) [119].

In addition, owing to its anti-HIV activity and pH-sensitive properties, CAP has been used to formulate nanoparticle formulations as a topical PrEP product. CAP nanoparticles loaded with Dolutegravir (DTG), an integrase strand transfer inhibitor, were prepared and incorporated into a thermosensitive gel for intravaginal delivery [115]. The DTG-loaded CAP nanoparticles synthesized by Mandal et al. were noncytotoxic to vaginal cells and demonstrated synergistic effects, combining the anti-HIV activities of DTG and CAP with the pH-sensitive properties of CAP. The formulated nanoparticles remained stable at vaginal pH 4.2, but quickly dissolved and released over 80% DTG at pH 7.4 of seminal fluid within one hour [115].

Moreover, cellulose derivatives, especially HPMC, have been used as an excipient in the formulation of intravaginal films for the delivery of several antiretroviral agents, including dapivirine [43], IQP-0528 [46], UAMC01398 [45], 5-chloro-3-[phenylsulfonyl] indole-2-carboxamide (CSIC) [44], EFdA [47, 48], and tenofovir [106]. Grammen et al. [45] developed a quick-dissolving (<10 minutes) vaginal film consisting of HPMC and polyethylene glycol (PEG) at varying proportions by a solvent evaporation method. Akil et al. [43] developed a vaginal film whose polymer matrix consisting of polyvinyl alcohol, HPMC, and PEG for the delivery of a non-NRTI, dapivirine. The film inhibited HIV-1 infection in vitro and ex vivo and rapidly released over half of the loaded dapivirine within 10 minutes of aqueous exposure [43]. The film also demonstrated safety and favorable PK and PD properties in Phase I clinical study [147]. Moreover, a vaginal film formulation that can provide sustained drug release has recently been developed by combining HPMC and zein [106]. The formulated vaginal film was found to provide a sustained release of the encapsulated tenofovir for 120 hours, demonstrating the promise of the film as an extended-release platform.

8.4.2 Chitosan

Chitosan is a naturally occurring cationic linear polysaccharide containing randomly distributed units of N-acetyl-D-glucosamine (acetylated unit) and β-(1–4)-linked D-glucosamine (deacetylated unit), produced by the partial deacetylation of chitin [148, 149]. To synthesize chitosan, the chitin (a N-acetylglucosamine polymer) shells of shrimp and other crustaceans are treated with an alkaline compound such as sodium hydroxide in the alkaline deacetylation reaction [148]. The degree of deacetylation, determined by the proportion of D-glucosamine and N-acetyl-D-glucosamine, can affect the physicochemical and biological properties of chitosan [149]. Chitosan was shown to possess useful properties including biocompatibility, biodegradability, low immunogenicity, and allergenicity, as well as a good anticoagulant, antimicrobial, and antioxidant activities [150]. In addition, chitosan is a cationic polymer possessing a high positive-charge density at pH<6.5 [151]. This property makes chitosan suitable for mucosal drug delivery, especially to the vaginal mucosa that has an acidic environment. In fact, chitosan has demonstrated good mucoadhesive properties for various vaginal drug delivery systems [152]. Because of these outstanding properties, chitosan has proved to be an attractive bio-based polymer for use in a number of biomedical and pharmaceutical applications [149, 150, 153].

Likewise, in the field of HIV prophylaxis, chitosan has been investigated for its potential use in a number of topical PrEP products. Chitosan nanoparticles loaded with tenofovir were synthesized by ionic gelation with drug-loading efficiency of 5.83% in aqueous conditions [54]. The formulated tenofovir-loaded chitosan nanoparticles were found to be noncytotoxic in vitro to vaginal epithelial cells and *Lactobacillus crispatus*, and demonstrated mucoadhesion to vaginal tissue that increased with a decrease in the nanoparticle size. The same group also prepared tenofovir-loaded chitosan-thioglycolic acid-conjugated nanoparticles with optimal formulations ranging in sizes from 240 to 252 nm [55]. The thiolated chitosan nanoparticles demonstrated superior properties to the non-thiolated ones, including an improved encapsulation efficiency (22.60% maximum) and a higher mucoadhesion by 4–5 times (up to 65% in two hours) [55].

Chitosan has also been formulated with PLGA to synthesize hybrid chitosan/PLGA nanoparticles loaded with tenofovir using a multiple emulsion method in which chitosan was added in the inner phase of the primary emulsion [52]. Belletti et al. [52] found that tenofovir release was limited at pH 7.4, while at the acidic pH of 4.6, a burst release followed by a sustained release over several hours was observed, suggesting potential use of the chitosan/PLGA formulation for sustained antiretrovial drug delivery to the vaginal mucosa.

In addition, chitosan has been used to formulate mucoadhesive vaginal tablets to provide vaginal mucoadhesion and a sustained release of ARVs over several days. The mucoadhesive tablets were formulated with varying proportions of natural bio-based polymers (chitosan, HPMC, pectin, and locust bean gum) using various techniques including direct compression, melt granulation, and hot-melt extrusion [110–114]. Cazorla-Luna et al. [114] found that due to the polyelectrolyte complexes formed by pectin and chitosan, the tablets formulated with these polymers provided a controlled and sustained release of tenofovir and remained adhered to the vaginal mucosa for up to four days.

Chitosan derivatives have also been employed to formulate pH-sensitive multilayer vaginal films. Chitosan lactate, chitosan tartate, and chitosan citrate, combined with Eudragit® S100, a polyanionic copolymer, were used to fabricate a vaginal film encapsulated with tenofovir using the layer-by-layer technique [105]. Chitosan derivatives and Eudragit form polyelectrolyte complexes that can provide an extended release of loaded drugs [154]. The film combining chitosan citrate and Eudragit S100 demonstrated a high mucosal adhesiveness with very moderate swelling [105]. It exhibited sustained release profiles of tenofovir in SVF for up to five days, but instantaneously released all encapsulated drugs in less than four hours in a mixture of simulated vaginal and semen fluids [105]. This pH-sensitive property of the film renders it a suitable candidate as an extended release topical PrEP product.

8.4.3 Polylactic Acid

PLA is a bio-based thermoplastic aliphatic polyester produced from renewable resources, such as sugars in corn and potato starch, cassava, or sugarcane [155, 156]. PLGA is a copolymer synthesized from lactic acid and glycolic acid [157]. PLA and PLGA have various uses in the medical and pharmaceutical industries, including in medical device production, tissue engineering, and wound-healing products (e.g. grafts, sutures, and prosthetics), and drug delivery systems, owing to their outstanding biocompatibility, biodegradability, and mechanical properties [158]. PLA has been utilized to produce biodegradable scaffolds and implants for bone and cartilage tissue engineering [159]. Due to its excellent biocompatibility and biodegradability, PLGA has been investigated in several studies in sustained and targeted delivery vehicles for small-molecule drugs, proteins, peptides, and plasmid DNA [157, 158]. An interesting feature of the PLGA carriers is that their erosion times and drug release rates are tunable by adjusting the hydrophilicity, crystallinity, and chemical interactions of the hydrolytic groups of PLGA [157]. For example, a highly hydrophilic amorphous polymer is suitable for a short-term release, whereas a semicrystalline PLGA polymer with a high percent crystallinity should be used for very long-term release [157]. These tunable degradability, release mechanisms, and mechanical properties render PLGA particularly attractive for controlled release applications.

In anti-HIV drug delivery, PLGA polymers have also been utilized to formulate various types of drug delivery systems for the delivery of several antiretroviral agents. Ham et al. [25] synthesized PLGA nanoparticles loaded with *N*-nonanoyl, des-Ser[1][L-thioproline[2], L--cyclohexylglycine[3]]-RANTES(2-68) (PSC-RANTES), a CCR5 chemokine receptor inhibitor, by a double-emulsion solvent evaporation method. The PLGA nanoparticles encapsulated with PSC-RANTES were biocompatible and nontoxic to human vaginal and ectocervical tissues and had a 4.8 times higher tissue uptake and permeability than that of the free PSC-RANTES during a four-hour exposure time [25]. The nanoparticles were also found to be able to penetrate deeper into the basal layers of the cervical epithelium, while the free drug could only permeate the outermost superficial layers of the tissue.

Woodrow et al. [160] formulated PLGA nanoparticles loaded with siRNA by a double-emulsion solvent-evaporation technique, achieving sustained intravaginal gene silencing effect in tissue for several days. In vivo studies showed that the siRNA-loaded PLGA nanoparticles appeared throughout the reproductive tract in the vaginal tract, cervix, and uterine horns and penetrated deep within the vaginal tissue, resulting in sustained gene silencing throughout the reproductive tract for 14 days [160].

In addition, formulations of pH-responsive PLGA nanoparticles have been developed for vaginal ARV delivery. Zhang et al. [117] synthesized pH-responsive nanoparticles of 250 nm in average size loaded with tenofovir or tenofovir disoproxil fumarate using a combination of PLGA and a pH-responsive polymer, methacrylic acid copolymer. The nanoparticles showed no toxicity for 48 hours to vaginal cells and demonstrated a significant pH-sensitive drug release upon contact with human semen fluid simulant [117].

Recently, Yang et al. [56] synthesized core-shell PLGA-based nanoparticles of size 180–200 nm encapsulated with dapivirine and griffithsin, an HIV entry inhibitor protein drug, using a double emulsion-solvent evaporation process. The PLGA nanoparticles were nontoxic, provided biphasic release profiles of both drugs (an initial burst release followed by a sustained release), and showed a strong synergistic effect in preventing HIV infection in vitro [56]. This study demonstrated a successful development of a PLGA nanoparticle that can simultaneously deliver both potent ARVs, providing a sustained release of both drugs while maintaining their anti-HIV effectiveness.

PLA has also been used to formulate electrospun nanofibers as a topical PrEP product. Ball et al. [122] synthesized nanofiber meshes from a combination of poly-L-lactic acid (PLLA) and polyethylene oxide (PEO), encapsulated with maraviroc, a potent CCR5-mediated HIV fusion inhibitor, or 3′-azido-3′deoxythymidine (AZT), a reverse transcriptase inhibitor. The drug-loaded PLLA/PEO fibers were nontoxic to TZM-bL cells and macaque ectocervical explants, demonstrated comparable anti-HIV efficacy to the free drugs, and exhibited burst release of maraviroc and AZT within one hour in vaginal fluid simulant. Both PLLA/PEO and poly-(D,L)-lactic acid (PDLLA)/PLLA-blended fibers provided sustained release profiles of maraviroc over several days [122]. In addition, Tyo et al. [123] synthesized electrospun fibers loaded with griffithsin for pH-sensitive vaginal drug delivery, combining methoxypolyethylene glycol-b-PLGA and poly(n-butyl acrylate-co-acrylic acid) in varying weight proportions. The developed GRFT-loaded fibers were noncytotoxic to vaginal cell lines and demonstrated pH-sensitive drug release upon contact with stimulated semen fluid [123].

PLA has also been used as a pod-coating polymer in the formulation of pod-inserted IVRs for the delivery of antiretroviral agents. In the pod vaginal rings, multiple pods – each can be loaded with a different API – are incorporated into the polymer matrix, usually a silicone elastomer [81]. The release rate of each API can be individually controlled by adjusting the amount of the coated polymer and the size of the delivery channel [57, 81]. The advantage of this pod-style IVR is that it allows for the simultaneous delivery of multiple drugs in a controlled manner. The pod silicone-based vaginal rings with PLA-coated pellets have been reported for the delivery of several ARVs including tenofovir [125], tenofovir disoproxil fumarate [22], as well as tenofovir in combination with tenofovir disoproxil fumarate [126] and acyclovir [57, 60]. Over the 28 days' in vivo studies, the rings exhibited sustained delivery with release rates that could be independently controlled for each drug [60, 125, 126], demonstrating that the pod-based IVRs are promising as sustained and controlled release topical PrEP products.

Recently, Benhabbour et al. [127] developed a biodegradable long-acting in situ forming implant using PLGA for the controlled and sustained delivery of multiple antiretroviral agents for up to one year. Unlike nonbiodegradable solid implants, the biodegradable implant formulated with PLGA does not require surgical removal of the used implant since PLGA can slowly degrade to nontoxic glycolic acid and lactic acid molecules. Results showed that the developed PLGA implant could be formulated with a single or multiple drugs and was able to sustain the plasma concentration of each individual drug over several weeks to up to a year [127].

8.4.4 Carrageenan

Carrageenan is a naturally occurring sulfated polysaccharide extracted from edible red algae, called Carrageen Moss or Carraigin, of the Rhodophyceae class [161]. It is comprised of alternating units of D-galactose and 3,6-anhydrogalactose, linked by α-(1,3) and β-(1,4)-glycosidic linkage [161]. It has an antiretroviral activity against human papillomavirus (HPV) [162] and herpes simplex virus (HSV) [163]. In addition, carrageenan has demonstrated anticoagulant, anticancer, and antihyperlipidemic properties [164]. It is found safe and nontoxic for food use [165] and is regarded nontoxic for non-parenteral pharmaceutical uses [164]. Thus, carrageenan has been used in industrial food products as gelling, emulsifying, and stabilizing agents, and cosmetic products such as toothpastes, shampoos, and cosmetic creams, as well as pharmaceutical formulations [161]. Carrageenan has been increasingly used as an excipient in drug delivery systems, including oral extended-release tablets, pellets, microcapsules/spheres, nanoparticle formulations, and controlled release formulations [164]. Due to its gelling properties, carrageenan has been widely used in topical gel formulations for vaginal, transscleral, buccal, and transdermal drug delivery applications that need prolonged retention time [164].

In the field of HIV prophylaxis, several formulations of carrageenan gels have been developed for intravaginal protection against HIV sexual transmission. Carrageenan gel formulations demonstrated efficacy against HIV [166] and other STIs including HPV [167] and HSV type 2 (HSV-2) [168]. Carrageenan has been used in the development of a gel formulation called Carraguard, by the Population Council. Later, the Carraguard-based formulation was combined with a potent non-nucleoside reverse transcriptase inhibitor (NNRTI), Medivir-150

(MIV) 150, which was found to be more efficacious than Carraguard alone [36]. Carraguard demonstrated safety for vaginal use in Phase I and II clinical studies [169, 170]. Although a Phase III trial did not show the efficacy of Carraguard in preventing HIV vaginal transmission, the study demonstrated that Carraguard was safe for vaginal use [72]. Thus, the carrageenan gel appears to be a safe and promising candidate as an anti-HIV drug delivery system and should be explored further for its potential use in combination with potent antiretroviral agents [73].

Moreover, carrageenan was used to formulate semisoft suppositories loaded with tenofovir for vaginal ARV delivery [21, 51]. In vitro release results showed that within the first two hours, approximately 40–50% of loaded tenofovir was released by diffusion out of the polymer matrix into the simulated vaginal and seminal fluids [51].

In addition, carrageenan has been used as an active pharmaceutical ingredient (API) in IVRs to provide extended protection against the sexual transmission of HIV-1 and other STIs. Ugaonkar et al. [61] developed a core-matrix IVR by formulating carrageenan with the NNRTI MIV 150, zinc acetate and levonorgestrel, a progestogen used in contraception. The formulated IVR simultaneously provides antiviral effects against HIV-1 (by MIV-150 and zinc acetate), HSV-2 (by carrageenan and zinc acetate), and HPV (by carrageenan), combined with the contraceptive effects exerted by levonorgestrel. The group formulated the IVR by loading the hydrophobic MIV-150 and levonorgestrel into the matrix of the IVR body, while incorporating the hydrophilic zinc acetate and carrageenan into the hydrophilic core of the IVR. Each API was released by diffusion from the IVR in a sustained manner over 94 days in vitro and 28 days in vivo in macaques [61].

Recently, carrageenan has been used as an API in combination with griffithsin, an HIV entry inhibitor, to develop vaginal fast-dissolving inserts as a precoital PrEP product for the prevention of HIV and sexually transmitted diseases [19, 49]. Griffithsin is a lectin derived from red algae and has shown broad-spectrum antiretroviral activity against several pathogens, including HIV [171, 172] and HSV-2 [173]. The study found that the carrageenan/griffithsin fast-dissolving inserts dissolved into semisolid gels in SVF in less than 60 seconds [49]. The inserts also demonstrated vaginal protection in vivo against simian-human immunodeficiency virus (SHIV) in rhesus macaques and against HSV-2 and HPV pseudovirus in mice [19].

8.4.5 Alginate

Alginate is a natural and biodegradable anionic polysaccharide, found mainly in the cell walls of brown algae of the Phaeophyceae class in species such as ascophyllum, durvillaea, laminaria, and sargassum [174, 175]. The linear polymer chain of alginate consists of alternating units of β-(1,4)-linked D-mannuronic acid and α-(1,4)-linked L-guluronic acid [176]. Due to its excellent biocompatibility and biodegradability, alginate has been used as a biomaterial in a number of biomedical applications including tissue engineering, tissue regeneration, wound dressings, and drug delivery [175–177]. Alginate is widely used as a thickening, gel-forming, and stabilizing excipient in drug delivery systems, including controlled-release products [178].

In the anti-HIV drug delivery field, alginate has been used in the formulations of microparticles and intravaginal films. Damelin et al. [118] developed alginate microbeads loaded with silver complexes. The developed alginate microbeads demonstrated pH-responsive behavior, being stable in the acidic vaginal simulated fluid but dissolving in simulated seminal fluid (pH ~ 7.2–7.8) [118]. This property is ideal as the alginate microbead vehicle can protect the silver complexes before intercourse while rapidly dissolving and releasing the encapsulated compounds following ejaculation.

Due to the anionic nature of alginate, it can form strong complexes with polycation polymers such as chitosan. Chitosan can be incorporated into the alginate polymer matrix in order to improve the controlled release properties of the alginate particles. Meng et al. [53] developed multilayer microparticles made of sodium alginate coated with thiolated chitosan for vaginal delivery of tenofovir. To prepare the multilayer microparticles, the alginate microparticle cores were synthesized by spray drying and coated with thiolated chitosan by a layer-by-layer method [53]. The developed thiolated chitosan-coated alginate microparticles showed a burst release of tenofovir followed by a sustained release, while the ones without chitosan coating only exhibited a burst release. The study also found that the microparticles had 20–50-fold higher mucoadhesion than the non-coated ones. Thus, the use of thiolated chitosan helped to enhance the controlled release properties and the mucoadhesiveness of the alginate microparticles.

Besides application in formulating microparticles, alginate has been used in the development of intravaginal films for protection against HIV sexual transmission. Ghosal et al. [104] formulated bioadhesive intravaginal films, encapsulated with abacavir, using sodium alginate as the main polymer combined with HPMC and glycerol. The films were prepared by solvent evaporation and demonstrated favorable physicochemical properties. The films formulated with alginate alone showed a quicker drug release, while the ones formulated with different combinations of alginate and HPMC demonstrated sustained drug release profiles over four hours [104].

Therefore, alginate has found several uses in anti-HIV drug delivery systems, both alone and together with other bio-based polymers such as chitosan and cellulose derivatives, being capable of forming both rapid release drug delivery vehicles and ones with controlled release characteristics.

8.4.6 Hyaluronic Acid

Hyaluronic acid is a natural anionic non-sulfated polysaccharide whose linear polymer chain consists of alternating units of D-glucuronic acid and N-acetyl-D-glucosamine linked by glucuronidic β-(1,3) or β-(1,4) bonds [179, 180]. Hyaluronic acid has favorable properties for biomedical applications including biocompatibility, biodegradability, and nontoxicity. It also has a high water affinity and has been widely used to form hydrogels in controlled release and targeted drug delivery systems [181] as well as in tissue engineering as hydrogel scaffolds [179]. Hyaluronic acid is hydrolysable and can be degraded by the hyaluronidase enzyme [179], which is abundantly found in human semen [182], cells, and tissues [183]. Thus, previous studies have explored the synthesis of hyaluronidase-sensitive drug delivery systems made of hyaluronic acid, enabling controlled and targeted drug delivery in the presence of the hyaluronidase enzyme at the diseased sites [184, 185].

In anti-HIV drug delivery, hyaluronic acid has been used in formulating stimuli-responsive nanofibers and nanoparticles. Agrahari et al. [116] synthesized hyaluronic acid nanoparticles that were triggered by the hyaluronidase enzyme to release the encapsulated tenofovir. The same group also synthesized tenofovir-loaded electrospun nanofibers made of thiolated hyaluronic acid [121]. Thiolated hyaluronic acid, formed by attaching sulfhydryl ligands onto hyaluronic acid, has been found to improve the mucoadhesive characteristics, permeation, and gelling properties of hyaluronic acid [186]. Both hyaluronic formulations were noncytotoxic to human vaginal cell lines and exhibited triggered release of the encapsulated tenofovir in the presence of seminal hyaluronidase enzyme [116, 121]. Moreover, the thiolated hyaluronic acid nanofibers improved tenofovir retention and bioavailability in the vaginal mucosa, as compared to the standard 1% tenofovir gel [121]. These studies demonstrated potential applications of hyaluronic acid in semen-triggered anti-HIV drug delivery systems.

8.4.7 Pectin

Pectin is a natural heteropolysaccharide extracted from citrus peels and apple pomace, consisting of α-1,4-D-galacturonic acid residues [187]. Pectin has been used widely in the pharmaceutical industry as a thickening, gelling, and binding agent, and as a polymer matrix in small-drug, protein, and cell delivery [187, 188]. Pectin derivatives that contain primary amine groups have favorable mucoadhesive properties for mucosal drug delivery, while the more hydrophobic ones that contain highly esterified galacturonic acid residues are more suitable for sustained and extended release [188].

In anti-HIV drug delivery, pectin has been used to formulate sustained-release vaginal tablets and bigels loaded with ARVs. Pectin-based bioadhesive vaginal tablets developed by Cazorla-Luna et al. [113] demonstrated a moderately prolonged swelling in SVF and provided a sustained release of dapivirine for up to five days. In addition, tenofovir-loaded freeze-dried bigels containing pectin as a polymer matrix have recently been developed by Martín-Illana et al. [124] for vaginal use. The group prepared nine different bigel formulations containing varying amounts of either pectin, chitosan, or HPMC. Results from mechanical tests, however, showed that the bigels with 3% pectin demonstrated superior mechanical properties over the other bigel formulations, having the highest hardness and deformation resistance with good mucoadhesiveness [124]. The study found that the release of tenofovir out of the pectin bigels was sustained for approximately 72 hours, although the release was accelerated in the presence of simulated semen fluid. This behavior was due to the acidic nature and pH-responsive behavior of pectin. The polymer matrix structure was stable at the acidic pH of vaginal fluid but swelled and released drug faster in the presence of seminal fluid (pH 7.2–7.8), rendering the pectin-based bigels a smart semen-triggered carrier for vaginal ARV delivery [124]. Therefore, pectin has shown promise as a potential bio-based polymer in the formulation of controlled and sustained release topical PrEP dosage forms.

8.5 Conclusion

Topically acting anti-HIV drug delivery systems represent an epidemiological alternative to oral PrEP, which requires user adherence to daily dosing to be fully effective. Bio-based materials have been widely used in the fabrication of anti-HIV drug delivery systems due to their outstanding biological, chemical, and physiological properties. Besides their biocompatibility and biodegradability, several bio-based polymers have favorable mucoadhesive properties (e.g. cellulose derivatives, alginate, chitosan, carrageenan, and pectin), pH-responsive characteristics (e.g. CAP, PLA, alginate, and pectin), and antimicrobial (e.g. chitosan) and antiviral (e.g. carrageenan) properties.

Earlier strategies for anti-HIV drug delivery systems were focused upon on-demand, fast-acting products. The latest approaches have moved toward extended-release dosage forms. Bio-based materials have been used to formulate topical anti-HIV drug delivery systems including both on-demand, faster-acting dosage forms, and controlled- and sustained-release, longer-acting products. However, their use in the fabrication of systemic PrEP, including subdermal implants and long-acting injectables, merits further exploration. There is a growing need for biodegradable implants that do not need surgical removal of the implant after the drug is fully released. Several bio-based polymers (e.g. PLA) offer outstanding biodegradability and tunable controlled-release kinetics and thus should be further explored for use in the formulation of long-acting biodegradable implants. It is important to note that anti-HIV drug delivery systems should be cheap and easily accessible to at-risk groups, especially in parts of the developing world where HIV/AIDS prevalence is still high. Bio-based materials are suitable for this purpose as they are derived from renewable feedstocks, rendering them cost-effective and economically attractive for manufacturing. With varying social and economic conditions and specific personal preferences, the development of multiple anti-HIV drug delivery strategies remains necessary to ensure compliance and efficacy. Bio-based materials have been and will be playing a significant role in the formulation and manufacturing of current and next-generation anti-HIV drug delivery systems that are developed in an effort to offer long-term global prevention of HIV/AIDS.

References

1. Fauci, A.S. and Folkers, G.K. (2012). Toward an AIDS-free generation. *The Journal of the American Medical Association* 308 (4): 343–344.
2. Gallo, R.C. and Montagnier, L. (2003). The discovery of HIV as the cause of AIDS. *New England Journal of Medicine* 349 (24): 2283–2285.
3. The Joint United Nations Programme on HIV/AIDS (2020). Global HIV & AIDS statistics – 2020 fact sheet http://unaids.mio.guru/en/resources/fact-sheet (accessed 10 January 2020).
4. Ganser-Pornillos, B.K., Yeager, M., and Sundquist, W.I. (2008). The structural biology of HIV assembly. *Current Opinion in Structural Biology* 18 (2): 203–217.
5. Shaw, G.M. and Hunter, E. (2012). HIV transmission. *Cold Spring Harbor Perspectives in Medicine* 2 (11): a006965.
6. McMichael, A.J. and Rowland-Jones, S.L. (2001). Cellular immune responses to HIV. *Nature* 410 (6831): 980–987.
7. Levy, J.A. (2009). HIV pathogenesis: 25 years of progress and persistent challenges. *AIDS* 23 (2): 147–160.
8. Simon, V., Ho, D.D., and Karim, Q.A. (2006). HIV/AIDS epidemiology, pathogenesis, prevention, and treatment. *The Lancet* 368 (9534): 489–504.
9. Desai, M., Iyer, G., and Dikshit, R.K. (2012). Antiretroviral drugs: critical issues and recent advances. *Indian Journal of Pharmacology* 44 (3): 288.
10. Sarkar, N.N. (2008). Barriers to condom use. *The European Journal of Contraception and Reproductive Health Care* 13 (2): 114–122.
11. Kalichman, S.C., Williams, E.A., Cherry, C. et al. (1998). Sexual coercion, domestic violence, and negotiating condom use among low-income African American women. *Journal of Women's Health* 7 (3): 371–378.
12. Wingood, G.M. and DiClemente, R.J. (1997). The effects of an abusive primary partner on the condom use and sexual negotiation practices of African-American women. *American Journal of Public Health* 87 (6): 1016–1018.
13. Paz-Bailey, G., Koumans, E.H., Sternberg, M. et al. (2005). The effect of correct and consistent condom use on chlamydial and gonococcal infection among urban adolescents. *Archives of Pediatrics and Adolescent Medicine* 159 (6): 536–542.
14. Weller, S.C. and Davis-Beaty, K. (2002). Condom effectiveness in reducing heterosexual HIV transmission. *Cochrane Database of Systematic Reviews* (1): 1–22.
15. Haynes, B.F. and Burton, D.R. (2017). Developing an HIV vaccine. *Science* 355 (6330): 1129–1130.
16. Kartikeyan, S., Bharmal, R.N., Tiwari, R.P., and Bisen, P.S. (2007). *HIV and AIDS: Basic Elements and Priorities*, XIV, 418. Dordrecht: Springer.
17. De Clercq, E. (2013). The acyclic nucleoside phosphonates (ANPs): antonín Holý's legacy. *Medicinal Research Reviews* 33 (6): 1278–1303.
18. Rosenberg, Z.F. and Devlin, B. (2012). Future strategies in microbicide development. *Best Practice and Research: Clinical Obstetrics and Gynaecology* 26 (4): 503–513.
19. Derby, N., Lal, M., Aravantinou, M. et al. (2018). Griffithsin carrageenan fast dissolving inserts prevent SHIV HSV-2 and HPV infections in vivo. *Nature Communications* 9 (1): 3881.

20. Garg, S., Goldman, D., Krumme, M. et al. (2010). Advances in development, scale-up and manufacturing of microbicide gels, films, and tablets. *Antiviral Research* 88: S19–S29.
21. Zaveri, T., Primrose, R.J., Surapaneni, L. et al. (2014). Firmness perception influences women's preferences for vaginal suppositories. *Pharmaceutics* 6 (3): 512–529.
22. Baum, M.M., Butkyavichene, I., Churchman, S.A. et al. (2015). An intravaginal ring for the sustained delivery of tenofovir disoproxil fumarate. *International Journal of Pharmaceutics* 495 (1): 579–587.
23. Fan, M.D., Kramzer, L.F., Hillier, S.L. et al. (2017). Preferred physical characteristics of vaginal film microbicides for HIV prevention in Pittsburgh women. *Archives of Sexual Behavior* 46 (4): 1111–1119.
24. Blakney, A.K., Jiang, Y., and Woodrow, K.A. (2017). Application of electrospun fibers for female reproductive health. *Drug Delivery and Translational Research* 7 (6): 796–804.
25. Ham, A.S., Cost, M.R., Sassi, A.B. et al. (2009). Targeted delivery of PSC-RANTES for HIV-1 prevention using biodegradable nanoparticles. *Pharmaceutical Research* 26 (3): 502–511.
26. Owen, A. and Rannard, S. (2016). Strengths, weaknesses, opportunities and challenges for long acting injectable therapies: insights for applications in HIV therapy. *Advanced Drug Delivery Reviews* 103: 144–156.
27. Johnson, L.M., Krovi, S.A., Li, L. et al. (2019). Characterization of a reservoir-style implant for sustained release of tenofovir alafenamide (TAF) for HIV pre-exposure prophylaxis (PrEP). *Pharmaceutics* 11 (7): 315.
28. Plosker, G.L. (2013). Emtricitabine/tenofovir disoproxil fumarate: a review of its use in HIV-1 pre-exposure prophylaxis. *Drugs* 73 (3): 279–291.
29. Grant, R.M., Lama, J.R., Anderson, P.L. et al. (2010). Preexposure chemoprophylaxis for HIV prevention in men who have sex with men. *New England Journal of Medicine* 2010 (363): 2587–2599.
30. Thigpen, M.C., Kebaabetswe, P.M., Paxton, L.A. et al. (2012). Antiretroviral preexposure prophylaxis for heterosexual HIV transmission in Botswana. *New England Journal of Medicine* 367 (5): 423–434.
31. Baeten, J.M., Donnell, D., Ndase, P. et al. (2012). Antiretroviral prophylaxis for HIV prevention in heterosexual men and women. *New England Journal of Medicine* 367 (5): 399–410.
32. Sidebottom, D., Ekström, A.M., and Strömdahl, S. (2018). A systematic review of adherence to oral pre-exposure prophylaxis for HIV – how can we improve uptake and adherence? *BMC Infectious Diseases* 18 (1): 581.
33. Traeger, M.W., Cornelisse, V.J., Asselin, J. et al. (2019). Association of HIV preexposure prophylaxis with incidence of sexually transmitted infections among individuals at high risk of HIV infection. *The Journal of the American Medical Association* 321 (14): 1380–1390.
34. Anton, P.A., Saunders, T., Elliott, J. et al. (2011). First phase 1 double-blind, placebo-controlled, randomized rectal microbicide trial using UC781 gel with a novel index of ex vivo efficacy. *PLoS One* 6 (9): 1–15.
35. El-Sadr, W.M., Mayer, K.H., Maslankowski, L. et al. (2006). Safety and acceptability of cellulose sulfate as a vaginal microbicide in HIV-infected women. *AIDS* 20 (8): 1109–1116.
36. Fernández-Romero, J.A., Thorn, M., Turville, S.G. et al. (2007). Carrageenan/MIV-150 (PC-815), a combination microbicide. *Sexually Transmitted Diseases* 34 (1): 9–14.
37. Jespers, V.A., Van Roey, J.M., Beets, G.I., and Buvé, A.M. (2007). Dose-ranging phase 1 study of TMC120, a promising vaginal microbicide, in HIV-negative and HIV-positive female volunteers. *Journal of Acquired Immune Deficiency Syndromes* 44 (2): 154–158.
38. Karim, Q.A., Karim, S.S.A., Frohlich, J.A. et al. (2010). Effectiveness and safety of tenofovir gel, an antiretroviral microbicide, for the prevention of HIV infection in women. *Science* 329 (5996): 1168–1174.
39. Low-Beer, N., Gabe, R., McCormack, S. et al. (2002). Dextrin sulfate as a vaginal microbicide: randomized, double-blind, placebo-controlled trial including healthy female volunteers and their male partners. *Journal of Acquired Immune Deficiency Syndromes* 31 (4): 391–398.
40. Mahalingam, A., Simmons, A.P., Ugaonkar, S.R. et al. (2011). Vaginal microbicide gel for delivery of IQP-0528, a pyrimidinedione analog with a dual mechanism of action against HIV-1. *Antimicrobial Agents and Chemotherapy* 55 (4): 1650–1660.
41. Manson, K.H., Wyand, M.S., Miller, C., and Neurath, A.R. (2000). Effect of a cellulose acetate phthalate topical cream on vaginal transmission of simian immunodeficiency virus in rhesus monkeys. *Antimicrobial Agents and Chemotherapy* 44 (11): 3199–3202.
42. Rohan, L.C., Moncla, B.J., Kunjara Na Ayudhya, R.P. et al. (2010). in vitro and ex vivo testing of tenofovir shows it is effective as an HIV-1 microbicide. *PLoS One* 5 (2): e9310.

43. Akil, A., Parniak, M.A., Dezzutti, C.S. et al. (2011). Development and characterization of a vaginal film containing dapivirine, a non-nucleoside reverse transcriptase inhibitor (NNRTI), for prevention of HIV-1 sexual transmission. *Drug Delivery and Translational Research* 1 (3): 209–222.
44. Gong, T., Zhang, W., Parniak, M.A. et al. (2017). Preformulation and vaginal film formulation development of microbicide drug candidate CSIC for HIV prevention. *Journal of Pharmaceutical Innovation* 12 (2): 142–154.
45. Grammen, C., Van den Mooter, G., Appeltans, B. et al. (2014). Development and characterization of a solid dispersion film for the vaginal application of the anti-HIV microbicide UAMC01398. *International Journal of Pharmaceutics* 475 (1–2): 238–244.
46. Ham, A.S., Rohan, L.C., Boczar, A. et al. (2012). Vaginal film drug delivery of the pyrimidinedione IQP-0528 for the prevention of HIV infection. *Pharmaceutical Research* 29 (7): 1897–1907.
47. Zhang, W., Hu, M., Shi, Y. et al. (2015). Vaginal microbicide film combinations of two reverse transcriptase inhibitors, EFdA and CSIC, for the prevention of HIV-1 sexual transmission. *Pharmaceutical Research* 32 (9): 2960–2972.
48. Zhang, W., Parniak, M.A., Sarafianos, S.G. et al. (2014). Development of a vaginal delivery film containing EFdA, a novel anti-HIV nucleoside reverse transcriptase inhibitor. *International Journal of Pharmaceutics* 461 (1–2): 203–213.
49. Lal, M., Lai, M., Ugaonkar, S. et al. (2018). Development of a vaginal fast-dissolving insert combining griffithsin and carrageenan for potential use against sexually transmitted infections. *Journal of Pharmaceutical Sciences* 107 (10): 2601–2610.
50. Ham, A.S. and Buckheit, R.W. Jr. (2017). Designing and developing suppository formulations for anti-HIV drug delivery. *Therapeutic Delivery* 8 (9): 805–817.
51. Zaveri, T., Hayes, J.E., and Ziegler, G.R. (2014). Release of tenofovir from carrageenan-based vaginal suppositories. *Pharmaceutics* 6 (3): 366–377.
52. Belletti, D., Tosi, G., Forni, F. et al. (2012). Chemico-physical investigation of tenofovir loaded polymeric nanoparticles. *International Journal of Pharmaceutics* 436 (1–2): 753–763.
53. Meng, J., Agrahari, V., Ezoulin, M.J. et al. (2017). Spray-dried thiolated chitosan-coated sodium alginate multilayer microparticles for vaginal HIV microbicide delivery. *The AAPS Journal* 19 (3): 692–702.
54. Meng, J., Sturgis, T.F., and Youan, B.-B.C. (2011). Engineering tenofovir loaded chitosan nanoparticles to maximize microbicide mucoadhesion. *European Journal of Pharmaceutical Sciences* 44 (1–2): 57–67.
55. Meng, J., Zhang, T., Agrahari, V. et al. (2014). Comparative biophysical properties of tenofovir-loaded, thiolated and nonthiolated chitosan nanoparticles intended for HIV prevention. *Nanomedicine* 9 (11): 1595–1612.
56. Yang, H., Li, J., Patel, S.K. et al. (2019). Design of poly (lactic-*co*-glycolic acid)(PLGA) nanoparticles for vaginal co-delivery of griffithsin and dapivirine and their synergistic effect for HIV prophylaxis. *Pharmaceutics* 11 (4): 184.
57. Baum, M.M., Butkyavichene, I., Gilman, J. et al. (2012). An intravaginal ring for the simultaneous delivery of multiple drugs. *Journal of Pharmaceutical Sciences* 101 (8): 2833–2843.
58. Gupta, K.M., Pearce, S.M., Poursaid, A.E. et al. (2008). Polyurethane intravaginal ring for controlled delivery of dapivirine, a nonnucleoside reverse transcriptase inhibitor of HIV-1. *Journal of Pharmaceutical Sciences* 97 (10): 4228–4239.
59. Johnson, T.J., Gupta, K.M., Fabian, J. et al. (2010). Segmented polyurethane intravaginal rings for the sustained combined delivery of antiretroviral agents dapivirine and tenofovir (vol 39, pg 203, 2010). *European Journal of Pharmaceutical Sciences* 41 (5): 736–736.
60. Moss, J.A., Malone, A.M., Smith, T.J. et al. (2012). Simultaneous delivery of tenofovir and acyclovir via an intravaginal ring. *Antimicrobial Agents and Chemotherapy* 56 (2): 875–882.
61. Ugaonkar, S.R., Wesenberg, A., Wilk, J. et al. (2015). A novel intravaginal ring to prevent HIV-1, HSV-2, HPV, and unintended pregnancy. *Journal of Controlled Release* 213: 57–68.
62. Woolfson, A.D., Malcolm, R.K., Morrow, R.J. et al. (2006). Intravaginal ring delivery of the reverse transcriptase inhibitor TMC 120 as an HIV microbicide. *International Journal of Pharmaceutics* 325 (1–2): 82–89.
63. Clement, M.E., Kofron, R., and Landovitz, R.J. (2020). Long-acting injectable cabotegravir for the prevention of HIV infection. *Current Opinion in HIV and AIDS* 15 (1): 19–26.
64. Jackson, A. and McGowan, I. (2015). Long-acting rilpivirine for HIV prevention. *Current Opinion in HIV and AIDS* 10 (4): 253–257.

65. Trezza, C., Ford, S.L., Spreen, W. et al. (2015). Formulation and pharmacology of long-acting cabotegravir. *Current Opinion in HIV and AIDS* 10 (4): 239.
66. Gunawardana, M., Remedios-Chan, M., Miller, C.S. et al. (2015). Pharmacokinetics of long-acting tenofovir alafenamide (GS-7340) subdermal implant for HIV prophylaxis. *Antimicrobial Agents and Chemotherapy* 59 (7): 3913–3919.
67. Matthews, R.P., Barrett, S.E., Patel, M. et al. (2019). First-in-human trial of MK-8591-eluting implants demonstrates concentrations suitable for HIV prophylaxis for at least one year. The 10th IAS Conference on HIV Science (IAS 2019), Mexico City, Mexico (21–24 July 2019).
68. Schlesinger, E., Johengen, D., Luecke, E. et al. (2016). A tunable, biodegradable, thin-film polymer device as a long-acting implant delivering tenofovir alafenamide fumarate for HIV pre-exposure prophylaxis. *Pharmaceutical Research* 33 (7): 1649–1656.
69. Cummins, J.E. Jr. and Doncel, G.F. (2009). Biomarkers of cervicovaginal inflammation for the assessment of microbicide safety. *Sexually Transmitted Diseases* 36 (3): S84–S91.
70. Gao, Y., Yuan, A., Chuchuen, O. et al. (2015). Vaginal deployment and tenofovir delivery by microbicide gels. *Drug Delivery and Translational Research* 5 (3): 279–294.
71. Singer, R., Derby, N., Rodriguez, A. et al. (2011). The nonnucleoside reverse transcriptase inhibitor MIV-150 in carrageenan gel prevents rectal transmission of simian/human immunodeficiency virus infection in macaques. *Journal of Virology* 85 (11): 5504–5512.
72. Skoler-Karpoff, S., Ramjee, G., Ahmed, K. et al. (2008). Efficacy of Carraguard for prevention of HIV infection in women in South Africa: a randomised, double-blind, placebo-controlled trial. *The Lancet* 372 (9654): 1977–1987.
73. Turville, S.G., Aravantinou, M., Miller, T. et al. (2008). Efficacy of Carraguard®-based microbicides in vivo despite variable in vitro activity. *PLoS One* 3 (9): e3162.
74. Horwood, J. (2007). Cellulose sulphate microbicide trial halted. *The Lancet Infectious Diseases* 7 (3): 183.
75. Ham, A.S., Nugent, S.T., Peters, J.J. et al. (2015). The rational design and development of a dual chamber vaginal/rectal microbicide gel formulation for HIV prevention. *Antiviral Research* 120: 153–164.
76. Malcolm, R.K., Forbes, C.J., Geer, L. et al. (2013). Pharmacokinetics and efficacy of a vaginally administered maraviroc gel in rhesus macaques. *Journal of Antimicrobial Chemotherapy* 68 (3): 678–683.
77. Guthrie, K.M., Dunsiger, S., Vargas, S.E. et al. (2016). Perceptibility and the "choice experience": user sensory perceptions and experiences inform vaginal prevention product design. *AIDS Research and Human Retroviruses* 32 (10–11): 1022–1030.
78. Morrow, K.M., Fava, J.L., Rosen, R.K. et al. (2014). Designing preclinical perceptibility measures to evaluate topical vaginal gel formulations: relating user sensory perceptions and experiences to formulation properties. *AIDS Research and Human Retroviruses* 30 (1): 78–91.
79. Rohan, L.C., Devlin, B., and Yang, H. (2013). Microbicide dosage forms. In: *Microbicides for Prevention of HIV Infection* (ed. J. Nuttall), 27–54. Berlin, Heidelberg: Springer.
80. Malcolm, R.K. (2003). The intravaginal ring. *Drugs and the Pharmaceutical Sciences* 126: 775–790.
81. Friend, D.R. (2011). Intravaginal rings: controlled release systems for contraception and prevention of transmission of sexually transmitted infections. *Drug Delivery and Translational Research* 1 (3): 185–193.
82. Kiser, P.F., Johnson, T.J., and Clark, J.T. (2012). State of the art in intravaginal ring technology for topical prophylaxis of HIV infection. *AIDS Reviews* 14 (1): 62–77.
83. Malcolm, R.K., Edwards, K.-L., Kiser, P. et al. (2010). Advances in microbicide vaginal rings. *Antiviral Research* 88: S30–S39.
84. Landovitz, R.J., Kofron, R., and McCauley, M. (2016). The promise and pitfalls of long acting injectable agents for HIV prevention. *Current Opinion in HIV and AIDS* 11 (1): 122.
85. Weld, E.D. and Flexner, C. (2020). Long-acting implants to treat and prevent HIV infection. *Current Opinion in HIV and AIDS* 15 (1): 33–41.
86. Chen, J., Walters, K., and Ashton, P. (2005). Correlation of in vitro-in vivo release rates for sustained release nevirapine implants in rats. *Journal of Controlled Release: Official Journal of the Controlled Release Society* 101 (1–3): 357.
87. Gao, S., Tang, G., Hua, D. et al. (2019). Stimuli-responsive bio-based polymeric systems and their applications. *Journal of Materials Chemistry B* 7 (5): 709–729.
88. Bhattarai, N., Gunn, J., and Zhang, M. (2010). Chitosan-based hydrogels for controlled, localized drug delivery. *Advanced Drug Delivery Reviews* 62 (1): 83–99.

89. Luo, Y., Kirker, K.R., and Prestwich, G.D. (2000). Cross-linked hyaluronic acid hydrogel films: new biomaterials for drug delivery. *Journal of Controlled Release* 69 (1): 169–184.
90. Basavaraj, K.H., Johnsy, G., Navya, M.A., and Rashmi, R. (2010). Biopolymers as transdermal drug delivery systems in dermatology therapy. *Critical Reviews™ in Therapeutic Drug Carrier Systems* 27 (2): 155–185.
91. Sharma, K., Singh, V., and Arora, A. (2011). Natural biodegradable polymers as matrices in transdermal drug delivery. *International Journal of Drug Development and Research* 3: 85–103.
92. Torres-Martínez, E.J., Cornejo Bravo, J.M., Serrano Medina, A. et al. (2018). A summary of electrospun nanofibers as drug delivery system: drugs loaded and biopolymers used as matrices. *Current Drug Delivery* 15 (10): 1360–1374.
93. Park, J., Ye, M., and Park, K. (2005). Biodegradable polymers for microencapsulation of drugs. *Molecules* 10 (1): 146–161.
94. Sailaja, A.K., Amareshwar, P., and Chakravarty, P. (2011). Different techniques used for the preparation of nanoparticles using natural polymers and their application. *International Journal of Pharmacy and Pharmaceutical Sciences* 3 (2): 45–50.
95. Chen, M.-C., Ling, M.-H., Lai, K.-Y., and Pramudityo, E. (2012). Chitosan microneedle patches for sustained transdermal delivery of macromolecules. *Biomacromolecules* 13 (12): 4022–4031.
96. Xu, J., Jiao, Y., Shao, X., and Zhou, C. (2011). Controlled dual release of hydrophobic and hydrophilic drugs from electrospun poly (l-lactic acid) fiber mats loaded with chitosan microspheres. *Materials Letters* 65 (17–18): 2800–2803.
97. Treesuppharat, W., Rojanapanthu, P., Siangsanoh, C. et al. (2017). Synthesis and characterization of bacterial cellulose and gelatin-based hydrogel composites for drug-delivery systems. *Biotechnology Reports* 15: 84–91.
98. Macri, L.K., Sheihet, L., Singer, A.J. et al. (2012). Ultrafast and fast bioerodible electrospun fiber mats for topical delivery of a hydrophilic peptide. *Journal of Controlled Release* 161 (3): 813–820.
99. Zhao, W., Jin, X., Cong, Y. et al. (2013). Degradable natural polymer hydrogels for articular cartilage tissue engineering. *Journal of Chemical Technology and Biotechnology* 88 (3): 327–339.
100. Venkatesan, J. and Kim, S.-K. (2010). Chitosan composites for bone tissue engineering – an overview. *Marine Drugs* 8 (8): 2252–2266.
101. Shishatskaya, E.I., Volova, T.G., Puzyr, A.P. et al. (2004). Tissue response to the implantation of biodegradable polyhydroxyalkanoate sutures. *Journal of Materials Science: Materials in Medicine* 15 (6): 719–728.
102. Pok, S. and Jacot, J.G. (2011). Biomaterials advances in patches for congenital heart defect repair. *Journal of Cardiovascular Translational Research* 4 (5): 646–654.
103. Mogoşanu, G.D. and Grumezescu, A.M. (2014). Natural and synthetic polymers for wounds and burns dressing. *International Journal of Pharmaceutics* 463 (2): 127–136.
104. Ghosal, K., Ranjan, A., and Bhowmik, B.B. (2014). A novel vaginal drug delivery system: anti-HIV bioadhesive film containing abacavir. *Journal of Materials Science: Materials in Medicine* 25 (7): 1679–1689.
105. Cazorla-Luna, R., Martín-Illana, A., Notario-Pérez, F. et al. (2020). Vaginal polyelectrolyte layer-by-layer films based on chitosan derivatives and Eudragit® S100 for pH responsive release of tenofovir. *Marine Drugs* 18 (1): 44.
106. Notario-Pérez, F., Martín-Illana, A., Cazorla-Luna, R. et al. (2019). Development of mucoadhesive vaginal films based on HPMC and zein as novel formulations to prevent sexual transmission of HIV. *International Journal of Pharmaceutics* 570: 118643.
107. Gupta, J., Othman, A., Qihai Tao, J., and Garg, S. (2011). Development and validation of a HPLC method for simultaneous determination of dapivirine and DS003 in combination microbicide tablet. *Current Pharmaceutical Analysis* 7 (1): 21–26.
108. Clark, M., Peet, M., Davis, S. et al. (2014). Evaluation of rapidly disintegrating vaginal tablets of tenofovir, emtricitabine and their combination for HIV-1 prevention. *Pharmaceutics* 6 (4): 616–631.
109. McConville, C., Friend, D.R., Clark, M.R., and Malcolm, K. (2013). Preformulation and development of a once-daily sustained-release tenofovir vaginal tablet containing a single excipient. *Journal of Pharmaceutical Sciences* 102 (6): 1859–1868.

110. Notario-Pérez, F., Cazorla-Luna, R., Martín-Illana, A. et al. (2018). Optimization of tenofovir release from mucoadhesive vaginal tablets by polymer combination to prevent sexual transmission of HIV. *Carbohydrate Polymers* 179: 305–316.
111. Notario-Pérez, F., Cazorla-Luna, R., Martín-Illana, A. et al. (2019). Tenofovir hot-melt granulation using Gelucire® to develop sustained-release vaginal systems for weekly protection against sexual transmission of HIV. *Pharmaceutics* 11 (3): 137.
112. Notario-Pérez, F., Martín-Illana, A., Cazorla-Luna, R. et al. (2018). Improvement of Tenofovir vaginal release from hydrophilic matrices through drug granulation with hydrophobic polymers. *European Journal of Pharmaceutical Sciences* 117: 204–215.
113. Cazorla-Luna, R., Martín-Illana, A., Notario-Pérez, F. et al. (2019). Dapivirine bioadhesive vaginal tablets based on natural polymers for the prevention of sexual transmission of HIV. *Polymers* 11 (3): 483.
114. Cazorla-Luna, R., Notario-Pérez, F., Martín-Illana, A. et al. (2019). Chitosan-based mucoadhesive vaginal tablets for controlled release of the anti-HIV drug tenofovir. *Pharmaceutics* 11 (1): 20.
115. Mandal, S., Khandalavala, K., Pham, R. et al. (2017). Cellulose acetate phthalate and antiretroviral nanoparticle fabrications for HIV pre-exposure prophylaxis. *Polymers* 9 (9): 423.
116. Agrahari, V., Zhang, C., Zhang, T. et al. (2014). Hyaluronidase-sensitive nanoparticle templates for triggered release of HIV/AIDS microbicide in vitro. *The AAPS Journal* 16 (2): 181–193.
117. Zhang, T., Sturgis, T.F., and Youan, B.-B.C. (2011). pH-responsive nanoparticles releasing tenofovir intended for the prevention of HIV transmission. *European Journal of Pharmaceutics and Biopharmaceutics* 79 (3): 526–536.
118. Damelin, L.H., Fernandes, M.A., and Tiemessen, C.T. (2015). Alginate microbead-encapsulated silver complexes for selective delivery of broad-spectrum silver-based microbicides. *International Journal of Antimicrobial Agents* 46 (4): 394–400.
119. Huang, C., Soenen, S.J., van Gulck, E. et al. (2012). Electrospun cellulose acetate phthalate fibers for semen induced anti-HIV vaginal drug delivery. *Biomaterials* 33 (3): 962–969.
120. Ball, C., Chou, S.-F., Jiang, Y., and Woodrow, K.A. (2016). Coaxially electrospun fiber-based microbicides facilitate broadly tunable release of maraviroc. *Materials Science and Engineering: C* 63: 117–124.
121. Agrahari, V., Meng, J., Ezoulin, M.J.M. et al. (2016). Stimuli-sensitive thiolated hyaluronic acid based nanofibers: synthesis, preclinical safety and in vitro anti-HIV activity. *Nanomedicine* 11 (22): 2935–2958.
122. Ball, C., Krogstad, E., Chaowanachan, T., and Woodrow, K.A. (2012). Drug-eluting fibers for HIV-1 inhibition and contraception. *PLoS One* 7 (11): e49792.
123. Tyo, K.M., Duan, J., Kollipara, P. et al. (2019). pH-responsive delivery of Griffithsin from electrospun fibers. *European Journal of Pharmaceutics and Biopharmaceutics* 138: 64–74.
124. Martín-Illana, A., Notario-Pérez, F., Cazorla-Luna, R. et al. (2019). Smart freeze-dried bigels for the prevention of the sexual transmission of HIV by accelerating the vaginal release of tenofovir during intercourse. *Pharmaceutics* 11 (5): 232.
125. Moss, J.A., Malone, A.M., Smith, T.J. et al. (2012). Safety and pharmacokinetics of intravaginal rings delivering tenofovir in pig-tailed macaques. *Antimicrobial Agents and Chemotherapy* 56 (11): 5952–5960.
126. Moss, J.A., Baum, M.M., Malone, A.M. et al. (2012). Tenofovir and tenofovir disoproxil fumarate pharmacokinetics from intravaginal rings. *AIDS* 26 (6): 707–710.
127. Benhabbour, S.R., Kovarova, M., Jones, C. et al. (2019). Ultra-long-acting tunable biodegradable and removable controlled release implants for drug delivery. *Nature Communications* 10 (1): 1–12.
128. Sczostak, A. (2009). Cotton linters: an alternative cellulosic raw material. *Macromolecular Symposia* 280 (1): 45–53.
129. Yilmaz, N.D., Çilgi, G.K., and Yilmaz, K. (2015). Natural polysaccharides as pharmaceutical excipients. In: *Handbook of Polymers for Pharmaceutical Technologies: Biodegradable Polymers, Volume 3* (ed. V.K. Thakur and M.K. Thakur), 483–516. Hoboken: Wiley Scrivener.
130. Shokri, J. and Adibkia, K. (2013). Application of cellulose and cellulose derivatives in pharmaceutical industries. In: *Cellulose-Medical, Pharmaceutical and Electronic Applications* (ed. T. van de Ven and L. Godbout), 47–66. London: IntechOpen.

131. Baumgartner, S., Kristl, J., and Peppas, N.A. (2002). Network structure of cellulose ethers used in pharmaceutical applications during swelling and at equilibrium. *Pharmaceutical Research* 19 (8): 1084–1090.
132. Edgar, K.J. (2007). Cellulose esters in drug delivery. *Cellulose* 14 (1): 49–64.
133. Kristmundsdóttir, T., Sigurdsson, P., and Thormar, H. (2003). Effect of buffers on the properties of microbicidal hydrogels containing monoglyceride as the active ingredient. *Drug Development and Industrial Pharmacy* 29 (2): 121–129.
134. Valenta, C. (2005). The use of mucoadhesive polymers in vaginal delivery. *Advanced Drug Delivery Reviews* 57 (11): 1692–1712.
135. Lai, B.E., Geonnotti, A.R., DeSoto, M.G. et al. (2010). Semi-solid gels function as physical barriers to human immunodeficiency virus transport in vitro. *Antiviral Research* 88 (2): 143–151.
136. Marrazzo, J., Ramjee, G., Nair, G. et al. (2013). Pre-exposure prophylaxis for HIV in women: daily oral tenofovir, oral tenofovir/emtricitabine, or vaginal tenofovir gel in the VOICE study (MTN 003). The 20th Conference on Retroviruses and Opportunistic Infections, Atlanta, GA (3–6 March 2013).
137. Rees, H., Delany-Moretlwe, S., Lombard, C. et al. (2015). FACTS 001 phase III trial of pericoital tenofovir 1% gel for HIV prevention in women. Conference on Retroviruses and Opportunistic Infections (CROI), Seattle, WA (23–26 February 2015).
138. Marrazzo, J.M., Ramjee, G., Richardson, B.A. et al. (2015). Tenofovir-based preexposure prophylaxis for HIV infection among African women. *New England Journal of Medicine* 372 (6): 509–518.
139. Dai, J.Y., Hendrix, C.W., Richardson, B.A. et al. (2016). Pharmacological measures of treatment adherence and risk of HIV infection in the VOICE study. *Journal of Infectious Diseases* 213 (3): 335–342.
140. Vermund, S.H. and Walker, A.S. (2015). Use of pharmacokinetic data in novel analyses to determine the effect of topical microbicides as preexposure prophylaxis against HIV infection. *Journal of Infectious Diseases* 213 (3): 329–331.
141. Chuchuen, O., Henderson, M.H., Sykes, C. et al. (2013). Quantitative analysis of microbicide concentrations in fluids, gels and tissues using confocal Raman spectroscopy. *PLoS One* 8 (12): e85124.
142. Chuchuen, O., Maher, J.R., Henderson, M.H. et al. (2017). Label-free analysis of tenofovir delivery to vaginal tissue using co-registered confocal Raman spectroscopy and optical coherence tomography. *PLoS One* 12 (9): 1–5.
143. Chuchuen, O., Maher, J.R., Simons, M.G. et al. (2017). Label-free measurements of tenofovir diffusion coefficients in a microbicide gel using Raman spectroscopy. *Journal of Pharmaceutical Sciences* 106 (2): 639–644.
144. Maher, J.R., Chuchuen, O., Henderson, M.H. et al. (2015). Co-localized confocal Raman spectroscopy and optical coherence tomography (CRS-OCT) for depth-resolved analyte detection in tissue. *Biomedical Optics Express* 6 (6): 2022–2035.
145. Presnell, A.L., Chuchuen, O., Simons, M.G. et al. (2018). Full depth measurement of tenofovir transport in rectal mucosa using confocal Raman spectroscopy and optical coherence tomography. *Drug Delivery and Translational Research* 8 (3): 843–852.
146. Neurath, A.R., Strick, N., Li, Y.-Y., and Debnath, A.K. (2001). Cellulose acetate phthalate, a common pharmaceutical excipient, inactivates HIV-1 and blocks the coreceptor binding site on the virus envelope glycoprotein gp120. *BMC Infectious Diseases* 1 (1): 17.
147. Bunge, K.E., Dezzutti, C.S., Rohan, L.C. et al. (2016). A Phase 1 trial to assess the safety, acceptability, pharmacokinetics and pharmacodynamics of a novel dapivirine vaginal film. *Journal of Acquired Immune Deficiency Syndromes (1999)* 71 (5): 498.
148. Arafat, A., Samad, S.A., Masum, S.M., and Moniruzzaman, M. (2015). Preparation and characterization of chitosan from shrimp shell waste. *International Journal of Scientific and Engineering Research* 6 (5): 538–541.
149. Cheung, R., Ng, T., Wong, J., and Chan, W. (2015). Chitosan: an update on potential biomedical and pharmaceutical applications. *Marine Drugs* 13 (8): 5156–5186.
150. Zhao, D., Yu, S., Sun, B. et al. (2018). Biomedical applications of chitosan and its derivative nanoparticles. *Polymers* 10 (4): 462.
151. Sogias, I.A., Williams, A.C., and Khutoryanskiy, V.V. (2008). Why is chitosan mucoadhesive? *Biomacromolecules* 9 (7): 1837–1842.

152. Acarturk, F. (2009). Mucoadhesive vaginal drug delivery systems. *Recent Patents on Drug Delivery and Formulation* 3 (3): 193–205.
153. Dash, M., Chiellini, F., Ottenbrite, R.M., and Chiellini, E. (2011). Chitosan – a versatile semi-synthetic polymer in biomedical applications. *Progress in Polymer Science* 36 (8): 981–1014.
154. Hamman, J.H. (2010). Chitosan based polyelectrolyte complexes as potential carrier materials in drug delivery systems. *Marine Drugs* 8 (4): 1305–1322.
155. Avérous, L. (2008). Polylactic acid: synthesis, properties and applications. In: *Monomers, Polymers and Composites from Renewable Resources* (ed. M.N. Belgacem and A. Gandini), 433–450. Kidlington: Elsevier Ltd.
156. Drumright, R.E., Gruber, P.R., and Henton, D.E. (2000). Polylactic acid technology. *Advanced Materials* 12 (23): 1841–1846.
157. Makadia, H.K. and Siegel, S.J. (2011). Poly lactic-*co*-glycolic acid (PLGA) as biodegradable controlled drug delivery carrier. *Polymers* 3 (3): 1377–1397.
158. Alsaheb, R.A.A., Aladdin, A., Othman, N.Z. et al. (2015). Recent applications of polylactic acid in pharmaceutical and medical industries. *Journal of Chemical and Pharmaceutical Research* 7 (12): 51–63.
159. Lopes, M.S., Jardini, A.L., and Maciel Filho, R. (2012). Poly (lactic acid) production for tissue engineering applications. *Procedia Engineering* 42: 1402–1413.
160. Woodrow, K.A., Cu, Y., Booth, C.J. et al. (2009). Intravaginal gene silencing using biodegradable polymer nanoparticles densely loaded with small-interfering RNA. *Nature Materials* 8 (6): 526–533.
161. Necas, J. and Bartosikova, L. (2013). Carrageenan: a review. *Veterinarni Medicina* 58 (4): 187–205.
162. Buck, C.B., Thompson, C.D., Roberts, J.N. et al. (2006). Carrageenan is a potent inhibitor of papillomavirus infection. *PLoS Pathogens* 2 (7): e69.
163. Gomaa, H.H.A. and Elshoubaky, G.A. (2016). Antiviral activity of sulfated polysaccharides carrageenan from some marine seaweeds. *International Journal of Current Pharmaceutical Review and Research* 7 (1): 34–42.
164. Li, L., Ni, R., Shao, Y., and Mao, S. (2014). Carrageenan and its applications in drug delivery. *Carbohydrate Polymers* 103: 1–11.
165. Kirsch, P.P. (2002). Carrageenan: a safe additive. *Environmental Health Perspectives* 110 (6): A288.
166. Pearce-Pratt, R. and Phillips, D.M. (1996). Sulfated polysaccharides inhibit lymphocyte-to-epithelial transmission of human immunodeficiency virus-1. *Biology of Reproduction* 54 (1): 173–182.
167. Rodríguez, A., Kleinbeck, K., Mizenina, O. et al. (2014). in vitro and in vivo evaluation of two carrageenan-based formulations to prevent HPV acquisition. *Antiviral Research* 108: 88–93.
168. Zacharopoulos, V.R. and Phillips, D.M. (1997). Vaginal formulations of carrageenan protect mice from herpes simplex virus infection. *Clinical and Diagnostic Laboratory Immunology* 4 (4): 465–468.
169. Kilmarx, P.H., van de Wijgert, J.H.H.M., Chaikummao, S. et al. (2006). Safety and acceptability of the candidate microbicide Carraguard in Thai women: findings from a phase II clinical trial. *Journal of Acquired Immune Deficiency Syndromes* 43 (3): 327–334.
170. van de Wijgert, J.H.H.M., Braunstein, S.L., Morar, N.S. et al. (2007). Carraguard vaginal gel safety in HIV-positive women and men in South Africa. *Journal of Acquired Immune Deficiency Syndromes* 46 (5): 538–546.
171. Mori, T., O'Keefe, B.R., Sowder, R.C. et al. (2005). Isolation and characterization of griffithsin, a novel HIV-inactivating protein, from the red alga *Griffithsia* sp. *Journal of Biological Chemistry* 280 (10): 9345–9353.
172. O'Keefe, B.R., Vojdani, F., Buffa, V. et al. (2009). Scaleable manufacture of HIV-1 entry inhibitor griffithsin and validation of its safety and efficacy as a topical microbicide component. *Proceedings of the National Academy of Sciences of the United States of America* 106 (15): 6099–6104.
173. Levendosky, K., Mizenina, O., Martinelli, E. et al. (2015). Griffithsin and carrageenan combination to target herpes simplex virus 2 and human papillomavirus. *Antimicrobial Agents and Chemotherapy* 59 (12): 7290–7298.
174. Alihosseini, F. (2016). Plant-based compounds for antimicrobial textiles. In: *Antimicrobial Textiles* (ed. G. Sun), 155–195. Cambridge: Woodhead Publishing.
175. Lee, K.Y. and Mooney, D.J. (2012). Alginate: properties and biomedical applications. *Progress in Polymer Science* 37 (1): 106–126.

176. Paul, W. and Sharma, C.P. (2015). Alginates: wound dressings. In: *Encyclopedia of Biomedical Polymers and Polymeric Biomaterials, 11 Volume Set* (ed. M. Mishra), 134–146. Boca Raton, FL: CRC Press.
177. Augst, A.D., Kong, H.J., and Mooney, D.J. (2006). Alginate hydrogels as biomaterials. *Macromolecular Bioscience* 6 (8): 623–633.
178. Tønnesen, H.H. and Karlsen, J. (2002). Alginate in drug delivery systems. *Drug Development and Industrial Pharmacy* 28 (6): 621–630.
179. Burdick, J.A. and Stevens, M.M. (2005). Biomedical hydrogels. In: *Biomaterials, Artificial Organs and Tissue Engineering* (ed. L.L. Hench and J.R. Jones), 107–115. Cambridge: Woodhead Publishing.
180. Kumar, S., Ali, J., and Baboota, S. (2019). Polysaccharide nanoconjugates for drug solubilization and targeted delivery. In: *Polysaccharide Carriers for Drug Delivery* (ed. S. Maiti and S. Jana), 443–475. Cambridge: Woodhead Publishing.
181. Yadav, A.K., Mishra, P., and Agrawal, G.P. (2008). An insight on hyaluronic acid in drug targeting and drug delivery. *Journal of Drug Targeting* 16 (2): 91–107.
182. Gmachl, M., Sagan, S., Ketter, S., and Kreil, G. (1993). The human sperm protein PH-20 has hyaluronidase activity. *FEBS Letters* 336 (3): 545–548.
183. Bollet, A.J., Bonner, W.M., and Nance, J.L. (1963). The presence of hyaluronidase in various mammalian tissues. *Journal of Biological Chemistry* 238 (3522): 7.
184. Yoon, H.Y., Koo, H., Choi, K.Y. et al. (2012). Tumor-targeting hyaluronic acid nanoparticles for photodynamic imaging and therapy. *Biomaterials* 33 (15): 3980–3989.
185. Zhang, M., Xu, C., Wen, L. et al. (2016). A hyaluronidase-responsive nanoparticle-based drug delivery system for targeting colon cancer cells. *Cancer Research* 76 (24): 7208–7218.
186. Griesser, J., Hetényi, G., and Bernkop-Schnürch, A. (2018). Thiolated hyaluronic acid as versatile Mucoadhesive polymer: from the chemistry behind to product developments – what are the capabilities? *Polymers* 10 (3): 243.
187. Sriamornsak, P. (2003). Chemistry of pectin and its pharmaceutical uses: a review. *Silpakorn University International Journal* 3 (1–2): 206–228.
188. Liu, L., Fishman, M.L., and Hicks, K.B. (2007). Pectin in controlled drug delivery – a review. *Cellulose* 14 (1): 15–24.

9

Chitin – A Natural Bio-feedstock and Its Derivatives: Chemistry and Properties for Biomedical Applications

Anu Singh[1], Shefali Jaiswal[1], Santosh Kumar[2] and Pradip K. Dutta[1]

[1]Polymer Research Group, Department of Chemistry, Motilal Nehru National Institute of Technology, Allahabad, Prayagraj, India
[2]Department of Organic and Nano System Engineering, Konkuk University, Seoul, South Korea

9.1 Bio-feedstocks

The population in India has increased at a very high rate, and therefore waste material has also rapidly increased. Such waste materials contain cooked food, exoskeletons of animals, seafood, peels, and disposable materials. These materials are also produced by living organisms such as fungi and algae in natural environment behaviors [1]. All living organisms from plant sources and animal sources present in oceans such as crabs, lobsters, shrimps, and seafood [2, 3] produce biopolymers such as chitin, glucan, and chitin–glucan complex (ChGC). All these naturally occurring polymers therefore possess biodegradable, biocompatible and nontoxic properties. Some of these biopolymers are also present in the cell walls of insects and mushrooms.

In Asian countries, many bio-feedstock-derived materials are used for food or food-application purposes. The Indian mushroom (*Agaricus bisporus*) is one of the important sources of chitin-based biopolymers and is also good for human health [4]. Mushroom holds

---

High-Performance Materials from Bio-based Feedstocks, First Edition. Edited by Andrew J. Hunt, Nontipa Supanchaiyamat, Kaewta Jetsrisuparb and Jesper T.N. Knijnenburg.

Table 9.1 Various bio-feedstocks for obtaining biopolymers.

Sea animals	Insects	Microorganisms
Shrimp	Ladybug	Fungi
Crab	Silkworm	Yeast
Lobster	Butterfly	Algae
Molluscs	Waxworm	Brown algae

rich sources of proteins, vitamins, minerals, and essential amino acids. The polymers derived from mushrooms mostly exhibit biomedically related properties. Consequently, materials derived from mushrooms and other bio-feedstocks are used to prevent many diseases such as cancer, bacterial infections, and hypertension [5–7]. Some bio-feedstocks available in nature used for obtaining biopolymers are given in Table 9.1.

9.1.1 Chitin

Chitin is the second important plentiful naturally occurring polysaccharide in the world after cellulose. It consists of amino sugars and belongs to a long-chain polymer of N-acetyl glucosamine (β [1→4]-linked 2-acetamido-2-deoxy-β-D-glucose) [8, 9], which is similar to the glucose units that form the principal structural part of green plants like cellulose. Chitin can be readily obtained from the cell walls of crabs or shrimps and fungi [10–12]. Initially, the production of chitin is associated with food industries such as shrimp canning [13]. Chitin is found as a complex network with the constituents of minerals and proteins. In industrial processing, chitin is extracted by demineralization with hydrochloric acid followed by the alkaline solution to dissolve proteins [14]. Chitins obtained from different sources differ somewhat in their structure and percentage of their content.

In 1811, chitin was first discovered by a French professor of natural history and it was first obtained from mushrooms. In 1843, Lassaigne showed the presence of nitrogen in chitin. In 1878, Ledderhose determined chitin to be made of glucosamine and acetic acid [15].

Chitin is one of the most carbon-containing organic semitransparent polysaccharide materials. After deacetylation, chitin is transformed into Cs, which is a copolymer of glucosamine and N-acetyl glucosamine. Chitin is soluble in slightly acidic medium pH ranging from 5 to 7. Three forms of chitin are known, i.e. α-, β-, and γ-forms. In α-chitin, four crystalline peaks are observed by X-ray diffraction, in β-chitin, two crystalline peaks are observed, and in γ-chitin, antiparallel and parallel structures are formed, which is similar to α-chitin by X-ray diffraction. The production of chitin by natural organisms is estimated to be approximately 10^{11} tons year^{-1} [16]. Chitin is a biocompatible, biodegradable, and nontoxic biomaterial that exhibits excellent properties such as cell viability, blood compatibility, antitumor activity, and antithrombogenesis [17–19].

9.1.2 Chitosan

Chitosan (Cs) is the N-deacetylated derivative of chitin obtained from partial deacetylation of chitin in an alkaline medium [20, 21]. After deacetylation, each Cs subunit consists of amine and hydroxyl groups, which can be chemically altered according to the application. Cs is also isolated from various endophytic fungal species [22]. It is a biocompatible, biodegradable, nontoxic, and cationic biopolymer that has been used in many diverse fields such as wastewater treatment, tissue engineering, gene delivery, food processing, medicine, cosmetics, wound healing, drug delivery, and biotechnology [23–26]. Cs can be easily modified into different forms such as nanofibers, beads, membranes, scaffolds, nanofibrils, nanoparticles (NPs), microparticles, gels, and sponge-like structures. Cs is soluble in acid solution, i.e. at $pH<6.5$. In addition to pH, Cs solubility also depends upon the degree of deacetylation (DD) and molecular weight (MW) [27–30]. It is soluble in some aqueous dilute acetic, formic, hydrochloric, malic, lactic, and succinic acids due to the presence of amino groups in its molecular structure. However, it is insoluble or very difficult to dissolve in water, organic solvents, and alkaline solution. This is due to the formation of intermolecular hydrogen bonds.

9.1.3 Glucan

Glucan molecule is a heterogeneous polysaccharides group of D-glucose polymers linked by glycosidic bonds. It occurs naturally in the cell walls of bacteria, yeast, fungi, and cereals such as oats and barley [31–34]. According to the linkage patterns, glucan exists in two forms, namely α-glucan and β-glucan. The α-1,6-glucan with α-1,3-branches is present in dextran, whereas α-1,4- and α-1,6-glucan are present in floridean starch, glycogen, and pullulan. Starch consists of a mixture of amylose and amylopectin; amylose is branched with α-1,4-glucan, and amylopectins are branched with α-1,4 and α-1,6-glucans [35, 36]. Cellulose, chrysolaminarin, curdlan, laminarin, lentinan, lichenin, oat β-glucan, pleuran, and zymosan are various types of β-glucan.

9.1.4 Chitin–Glucan Complex

ChGC is a fungal copolymer consisting of chitin and β-glucan and is found in the cell walls of fungi and yeast. ChGC can be obtained from various sources such as *Mycothallus* [37, 38], *Komagataella pastoris* [39], *Schizophyllum commune* [40], *Gongronella butleri* [41], *Candida albicans* [42], *Armillariella mellea* [43], and *Agaricus bisporus* [44]. The chemical structure and physicochemical properties of ChGC from different sources have not been studied systematically, but, in general, the structure of ChGC may be represented by the formula $[C_6H_7O_2(OH)_2NHCOCH_3]_n\ [C_6H_7O_2(OH)_3]_m$ [45]. The nature of the chitin–glucan bonds in ChGC remains disputable. According to the published data [46, 47], ChGC is a complex system, which differs in the concentration ratio of chitin and glucan in different species. The free glucan can also be present in native complexes. The chemical composition of fungus cell walls and the chitin–to–glucan ratio vary with types of fungi, soil composition, climatic, ecological conditions under which the fungi of a given type grow and the conditions of cultivation [48].

9.1.5 Polyphenols

In recent years, polyphenols have received significant interest from scientists because of their importance in the human diet and occurrence in the natural form [49]. Polyphenols can be found in trees, fruits, plants, vegetables, tea, wine, and coffee [50–52] where they act as secondary metabolites for plant growth. Their antioxidant, anticancer, antiallergic, antidiabetic, antibacterial, antiviral, and anti-inflammatory activities make polyphenols very useful in biomedical applications [53–55]. Polyphenols possess a better antioxidant activity in comparison to other chemical substances, and therefore play a major role in cancer research. The antioxidant activity of polyphenols mainly depends on their structure. Polyphenols form a phenoxy radical by donating a hydrogen atom and these phenoxy radicals are stabilized by resonance. Therefore, polyphenols are used in scavenging activity and may act as chain breakers [56].

9.2 Synthetic Route

Chemical and enzymatic routes are available for the isolation of chitin and ChGC. Herein, we describe the chemical synthetic route for the isolation of ChGC.

9.2.1 Isolation of ChGC

Many researchers have used different techniques to isolate polysaccharides from fungi, yeasts, plants, and crustacean shells. The schematic preparation methods for chitin and Cs are shown in Figure 9.1. Herein, isolation of ChGC is performed in an aqueous solvent and the yield depends on the fungal materials, temperature, and different environments of soil materials. We have reported the synthetic route of ChGC by using *S. commune* [40]. Briefly, biomass was digested with different concentrations of hot alkali sodium hydroxide for two hours. Thereafter, biomass was filtered and washed with distilled water up to three times and the resulting ChGC was dried. Further, the dried sample was heated with 25–60% NaOH at 90 °C for six hours. The product was first filtered and washed with distilled water until neutral pH followed by drying at 50 °C. The alkali-insoluble material of ChGC was obtained.

9.2.2 Derivatives of ChGC and Its Modified Polymers

Some derivatives of ChGC are reported in the literature. Machova et al. [57] modified ChGC to form a carboxymethylated ChGC. Briefly, ChGC was suspended in a mixture of NaOH and isopropanol with stirring for one hour. Then sodium salt of monochloroacetic acid was added to the reaction mixture and maintained at 70 °C for two hours. The resulting reaction mixture was neutralized with 6N HCl and dialyzed with a dialysis membrane to remove untreated particles. Thereafter, the sample was dried, dissolved in water, separated by centrifugation and freeze-dried [58]. The carboxymethylated substituted ChGC was formed.

Chorvatovicova et al. [59] and Nudga et al. [60] modified ChGC by sulfoethyl ChGC. Briefly, the complex was suspended in a mixture of NaOH and isopropanol and stirred for one hour at 10 °C. Thereafter, the suspension was heated at 70 °C for three hours followed by the addition of sodium β-chlorethylsulfonate. After that, the solution was cooled and neutralized with ethanol/acetic acid solution. The product was dialyzed to remove unreacted particles. Finally, the modified ChGC product was freeze-dried and stored.

Skorik et al. [61] prepared N-(2-carboxyethyl)chitosan–glucan complex (CE-CsGC) by deacetylation of ChGC. First, ChGC was reacted with 25–60% of NaOH at 90 °C for six hours. The sample was filtered, washed with distilled water until pH 7, and the resulting CsGC was dried. Another method for the preparation of CsGC was done by the treatment of ChGC with hydrazine hydrate. Further, CsGC was modified in CE-CsG by treatment with acrylic acid at 90 °C for 48 hours.

Recently, Singh et al. [62, 63] synthesized chitin–glucan–quercetin conjugate and chitin–glucan–gallic acid conjugate by free radical methods. Briefly, ChGC was dissolved in an acidic medium and entered a hydrogen peroxide/ascorbic acid redox pair system for the generation of free radicals in hydroxyl, amino, α-methylene groups on ChGC. Thereafter, polyphenols (quercetin/gallic acid) were added to the solution and covalently inserted in ChGC radicals. The resulting sample was dialyzed using a membrane and finally freeze-dried.

On the other hand, Singh et al. [64] reported a novel chitin–glucan aldehyde–quercetin conjugate via a condensation reaction. Sodium periodate reagent was used to oxidize ChGC into chitin–glucan–aldehyde complex. Then the sample was lyophilized and further treated

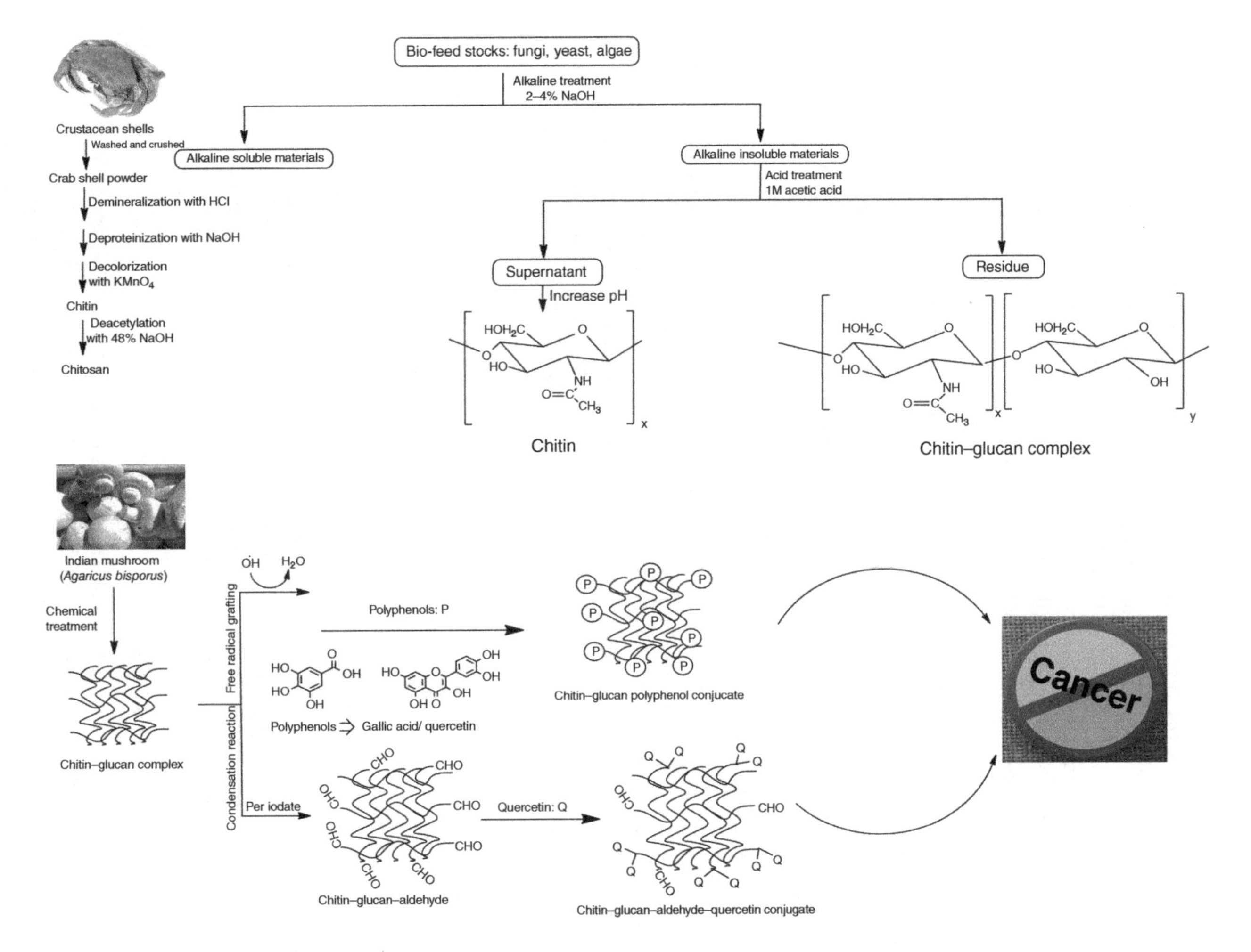

Figure 9.1 *Schematic representation of the cell-wall fractionation and sea animal-based crustacean shell chemical methods for recovery of polymers from bio-feedstocks and synthetic routes of modified polymers via free radical grafting and condensation method.*

with quercetin in the presence of acetic acid and dimethyl sulfoxide (DMSO). Thereafter, the resulting solution was dialyzed and then freeze-dried. The preparation methods of modified ChGC are shown in Figure 9.1.

9.2.3 Preparation of D-Glucosamine from Chitin/Chitosan–Glucan

Preparation of D-glucosamine was performed by a chemical method using acid hydrolysis of chitin, ChGC, and CsGC. Herein, concentrated acids like hydrochloric acid and sulfuric acid are used for the degradation of biopolymers to form D-glucosamine. Shabrukova et al. [65] synthesized D-glucosamine hydrochloride, N-acetyl-D-glucosamine, chitooligosaccharide, glucose, fructose, and ammonia by acid hydrolysis of ChGC and CsGC.

9.3 Properties of Chitin, ChGC, and Its Derivatives for Therapeutic Applications

The preparation of chitin, ChGC, and their derivatives is not sufficient unless their properties adhere to therapeutic applications. In this section, some of the important properties like antibacterial, anticancer, and antioxidant activities are discussed.

9.3.1 Antibacterial Activity

The antibacterial activities of chitin, ChGC, and their derivatives depend on various factors including the variety and species of microorganism, MW, degree of polymerization, degree of acetylation, chemical structure, functionalization/derivatization, pH, and solubility [66, 67]. These polymers inhibited the growth of bacteria and this activity was shown through the electrostatic interaction between positive and negative ion moieties. The antibacterial activity may also depend on the pH value of polymers, with a higher antibacterial activity at a lower pH value [68].

9.3.2 Anticancer Activity

Nowadays, cancer is a highly alarming disease. Chemotherapy drugs are not that effective due to their toxicity and selectivity issues. To eliminate these problems, researchers have developed new materials for the treatment of cancer [69]. Herein, we have focused on ChGC and its derivatives for the treatment of cancer and many other diseases [70]. These polymers are beneficial for some cancer cell lines like lung, colon, HeLa, microphage cancer, leukemia, and breast by in vitro and in vivo methods. Various research studies have shown that the efficacy of natural polymers is dose- and time-dependent for cancerous cell lines without affecting the normal cell lines. Recently, polyphenols have also played a major role in the treatment of cancer [15].

9.3.3 Antioxidant Activity

A wide variety of diseases such as rheumatoid arthritis and carcinogenesis is caused by free radicals such as hydroxyl (OH) and superoxide (O_2^-) radicals produced by oxidation of lipoproteins. To prevent oxidative damage, organisms and living systems are well protected by materials that inhibit the formation of free radicals, and such materials are known as antioxidants [71]. Polyphenols and many other polymers contain antioxidant activity. Herein, we analyzed the antioxidant activity of ChGC and its derivatives which can be attributed to in vitro and in vivo free radical scavenging activities. This activity is very useful for biomedical applications. For example, it protects the skin against oxidative damage, and its use as additives may prevent the oxidation of other active ingredients such as essential oils or vitamins. Antioxidant activity is measured by different methods such as alkyl, superoxide, hydroxyl, ABTS (2,2′-azino-bis-3-ethylbenzothiazoline-6-sulfonic acid) and DPPH (2,2-diphenyl-1-picrylhydrazyl) radical scavenging activity assay. The mechanism is still unclear but it is assumed to take place via chelation of free metal ions by the hydroxyl and amino groups of polysaccharide, which lead to the formation of a stable system.

9.3.4 Therapeutic Applications

The design of a biomedical biopolymer system is very useful for cancer therapeutic applications. Polyphenols are the major sources of antioxidants making them attractive in biomedical applications such as osteoarthritis, cartilage, and cancer diseases. Characteristic properties like the degree of polymerization [72, 73] of the polyphenol, its biocompatibility, and biodegradability along with the wide range of features of a particular biopolymer and its low cost make it suitable to utilize in the field of therapeutic applications [74].

Polyphenol–polymer conjugates are studied as novel materials for curing cancer and its related diseases. An enormous number of biopolymers are today available on earth. Proper selection of those biomaterials makes them suitable for biomedical applications. A large number of biopolymers have been developed with polyphenols by using a grafting technique, where phenolic compounds were covalently bonded to amino and hydroxyl groups of polymers [75]. Therefore, intra- and intermolecular hydrogen bonds are destructed and crystallinity was reduced. After grafting, biomedical properties were enhanced in comparison to native polymers. Recently, researchers of the authors' laboratory [62, 64] synthesized chitin–glucan–quercetin conjugate and chitin–glucan–gallic acid conjugate to investigate therapeutic properties such as antioxidant, anticancer, and antibacterial in comparison to native ChGC. The anticancer activity of both conjugates was checked by microphage cancer cell lines and its biocompatibility was confirmed with peripheral blood mononuclear cells (PBMCs). Upon increasing the concentration of chitin–glucan–quercetin/gallic acid conjugates, the cell death in cancer cell lines was increased, whereas normal cell lines remained unchanged. These results showed that the loading of polyphenols in ChGC increases anticancer activity. On the other hand, chitin–glucan–aldehyde–quercetin conjugate [64] showed better antioxidant and anticancer activities in comparison to native materials. The prepared conjugates protect human hepatocellular carcinoma (HepG2) and epithelial cervical cancer (HeLa) cells against oxidative DNA lesions and DNA repair of lesions stimulate as induced by alkylating agents. The derivatives of ChGC contain a potential source of antioxidants for cancer therapeutic applications.

9.4 Gene Therapy – A Biomedical Approach

Gene therapy is among the most promising and vigorously developing trends of modern science. It has gained enormous consideration over the past two decades to cure various diseases such as inherited genetic diseases [76–78], cancer [74, 79–83], infections [84–86], and obesity [87].

Gene therapy involves the delivery of various nucleic acids such as pDNA, mRNA, siRNA, and antisense oligonucleotides (ODNs) into various cells [88]. These nucleic acids present qualities that encode a functional protein, which is essential in avoiding the progression of diseases [89]. However, naked nucleic acids cannot be delivered directly into the cell because of their inability to cross the cell membrane and are easily damaged by serum nucleases [90]. Therefore, the main task in the field of gene therapy is to develop safe and efficient gene transfection vectors. Broadly, gene vectors are divided into two classes: viral vectors and nonviral vectors. Viral vectors involve both adenoviruses [91], and retro-related viruses [92].

The initial research has focused on utilizing viral vectors because of their ability to display high effectiveness and long-term gene expression [93]. However, multiple hurdles related to viral vector frameworks such as random recombination, immunogenicity, fatal inflammation, and insertional mutagenesis [94, 95] limit their quick advancement. Therefore, researchers shifted their interest to using nonviral vectors with a better safety profile that compact and protect ODNs for gene therapy. Recently, the nonviral vectors have gained new momentum for gene delivery because of the need for effective delivery systems. Nonviral vectors have an optimum security profile with low immune response, reduced immunotoxicity and biocompatibility [96]. In addition to this, nonviral vectors can be produced on a large scale at a low cost. They have the ability to show high transfection efficiency, and they are more flexible for chemical structure modification for effective gene delivery [97, 98]. Nonviral vectors have been widely used in various studies such as gene editing [76, 99–101] and CAR-T cell therapy [102], and they have been proven effective in wound healing [103], cancer, and neurodegenerative disorders [104].

Nonviral vectors consist of inorganic NPs such as metals, ceramics, magnetic materials and quantum dots as well as organic NPs such as dendrimers, polymers, liposomes, carbon nanotubes, graphene and micelles. Few organic NPs have low transfection rates [105], however, they show promising results in light of the fact that they have little oncogenic or dangerous potential if selected properly [106]. On the other hand, inorganic NPs have indicated fewer effective results for gene therapy regardless of numerous in vitro and preclinical studies. Among various organic nonviral vectors, the most attractive are the cationic lipids (lipoplexes) and polymer complexes [107–109]. Complexes based on synthetic polymers show efficient gene transfection but possess high toxicity leading to cell death. Therefore, there is an urgent need to find efficient vectors based on less toxic natural polymers. Cs holds a prominent place among natural polymers and possesses a unique set of physicochemical and biological properties. It has gained much consideration from researchers to construct gene delivery vectors based on it. Numerous researches have constructed Cs-based gene vectors. This section will mainly focus on the organic modifications made on Cs to enhance the gene delivery process.

9.5 Cs: Properties and Factors Affecting Gene Delivery

Cs is a biocompatible, biodegradable, nontoxic, and cationic polysaccharide which has helped it become one of the safest nonviral vectors for gene delivery in comparison to other nonviral vectors such as polycationic polymers and cationic lipids. The polycationic characteristics empower Cs to easily form complexes with negatively charged nucleic acids through electrostatic interactions, thereby protecting DNA from DNase I and II degradation [110, 111] as well as an increase in the cell uptake of the complex [112]. Further, the amino-functional groups of Cs contribute to the endosome escape of the complex since it could trigger a "proton sponge effect" like polyethylenimine (PEI) [112, 113]. The MW of Cs can alter the stability as well as the size of nanocomplexes, cellular uptake, and the release rate of the gene, which finally influences transfection efficiency [114–120]. Along with MW, N/P charge ratio [121, 122], pH, DD [118, 123, 124], and cell type also affect the transfection efficiency. Numerous in vitro studies have been carried out related to transfection of Cs/DNA complexes to various cell lines such as human lung carcinoma cells (A549), human osteosarcoma cell line (MG63), CV-1 in Origin with SV40 gene (COS-7) cells, human HeLa, and liver HepG2.

In addition to the aforementioned advantages, Cs exhibits inherent mucous adhesive property as well as a wound-healing activity which proves Cs to be a good candidate for designing gene delivery carriers for wound-healing application and mucosal tumor treatment [125–129]. It also exhibits potential for the delivery of clustered regularly interspaced short palindromic repeats (CRISPR)/Cas9 ribonucleoproteins [130].

Due to its versatile properties, Cs has been widely used as a gene delivery vector. However, native Cs has several limitations for this application such as low solubility at physiological pH, which, in turn, influences low transfection efficiency. Therefore, the chemical modification of Cs makes it possible to obtain derivatives with the desired properties related to gene delivery.

9.6 Organic Modifications of Cs Backbone for Enhancing the Properties of Cs Associated with Gene Delivery

A large number of organic modifications in the Cs backbone were studied for enhancing the properties related to the gene delivery process by incorporating various groups/ligands such as hydrophilic and hydrophobic groups, cationic substituents and target ligands.

9.6.1 Modification of Cs with Hydrophilic Groups

The key benefit of attaching hydrophilic moiety to Cs is to enhance its solubility, which includes quaternization and PEG modification. One of the most popular methods that are intensively studied in gene delivery applications is the complete quaternization of amino groups in Cs by alkylation [131–135]. The major advantages of trimethylated chitosan (TMC) are that it exhibits a greater solubility at physiological pH when it achieved a 40% degree of quaternization [136]. Moreover, TMC NPs strongly reduce the aggregation tendency and pH dependency of genetic material complexation as compared to unmodified Cs NPs [133]. Germershaus et al. [133] reported that cellular uptake of TMC/pDNA was far more efficient compared to Cs polyplexes in NIH/3T3 cells. However, quaternization of Cs tends to increase the cytotoxicity, which was the major shortcoming of these carriers. To eliminate this drawback, quaternized Cs was additionally modified with PEG, which resulted in a decrease in the cytotoxic effects and further enhanced the solubility. Finally, PEG-TMC NPs improved the transfection efficiency by 10-fold when compared with non-grafted TMC [133]. The introduction of folate groups onto TMC also significantly increases the gene transfection efficiency of pDNA in SKOV3, is an ovarian cancer cell line, and KB cells which overexpress folate receptors as reported by Zheng et al. [135]. In a recent study, P(PEGMA-co-DMAEMA) and folate groups were attached onto allyl-TMC to achieve TMC-g-P(PEGMA-co-DMAEMA)-FA polymer. These modifications contribute to superior water solubility and stronger pDNA complexation ability and finally improve gene transfection as compared to pristine Cs [137]. All these studies have demonstrated that modification of TMC provides a successful gene transfection system by improving various factors such as particle size, stability, targeting ability, and reduced toxicity. Further modification of quaternized Cs was used to design a gene delivery carrier with enhanced properties [138, 139]. Quaternized derivatives could also be prepared by introducing a quaternary nitrogen atom into Cs [140, 141].

Another important method affecting the solubility of Cs is PEGylation. The molecular length and degree of substitution of PEG are the two main factors that affect the solubility of Cs, which generally increases with the degree of PEG substitution [142, 143]. Nanocomplexes designed by PEGylated Cs were found to have a prolonged circulation time in blood [144], which, in turn, delays tumor accumulation, leading to the loss of siRNA and exhibiting poor endosomal escape efficiency [145]. However, this shortcoming could be overcome by attaching a targeting ligand such as folic acid to decrease the targeting time [144, 146]. Examples of other hydrophilic modifications of Cs for gene delivery include poly (vinyl pyrrolidone), poly (β-malic acid), poly (aspartic acid), and dextran [147].

9.6.2 Modification in Cs by Hydrophobic Groups

The modification of Cs with hydrophobic substituents has several advantages compared to native Cs polyplexes. Hydrophobic groups assist the dissociation of DNA inside the cell, provide effective shelter from enzymatic degradation, and improve the penetrating tendency across the cell membrane [148, 149]. The introduction of hydrophobic moieties into the Cs backbone gives an amphiphilic polymer [150]. For example, a stearic acid (SA)-modified Cs by the carbodiimide method gives an amphiphilic polymer with low cell toxicity and high transfection efficiency [151]. Meng et al. [152] further modified Cs-SA with spermine to construct a novel gene delivery vector. The spermine-SA-Cs/pDNA nanocomplex was found to have high DNA-binding ability, high gene knock-down, and reduced toxicity. The other hydrophobic moieties include caproic and alkyl halides [153, 154], deoxycholic acid [155, 156], and 5β-cholanic acid [157].

A series of fatty acid-grafted Cs were synthesized by [154] as potential nonviral gene delivery vectors and these nanomicelles showed excellent hemo- and cytocompatibility. The authors also demonstrated that grafting of palmitic acid onto Cs significantly increased the transfection efficiency of Cs. Caproic acid-modified chitosan (CCS) was prepared to analyze the effect of different degrees of amino substitutions of caproic acid in terms of DNA-binding capacity, stability, cellular uptake, unpacking ability, and transfection efficiency. CCS polymers revealed good pDNA-binding ability and shows protection against pDNA from DNase I. The spherical NPs exhibit good stability in serum and had a higher cellular uptake than Cs/pDNA polyplexes. The CCS-15 polymer revealed a 31-fold higher gene expression as compared to Cs [158]. Alkylated Cs derivatives synthesized by N-alkylation of butyl, hexadecyl, dodecyl, and octyl bromide revealed a high transfection efficiency [159]. Muddineti et al. [160] reported that Cs modified cholesterol micelles (CHCS) have the ability to condense siRNA which was optimum at N/P ratio of 40. CHCS/siRNA show efficient cell uptake in a time-dependent manner via clathrin-mediated endocytosis. Hashem et al. [161] synthesized pDNA gene-loaded Cs-modified sodium deoxycholate (Cs-DS) NPs. It showed a sustained release of pDNA and exhibited more transfection efficiency than the naked p53 gene. According to polymerase chain reaction (PCR) and enzyme-linked immunosorbent assay (ELISA), gene transfection revealed a direct association between the concentration of pDNA and gene expression. The highest expression of the gene was 5.7 times more compared to naked pDNA at the same concentration ($9\,\mu g\,ml^{-1}$).

9.6.3 Modification by Cationic Substituents

The most efficient nucleic acid delivery can be achieved by cationic nanocarriers. This can be elucidated by mainly two factors. First, the proper interaction between the positively charged polymer and the negatively charged nucleic acid protects them from nuclease degradation. Second, cationic nanocarriers possess a net cationic charge to improve the interaction of the cationic complex with the negative charge on the surface of the cell [162]. However, the shortcoming associated with Cs is its low cationic charge density [157, 163, 164]. Therefore, an increase in the cationic charge density is the crucial factor affecting endosomal rupture, which in turn increases transfection efficiency. Several types of modifications in Cs have been reported to increase the cationic charge density. The increase in cationic charge density was attained by attaching Cs to different moieties such as PEI [165], polylysine [166], spermine [167], and arginine [168, 169].

PEI is the most promising cationic nonviral vector which is readily soluble in water, possesses a high buffering capacity, and facilitates the endosomal escape of DNA. However, its toxicity and non-biodegradability restrict its applications in gene delivery. Fortunately, many experimental analyses have demonstrated that PEI conjugation with Cs improves the transfection efficacy and decreases the toxicity. Huad et al. [170] designed chitosan-grafted-PEI (CP)/DNA NPs with better transfection efficiency than that of Cs/DNA and PEI (25 kDa)/DNA NPs [171]. Zhang et al. [172] synthesized a Cs-PEI copolymer with arginine (Arg) which could strongly bind pDNA and nanocomplexes with sizes of approximately 170 nm were formed. In vitro cytotoxicity assay of Cs-PEI-Arg showed lower cytotoxicity compared with Cs-PEI, which is similar to Cs and far below that of PEI. In addition, the prepared Cs-PEI-Arg nanocomplexes had better transfection efficiency than Cs-PEI, which is greater than that of PEI. Polylysine is another cationic nonviral vector used for gene delivery due to its biodegradable properties. However, it exhibits high toxicity and low transfection efficacy. Yu et al. [166] found that poly-L-lysine grafted onto the Cs backbone displayed greater gene transfection capacity than pure Cs.

Another strategy that is used to increase the cationic charge density is Cs modification with imidazole derivatives [173–177], such as histidine and urocanic acid, which possess the proton sponge ability that is very crucial for gene delivery carriers. Therefore, modifying Cs with imidazole moieties would enhance gene transfection efficiency without compromising biocompatibility and biodegradability. Recently, Hsueh et al. [178] modified Cs with urocanic acid (via carbodiimide method) as a promising gene carrier due to its better biocompatibility, DNA protection ability, and high transfection efficiency.

9.6.4 Modification by Target Ligands

Ligand-conjugated Cs is a promising method to deliver actives such as genes, proteins, and therapeutic agents, specifically and preferentially within the target sites while minimizing non-targeted delivery [179]. These modifications include biologically active ligands such as mannose [180–184], galactose [168, 185–189], proteins, peptides, and antibodies to achieve target delivery. Furthermore, aptamers are also combined with therapeutic molecules and vectors to design targeting gene or drug delivery systems for tumor-targeted therapies [190–193]. Further, it has been reported that targeting ligands can enhance specific delivery, improve cellular uptake, and reduce toxicity [194–196]. Among various ligands, folate- and galactose-derived Cs nanocarriers were considered the most promising in cancer therapy due to the presence of folate and galactose receptors in cancer cells [100]. Galactosylated Cs is a common carrier used in gene or drug delivery to target the HepG2 cells ([100, 185, 197, 198]), HEK293 [188], SMMC7721 [199], and macrophages [103]. Han et al. [200] prepared a series of galactose-grafted trimethyl chitosan–cysteine (GTC) conjugates to study the effect of galactosylation degrees on the entire process of oral gene delivery to optimize the antitumor efficacy. The results of both in vitro and in vivo studies demonstrated that GTC NPs were excellent in enhancing cellular uptake, intracellular distribution, silencing target genes, and finally, lead to the inhibition of cell growth. Recently, Jin et al. [201] introduced methoxy poly (ethylene glycol) (mPEG) as a hydrophilic group on GCS which showed strong interaction with pDNA and excellent protection of pDNA from DNase I and serum degradation. These nanocomplexes demonstrated a high loading efficiency and sustainable release of pDNA. In addition to this, the nanocomplexes offered a nontoxic effect on L929 cells but showed inhibitory effects on Bel-7402 cells. Chitosan-grafted folic acid (CSF) is one of the most promising carriers for targeting specific cancer cells during gene delivery due to the presence of folate receptors expressed on the surface. More chemical modifications based on CSF NPs and their derivatives for drug and gene delivery were reported by John et al. [202]. Reports based on Cs modified by peptides, aptamers, and antibody and their relevant outcomes are summarized in Table 9.2.

9.7 Multifunctional Modifications of Cs

Regardless of the wide variety of applications of gene delivery carriers based on Cs, most in vitro studies have been performed in the therapeutic field and only limited in vivo studies have achieved promising results to enter the clinical phases [211].

Even after the enormous growth and assessment, Cs-based gene delivery systems still have numerous limitations such as their weak buffering capacity, lack of cell specificity, poor stability at physiological pH, and loss of positive charge (insufficient to ensure the endosomal escape). To overcome these limitations, researchers have been encouraged to develop multifunctional derivatives of Cs that improve the properties of the delivery systems. Representative examples of multifunctional Cs to enhance the efficiency of siRNA delivery are shown in Table 9.3.

9.8 Miscellaneous

Until now, there have been many other works on Cs-based nanocarriers in gene delivery for multipurpose applications reported by various researchers. Very recently, the authors' laboratory has reported methyl methacrylate modified chitosan conjugate (CSMMA) for efficient gene transfection and drug delivery [215] (Figure 9.2). In all three cancer cell lines (A549, HeLa and HepG2), in vitro cytotoxicity assay and transfection tests demonstrated improved cell viability and transfection efficiency of CSMMA over unmodified Cs and lipofectamine®. In vitro drug release study of curcumin-loaded CSMMA NPs verified that pH 5.0 was more favorable than physiological pH for the drug release. These results demonstrate that CSMMA NPs can prove to be a good nonviral vector for a drug as well as gene delivery.

Some reports on Cs modification for gene delivery that could not be included in accordance with the classification given in this section are summarized in Table 9.4.

9.9 Conclusion

In this chapter, we describe the pivotal role of chitin, ChGC, and their derivatives in antibacterial, antioxidant, and anticancer applications, as well as Cs as the main matrix for effective gene therapy. Herein, we have seen that the derivatives showed excellent biomedical applications

Table 9.2 *Summary of special characteristics of target ligand-modified Cs-based nanocarriers in gene delivery.*

Modification by target ligands		Characteristics	Active agents	Cell type	In vitro/ in vivo application	Reference
Antibody	Cs-modified transferrin antibody and bradykinin B2 antibody	• Average size 235.7 ± 10.2 nm, zeta potential 22.88 ± 1.78 mV • Improved cellular accumulation and gene-silencing efficiency compared to single-antibody Cs-NPs and non-modified Cs-NPs	siRNA	U138-MG and hCMEC/D3	In vitro	[203]
	Cs-modified CD7-specific single-chain antibody (scFvCD7)	• Average diameter nearly 320 nm, zeta potential +17 mV • Increased cellular association and gene-silencing efficiency compared to native Cs NPs	siRNA	CD_4 + T	In vitro	[204]
Peptides	(EGFR)-targeted peptide-modified Cs	• Targeted NPs displayed greater and more selective intracellular uptake than non-targeted NPs • Effective depletion of Mad2 expression resulting in cell death	siRNA	A549 cells	In vitro	[205]
	PEI modified with N-octyl-N-quaternary Cs (OTMCS) and trifunctional peptide (RGDC-TAT-NLS)	• Greater cell viability and transfection efficacy of synthesized NPs compared to PEI, OTMCS-PEI, and OTMCS-PEI-R13	pDNA	Hela cells	In vitro and in vivo	[196]
	PEI-g-Cs with RGD dendrimer peptide (PEI-Cs-RGD)	• Targeted gene vector was biocompatible and significantly controls tumor growth in PC3-xenograft mouse model by silencing *BCL2* mRNA	pDNA	HEK293, MDA-MB-231, and PC3	In vivo and in vitro	[206]
	RGD peptidomi-metic (RGDp) modified PEG-Cs	• Therapeutic efficacy of targeted NPs (targets MCT1 and ASCT2 transporters) could be efficiently internalized into αvβ3 integrin-expressing cells. • Attaching a lipophilic linker to the targeted NPs enhances delivery efficacy more than a hydrophilic linker	siRNA	H1299 cell	In vitro and in vivo	[195]

(*Continued*)

Table 9.2 *(Continued)*

Modification by target ligands		Characteristics	Active agents	Cell type	In vitro/ in vivo application	Reference
	Poly(histidine-arginine)$_6$ (H6R6) peptide grafted CS	• in vitro: NPs displayed greater transfection efficiency and superior endosomal escape capacity than unmodified Cs NPs • in vivo: Significant suppression of tumor cell growth and metastases	siRNA	4T1	In vitro and in vivo	[207]
Aptamers	AS1411-Cs-ss-PEI -urocanic acid (ACPU) micelle	• Dual pH/ redox sensitivity and targeting effects for tumor therapy • Excellent tumor penetrating efficacy and low toxicity	Doxorubicin (Dox) and Toll-like receptor 4 siRNA	A549 cells	In vitro and in vivo	[208]
	Cs-PEG- transferrin (Cs-PEG-TfP)	• Size 165–248 nm, surface charge 3–4.5 mV • in vitro: Great antitumor activities by synergistic effect • in vivo: Transfection efficiency displayed higher expression in HCT119 cells than HEK293 cells	psiRNA-bcl2 and (KLA)$_4$ peptide	HCT119 cells, HEK293 cells	In vitro	[209]
Peptide + receptors	TAT-LHRH-chitosan conjugate (TLC)	• TLC NPs (size 70–85 nm, zeta potential +30 mV) has a strong DNA-binding affinity • NPs were nontoxic and highly selective for BEL-7402 cells	pDNA	BEL-7402	In vitro	[210]

Table 9.3 Summary of special features/application in multifunctional modification of Cs-based nanocarriers in gene delivery.

Multifunctional modification	Special features/application after modification	Cell lines	In vitro/ in vivo	Reference
Mannose-g-trimethylchitosan-cysteine (MTC) NPs	• MTC NPs were created by the ionic gelation method • Improved stability as well as the quality of transfection by overcoming extracellular and intracellular barriers such as for oral delivery of TNF-α siRNA	Raw 264.7	In vitro and in vivo both	[212]
Folate-targeted PEG-modified amino acid-functionalized chitosan (Cs-PFA)	• Cs-PFA NPs showed hemocompatibility and increased gene expression, higher affinity toward cancer cells, and penetrated into 3D spheroids to a higher extent than non-targeted NPs • Decrease in tumor-spheroids volume was promoted by Cs-PFA-mediated delivery of p53 tumor suppressor	HeLa cells	In vitro	[213]
TMC modified with poly(ε-caprolactone) (PCL)	• TMCSP-g-PCL NPs could effectively bind pDNA • Polyplexes were stable in physiological salt condition and showed high uptake efficiency	HEK293T	In vitro	[139]
Tat-tagged and folate-modified N-succinyl chitosan (Tat-Suc-FA-Cs)	• Tat-Suc-FA polymers possessed much lower cytotoxicity than CS • Tat-Suc-FA polymers could form stable NPs (54–106 nm, positive charge 3–44 mV) with pDNA • Low-toxicity Tat-Suc-FA-Cs polymers could be used as gene delivery vectors	K562	In vitro	[214]
N-succinyl chitosan–poly-L-lysine–palmitic acid (NSC–PLL–PA)	• Triblock copolymer micelle was designed to co-deliver Dox and siRNA–P-glycoprotein (P-gp) • The NPs (170 nm) were less stable at pH 5.3 thus improving the intracellular release of encapsulated drug and siRNA and preventing tumor growth • Improved antitumor efficacy due to downregulation of P-gp expression and increasing intracellular Dox concentration	HepG2 and HepG2/ ADM	In vitro and in vivo	[74]

because of their increased stability and solubility in comparison to native material. Overall, in this article, an attempt has been made to understand the importance of chitin, a natural bio-feedstock and its derivatives, by describing its chemistry and properties for biomedical applications. It is expected that future research will motivate biotechnologists, industrialists, entrepreneurs, and researchers to take the challenges and opportunities for chitin-based natural bio-feedstocks for biomedical applications.

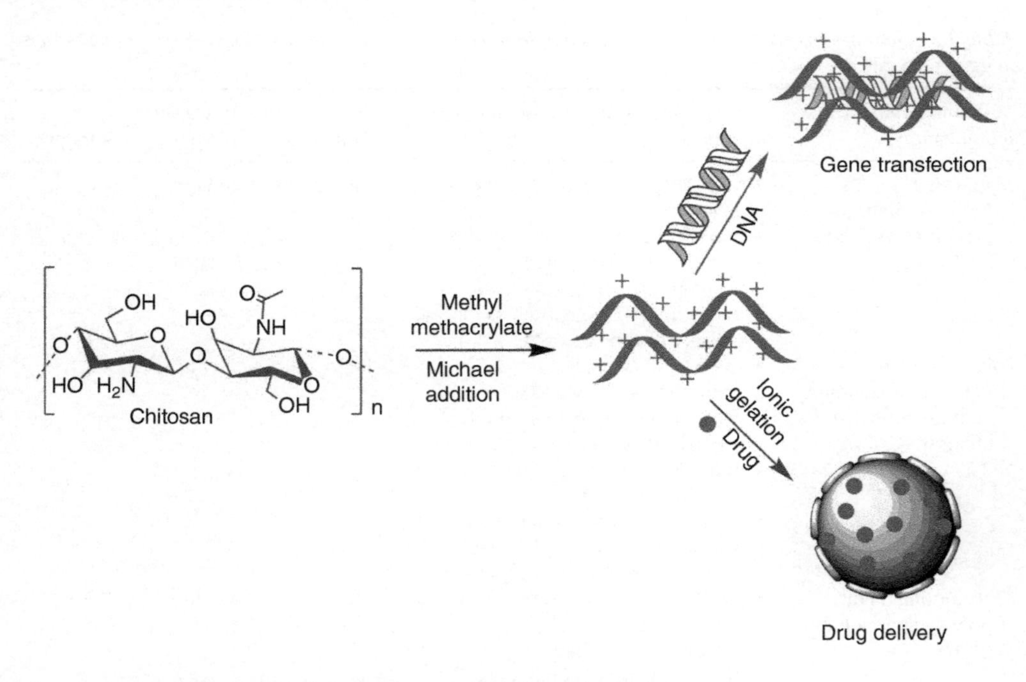

Figure 9.2 *CSMMA NPs for efficient gene transfection and drug delivery.*

Table 9.4 *Summary of special features/biomedical application in Cs-based nanocarriers in gene delivery.*

Cs modification	Special features/application after modification	Cell types	In vitro/ in vivo application	Reference
Cs grafted with alkylamine (Cs-AA) such as propylamine (PA), (diethylamino) propylamine (DEAPA), and N,N-dimethyl-dipropylenetriamine (DMAPAPA)	• In vitro: Cs-AA (especially, Cs-DMAPAPA) showed strong binding capacity with gene and excellent biocompatibility • Cs-DMAPAPA displayed 1.58 times greater buffering capacity than Cs and enhanced transfection performance • In vivo: DMAPAPA-Cs/pDNA complexes exhibited superior transfection behavior and tumorous penetrability over other Cs-AA complexes and Cs	A549	In vitro and in vivo	[216]
Cs-g-Piperazine	• Improved water solubility at physiological pH with low cytotoxicity and efficient combination with siRNA • In vivo: significant tumor reduction and no adverse effects of complexes	A549-luc	In vitro and in vivo	[217]

Table 9.4 (Continued)

Cs modification	Special features/application after modification	Cell types	In vitro/ in vivo application	Reference
Triphenylamine modified Cs (TPAC)	• Improved solubility, biodegradability, and low viscosity at biological pH • Enhanced availability of delivered pDNA to the nucleus • Best transfection efficacy at 1 : 5 ratio of pGL3 to TPAC • In vivo: higher luciferase gene expression of pGL3/TPAC nanocomplexes in Balb/c mice in contrast to pGL3/Cs	A549, HeLa and HepG2	In vitro and in vivo	[218]
2-Acrylamido-2-methylpropane coupled Cs (CsAMP)	• In vitro: CsAMP/DNA exhibited a higher buffering capacity leading to a greater transfection efficiency than Cs and lipofectamines • In vivo: higher luciferase expression than Cs	A549, HeLa and HepG2	In vitro and in vivo	[219]

Acknowledgments

The authors are grateful to Director Prof. Rajeev Tripathi, Motilal Nehru National Institute of Technology (MNNIT) Allahabad, Prayagraj (India) for providing stipend/infrastructure to AS & SJ. The authors are thankful to Technical Education Quality Improvement Programme (TEQIP) for providing a research-related fund. The authors are also thankful to Centre for Interdisciplinary Research (CIR), MNNIT Allahabad for providing characterization facilities.

References

1. Yadav, M., Goswami, P., Paritosh, K. et al. (2019). Seafood waste: a source for preparation of commercially employable chitin/chitosan materials. *Bioresources and Bioprocessing* 6: 8–27. https://doi.org/10.1186/s40643-019-0243-y.
2. Kaur, S., Dhillon, G.S. (2014). The versatile biopolymer chitosan: potential sources, evaluation of extraction methods and applications. *Critical Reviews in Microbiology* 40(2): 155–175. DOI: https://doi.org/10.3109/1040841X.2013.770385.
3. Khanafari, A., Marandi, R., and Sanatei, S. (2008). Recovery of chitin and chitosan from shrimp waste by chemical and microbial methods. *Journal of Environmental Health Science & Engineering* 5 (1): 19–24.
4. Dutta, P.K., Dutta, J., Chattopadhyaya, M.C. et al. (2004). Chitin and chitosan: novel biomaterials waiting for future developments. *Journal of Polymer Materials* 21 (3): 321–333.
5. Bilal, M., Iqbal, H.M.N. (2019). Naturally-derived biopolymers: potential platforms for enzyme immobilization. *International Journal of Biological Macromolecules* 130: 462–482. DOI: https://doi.org/10.1016/j.ijbiomac.2019.02.152.
6. Dutta, J., Tripathi, S., Dutta, P.K. (2012). Progress in antimicrobial activities of chitin, chitosan and its oligosaccharides: a systematic study needs for food applications. *Food Science and Technology International* 18(1): 3–34. DOI: https://doi.org/10.1177/1082013211399195.
7. Dutta, P.K., Kumar, M.N.V.R., and Dutta, J. (2002). Chitin and chitosan for versatile applications. *Journal of Macromolecular Science, Part C: Polymer Reviews* 42: 307–354. https://doi.org/10.1081/MC-120006451.

8. Prashanth, K.V.H. and Tharanathan, R.N. (2007). Chitin/chitosan: modifications and their unlimited application potential – an overview. *Trends in Food Science and Technology* 18 (3): 117–131.
9. Rinaudo, M. (2006). Chitin and chitosan: properties and applications. *Progress in Polymer Science* 31: 603–632. http://dx.doi.org/10.1016/j.progpolymsci.2006.06.001.
10. Anitha, A., Sowmya, S., Sudheesh Kumar, P.T. et al. (2014). Chitin and chitosan in selected biomedical applications. *Progress in Polymer Science* 39 (9): 1644–1667. http://dx.doi.org/10.1016/j.progpolymsci.2014.02.008.
11. Fairbridge, R.W. and Jablonski, D. (1979). *The Encyclopedia of Paleontology*, 186–189. Berlin, Germany: Springer.
12. Khor, E. and Lim, L.Y. (2003). Implantable applications of chitin and chitosan. *Biomaterials* 24 (13): 2339–2349. https://doi.org/10.1016/S0142-9612(03)00026-7.
13. Hudson, S.M. and Jenkins, D.W. (2001). *Chitin and Chitosan. Encyclopedia of Polymer Science and Technology*, 805–827. New York: John Wiley & Sons.
14. Jang, M.K., Kong, G., Jeong, Y. et al. (2004). Physicochemical characterization of α-chitin, α-chitin, and γ-chitin separated from natural resources. *Journal of Polymer Science, Part A: Polymer Chemistry* 42: 3423–3432. https://doi.org/10.1002/pola.20176.
15. Adhikari, H.S. and Yadav, P.N. (2018). Anticancer activity of chitosan, chitosan derivatives, and their mechanismofaction.*InternationalJournalofBiomaterials*2018:29.https://doi.org/10.1155/2018/2952085.
16. Elieh-Ali-Komi, D. and Hamblin, M.R. (2016). Chitin and chitosan: production and application of versatile biomedical nanomaterials. *International Journal of Advanced Research* 4 (3): 411–427.
17. Aranaz, I., Mengíbar, M., Harris, R. et al. (2009). Functional characterization of chitin and chitosan. *Current Chemical Biology* 3:203–230. DOI: https://doi.org/10.2174/2212796810903020203.
18. Dutta, P.K., Dutta, J., and Tripathi, V.S. (2004). Chitin and chitosan: chemistry, properties and applications. *Journal of Scientific and Industrial Research* 63: 20–31. http://nopr.nisca ir.res.in/handle/123456789/5397.
19. Honarkar, H., Barikani, M., Honarkar, H. et al. (2009). Applications of biopolymers I: chitosan. *Monatshefte für Chemie – Chemical Monthly* 140: 1403–1420.
20. Andres, E., Albesa-Jove, D., Biarnes, X. et al. (2014). Structural basis of chitin oligosaccharide deacetylation. *Angewandte Chemie International Edition* 53 (27): 6882–6887.
21. Liu, L., He, Y., Shi, X. et al. (2018). Phosphocreatine-modified chitosan porous scaffolds promote mineralization and osteogenesis in vitro and in vivo. *Applied Materials Today* 12: 21–33.
22. Dutta, J. (2013). Engineering of polysaccharides via nanotechnology. In: *Multifaceted Development and Application of Biopolymers for Biology, Biomedicine and Nanotechnology* (ed. P.K. Dutta and J. Dutta), 87–134. New York: Springer.
23. Dash, M., Chiellini, F., Ottenbrite, R.M. et al. (2011). Chitosan a versatile semi-synthetic polymer in biomedical applications. *Progress in Polymer Science* 36: 981–1014. https://doi.org/10.1016/j.progpolymsci.2011.02.001.
24. Hein, S., Wang, K., Stevens, W.F. et al. (2008). Chitosan composites for biomedical applications: status, challenges and perspectives. *Materials Science and Technology* 24 (9): 1053–1061. https://doi.org/10.1179/174328408X341744.
25. Shi, C., Zhu, Y., Ran, X. et al. (2006). Therapeutic potential of chitosan and its derivatives in regenerative medicine. *Journal of Surgical Research* 133(2):185–192. DOI: https://doi.org/10.1016/j.jss.2005.12.013.
26. Xia, W., Liu, P., Liu, J. (2008). Advance in chitosan hydrolysis by non-specific cellulases. *Bioresource Technology* 99 (15): 6751–6762. DOI: https://doi.org/10.1016/j.biortech.2008.01.011.
27. Alameh, M., Lavertu, M., Tran-Khanh, N. et al. (2017). siRNA delivery with chitosan: influence of chitosan molecular weight, degree of deacetylation, and amine to phosphate ratio on in vitro silencing efficiency, hemocompatibility, biodistribution, and in vivo efficacy. *Biomacromolecules* 19 (1): 112–131.
28. Kumar, B.A., Varadaraj, M.C., Tharanathan, R.N. (2007). Low molecular weight chitosan – preparation with the aid of pepsin, characterization, and its bactericidal activity. *Biomacromolecules* 8(2): 566–572. DOI: https://doi.org/10.1021/bm060753z.
29. Opanasopit, P., Tragulpakseerojn, J., Apirakaramwong, A. et al. (2011). The development of poly-l-arginine-coated liposomes for gene delivery. *International Journal of Nanomedicine* 6: 2245–2252.
30. Park, J.K., Chung, M.J., Choi, H.N. et al. (2011). Effects of the molecular weight and the degree of deacetylation of chitosan oligosaccharides on antitumor activity. *International Journal of Molecular Sciences* 12 (1): 266–277. DOI: https://doi.org/10.3390/ijms12010266.

31. Di Luzio, N.R., Williams, D.L., McNamee, R.B. et al. (1979). Comparative tumor-inhibitory and anti-bacterial activity of soluble and particulate glucan. *International Journal of Cancer* 24: 773–779. DOI: https://doi.org/10.1002/ijc.2910240613.
32. Sandula, J., Machová, E., Hríbalová, V. (1995). Mitogenic activity of particulate yeast β-(1→3)-d-glucan and its water-soluble derivatives. *International Journal of Biological Macromolecules* 17:323–326. DOI: https://doi.org/10.1016/0141-8130(96)81839-3.
33. Tohamy, A.A., El-Ghor, A.A., El-Nahas, S.M. et al. (2003).β-Glucan inhibits the genotoxicity of cyclo-phosphamide, adramycin and cisplatin. *Mutation Research* 541(1–2): 45–53. DOI: https://doi.org/10.1016/s1383-5718(03)00184-0.
34. Zeković, D.B., Kwiatkowski, S., Vrvić, M.M. et al. (2005). Natural modified (1→3)-β-d-glucans in health promotion and disease alleviation, *Critical Reviews in Biotechnology* 25: 205–230. DOI: https://doi.org/10.1080/07388550500376166.
35. Lowman, D.W., West, L.J., Bearden, D.W. et al. (2011). New insights into the structure of (1–3,1–6)-β-d glucan side chains in the Candida glabrata cell wall. *PLoS One* 6 (11): 2011–27614. https://doi.org/10.1371/journal.pone.0027614.
36. Synytsya, A., Novak, M. (2014). Structural analysis of glucans. *Annals of Translational Medicine* 2(2):17–30. DOI: https://doi.org/10.3978/j.issn.2305-5839.2014.02.07.
37. Ivshina, T.N., Artamonova, S.D., Ivshin, V.P. et al. (2009). Isolation of the ChGC from the fruiting bodies of mycothallus. *Applied Biochemistry and Microbiology* 45: 313–318.
38. Pestov, A.V., Drachuk, S.V., Koryakova, O.V. et al. (2009). Isolation and characterization of chitin-glucan complexes from the mycothallus of fungi belonging to Russula genus. *Chemistry for Sustainable Development* 17: 281–287.
39. Farinha, I., Duarte, P., Pimentel, A. et al. (2015). Chitin–glucan complex production by *Komagataellapastoris*: downstream optimization and product characterization. *Carbohydrate Polymers* 130: 455–464. DOI: https://doi.org/10.1016/j.carbpol.2015.05.034.
40. Abdel-Mohsen, A.M., Jancar, J., Massoud, D. et al. (2016). Novel chitin/chitosan-glucan wound dressing: isolation, characterization, antibacterial activity and wound healing properties. *International Journal of Pharmaceutics* 510(1):86–99. DOI: https://doi.org/10.1016/j.ijpharm.2016.06.003.
41. Nwea, N., Stevens, W.F., Tokura, S. et al. (2008). Characterization of chitosan and chitosan–glucan complex extracted from the cell wall of fungus *Gongronellabutleri* USDB 0201 by enzymatic method. *Enzyme and Microbial Technology* 42: 242–251. DOI: https://doi.org/10.1016/j.enzmictec.2007.10.001.
42. Estrada-Mata, E. Navarro-Arias, M.J., Pérez-García, L.A. et al. (2015). Members of the *Candida parapsilosis* complex and *Candida albicans* are differentially recognized by human peripheral blood mononuclear cells. *Frontiers in Microbiology* 6: 1527–1538.DOI: https://doi.org/10.3389/fmicb.2015.01527.
43. Ivshin, V.P., Artamonova, S.D., Ivshina, T.N. et al. (2007). Methods for isolation of chitin–glucan complexes from higher fungi native biomass, *Polymer Science* 49(11–12): 2215–2222. DOI: https://doi.org/10.1134/S1560090407110097.
44. Singh, A. and Dutta, P.K. (2017). Extraction of chitin-glucan complex from *Agaricus bisporus*: characterization and antibacterial activity. *Journal of Polymer Materials* 34 (1): 1–9.
45. Hong, Y., Ying, T. (2019). Characterization of a chitin-glucan complex from the fruiting body of Termitomyces albuminosus (Berk.) Heim. *International Journal of Biological Macromolecules* 134: 131–138. DOI: https://doi.org/10.1016/j.ijbiomac.2019.04.198.
46. Kollar, R., Petracova, E., Ashwell, G. et al. (1997). Architecture of the yeast cell wall β(1→6)-glucan interconnects mannoprotein, β(1→3)-glucan, and chitin. *The Journal of Biological Chemistry* 272(28): 17762–17775. DOI: https://doi.org/10.1074/jbc.272.28.17762.
47. Siestma, J.H. and Wessels, J.G.H. (1979). Evidence for covalent linkages between chitin and β-glucan in a fungal wall. *Journal of General Microbiology* 114: 99–108. https://doi.org/10.1099/00221287-114-1-99.
48. Alessandri, G., Milani, C., Duranti, S. et al. (2019). Ability of bifidobacteria to metabolize chitin-glucan and its impact on the gut microbiota. *Scientific Reports* 9: 5755–5769.
49. Strack, D. (1997). Phenolic metabolism. In: *Plant Biochemistry* (ed. P.M. Dey and J.B. Harborne), 387–416. London, UK: Academic Press.
50. Higdon, J.V., Frei, B. (2003). Tea catechins and polyphenols: health effects, metabolism, and antioxidant functions. *Critical Reviews in Food Science and Nutrition* 43: 89–143. DOI: https://doi.org/10.1080/10408690390826464.

51. Nanditha, B. and Prabhasankar, P. (2008). Antioxidants in bakery products: a review. *Critical Reviews in Food Science and Nutrition* 49: 1–27. https://doi.org/10.1080/10408390701764104.
52. Vuong, V., Stathopoulos, C.E., Nguyen, M.H. et al. (2011). Isolation of green tea catechins and their utilization in the food industry. *Food Reviews International* 27 (3): 227–247. https://doi.org/10.1080/87559129.2011.563397.
53. Crozier, A., Jaganath, I.B., Clifford, M.N. (2009). Dietary phenolics: chemistry, bioavailability and effects on health. *Natural Product Reports* 26: 1001–1043. DOI: https://doi.org/10.1039/b802662a.
54. Rice-Evans, C.A., Miller, N.J., Paganga, G. (1996). *Free Radical Biology and Medicine* 20 (7): 933–956. DOI: https://doi.org/10.1016/0891-5849(95)02227-9.
55. Singh, A., Holvoet, S., Mercenier, A. (2011). Dietary polyphenols in the prevention and treatment of allergic diseases. *Clinical & Experimental Allergy* 41: 1346–1359.DOI: https://doi.org/10.1111/j.1365-2222.2011.03773.x.
56. Oliver, S., Vittorio, O., Cirillo, G. et al. (2016). Enhancing the therapeutic effects of polyphenols with macromolecules. *Polymer Chemistry* 7: 1529–1544. DOI:https://doi.org/10.1039/C5PY01912E.
57. Machova, E., Kogan, G., Soltes, L. et al. (1999). Ultrasonic depolymerization of the chitin–glucan isolated from *Aspergillus niger*. *Reactive & Functional Polymers* 42: 265–271. https://doi.org/10.1016/S1381-5148(98)00085-6.
58. Slamenová, D., Kovácíková, I., Horváthová, E. et al. (2010). Carboxymethyl chitin-glucan (CM-CG) protects human HepG2 and HeLa cells against oxidative DNA lesions and stimulates DNA repair of lesions induced by alkylating agents. *Toxicology in vitro* 24: 1986–1992. DOI: https://doi.org/10.1016/j.tiv.2010.08.015.
59. Chorvatovicova, D., Kovacikova, Z., Sandula, J. et al. (1993). Protective effect of sulfoethylglucan against hexavalent chromium. *Mutation Research Letters* 302: 207–211. https://doi.org/10.1016/0165-7992(93)90106-6.
60. Nudga, L.A., Petrova, V.A., Benkovich, A.D. et al. (2001). Comparative study of reactivity of cellulose, chitosan, and chitin-glucan complex in sulfoethylation. *Russian Journal of Applied Chemistry* 74 (1): 145–148.
61. Skorik, Y.A., Pestov, A.V., Yatluk, Y.G. (2010). Evaluation of various chitin-glucan derivatives from *Aspergillus niger* as transition metal adsorbents. *Bioresource Technology* 101 (6): 1769–1775. DOI: https://doi.org/10.1016/j.biortech.2009.10.033.
62. Singh, A., Kumar, H., Kureel, A.K. et al. (2019). Improved antibacterial and antioxidant activities of gallic acid grafted chitin-glucan complex. *Journal of Polymer Research* 26: 234–244. https://doi.org/10.1007/s10965-019-1893-3.
63. Singh, A., Lavkush, Kureel, A.K. et al. (2018). Curcumin loaded chitin-glucan quercetin conjugate: synthesis, characterization, antioxidant, in vitro release study, and anticancer activity. *International Journal of Biological Macromolecules* 110: 234–244. DOI: https://doi.org/10.1016/j.ijbiomac.2017.11.002.
64. Singh, A., Kumar, H., Kureel, A.K. et al. (2018). Synthesis of chitin-glucan-aldehyde-quercetin conjugate and evaluation of anticancer and antioxidant activities. *Carbohydrate Polymers* 193: 99–107. https://doi.org/10.1016/j.carbpol.2018.03.092.
65. Shabrukova, N.V., Shestakova, L.M., Zainetdinova, D.R. et al. (2002). Gamayurova research of acid hydrolyses of chitin-glucan and chitosan-glucan complexes. Thematic course: natural polysaccharides. Part III. Chemistry and computational simulation. *Butlerov Communications* 2 (8): 57–60.
66. Ghosh, S., Haldar, J. (2020). Cationic polymer-based antibacterial smart coatings. Advances. In: Smart Coatings and Thin Films for Future Industrial and Biomedical Engineering Applications, Makhlouf, A. S. H., Abu-Thabit, N. Y., Eds., Amsterdam, The Netherland: Elsevier. 557–582. https://doi.org/10.1016/B978-0-12-849870-5.00011-2.
67. Tachaboonyakiat, W. (2017). Antimicrobial applications of chitosan. *Chitosan Based Biomaterials Volume 2 Tissue Engineering and Therapeutics* 2: 245–274. http://dx.doi.org/10.1016/B978-0-08-100228-5.00009-2.
68. Bakshi, P.S., Selvakumar, D., Kadirvelu, K. et al. (2018). Comparative study on antimicrobial activity and biocompatibility of N-selective chitosan derivatives. *Reactive and Functional Polymers* 124: 149–155. https://doi.org/10.1016/j.reactfunctpolym.2018.01.016.
69. Zhang, H., Wu, F., Li, Y. et al. (2016). Chitosan-based nanoparticles for improved anticancer efficacy and bioavailability of mifepristone. *Beilstein Journal of Nanotechnology* 7: 1861–1870.

70. Kothari, D., Patel, S., Kim, S.K. (2018). Anticancer and other therapeutic relevance of mushroom polysaccharides: a holistic appraisal. *Biomedicine & Pharmacotherapy* 105: 377–394. DOI: https://doi.org/10.1016/j.biopha.2018.05.138.
71. Neha, K., Haider, M.R., Pathak, A. et al. (2019). Medicinal prospects of antioxidants: a review. *European Journal of Medicinal Chemistry* 178: 687–704. https://doi.org/10.1016/j.ejmech.2019.06.010.
72. Mitjans, M., Ugartondo, V., Martinez, V. et al. (2011). Role of galloylation and polymerization in cytoprotective effects of polyphenolic fractions against hydrogen peroxide insult. *Journal of Agricultural and Food Chemistry* 59 (5): 2113–2119.
73. Valko, M., Leibfritz, D., Moncol, J. et al. (2007). Free radicals and antioxidants in normal physiological functions and human disease. *International Journal of Biochemistry & Cell Biology* 39 (1): 44–84.
74. Zhang, C.G., Zhu, W., Liu, Y. et al. (2016). Novel polymer micelle mediated co-delivery of doxorubicin and P-glycoprotein siRNA for reversal of multidrug resistance and synergistic tumor therapy. *Scientific Reports* 6 (1): 1–12.
75. Bhattacharya, A. and Misra, B.N. (2004). Grafting: a versatile means to modify polymers techniques, factors and applications. *Progress in Polymer Science* 29: 767–814. https://doi.org/10.1016/j.progpolymsci.2004.05.002.
76. Gao, X., Tao, Y., Lamas, V. et al. (2018). Treatment of autosomal dominant hearing loss by in vivo delivery of genome editing agents. *Nature* 553 (7687): 217–221.
77. Mendell, J.R., Al-Zaidy, S., Shell, R. et al. (2017). Single-dose gene-replacement therapy for spinal muscular atrophy. *New England Journal of Medicine* 377 (18): 1713–1722.
78. Portier, I., Vanhoorelbeke, K., Verhenne, S. et al. (2018). High and long-term von Willebrand factor expression after Sleeping Beauty transposon-mediated gene therapy in a mouse model of severe von Willebrand disease. *Journal of Thrombosis and Haemostasis* 16 (3): 592–604.
79. Wang, P., Zhang, L., Xie, Y. et al. (2017). Genome editing for cancer therapy: delivery of cas9 protein/sgRNA plasmid via a gold nanocluster/lipid core shell nanocarrier. *Advanced Science* 4 (11): 1700175–1700185.
80. Zhang, L., Zheng, W., Tang, R. et al. (2016). Gene regulation with carbon-based siRNA conjugates for cancer therapy. *Biomaterials* 104: 269–278.
81. Zhang, M.Z., Zhou, X.J., Wang, B. et al. (2013). Lactosylated gramicidin-based lipid nanoparticles (lac-GLN) for targeted delivery of anti-miR-155 to hepatocellular carcinoma. *Journal of Controlled Release* 168 (3): 251–261.
82. Zhao, W., Yang, Y., Song, L. et al. (2017). A vesicular stomatitis virus inspired DNA nanocomplex for ovarian cancer therapy. *Advanced Science* 5 (3): 1–9.
83. Zhao, W., Yang, Y., Song, L. et al. (2018). A vesicular stomatitis virus-inspired DNA nanocomplex for ovarian cancer therapy. *Advanced Science* 5 (3): 1700263.
84. Falkenhagen, A., Singh, J., Asad, S. et al. (2017). Control of HIV infection in vivo using gene therapy with a secreted entry inhibitor. *Molecular Therapy – Nucleic Acids* 9: 132–144.
85. Hetzel, M., Mucci, A., Blank, P. et al. (2018). Hematopoietic stem cell gene therapy for IFN gamma R1 deficiency protects mice from mycobacterial infections. *Blood* 131 (5): 533–545.
86. Mitsuyasu, R.T., Merigan, T.C., Carr, A. et al. (2009). Phase 2 gene therapy trial of an anti-HIV ribozyme in autologous CD34+ cells. *Nature Medicine* 15 (3): 285–292.
87. Won, Y.W., Adhikary, P.P., Lim, K.S. et al. (2014). Oligopeptide complex for targeted non-viral gene delivery to adipocytes. *Nature Materials* 13 (12): 1157–1164.
88. Wirth, T., Parker, N., and Ylä-Herttuala, S. (2013). History of gene therapy. *Gene* 525: 162–169.
89. Wong, J.K., Mohseni, R., Hamidieh, A.A. et al. (2017). Will nanotechnology bring new hope for gene delivery. *Trends in Biotechnology* 35 (5): 434–454.
90. Niven, R., Pearlman, R., Wedeking, T. et al. (1998). Biodistribution of radiolabeled lipid-DNA complexes and DNA in mice. *Journal of Pharmaceutical Sciences* 87 (11): 1292–1299.
91. Barnard, A.R., Groppe, M., and MacLaren, R.E. (2014). Gene therapy for choroideremia using an adeno-associated viral (AAV) vector. *Cold Spring Harbor Perspectives in Medicine* 5 (3): a017293.
92. Lundstrom, K. and Boulikas, T. (2003). Viral and non-viral vectors in gene therapy: technology development and clinical trials. *Technology in Cancer Research & Treatment* 2 (5): 471–485.
93. Anderson, M.J., Viars, C.S., Czekay, S. et al. (1998). Cloning and characterization of three human fork-head genes that comprise an FKHR-like gene subfamily. *Genomics* 47 (2): 187–199.

94. Dewey, R.A., Morrissey, G., Cowsill, C.M. et al. (1999). Chronic brain inflammation and persistent herpes simplex virus 1 thymidine kinase expression in survivors of syngeneic glioma treated by adenovirus-mediated gene therapy: implications for clinical trials. *Nature Medicine* 7 (6): 1256–1263.
95. Fox, J.L. (2000). Gene-therapy death prompts broad civil lawsuit. *Nature Biotechnology* 18 (11): 1136–1136.
96. Cao, Y., Tan, Y.F., Wong, Y.S. et al. (2019). Recent advances in chitosan-based carriers for gene delivery. *Marine Drugs* 17 (6): 381–402.
97. Cortez, M.A., Godbey, W.T., Fang, Y.L. et al. (2015). The synthesis of cyclic poly(ethyleneimine) and exact linear analogues: an evaluation of gene delivery comparing polymer architectures. *Journal of the American Chemical Society* 137 (20): 6541–6549.
98. Song, H., Yu, M.H., Lu, Y. et al. (2017). Plasmid DNA delivery: nanotopography matters. *Journal of the American Chemical Society* 139 (50): 18247–18254.
99. Lee, K., Conboy, M., Park, H.M. et al. (2017). Nanoparticle delivery of cas9 ribonucleoprotein and donor DNA in vivo induces homology-directed DNA repair. *Nature Biomedical Engineering* 1 (11): 889–901.
100. Wang, B., Chen, P., Zhang, J. et al. (2017). Self-assembled core-shell-corona multifunctional non-viral vector with AIE property for efficient hepatocyte-targeting gene delivery. *Polymer Chemistry* 8 (48): 7486–7498.
101. Yin, H., Song, C.Q., Suresh, S. et al. (2017). Structure-guided chemical modification of guide RNA enables potent non-viral in vivo genome editing. *Nature Biotechnology* 35 (12): 1179–1187.
102. Smith, T.T., Stephan, S.B., Moffett, H.F. et al. (2017). In situ programming of leukaemia-specific T cells using synthetic DNA nanocarriers. *Nature Nanotechnology* 12 (18): 813–820.
103. Zhang, J., Tang, C., and Yin, C.H. (2013). Galactosylated trimethyl chitosan-cysteine nanoparticles loaded with Map4k4 siRNA for targeting activated macrophages. *Biomaterials* 34: 3667–3677.
104. Mead, B.P., Kim, N., Miller, G.W. et al. (2017). Novel focused ultrasound gene therapy approach noninvasively restores dopaminergic neuron function in a rat Parkinson's disease model. *Nano Letters* 17 (6): 3533–3542.
105. Kirtane, A.R. and Panyam, J. (2013). Polymer nanoparticles: weighing up gene delivery. *Nature Nanotechnology* 8 (11): 805–806.
106. Sun, N.F., Liu, Z.A., Huang, W.B. et al. (2014). The research of nanoparticles as gene vector for tumor gene therapy. *Critical Reviews in Oncology/Hematology* 89 (3): 352–357.
107. Panizo, M.V., Domínguez-Bajo, A., Portolés, M.T. et al. (2019). Nonviral gene therapy: design and application of inorganic nanoplexes. In: Filice, M., Ruiz-Cabello, J. *Nucleic Acid Nanotheranostics*, 365–390. Amsterdam, The Netherlands: Elsevier.
108. Yin, H., Kanasty, R.L., Eltoukhy, A.A. et al. (2014). Non-viral vectors for gene-based therapy. *Nature Reviews Genetics* 15 (8): 541–555.
109. Yokoo, T., Kamimura, K., Kanefuji, T. et al. (2018). Nucleic acid-based therapy: development of a nonviral-based delivery approach. In: *in vivo and ex vivo Gene Therapy for Inherited and Non-Inherited Disorders* (ed. H. Bachtarzi), 3–20. Rijeka, Croatia: IntechOpen.
110. Garaiova, Z., Strand, S.P., Reitan, N.K. et al. (2012). Cellular uptake of DNA-chitosan nanoparticles: the role of clathrin- and caveolae-mediated pathways. *International Journal of Biological Macromolecules* 51 (5): 1043–1051.
111. Ma, L., Shen, C.A., Gao, L. et al. (2016). Anti-inflammatory activity of chitosan nanoparticles carrying NF-κB/p65 antisense oligonucleotide in RAW264.7 macropghage stimulated by lipopolysaccharide. *Colloids and Surfaces B: Biointerfaces* 142: 297–306.
112. Lin, J.T., Liu, Z.K., Zhu, Q.L. et al. (2017). Redox-responsive nanocarriers for drug and gene co-delivery based on chitosan derivatives modified mesoporous silica nanoparticles. *Colloids and Surfaces B: Biointerfaces* 155: 41–50.
113. Richard, I., Thibault, M., De Crescenzo, G. et al. (2013). Ionization behaviour of chitosan and chitosan-DNA polyplexes indicate that chitosan has a similar capability to induce a proton-sponge effect as PEI. *Biomacromolecules* 14 (6): 1732–1740.
114. Huang, M., Fong, C.W., and Khor, et al. (2005). Transfection efficiency of chitosan vectors: effect of polymer molecular weight and degree of deacetylation. *Journal of Controlled Release* 106 (3): 391–406.

115. Kong, F., Liu, G., and Sun, B. (2012). Phosphorylatable short peptide conjugated low molecular weight chitosan for efficient siRNA delivery and target gene silencing. *International Journal of Pharmaceutics* 422 (1–2): 445–453.
116. Liu, X., Howard, K.A., Dong, M. et al. (2007). The influence of polymeric properties on chitosan/siRNA nanoparticle formulation and gene silencing. *Biomaterials* 28 (6): 1280–1288.
117. MacLaughlin, F.C., Mumper, R.J., Wang, J. et al. (1998). Chitosan and depolymerized chitosan oligomers as condensing carriers for in vivo plasmid delivery. *Journal of Controlled Release* 56 (1–3): 259–272.
118. Malmo, J., Sørgård, H., Vårum, K.M. et al. (2012). siRNA delivery with chitosan nanoparticles: molecular properties favoring efficient gene silencing. *Journal of Controlled Release* 158 (2): 261–268.
119. Ragelle, H., Vandermeulen, G., Préat, V. et al. (2013). Chitosan-based siRNA delivery systems. *Journal of Controlled Release* 172 (1): 207–218.
120. Sato, T., Ishii, T., and Okahata, Y. (2001). In vitro gene delivery mediated by chitosan. Effect of pH, serum, and molecular mass of chitosan on the transfection efficiency. *Biomaterials* 22 (15): 2075–2080.
121. Nielsen, E.J., Nielsen, J.M., Becker, D. et al. (2010). Pulmonary gene silencing in transgenic EGFP mice using aerosolised chitosan/siRNA nanoparticles. *Pharmaceutical Research* 27 (12): 2520–2527.
122. Veilleux, D., Nelea, M., Biniecki, K. et al. (2016). Preparation of concentrated chitosan/DNA nanoparticle formulations by lyophilization for gene delivery at clinically relevant dosages. *Journal of Pharmaceutical Sciences* 105 (1): 88–96.
123. Köping-Höggård, M., Tubulekas, I., Guan, H. et al. (2001). Chitosan as a nonviral gene delivery system. Structure–property relationships and characteristics compared with polyethylenimine in vitro and after lung administration in vivo. *Gene Therapy* 8: 1108–1121.
124. Mao, S., Sun, W., and Kissel, T. (2010). Chitosan-based formulations for delivery of DNA and siRNA. *Advanced Drug Delivery Reviews* 62 (1): 12–27.
125. Lord, M.S., Ellis, A.L., Farrugia, B.L. et al. (2017). Perlecan and vascular endothelial growth factor-encoding DNA-loaded chitosan scaffolds promote angiogenesis and wound healing. *Journal of Controlled Release* 250: 48–61.
126. Martirosyan, A., Olesen, M.J., Fenton, R.A. et al. (2016). Mucin-mediated nanocarrier disassembly for triggered uptake of oligonucleotides as a delivery strategy for the potential treatment of mucosal tumours. *Nanoscale* 8 (25): 12599–12607.
127. Mei, L., Fan, R.R., Li, X.L. et al. (2017). Nanofibers for improving the wound repair process: the combination of a grafted chitosan and an antioxidant agent. *Polymer Chemistry* 8 (10): 1664–1171.
128. Mohammed, M., Syeda, J., Wasan, K. et al. (2017). An overview of chitosan nanoparticles and its application in non-parenteral drug delivery. *Pharmaceutics* 9 (4): 53–79.
129. Xu, J.K., Strandman, S., Zhu, J.X.X. et al. (2015). Genipin-crosslinked catechol chitosan mucoadhesive hydrogels for buccal drug delivery. *Biomaterials* 37: 395–404.
130. Qiao, J., Sun, W., Lin, S. et al. (2019). Cytosolic delivery of CRISPR/Cas9 ribonucleoproteins for genome editing using chitosan-coated red fluorescent protein. *Chemical Communications* 55 (32): 4707–4710.
131. Eivazy, P., Atyabi, F., Jadidi-Niaragh, F. et al. (2016). The impact of the codelivery of drug-siRNA by trimethyl chitosan nanoparticles on the efficacy of chemotherapy for metastatic breast cancer cell line (MDA-MB-231). *Artificial Cells Nanomedicine and Biotechnology* 45 (5): 889–896.
132. Gao, Y., Wang, Z.Y., Zhang, J. et al. (2014). RVG-peptide-linked trimethylated chitosan for delivery of siRNA to the brain. *Biomacromolecules* 15 (3): 1010–1018.
133. Germershaus, O., Mao, S., Sitterberg, J. et al. (2008). Gene delivery using chitosan, trimethyl chitosan or polyethylenglycol-graft-trimethyl chitosan block copolymers: establishment of structure–activity relationships in vitro. *Journal of Controlled Release* 125 (2): 145–154.
134. Zheng, H., Tang, C., Yin, C. et al. (2015). Exploring advantages/disadvantages and improvements in overcoming gene delivery barriers of amino acid modified trimethylated chitosan. *Pharmaceutical Research* 32 (6): 2038–2050.
135. Zheng, Y., Cai, Z., Song, X. et al. (2009). Receptor mediated gene delivery by folate conjugated N-trimethyl chitosan in vitro. *International Journal of Pharmaceutics* 382 (1–2): 262–269.

136. Jintapattanakit, A., Mao, S., Kissel, T. et al. (2008). Physicochemical properties and biocompatibility of N-trimethyl chitosan: effect of quaternization and dimethylation. *European Journal of Pharmaceutical Sciences* 70 (2): 563–571.
137. Ren, H.Q., Liu, S., Yang, J.X. et al. (2016). N,N,N-trimethyl chitosan modified with well-defined multifunctional polymer modules used as pDNA delivery vector. *Carbohydrate Polymers* 137: 222–230.
138. He, C., Yin, L., Song, Y. et al. (2015). Optimization of multifunctional chitosan-siRNA nanoparticles for oral delivery applications, targeting TNF-alpha silencing in rats. *Acta Biomaterialia* 17: 98–106.
139. Tang, S., Huang, Z., Zhang, H. et al. (2014). Design and formulation of trimethylated chitosan-graft-poly(ε-caprolactone) nanoparticles used for gene delivery. *Carbohydrate Polymers* 101: 104–112.
140. Faizuloev, E., Marova, A., Nikonova, A. et al. (2012). Water-soluble N-[(2-hydroxy-3-trimethylammonium) propyl] chitosan chloride as a nucleic acids vector for cell transfection. *Carbohydrate Polymers* 89 (4): 1088–1094.
141. Li, G.F., Wang, J.C., Feng, X.M. et al. (2015). Preparation and testing of quaternized chitosan nanoparticles as gene delivery vehicles. *Applied Biochemistry and Biotechnology* 175 (7): 3244–3257.
142. Jeong, Y.I., Kim, D.G., Jang, M.K. et al. (2008). Preparation and spectroscopic characterization of methoxy poly (ethylene glycol)-grafted water-soluble chitosan. *Carbohydrate Research* 343 (2): 282–289.
143. Sugimoto, M., Morimoto, M., Sashiwa, H. et al. (1998). Preparation and characterization of water-soluble chitin and chitosan derivatives. *Carbohydrate Polymers* 36 (1): 49–59.
144. Li, T.S., Yawata, T., and Honke, K. (2014). Efficient siRNA delivery and tumor accumulation mediated by ionically cross-linked folic acid-poly (ethylene glycol)-chitosan oligosaccharide lactate nanoparticles: for the potential targeted ovarian cancer gene therapy. *European Journal of Pharmaceutical Sciences* 52: 48–61.
145. Gutoaia, A., Schuster, L., Margutti, S. et al. (2016). Fine-tuned PEGylation of chitosan to maintain optimal siRNA-nanoplex bioactivity. *Carbohydrate Polymers* 143: 25–34.
146. Wang, F., Wang, Y., Ma, Q. et al. (2017). Development and characterization of folic acid-conjugated chitosan nanoparticles for targeted and controlled delivery of gemcitabine in lung cancer therapeutics. *Artificial Cells, Nanomedicine, and Biotechnology* 45 (8): 1530–1538.
147. Jiang, H.L., Xing, L., Luo, C.Q. et al. (2018). Chemical modification of chitosan as a gene transporter. *Current Organic Chemistry* 22 (7): 668–689.
148. Inamdar, N. and Mourya, V.K. (2011). *Chitosan and Anionic Polymers – Complex Formation and Applications*, 333–377. New York: Nova Science Publishers Inc.
149. Kritchenkov, A.S., Andranovitš, S., and Skorik, Y.A. (2017). Chitosan and its derivatives: vectors in gene therapy. *Russian Chemical Reviews* 86 (3): 231–239.
150. Kuhn, P.S., Levin, Y., Barbosa, M.C. et al. (1999). Charge inversion in DNA–amphiphile complexes: possible application to gene therapy. *Physica A: Statistical Mechanics and Its Applications* 274 (1–2): 8–18.
151. Yan, J., Du, Y.Z., Chen, F.Y. et al. (2013). Effect of proteins with different isoelectric points on the gene transfection efficiency mediated by stearic acid grafted chitosan oligosaccharide micelles. *Molecular Pharmacology* 10 (7): 2568–2577.
152. Meng, T., Wu, J., Yi, H. et al. (2016). A spermine conjugated stearic acid-g-chitosan oligosaccharide polymer with different types of amino groups for efficient p53 gene therapy. *Colloids and Surfaces. B, Biointerfaces* 145: 695–705.
153. Liu, W.G., Zhang, X., Sun, S.J. et al. (2013). N-alkylated chitosan as a potential nonviral vector for gene transfection. *Bioconjugate Chemistry* 14 (4): 782–789.
154. Sharma, D. and Singh, J. (2017). Synthesis and characterization of fatty acid grafted chitosan polymer and their nanomicelles for nonviral gene delivery applications. *Bioconjugate Chemistry* 28 (11): 2772–2783.
155. Chae, S.Y., Son, S., Lee, M. et al. (2005). Deoxycholic acid-conjugated chitosan oligosaccharide nanoparticles for efficient gene carrier. *Journal of Controlled Release* 109 (1–3): 330–334.
156. Zhu, D., Jin, X., Leng, X. et al. (2010). Local gene delivery via endovascular stents coated with dodecylated chitosan–plasmid DNA nanoparticles. *International Journal of Nanomedicine* 5: 1095–1102.
157. Yoo, H.S., Lee, J.E., Chung, H. et al. (2005). Self-assembled nanoparticles containing hydrophobically modified glycol chitosan for gene delivery. *Journal of Controlled Release* 103 (1): 235–243.

158. Layek, B. and Singh, J. (2013). Caproic acid grafted chitosan cationic nanocomplexes for enhanced gene delivery: effect of degree of substitution. *International Journal of Pharmaceutics* 447 (1–2): 182–191.
159. Liu, W.G., Zhang, X., Sun, S.J. et al. (2003). N-alkylated chitosan as a potential nonviral vector for gene transfection. *Bioconjugate Chemistry* 14 (4): 782–789.
160. Muddineti, O.S., Shah, A., Rompicharla, S.V.K. et al. (2018). Cholesterol-grafted chitosan micelles as a nanocarrier system for drug-siRNA co-delivery to the lung cancer cells. *International Journal of Biological Macromolecules* 118: 857–863.
161. Hashem, F.M., Nasr, M., Khairy, A. et al. (2019). in vitro cytotoxicity and transfection efficiency of pDNA encoded p53 gene-loaded chitosan-sodium deoxycholate nanoparticles. *International Journal of Nanomedicine* 14: 4123–4131.
162. Parraga, J.E., Zorzi, G.K., and Diebold, Y. (2014). Nanoparticles based on naturally-occurring biopolymers as versatile delivery platforms for delicate bioactive molecules: an application for ocular gene silencing. *International Journal of Pharmaceutics* 477 (1–2): 12–20.
163. Du, Y.Z., Lu, P., Zhou, J.P. et al. (2010). Stearic acid grafted chitosan oligosaccharide micelle as a promising vector for gene delivery system: factors affecting the complexation. *International Journal of Pharmaceutics* 39 (1–2): 260–266.
164. Hu, F.Q., Zhao, M.D., Yuan, H. et al. (2006). A novel chitosan oligosaccharide-stearic acid micelle for gene delivery: properties and in vitro transfection studies. *International Journal of Pharmaceutics* 315 (1–2): 158.
165. Gao, J.Q., Zhao, Q.Q., Lv, T.F. et al. (2010). Gene-carried chitosan-linked-PEI induced high gene transfection efficiency with low toxicity and significant tumor-suppressive activity. *International Journal of Pharmaceutics* 387 (1–2): 286–294.
166. Yu, H., Chen, X., Lu, T. et al. (2007). Poly(l-lysine)-graft-chitosan copolymers: synthesis, characterization, and gene transfection effect. *Biomacromolecules* 8 (5): 1425–1435.
167. Jiang, H.L., Lim, H.T., Kim, Y.K. et al. (2011). Chitosan-graft-spermine as a genecarrier in vitro and in vivo. *European Journal of Pharmaceutics and Biopharmaceutics* 77 (1): 36–42.
168. Gao, Y., Xu, Z., Chen, S. et al. (2008). Arginine–chitosan/DNA self-assemble nanoparticles for gene delivery: in vitro characteristics and transfection efficiency. *International Journal of Pharmaceutics* 359 (1–2): 241–246.
169. Plianwong, S., Opanasopit, P., Ngawhirunpat, T. et al. (2013). Chitosan combined with poly l-arginine as efficient, safe, and serum-insensitive vehicle with RNase protection ability for siRNA delivery. *BioMed Research International* 2013: 574136.
170. Lu, H., Dai, Y., Lv, L. et al. (2014). Chitosan-graft-polyethylenimine/DNA nanoparticles as novel non-viral gene delivery vectors targeting osteoarthritis. *Plos One* 9 (1): e84703.
171. Lu, H., Dai, Y., Lv, L. et al. (2014). Chitosan-graft-polyethylenimine/DNA nanoparticles as novel non-viral gene delivery vectors targeting osteoarthritis. *PLoS One* 9 (1): e84703.
172. Zhang, X., Duan, Y., Wang, D. et al. (2015). Preparation of arginine modified PEI-conjugated chitosan copolymer for DNA delivery. *Carbohydrate Polymers* 122: 53–59.
173. Dutta, P.K. (2016). *Chitin and Chitosan for Regenerative Medicine*. New Delhi, India: Springer.
174. Ghosn, B., Singh, A., Li, M. et al. (2010). Efficient gene silencing in lungs and liver using imidazole-modified CS as a nanocarrier for small interfering RNA. *Oligonucleotides* 20 (3): 163–172.
175. Moreira, C., Oliveira, H., Pires, L.R. et al. (2009). Improving chitosan-mediated gene transfer by the introduction of intracellular buffering moieties into the chitosan backbone. *Acta Biomaterialia* 5 (8): 2995–3006.
176. Shi, B., Zhang, H., Shen, Z. et al. (2013). Developing a chitosan supported imidazole Schiff-base for high-efficiency gene delivery. *Polymer Chemistry* 4 (3): 840–850.
177. Wang, W., Yao, J., Zhou, J.P. et al. (2008). Urocanic acid-modified chitosan mediated p53 gene delivery inducing apoptosis of human hepatocellular carcinoma cell line HepG2 is involved in its antitumor effect in vitro and in vivo. *Biochemical and Biophysical Research Communications* 377 (2): 567–572.
178. Hsueh, Y.S., Subramaniam, S., Tseng, Y.C. et al. (2017). in vitro and in vivo assessment of chitosan modified urocanic acid as gene carrier. *Materials Science & Engineering C: Materials for Biological Applications* 70 (1): 599–606.

179. Du, H., Liu, M., Yang, X. et al. (2015). The role of glycyrrhetinic acid modification on preparation and evaluation of quercetin-loaded CS-based self-aggregates. *Journal of Colloid and Interface Science* 460: 87–96.
180. Kim, T.H., Jin, H., Kim, H.W. et al. (2006). Mannosylated chitosan nanoparticle-based cytokine gene therapy suppressed cancer growth in BALB/c mice bearing CT-26 carcinoma cells. *Molecular Cancer Therapeutics* 5 (7): 1723–1732.
181. Peng, W., Du, Y., Luo, M. et al. (2013). Preventative vaccine-loaded mannosylated chitosan nanoparticles intended for nasal mucosal delivery enhance immune responses and potent tumor immunity. *Molecular Pharmaceutics* 10 (8): 2904–2914.
182. Peng, Y., Yao, W., Wang, B. et al. (2015). Mannosylated chitosan nanoparticles based macrophage-targeting gene delivery system enhanced cellular uptake and improved transfection efficiency. *Journal of Nanoscience and Nanotechnology* 15 (4): 2619–2627.
183. Shilakari Asthana, G., Asthana, A., Kohli, D.V. et al. (2014). Mannosylated chitosan nanoparticles for delivery of antisense oligonucleotides for macrophage targeting. *Biomedical Research International* 1–17.
184. Yao, W., Jiao, Y., Luo, J. et al. (2012). Practical synthesis and characterization of mannose-modified chitosan. *International Journal of Biological Macromolecules* 50 (3): 821–825.
185. Gao, S., Chen, J., Dong, L. et al. (2005). Targeting delivery of oligonucleotide and plasmid DNA to hepatocyte via galactosylated chitosan vector. *The European Journal of Pharmaceutics and Biopharmaceutics* 60 (3): 327–334.
186. Kim, T.H., Park, I.K., Nah, J.W. et al. (2004). Galactosylated chitosan/DNA nanoparticles prepared using water-soluble chitosan as a gene carrier. *Biomaterials* 25 (17): 3783–3792.
187. Lu, B., Wu, D.Q., Zheng, H. et al. (2010). Galactosyl conjugated N-succinyl-chitosan-graft-polyethylenimine for targeting gene transfer. *Molecular BioSystems* 6 (12): 2529–2538.
188. Song, B., Zhang, W., Peng, R. et al. (2008). Synthesis and cell activity of novel galactosylated chitosan as a gene carrier. *Colloids and Surfaces B: Biointerfaces* 70 (2): 181–186.
189. Xia, Y., Zhong, J., Zhao, M. et al. (2019). Galactose-modified selenium nanoparticles for targeted delivery of doxorubicin to hepatocellular carcinoma. *Drug Delivery* 26 (1): 1–11.
190. Li, L., Hu, X.Q., Zhang, M. et al. (2017). Dual tumor-targeting nanocarrier system for siRNA delivery based on pRNA and modified chitosan. *Molecular Therapy – Nucleic Acids* 8: 169–183.
191. Liu, Z., Duan, J.H., Song, Y.M. et al. (2012). Novel HER2 aptamer selectively delivers cytotoxic drug to HER2-positive breast cancer cells in vitro. *Journal of Translational Medicine* 10 (1): 148.
192. Thiel, K.W., Hernandez, L.I., Dassie, J.P. et al. (2012). Delivery of chemo-sensitizing siRNAs to HER2(+)-breast cancer cells using RNA aptamers. *Nucleic Acids Research* 40 (13): 6319–6337.
193. Wang, T., Gantier, M.P., Xiang, D. et al. (2015). EpCAM aptamer mediated survivin silencing sensitized cancer stem cells to doxorubicin in a breast cancer model. *Theranostics* 5 (12): 1456–1472.
194. Chen, C., Li, Y.Y., Yu, X.P. et al. (2018). Bone-targeting melphalan prodrug with tumor-microenvironment sensitivity: synthesis. In vitro and in vivo evaluation. *Chinese Chemical Letters* 29 (11): 1609–1612.
195. Corbet, C., Ragelle, H., Pourcelle, V. et al. (2016). Delivery of siRNA targeting tumor metabolism using non-covalent pegylated chitosan nanoparticles: identification of an optimal combination of ligand structure, linker and grafting method. *Journal of Controlled Release* 223: 53–63.
196. Hu, J., Zhu, M.M., Liu, K.H. et al. (2016). A biodegradable polyethylenimine-based vector modified by trifunctional peptide R18 for enhancing gene transfection efficiency in vivo. *PLoS One* 11 (12): 1–21.
197. Jiang, H.L., Kwon, J., Kim, Y. et al. (2007). Galactosylated chitosan-graft-polyethylenimine as a gene carrier for hepatocyte targeting. *Gene Therapy* 14 (19): 1389.
198. Xia, Y., Zhong, J., Zhao, M. et al. (2019). Galactose-modified selenium nanoparticles for targeted delivery of doxorubicin to hepatocellular carcinoma. *Drug Delivery* 26 (1): 1–11.
199. Jiang, H., Wu, H., Xu, Y.L. et al. (2011). Preparation of galactosylated chitosan/tripolyphosphate nanoparticles and application as a gene carrier for targeting SMMC7721 cells. *Journal of Bioscience and Bioengineering* 111 (6): 719–724.
200. Han, L., Tang, C., and Yin, C. (2014). Oral delivery of shRNA and siRNA via multifunctional polymeric nanoparticles for synergistic cancer therapy. *Biomaterials* 34 (14): 4589–4600.

201. Jin, J., Fu, W., Liao, M. et al. (2017). Construction and characterization of Gal–chitosan graft methoxy poly (ethylene glycol) (Gal-CS-mPEG) nanoparticles as efficient gene carrier. *Journal of Ocean University of China* 16 (5): 873–881.
202. John, A.A., Jaganathan, S., Manikandan, D.A. et al. (2017). Folic acid decorated chitosan nanoparticles and its derivatives for the delivery of drugs and genes to cancer cells. *Current Science* 113: 1530–1542.
203. Gu, J., Al-Bayati, K., Ho, K. et al. (2017). Development of antibody-modified chitosan nanoparticles for the targeted delivery of siRNA across the blood–brain barrier as a strategy for inhibiting HIV replication in astrocytes. *Drug Delivery and Translational Research* 7 (4): 497–506.
204. Lee, J., Yun, K.S., Choi, C.S. et al. (2012). T cell-specific siRNA delivery using antibody-conjugated chitosan nanoparticles. *Bioconjugate Chemistry* 23 (6): 1174–1180.
205. Nascimento, A.V., Singh, A., Bousbaa, H. et al. (2014). Mad2 checkpoint gene silencing using epidermal growth factor receptor-targeted chitosan nanoparticles in non-small cell lung cancer model. *Molecular Pharmacology* 11 (10): 3515–3527.
206. Kim, Y.M., Park, S.C., Jang, M.K. et al. (2017). Targeted gene delivery of polyethyleneimine-grafted chitosan with RGD dendrimer peptide in αvβ3 integrin-overexpressing tumor cells. *Carbohydrate Polymers* 174: 1059–1068.
207. Sun, P., Huang, W., Kang, L. et al. (2017). siRNA-loaded poly(histidine-arginine) 6-modified chitosan nanoparticle with enhanced cell-penetrating and endosomal escape capacities for suppressing breast tumor Metastasis. *International Journal of Nanomedicine* 12: 3221–3234.
208. Yang, S., Ren, Z., Chen, M. et al. (2018). Nucleolin-targeting AS1411-aptamer modified graft polymeric micelle with dual pH/redox sensitivity designed to enhance tumor therapy through the co-delivery of doxorubicin/TLR4 siRNA and suppression of invasion. *Molecular Pharmaceutics* 15: 314–325.
209. Jeong, G.-W., Park, S.-C., Choi, C. et al. (2015). Anticancer effect of gene/peptide co-delivery system using transferrin-grafted LMWSC. *International Journal of Pharmaceutics* 488: 165–173.
210. Liu, L., Dong, X., Zhu, D. et al. (2014). TAT-LHRH conjugated low molecular weight chitosan as a gene carrier specific for hepatocellular carcinomacells. *International Journal of Nanomedicine* 9: 2879–2889.
211. Woensel, V., Wauthoz, M., Rosière, N. et al. (2016). Development of siRNA-loaded chitosan nanoparticles targeting Galectin-1 for the treatment of glioblastoma multiforme via intranasal administration. *Journal of Controlled Release* 227: 71–81.
212. He, C., Yin, L., Tang, C. et al. (2013). Multifunctional polymeric nanoparticles for oral delivery of TNF-α siRNA to macrophages. *Biomaterials* 34: 2843–2854.
213. Gaspar, V.M., Costa, E.C., Queiroz, J.A. et al. (2015). Folate-targeted multifunctional amino acid–CS nanoparticles for improved cancer therapy. *Pharmaceutical Research* 32 (2): 562–577.
214. Yan, C.Y., Gu, J.W., Hou, D.P. et al. (2015). Synthesis of Tat tagged and folate modified *N*-succinyl-chitosan self-assembly nanoparticles as a novel gene vector. *International Journal of Biological Macromolecules* 72: 751–756.
215. Jaiswal, S., Dutta, P., Kumar, S. et al. (2019). Methyl methacrylate modified chitosan: synthesis, characterization and application in drug and gene delivery. *Carbohydrate Polymers* 211: 109–117.
216. Huang, G., Chen, Q., Wu, W. et al. (2020). Reconstructed chitosan with alkylamine for enhanced gene delivery by promoting endosomal escape. *Carbohydrate Polymers* 227: 115339–115350.
217. Capel, V., Vllasaliu, D., Watts, P. et al. (2018). Water-soluble substituted chitosan derivatives as technology platform for inhalation delivery of siRNA. *Drug Delivery* 25 (1): 644–653.
218. Garg, P., Kumar, S., Pandey, S. et al. (2013). Triphenylamine coupled chitosan with high buffering capacity and low viscosity for enhanced transfection in mammalian cells, in vitro and in vivo. *Journal of Materials Chemistry B* 1 (44): 6053–6065.
219. Kumar, S., Garg, P., Pandey, S. et al. (2015). Enhanced chitosan–DNA interaction by 2-acrylamido-2-methylpropane coupling for an efficient transfection in cancer cells. *Journal of Materials Chemistry B* 3 (17): 3465–3475.

10

Carbohydrate-Based Materials for Biomedical Applications

Chadamas Sakonsinsiri

Department of Biochemistry, Faculty of Medicine, Khon Kaen University, Khon Kaen, Thailand

10.1 Introduction

Carbohydrates are the most abundant biomolecules in nature. They serve as energy sources and structural elements, and play essential roles in the regulation of cellular functions. Carbohydrates are made up of monosaccharide building blocks, which can be linked in a variety of ways generating diverse structures to perform different functions. Simple carbohydrates (mono- or disaccharides) and more complex carbohydrate systems – polysaccharides – are considered a source of low-cost, biocompatible, and renewable materials. Extraction of natural carbohydrate-based biopolymers, *e.g.*, cellulose, starch, and chitin, can be achieved through different methods. On the other hand, the structural diversity-complex carbohydrates can be found on the outside surface of cells which form a sugar-rich layer called glycocalyx, which are key mediators in many biological and recognition events, including fertilization, inflammation, cancer metastasis, and host-pathogen adhesion [1–3]. For instance, some pathogens, such as bacteria and viruses, invade host cells through carbohydrate–protein interactions (Figure 10.1).

Individual carbohydrate–protein interactions are typically weak. However, multivalent bindings result in the improvement of affinity and specificity in such interactions. An example of natural multivalent interactions is found in the highly branched network of hairs in a gecko which create numerous contacts to the surface and allow them to climb and hang on ceilings. Similarly, the adhesion of cells, viruses, bacteria, and toxins uses multiple contacts

High-Performance Materials from Bio-based Feedstocks, First Edition. Edited by Andrew J. Hunt, Nontipa Supanchaiyamat, Kaewta Jetsrisuparb and Jesper T.N. Knijnenburg.

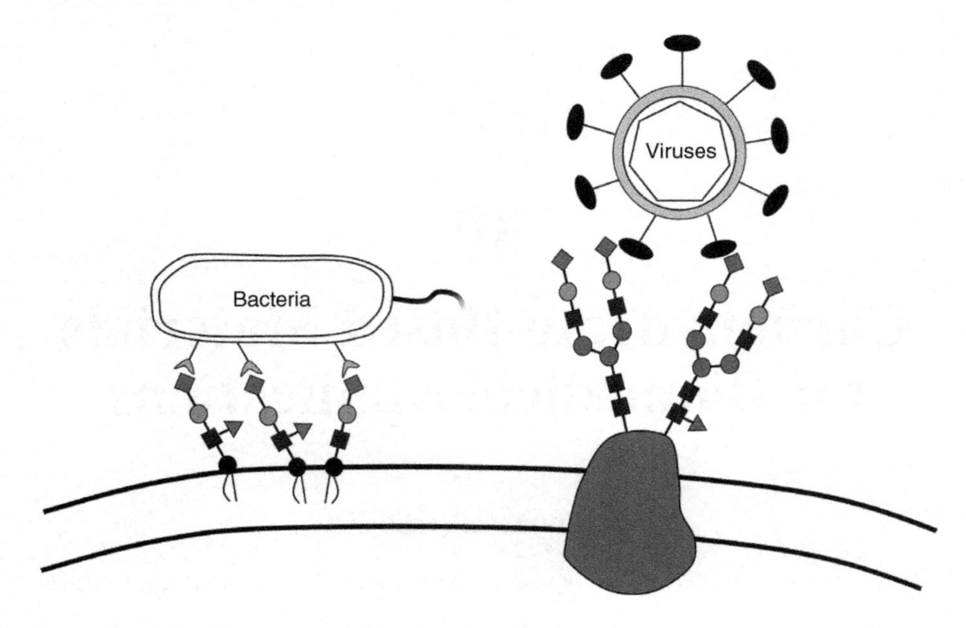

Figure 10.1 Carbohydrate–protein interactions between pathogenic agents and host cells which occur through multivalent binding.

for adhesion. Understanding the structures and interactions of the glycans leads to the design and synthesis of glycomimetics which ultimately can provide great benefits to prevent and treat diseases. Various glycomimetics which are molecules that mimic the key bioactive structural motifs of carbohydrates have been fabricated as probes or inhibitors, which may be beneficial for treating diseases, *e.g.*, inflammatory diseases, and viral and bacterial infections. The attachment of glycomimetics onto multivalent scaffolds to form multivalent glycoconjugates is an approach to increase their affinity to bind to target proteins. Several structures are used to present multiple copies of carbohydrate epitopes on to multivalent scaffolds, for instance, glycopolymers, glycoclusters, glycodendrimers, and glyconanoparticles (Figure 10.2).

Carbohydrate-based polymers can be derived from nature. Because of their nontoxic, biodegradable, and biocompatible properties, bio-based glycopolymers have attracted increasing attention in biomedical fields. In addition, synthetic materials bearing sugar residues and their interactions with lectins, which are carbohydrate-binding proteins, have been used for probing and inhibiting such interactions. The following sections outline selected bio-based glycopolymers and synthetic carbohydrate-functionalized materials and their uses in biomedical fields including biosensing, drug delivery, gene therapy, and wound-healing applications.

10.2 Bio-based Glycopolymers

10.2.1 Chitin and Chitosan

Chitin and chitosan are natural polysaccharides, which can be obtained from the shell waste of shrimps and crabs. Chitin is a linear polysaccharide consisting of β(1→4)-linked *N*-acetyl-D-glucosamine, whereas chitosan is generated by the deacetylation of chitin (Figure 10.3).

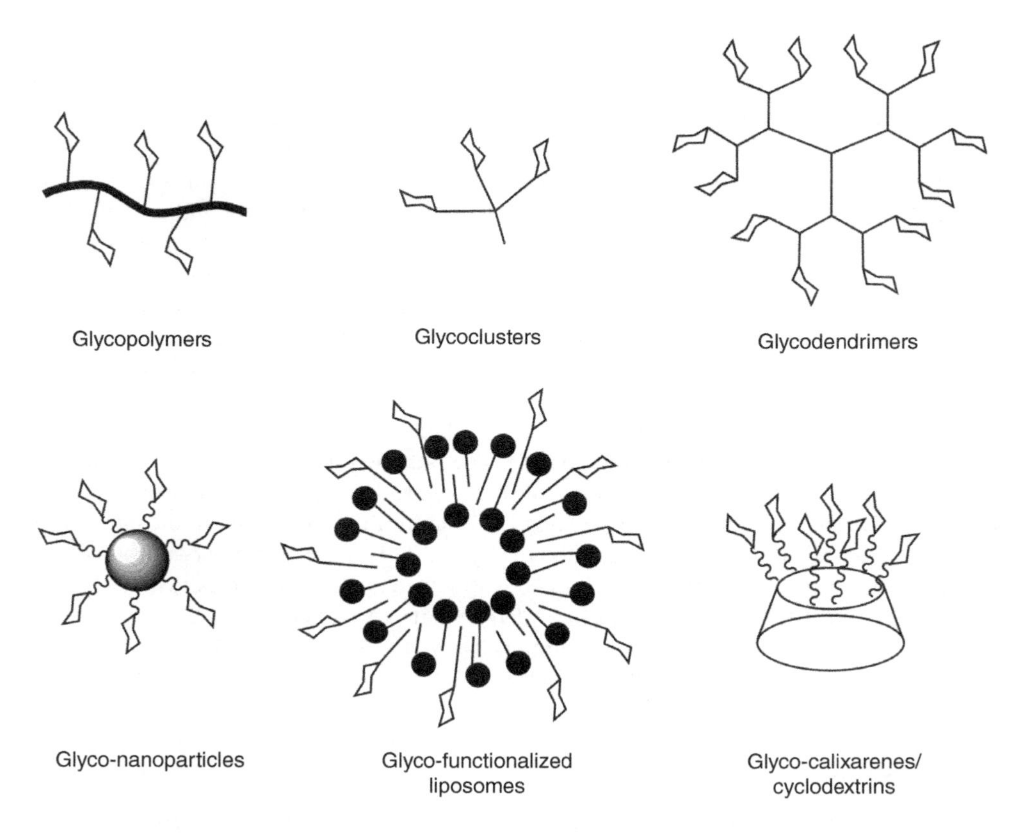

Figure 10.2 *Exemplified multivalent glycomaterials using different scaffolds.*

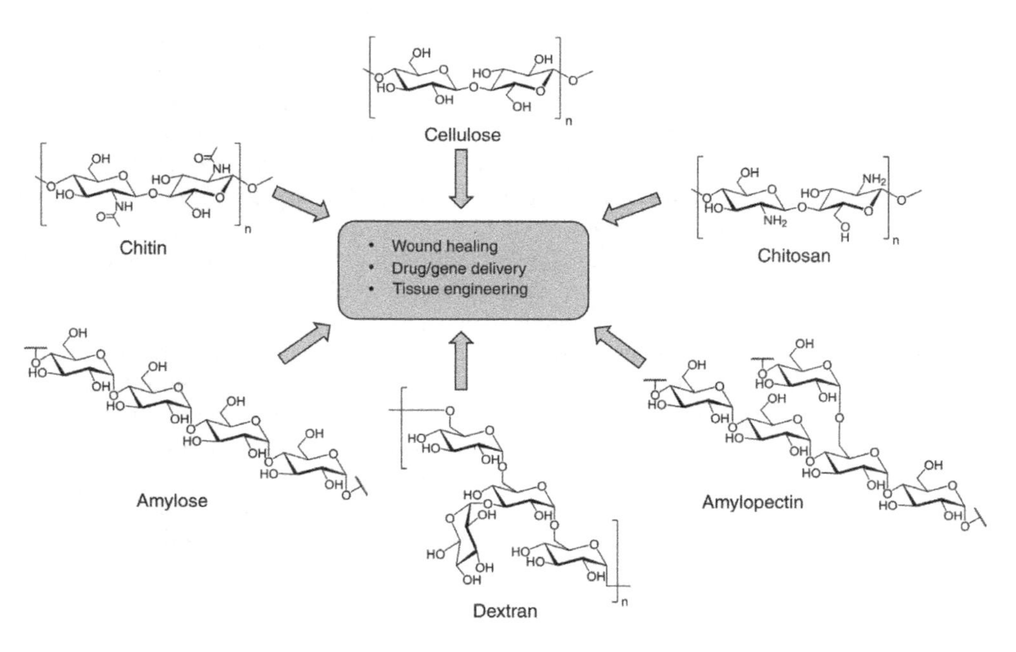

Figure 10.3 *Chemical structures and applications of cellulose, chitosan, amylopectin, dextran, amylose, and chitin.*

Structural modification of chitin and chitosan could enhance their solubility. Diverse derivatives of chitin and chitosan are generated from many modification approaches, such as grafting, degradation, and cross-linking. They have been employed as scaffolds for different applications, *e.g.*, wound-healing ointments and dressings, and drug delivery and gene delivery applications. Several studies have shown that the PEGylation of chitin/ chitosan using polyethylene glycols (PEGs) could overcome the issue of limited solubility. Recently, a chitosan-PEG scaffold which was fabricated through lyophilization and heparin conjugation was shown to increase the affinity of growth factors [4]. This system was used for the controlled delivery of growth factors to the injury site, permitting improved wound healing *in vitro* and *in vivo*. Chitin and chitosan could form polymeric nanoparticles for encapsulating and delivering drugs or active compounds to target cells. A grafted chitosan with PEG methacrylate has also been prepared *via* double cross-linking (ionic and covalent) reaction using a reverse emulsion method [5]. The obtained micro-/nanoparticles have the potential to be used as drug carriers for the controlled release of bevacizumab, an ophthalmic drug, *via* local injection close to the retina. Moreover, chitosan and its derivatives can be served as potential nonviral gene delivery systems. Their polycationic characteristics can form complexes with negatively charged nucleic acids, *e.g.*, DNA, small interfering RNA (siRNA), and microRNA (miRNA), resulting in the condensation and protection of nucleic acids. For example, Nguyen *et al.* [6] have developed chitosan nanoparticles for delivering miRNA to macrophages, causing reverse cholesterol transport and cholesterol efflux *in vivo*.

10.2.2 Cellulose

Cellulose is the most abundant biopolymer in nature. The cellulose polymer consists of long polymer chains of glucose monomers linked by β(1→4) glycosidic bonds. There are three main types of nanocellulose: (i) cellulose nanocrystals (CNCs); (ii) cellulose nanofibrils (CNFs); and (iii) bacterial cellulose (BC). Both CNCs and CNFs can be extracted from many sources, including wood, cotton, and wheat straws. Acid hydrolysis is a common method used for extracting CNCs, while a mechanically delaminating technique is employed for acquiring CNFs. In contrast, the production of BC can be achieved through the biosynthesis of bacteria, such as *Acetobacter xylinum* [7]. Due to its nontoxic and biodegradable properties, cellulose has been examined for drug delivery, wound healing, and tissue engineering. Cellulose contains many hydroxyl groups which can form polymer networks *via* hydrogen bonding with different functional materials. Preparation of cellulose-based hydrogels can be performed by using different approaches, such as radiation [8], graft copolymerization [9], and cross-linking strategies [10]. For example, You *et al.* [11] developed injectable hydrogels based on quaternized cellulose and rigid rod-like cationic CNCs using β-glycerophosphate as a linker. The cellulose-based hydrogels were loaded with doxorubicin, an anti-cancer drug, and used for the drug controlled released and *in vivo* studies of liver cancer tumor growth. A biocompatible double-membrane hydrogel system was fabricated from CNCs and alginate for the controlled release of two separate drug molecules [12]. Moreover, BC can be applied for wound dressings in skin treatment as exemplified by the work of Khalid *et al.* which demonstrated that the developed BC-zinc oxide nanocomposites can be used as a potential dressing material for burns.

10.2.3 Starch

Starch is the main carbohydrate reserve material in plants, which is presented in the form of insoluble granules of amylose and amylopectin. The major sources of starch are corn, rice, tapioca, wheat, and potatoes. Starches are extensively used in food and pharmaceutical industries where they are used for thickening, coating, and tablet formulation. Modification of starch through different methods such as structural reorganization, functionalization, or depolymerization can generate a variety of starches with different properties. In biomedical aspects, starch and starch-based hydrogels can be used as carriers for the controlled release of drugs, proteins, and other bioactive agents. As an example, Zhang and coworkers successfully prepared a pH-responsive copolymer constructed from poly(L-glutamic acid)-grafted starch nanoparticles and employed for controlled release of insulin, an oral drug-protein [13]. *In vitro* insulin-release experiments showed slow insulin release from the nanoparticles in artificial gastric juice was slower than that in artificial intestinal liquid, suggesting that the fabricated materials could protect the insulin against high pH conditions. Qi *et al.* [14] evaluated the encapsulation and release of indomethacin, an anti-inflammatory drug, as a drug model using spherical octenyl succinic anhydride starch-based nanoparticles. Waghmare and coworkers prepared starch-based nanofibrous scaffolds for wound-healing applications [15]. The addition of polyvinyl alcohol (PVA) was performed to facilitate the electrospinning process and decrease electrostatic repulsions in starch solutions. Another example of using starch-based materials as wound dressings was shown in the study of Hassen *et al.* [16] in which the antibacterial hydrogel membranes comprised of turmeric were prepared by cross-linking PVA with starch by using glutaraldehyde. The fabricated hydrogels exhibited antibacterial activities to both gram-positive and gram-negative bacteria.

10.2.4 Dextran

Dextran is a branched polymer composed of glucose residues which are linked by a linear 1,6-glycosidic bonds with branching at 1,3-linkages Dextran-based hydrogels can be synthesized through both physical and chemical cross-linking approaches. Liu and coworkers [17] prepared sericin-/dextran-injectable hydrogels for drug delivery and optical monitoring strategy. Sericin is a natural photoluminescent protein from silk, which allows the monitoring of the remaining drug real-time *in vivo*. The hydrogels were loaded with doxorubicin and significantly suppressed the tumor growth. Dextran hydrogels using human serum albumin (HSA) and PEG were successfully synthesized as both a drug delivery system and a crosslinker [18]. The established dextran/HSA/PAG hydrogens could be used for sustained delivery of doxorubicin to Michigan Cancer Foundation-7 (MCF-7) breast cancer cells. Sun *et al.* successfully prepared and used dextran hydrogel scaffolds for tissue repair for burn injuries in animal models [19, 20]. The authors concluded that dextran-based hydrogels could enhance skin healing and decrease the risk of skin graft loss. Recently, a dextran/hyaluronic acid hydrogel functionalized with the antimicrobial sanguinarine incorporated into gelatin microspheres has been produced [21]. The synthetic scaffolds have been shown to be a promising therapeutic option for infected burn treatment in a rat model of full-thickness burns.

10.3 Synthetic Carbohydrate-based Functionalized Materials

The involvement of carbohydrate–protein interactions and their multivalent binding in numerous biological and pathological processes have directed research into the rational design and synthesis of glycomimetics and multivalent glycomaterials to understand and mediate such interactions [22, 23]. Various glycomimetics have been developed as probes or inhibitors, which may be beneficial for treating diseases, such as inflammatory diseases and viral and bacterial infections. Both bio-derived and synthetic glycomaterials have been extensively studied for biorecognition applications. Examples of glycomimetics for lectin targeting and their multivalent scaffolds including dendrimers, polymers, and nanoparticles are highlighted.

10.3.1 Glycomimetics

Glycomimetics are molecules that mimic the key bioactive structural motifs of carbohydrates. Selectin inhibitors are among such glycomimetics that have been developed. Selectins, a family of three adhesion molecules (*i.e.*, E-selectin, L-selectin and P-selectin), play crucial roles in tethering and rolling of leukocytes on endothelial cells, contributing to a vaso-occlusive crisis (VOC) of sickle cell disease patients [24]. As an example, rivipansel (also called GMI-1070; Figure 10.4) has been shown to be a potent pan-selectin antagonist which could reverse VOC in mice [25]. This glycomimetic was rationally designed to mimic the biologically active conformation of the sialyl-Lewis^x of E-selectin and it currently is in Phase III clinical trial for sickle cell disease. In addition, to be used as a pan-selectin inhibitor, the compound was incorporated with an extended sulfated domain, which was required for binding to L- and P-selectins [25]. Besides, glycomimetic compounds can be effective drug candidates for the

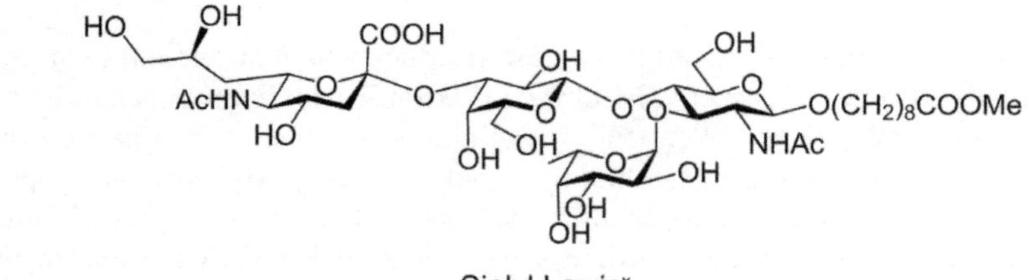

Sialyl Lewis^x

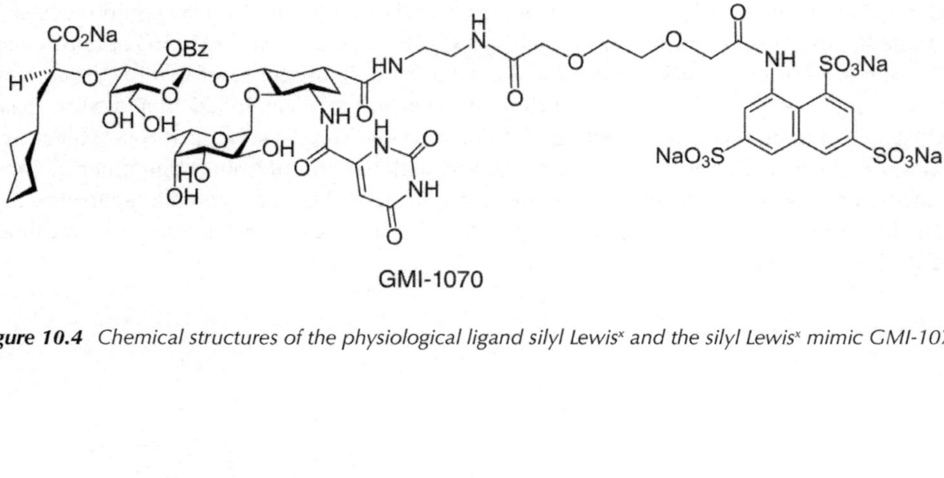

GMI-1070

Figure 10.4 *Chemical structures of the physiological ligand silyl Lewis^x and the silyl Lewis^x mimic GMI-1070.*

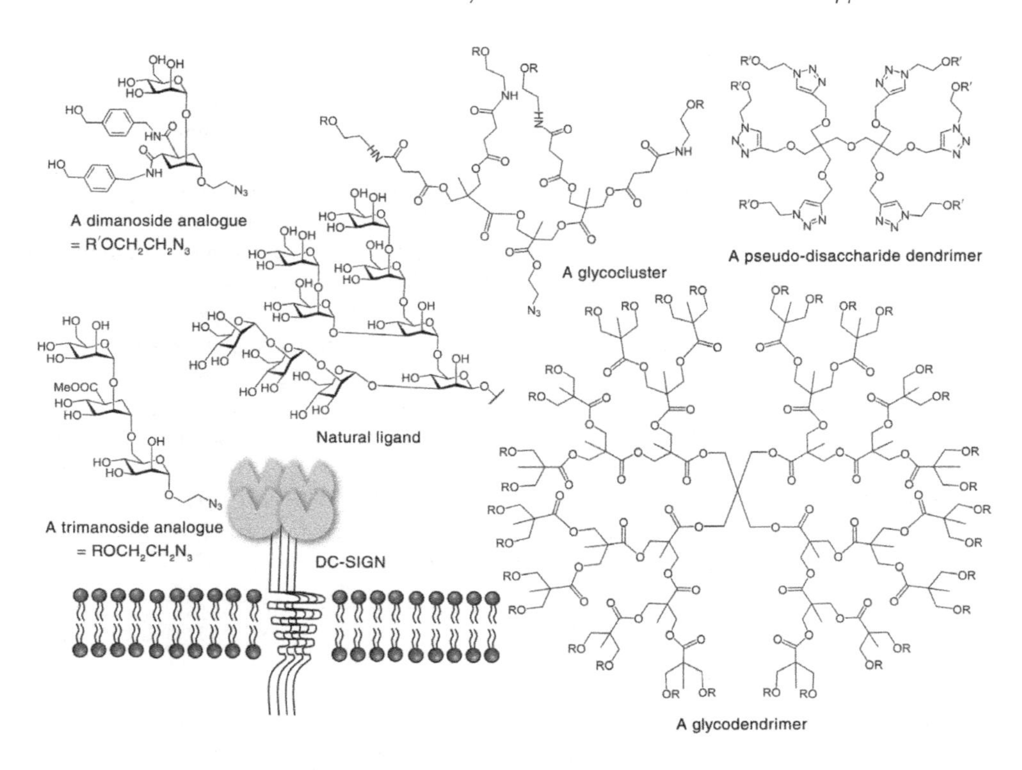

Figure 10.5 *The natural ligand Man$_9$ for DC-SIGN and selected mono- and multivalent analogues.*

inhibition of DC-SIGN (Dendritic Cell-Specific ICAM-3 Grabbing Nonintegrin), which is a C-type lectin involved in the recognition of viruses, such as the human immunodeficiency virus type 1 (HIV-1) and the Ebola virus [26]. The main carbohydrate epitope that is essential for the ligand binding is the two or three determinant mannose residues of the natural ligand (Figure 10.5) [27]. A tetravalent dendron bearing a linear trimannoside analogue has been developed and this compound exhibited anti-HIV activity in a cell-based assay [28]. Moreover, the dimannoside Manα-1,2-Man mimic containing a cyclohexane-based aglycone designed by Reina *et al.* showed anti-Ebola viral activity *in vitro* [29]. Another DC-SIGN dimannoside inhibitor development by Varga *et al.* was found to bind DC-SIGN with an enhanced selectivity [30].

10.3.2 Presentation of Glycomimetics in Multivalent Scaffolds

10.3.2.1 Glycodendrimers

Dendrimers are a class of hyperbranched nanopolymers with well-defined structures. Displaying the developed carbohydrate-based DC-SIGN antagonists, both pseudo-dimannosides and pseudo-trimannosides onto multivalent scaffolds to make glycoclusters/glycodendrimers led to enhanced binding and more efficient inhibition of DC-SIGN-mediated Ebola or HIV viral infections [28, 31, 32].

10.3.2.2 Glycopolymers

Polymeric materials have been widely used as multivalent scaffolds for mediating lectin-carbohydrate protein interactions. Focusing mainly on the development of DC-SIGN antagonists, several examples of developed glycopolymeric probes and inhibitors for carbohydrate–protein interactions are described as follows. Bacer and coworkers reported that their synthetic fully mannosylated polymer could inhibit the binding of the HIV viral envelope glycoprotein gp120 to DC-SIGN. The glycopolymers were prepared by the combination of copper-mediated living radical polymerization and click chemistry [33]. Yilmaz *et al.* have investigated the formation of single-chain glycopolymer folding using atom transfer radical coupling reaction, which resulted in augmented DC-SIGN-binding affinity than that of the unfolded single-chain glycopolymer [34]. Beyer *et al.* have recently developed a series of bottlebrush copolymers mimicking the architecture of ligands *via* a "grafting from" method for targeting two lectins, which are DC-SIGN and mannose-binding lectin. The authors concluded that following the binding studies using surface plasmon resonance, the brush polymers showed higher affinity against both lectins [35]. Another study performed by Jarvis *et al.* indicated that their fabricated mannose-containing polymers with various lengths could successfully investigate the effects of receptor clustering on DC-SIGN-mediated internalization and trafficking [36].

10.3.2.3 Glyconanoparticles

Different types of nanoparticles, such as gold nanoparticles and quantum dots, have emerged as important multivalent scaffolds for presenting carbohydrate epitopes. As an example, mannose-functionalized gold nanoparticles with different spacer lengths and variable carbohydrate density have been developed through direct glycosylation, peptide linkage, or thiourea linkage. The thiourea linkage was shown to be the most efficient method to construct glyconanoparticles which have been applied in the inhibition of the interactions between HIV-1 gp120 and DC-SIGN [37] (Figure 10.6). Conjugating α-fucosyl-β-alanyl amide to the surface of gold nanoparticles *via* specific linkers could bind and be internalized dendritic cells (DCs) through high binding affinities against DC-SIGN receptors which were expressed on DCs. This approach could be extended to cell imaging and antigen delivery to target cells [38]. Moreover, Chiodo and coworkers have developed gold nanoparticles bearing galactofuranose which triggered a pro-inflammatory response in DCs. *via* binding to DC-SIGN [39]. Recently, Budhadev *et al.* have successfully used the developed polyvalent glyco-gold nanoparticles as a new technique for the determination of K_d of the two closely related DC-SIGN and DC-SIGNR C-type *via* fluorescence quenching [40]. The glyco-gold nanoparticles had the potential to inhibit DC-SIGN-mediated Ebola virus surface glycoprotein-driven viral infections of host cells. Besides gold nanoparticles, quantum dots, which are fluorescent semiconductor nanocrystals, have been employed as multivalent structures to present different sugar moieties on the surface. Guo and coworkers successfully conjugated azido-functionalized mono- and dimannoside residues on the surface of quantum dots. The synthetic glyco-quantum dots have been employed as probes for the detection of carbohydrate-DC-SIGN interactions using Förster resonance energy transfer (FRET) and transmission electron microscopy imaging and as potential inhibitors of Ebola virus infection determined by cell-based assays [41, 42].

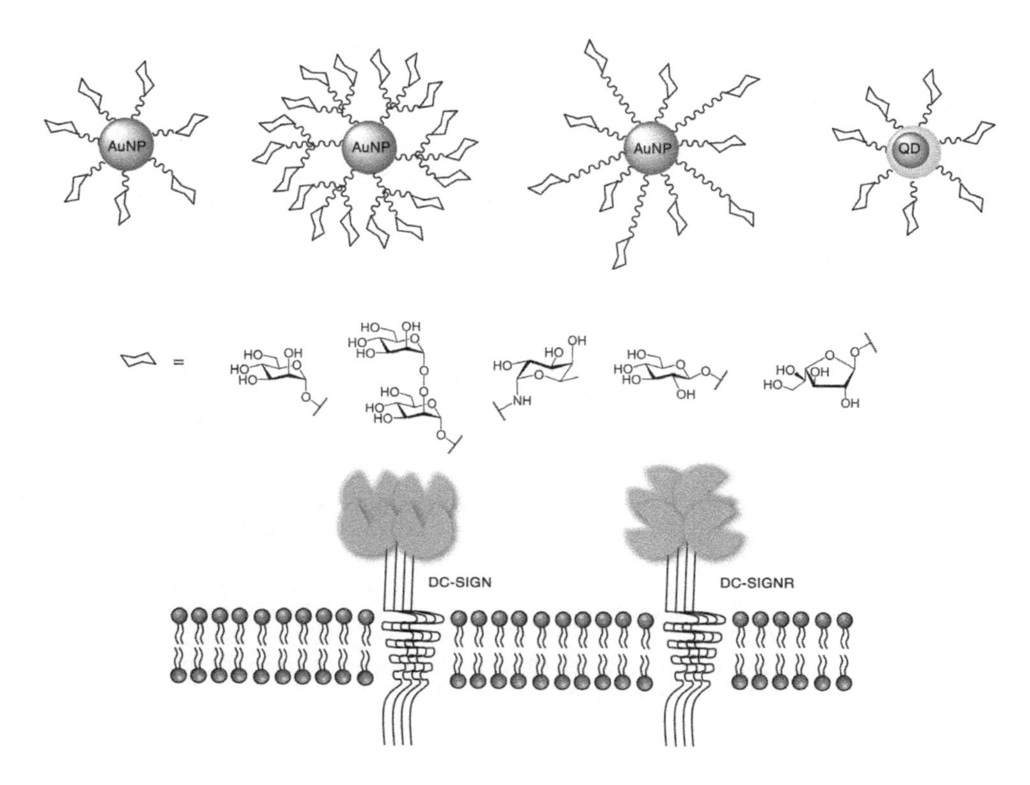

Figure 10.6 *Selected synthetic glyco-gold nanoparticles and glycol-quantum dots for probing carbohydrate–protein interactions.*

10.4 Conclusion

Carbohydrates act as an energy source, structural elements, and cell-to-cell communication molecules. Bio-based carbohydrate polymers can be obtained from a number of renewable sources, *e.g.*, plants and microorganisms. Starch, cellulose, and chitin and their derivatives can be applied in biomedical fields. Carbohydrates are also found in the cell surface of eukaryotes and are implicated in numerous pathological processes, *e.g.*, viral infection. Several nanomaterials have been extensively used as multivalent scaffolds to present multiple copies of carbohydrates to generate different glyconanomaterials, which can be used as probes or inhibitors for viral infection, such as DC-SIGN on DCs.

References

1. Diekman, A.B. (2003). Glycoconjugates in sperm function and gamete interactions: how much sugar does it take to sweet-talk the egg? *Cellular and Molecular Life Sciences* 60: 298–308.
2. Sacchettini, J.C., Baum, L.G., and Brewer, C.F. (2001). Multivalent protein-carbohydrate interactions. A new paradigm for supermolecular assembly and signal transduction. *Biochemistry* 40: 3009–3015.
3. Schröter, S., Osterhoff, C., Mcardle, W., and Ivell, R. (1999). The glycocalyx of the sperm surface. *Human Reproduction Update* 5: 302–313.

4. Vijayan, A., Sabareeswaran, A., and Kumar, G.S.V. (2019). PEG grafted chitosan scaffold for dual growth factor delivery for enhanced wound healing. *Scientific Reports* 9: 19165.
5. Savin, C.-L., Popa, M., Delaite, C. et al. (2019). Chitosan grafted-poly(ethylene glycol) methacrylate nanoparticles as carrier for controlled release of bevacizumab. *Materials Science and Engineering: C* 98: 843–860.
6. Nguyen, M.-A., Wyatt, H., Susser, L. et al. (2019). Delivery of microRNAs by chitosan nanoparticles to functionally alter macrophage cholesterol efflux in vitro and in vivo. *ACS Nano* 13: 6491–6505.
7. Brown, R.M.,.J., Willison, J.H., and Richardson, C.L. (1976). Cellulose biosynthesis in Acetobacter xylinum: visualization of the site of synthesis and direct measurement of the in vivo process. *Proceedings of the National Academy of Sciences of the United States of America* 73: 4565–4569.
8. Fekete, T., Borsa, J., Takács, E., and Wojnárovits, L. (2014). Synthesis of cellulose derivative based superabsorbent hydrogels by radiation induced crosslinking. *Cellulose* 21: 4157–4165.
9. Bao, Y., Ma, J., and Li, N. (2011). Synthesis and swelling behaviors of sodium carboxymethyl cellulose-g-poly(AA-co-AM-co-AMPS)/MMT superabsorbent hydrogel. *Carbohydrate Polymers* 84: 76–82.
10. Ciolacu, D., Rudaz, C., Vasilescu, M., and Budtova, T. (2016). Physically and chemically cross-linked cellulose cryogels: structure, properties and application for controlled release. *Carbohydrate Polymers* 151: 392–400.
11. You, J., Cao, J., Zhao, Y. et al. (2016). Improved mechanical properties and sustained release behavior of cationic cellulose nanocrystals reinforeced cationic cellulose injectable hydrogels. *Biomacromolecules* 17: 2839–2848.
12. Lin, N., Gèze, A., Wouessidjewe, D. et al. (2016). Biocompatible double-membrane hydrogels from cationic cellulose nanocrystals and anionic alginate as complexing drugs codelivery. *ACS Applied Materials & Interfaces* 8: 6880–6889.
13. Zhang, Z., Shan, H., Chen, L. et al. (2013). Synthesis of pH-responsive starch nanoparticles grafted poly (l-glutamic acid) for insulin controlled release. *European Polymer Journal* 49: 2082–2091.
14. Qi, L., Ji, G., Luo, Z. et al. (2017). Characterization and drug delivery properties of OSA starch-based nanoparticles prepared in [C_3OHmim]Ac-in-oil microemulsions system. *ACS Sustainable Chemistry & Engineering* 5: 9517–9526.
15. Waghmare, V.S., Wadke, P.R., Dyawanapelly, S. et al. (2017). Starch based nanofibrous scaffolds for wound healing applications. *Bioactive Materials* 3: 255–266.
16. Hassan, A., Niazi, M.B.K., Hussain, A. et al. (2018). Development of anti-bacterial PVA/starch based hydrogel membrane for wound dressing. *Journal of Polymers and the Environment* 26: 235–243.
17. Liu, J., Qi, C., Tao, K. et al. (2016). Sericin/dextran injectable hydrogel as an optically trackable drug delivery system for malignant melanoma treatment. *ACS Applied Materials & Interfaces* 8: 6411–6422.
18. Noteborn, W.E.M., Gao, Y., Jesse, W. et al. (2017). Dual-crosslinked human serum albumin-polymer hydrogels for affinity-based drug delivery. *Macromolecular Materials and Engineering* 302: 1700243.
19. Shen, Y.-I., Song, H.-H.G., Papa, A.E. et al. (2015). Acellular hydrogels for regenerative burn wound healing: translation from a porcine model. *Journal of Investigative Dermatology* 135: 2519–2529.
20. Sun, G., Zhang, X., Shen, Y.-I. et al. (2011). Dextran hydrogel scaffolds enhance angiogenic responses and promote complete skin regeneration during burn wound healing. *Proceedings of the National Academy of Sciences of the United States of America* 108: 20976–20981.
21. Zhu, Q., Jiang, M., Liu, Q. et al. (2018). Enhanced healing activity of burn wound infection by a dextran-HA hydrogel enriched with sanguinarine. *Biomaterials Science* 6: 2472–2486.
22. Bernardi, A., Jiménez-Barbero, J., Casnati, A. et al. (2013). Multivalent glycoconjugates as anti-pathogenic agents. *Chemical Society Reviews* 42: 4709–4727.
23. Jeong, W.-J., Bu, J., Kubiatowicz, L.J. et al. (2018). Peptide-nanoparticle conjugates: a next generation of diagnostic and therapeutic platforms? *Nano Convergence* 5: 38.
24. Turhan, A., Weiss, L.A., Mohandas, N. et al. (2002). Primary role for adherent leukocytes in sickle cell vascular occlusion: a new paradigm. *Proceedings of the National Academy of Sciences of the United States of America* 99: 3047–3051.
25. Chang, J., Patton, J.T., Sarkar, A. et al. (2010). GMI-1070, a novel pan-selectin antagonist, reverses acute vascular occlusions in sickle cell mice. *Blood* 116: 1779–1786.

26. Geijtenbeek, T.B., Kwon, D.S., Torensma, R. et al. (2000). DC-SIGN, a dendritic cell-specific HIV-1-binding protein that enhances trans-infection of T cells. *Cell* 100: 587–597.
27. Bernardi, A. and Cheshev, P. (2008). Interfering with the sugar code: design and synthesis of oligosaccharide mimics. *Chemistry (Weinheim an der Bergstrasse, Germany)* 14: 7434–7441.
28. Sattin, S., Daghetti, A., Thépaut, M. et al. (2010). Inhibition of DC-SIGN-mediated HIV infection by a linear trimannoside mimic in a tetravalent presentation. *ACS Chemical Biology* 5: 301–312.
29. Reina, J.J., Sattin, S., Invernizzi, D. et al. (2007). 1,2-Mannobioside mimic: synthesis, DC-SIGN interaction by NMR and docking, and antiviral activity. *ChemMedChem* 2: 1030–1036.
30. Varga, N., Sutkeviciute, I., Guzzi, C. et al. (2013). Selective targeting of dendritic cell-specific intercellular adhesion molecule-3-grabbing nonintegrin (DC-SIGN) with mannose-based glycomimetics: synthesis and interaction studies of bis(benzylamide) derivatives of a pseudomannobioside. *Chemistry – A European Journal* 19: 4786–4797.
31. Berzi, A., Varga, N., Sattin, S. et al. (2014). Pseudo-mannosylated DC-SIGN ligands as potential adjuvants for HIV vaccines. *Viruses* 6: 391–403.
32. Ordanini, S., Varga, N., Porkolab, V. et al. (2015). Designing nanomolar antagonists of DC-SIGN-mediated HIV infection: ligand presentation using molecular rods. *Chemical Communications* 51: 3816–3819.
33. Becer, C.R., Gibson, M.I., Geng, J. et al. (2010). High-affinity glycopolymer binding to human DC-SIGN and disruption of DC-SIGN interactions with HIV envelope glycoprotein. *Journal of the American Chemical Society* 132: 15130–15132.
34. Yilmaz, G., Uzunova, V., Napier, R., and Becer, C.R. (2018). Single-chain glycopolymer folding via host–guest interactions and its unprecedented effect on DC-SIGN binding. *Biomacromolecules* 19: 3040–3047.
35. Beyer, V.P., Monaco, A., Napier, R. et al. (2020). Bottlebrush glycopolymers from 2-oxazolines and acrylamides for targeting dendritic cell-specific intercellular adhesion molecule-3-grabbing nonintegrin and mannose-binding lectin. *Biomacromolecules* 21: 2298–2308.
36. Jarvis, C.M., Zwick, D.B., Grim, J.C. et al. (2019). Antigen structure affects cellular routing through DC-SIGN. *Proceedings of the National Academy of Sciences of the United States of America* 116: 14862–14867.
37. Martínez-Avila, O., Hijazi, K., Marradi, M. et al. (2009). Gold manno-glyconanoparticles: multivalent systems to block HIV-1 gp120 binding to the lectin DC-SIGN. *Chemistry* 15: 9874–9888.
38. Arosio, D., Chiodo, F., Reina, J.J. et al. (2014). Effective targeting of DC-SIGN by α-fucosylamide functionalized gold nanoparticles. *Bioconjugate Chemistry* 25: 2244–2251.
39. Chiodo, F., Marradi, M., Park, J. et al. (2014). Galactofuranose-coated gold nanoparticles elicit a pro-inflammatory response in human monocyte-derived dendritic cells and are recognized by DC-SIGN. *ACS Chemical Biology* 9: 383–389.
40. Budhadev, D., Poole, E., Nehlmeier, I. et al. (2020). Glycan-gold nanoparticles as multifunctional probes for multivalent lectin-carbohydrate binding: implications for blocking virus infection and nanoparticle assembly. *Journal of the American Chemical Society* 142: 18022–18034.
41. Guo, Y., Nehlmeier, I., Poole, E. et al. (2017). Dissecting multivalent lectin-carbohydrate recognition using polyvalent multifunctional glycan-quantum dots. *Journal of the American Chemical Society* 139: 11833–11844.
42. Guo, Y., Sakonsinsiri, C., Nehlmeier, I. et al. (2016). Compact, polyvalent mannose quantum dots as sensitive, ratiometric FRET probes for multivalent protein-ligand interactions. *Angewandte Chemie (International Edition in English)* 55: 4738–4742.

11

Organic Feedstock as Biomaterial for Tissue Engineering

Poramate Klanrit

Department of Biochemistry, Faculty of Medicine, Khon Kaen University, Khon Kaen, Thailand

11.1 Introduction

Langer and Vacanti described tissue engineering in the early 1990s as a "cross-disciplinary field," which combines the core principles of engineering and life sciences for the development of biological replacements that restore, or maintain tissue functions [1]. Tissue engineering basis aims at inducing specific regeneration processes, resulting in overcoming the downsides of organ transplantation, such as donor shortage and the need for immunosuppressive therapy. Many approaches have been reported recently with investigations via tissue engineering "triad" (Figure 11.1), including cells, (bio)materials as cell scaffold, and growth factors [2]. Two main approaches are currently applied in this area to create engineered tissue. First, biomaterial scaffolds are used as a "home" of cells seeded *in vitro*; cells are then encouraged to settle and grow the matrix to produce their foundations of tissue for transplantation. The second approach involves using the scaffold as a growth factor/drug delivery device. This strategy combines the scaffold and growth factors. So, upon implantation, cells from the body are recruited to the scaffold site and form tissue throughout the matrices [3].

Our organs and tissues are susceptible to severe damaging via several factors, including accident injuries, diseases such as cancers and organ failures, and congenital anomaly and defects [4]. Usually, these damages require surgical treatment and perhaps, with organ transplantation, in the way that pharmaceutical treatments are no more applicable. Organ and tissue transplantations are the only option to reconstruct the devastated tissues or organs.

High-Performance Materials from Bio-based Feedstocks, First Edition. Edited by Andrew J. Hunt, Nontipa Supanchaiyamat, Kaewta Jetsrisuparb and Jesper T.N. Knijnenburg.

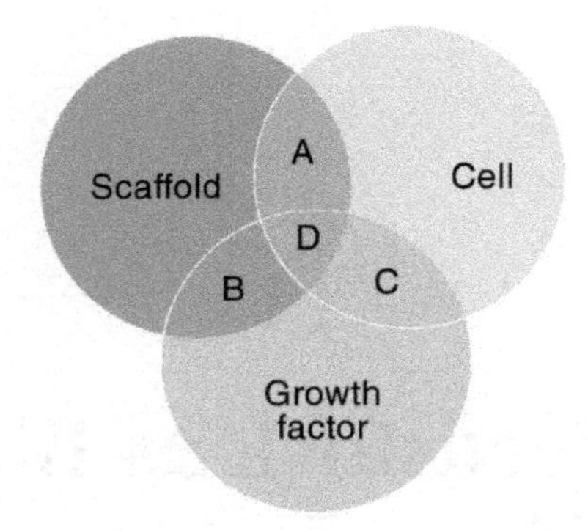

Figure 11.1 Tissue engineering triad showing the factors influenced in tissue engineering, where combinations of factors are always applied in research including cell-seeded scaffolds (A), functionalized scaffolds with biochemicals/growth factors (B), induced cells with designated characteristics (C), and all factors in combination (D). Source: Adapted from Langer and Vacanti [1].

However, the shortage of organ donation is still a worldwide limitation since there are few registered organ or tissue donors. Researchers are currently developing new artificial tissue and cell scaffolds, with the hope of producing representatives to overcome the shortage of donated organs.

The human body ingests food and water, digests them into smaller molecules, and uses those as precursors for cellular metabolisms. The molecules will be either broken down into smaller metabolites to obtain energy or built up to larger macromolecules and polymers to maintain or heal the structure of the cells, organs, and body. Tissue engineering aims at bypassing some of the build-up processes and providing a space for cells to migrate in and accelerate the regeneration process. Mostly, scientists look for the material that has similar properties or structure to the native tissue but can be obtained in a large amount for production. Natural organic feedstocks have been the superior choice as they have similar properties, structures, and compositions with a high chance of biocompatibility.

Organic feedstocks are a source of carbon-containing raw material, enriched in carbon and nitrogen sources that can be obtained from nature and have been used as biomaterials for biomedical applications. Natural biomaterials include substances that are natural in origin or certain chemical modifications of the materials used for the development of cell scaffolds or any other body implants. Such biomaterials can be commonly classified into two main categories, based on their nature of origin, protein-based and polysaccharide-based natural biomaterials. Therefore, there has been a focus on potential natural organic feedstocks which can be utilized and developed as biomaterials for cell scaffolds in tissue engineering applications (Figure 11.2).

11.2 Protein-based Natural Biomaterials

Protein-based biomaterials consist of unique proteins as a significant composition [5]. They involve specific mechanical properties and molecular motifs derived from natural proteins that provide structural support, cell attachment, and a guide to cell and tissue remodeling. The frequently used protein-based biomaterials are described as follows.

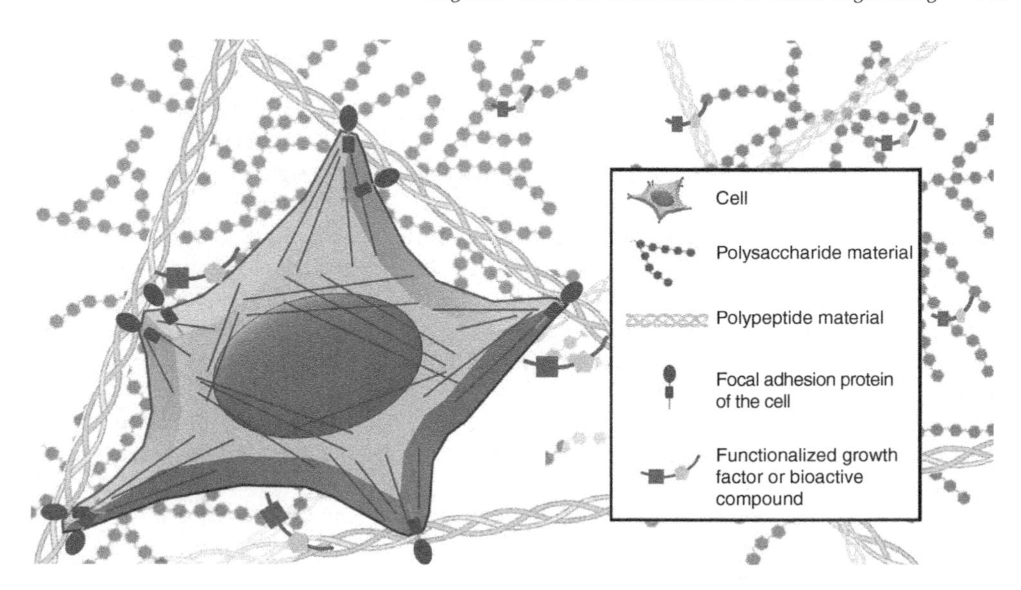

Figure 11.2 *Protein- or polysaccharide-based scaffold architecture mimics the natural extracellular matrix (ECM) for cells to adhere to (via focal adhesion proteins) and grow. The polypeptide or polysaccharide can be tailored with additional growth factors or bioactive compounds to improve cell adhesion, growth, and differentiation, which benefits the tissue engineering approach.*

11.2.1 Silk

Silk is a natural protein material, which is generally classified as fibrous proteins, composed of two major assembling polypeptidic components, fibroin and sericin [6]. Fibroin is an insoluble protein-based polymer presented in silk created by numerous insects such as spiders and mulberry silkworms. Silk in its raw state consists of two main proteins, sericin and fibroin, with a glue-like layer of sericin coating two singular filaments of fibroin called brins [7]. Silks are fibrous proteins possessing extraordinary mechanical properties and have been used as medical sutures for surgery practice [8, 9].

Recently, regenerated silk solutions from silk fibroin have been used to create a range of medical biomaterials, such as gels, sponges, and sheets. Silks can be chemically adjusted through amino acid side chains to functionalize the surface properties of the material or to immobilize specific growth factors [10]. Silk proteins have been shown to be promising biomaterials due to their unique biocompatibility, biodegradability, self-assembly, mechanical strength, and ability to tailor their composites. Silk protein blended with some organic/inorganic solvents and water-soluble gelatin materials have been extensively used for the production of 3D porous composite scaffolds for regeneration of bone, ligament, and skin tissues [11].

11.2.2 Collagen

Collagen is the protein that is found the most in our body, which provides strength and structural support to tissues, including blood vessels, skin, tendons, cartilage, and bone. Collagen has excellent biocompatibility and bioresorbable features [12, 13]. Bundles of collagen or collagen fibers are a significant part of the tissue extracellular matrix (ECM) that supports most cell structures from the outside.

The abundance of collagen in all connective tissues makes it one of the most studied biomolecules of the ECM. These fibrous protein species are the primary component of skin and bone, which accounts for approximately 25% of the total dry weight of mammals [14]. To date, 29 different collagen types have been discovered and characterized, and all display a typical triple helix structure [15]. Collagen types I, II, III, V, and XI are known to shape collagen fibers. Collagen molecules are composed of three α chains assembling due to their molecular structure. Every α chain is comprised of amino acids with the sequence based on -Gly-X-Y-. The existence of glycine is vital at every third amino acid position to allow tight packing of the three α chains in the tropocollagen molecule, and the X and Y positions are mostly proline and 4-hydroxyproline [16, 17].

There are around 25 different α chain conformations, each produced by their specific gene. The combination of these chains in the set of three assembles to form the 29 different types of collagen currently known [14].

Most soft and hard connective tissues (e.g. bone, cartilage, tendon, cornea, blood vessels, and skin) rely on collagen fibrils, and their networks function as ECM, organized in a three-dimensional (3D) architecture surrounded by various cells. Collagen plays a role in maintaining the biologic functions and structural integrity of ECM to maintain proper physiologic functions [18]. Therefore, the ultimate goal of tissue regeneration is to restore the native ECM, especially the complex collagen networks under normal physiologic regeneration. Only a few types are utilized to create collagen-based biomaterials. To date, type I collagen is the gold standard in the field of tissue engineering [18].

Animal-derived and recombinant collagens, especially type I, are recognized as one of the most potent biomaterials available and are now widely employed for tissue engineering, cosmetic surgery, and drug delivery systems [19–21]. They are utilized either in their native fibrillar forms or fabricated forms, such as collagen sponges and sheets. Butler and coworkers utilized type I collagen sponges to achieve patellar tendons in rabbits under various culture conditions [22–24]. The engineered neotissue reached about three–fourths of the mechanical properties of healthy tissue via the collagen sponge combined with bone marrow-derived mesenchymal stem cells (MSCs) [24]. Expression of collagen type I and type III by seeding cells was confirmed by gene-level expression assays [23]. Collagen can be cross-linked with some other materials such as hyaluronic acid, hydroxyapatite, and chitosan. Cross-linked hydroxyapatite-chitosan fibers and MSC seeding have been used to generate porous cell scaffolds for bone tissue engineering applications [25–28]. Cross-linked collagen-based composite scaffolds exhibited a minimal effect on inflammatory and non-fibrotic cellular growth [29]. Bovine collagen type I was proposed and examined as 3D scaffolds for re-establishing the collagen fibrillar like the structure of the skin. Such skin substitutes with cell-seeded collagens have been commercialized widely, such as Apligraf® from Organogenesis, Inc., and OrCel® from Ortec Int. [30].

The potential of using collagen as biomaterials for engineering collagen (type I) tissues in tendon and skin can be established, but the use of these animal-derived collagens in humans can sometimes cause possible allergic reactions and pathogen transmission [31], and the use of recombinant collagens may endure with the lack of biological activities, unlike the native tissue. This is because recombinant collagen does not go through significant post-translational modifications [23, 32]. The challenges remain for the development of precise amino acid residues in collagen molecules with higher yields but fewer allergic reactions. Besides, the combination of the material and stem cell technology will allow the cells to differentiate into target cell types with specific growth factors and cytokines. The scheme of these cell-scaffold systems developed into needed tissues will probably have a wide range of applications for tissue repair and regeneration.

11.2.3 Decellularized Skins

Skin is the largest organ of humans and serves a variety of functions. Being predominantly as a barrier protecting the internal organs from the external environment, it also serves other purposes such as controlling fluid homeostasis, sensory detection, vitamin D synthesis, and self-healing. Human skin consists of the epidermis anchored tightly to an underneath dermis, giving elasticity and mechanical resistance [33]. Damage or loss of integrity of the skin caused by wounds may ruin the functions of the skin to varying extents. Wounds can be affected by mechanical trauma, surgical procedures, reduced blood circulation, burns, or aging. Most skin wounds can heal naturally through dynamically remodeling processes, which involve soluble factors, blood elements, ECM components, and cells [34].

A decellularized skin (decellularized matrix) is an acellular tissue made by taking a section of the skin layer from a donor source, which is mostly a human cadaver, porcine, and bovine origins. In the case of human donors, the decellularized skin is screened for infectious agents, such as HIV, hepatitis, and syphilis [35]. The decellularization process involves very different protocols combining various agents such as enzymes, hyper-/hypotonic solutions, detergents, and organic solvents. However, the core principle is to achieve effectively the disruption of cells and the elimination of all DNA content in the ECM that may trigger allergic reactions [35]. This treatment will be repeated until the matrices meet the desired quality. The matrices can also be seeded with cells, so-called "re-cellularization," to enhance tissue remodeling and regeneration [36].

To date, decellularized skin has become increasingly popular for the coverage of soft-tissue open wounds. Although skin grafting is significantly less expensive, the use of decellularized skin can be a successful alternative to a painful and aesthetically undesirable donor site [35]. In the case of significant tissue injury or disease, tissue autografts are often regarded as the gold standard, but the limitations, including donor site morbidity, low availability, and troubling failure rates are observed [37]. Recently, engineering human skin is an interdisciplinary and active field of research, but replicating the full properties of the ECM is still a significant challenge [38, 39].

Tissue-engineered skin preparation usually includes cells and/or ECM [40]. An ideal decellularized skin should be sterile, act as a barrier, and have a low inflammatory response with no local toxicity. The engineered skins should also heal the surface rapidly, meet the required physical and mechanical properties, and possess controlled degradation [41]. They should also facilitate angiogenesis and have a long shelf-life with low storage requirements [42]. Various approaches have been implemented to develop engineered skin with mono- or multilayered cultures and 3D matrices for full-thickness models.

The clinical function-decellularized skins are widely held to be a consequence of the presence of specific ECM components in the transplant. Collagen types I and IV, Fibronectin, and Laminin are essential components of connective tissue and provide structure [43]. The initial testing of a new human tissue transplant using Epiflex®, a decellularized human dermis established from screened consenting donors, was described [44]. The transplant was approved as a drug in Germany and initial clinical investigations suggest that Epiflex can be utilized effectively in the treatment of burns, hypertrophic scars, and as a transplant seeded with autologous dermal fibroblasts for soft-tissue regeneration. While great developments have been made in reducing morbidity and mortality arising because of burn wounds, some of the challenges remain. Tissue-engineered skin substitutes with a combination of cell scaffolds and growth factors might be a more promising choice. The utilization of a multi-material strategy may help develop composite scaffolds more precisely to overcome limitations regarding inferior mechanical properties and biocompatibility [45]. The ultimate goal is to entirely restore skin anatomy and physiology using the automated fabrication of engineered tissues to increase precision, reduce costs, and optimize for individual patient needs using 3D bioprinting. Lastly, the technology may adopt a regenerative therapy scheme using stem cell technology for individually targeting pigmentation, wound closure, angiogenesis, and skin sensation [46].

11.2.4 Fibrin/Fibrinogen

Fibrin is a protein-based natural polymer from blood and is formed by the polymerization of fibrinogen, following the induction of the coagulation cascade. Thrombin activates fibrinogen, which binds to adjacent fibrinogen molecules and results in the construction of an insoluble fibrin matrix. This fibrin network is the primary protein factor in clots and consequently provides a scaffold that supports cell infiltration, angiogenesis, and promoting cell attachment and proliferation during tissue repair and wound healing [47]. Blood is composed of 55% plasma and 45% cells. Plasma contains mostly water (92%), including soluble proteins, electrolytes, and metabolic wastes. The major soluble component is fibrinogen or a clotting protein [48]. Centrifuging whole blood separates its components according to density. Erythrocytes were collected at the bottom side of the tube while the thin and white-tinted buffy coat stays at the top of the erythrocytes, and plasma forms the supernatant [49]. Varying centrifugation speed and duration can separate blood concentrate components. Anticoagulating agents or enzymatic supplements may be required to separate platelet-rich plasma (PRP) and platelet-poor plasma (PPP). The PRP will have enough platelets for therapeutic use, yet less abundant than the PPP at roughly 25 and 75% of the supernatant volume, respectively [50].

Because of the strong adhesive properties of fibrin, it is usually used as a medical sealant [50]. Medical sealant, also known as "fibrin glue" or "fibrin tissue sealant," is a kind of biodegradable biomaterial developed and used for decades. Fibrin sealants are currently used to boost local surgical hemostasis and to provide efficient tissue adherence. [51] A sealant forms a sealing wall that inhibits the outflow of gas or liquid from a surgical site. It automatically polymerizes by itself and is often most effective in a dry field. It can adhere structures together that make it function as both sealants and adhesives when applied to the leaking blood vessels. Fibrin sealant must be commercially available via FDA approval for clinical use in all of these categories: hemostats, sealants, and adhesives [52]. The elasticity, tensile strength, and tissue adhesiveness of fibrin glue have made it an important material in microsurgery, including cardiopulmonary bypass surgery, colostomy closure, and splenic injury repair and hepatic surgery [53, 54]. However, the development of more functions and applications of medical sealant is still required. In recent years, tissue engineering research has focused on the development of biomaterials to improve this healing process. González-Quevedo and coworkers reported nanostructured fibrin-agarose hydrogel composite revealed significantly improved biomechanical properties when compared with MatriDerm®. Moreover, high cell viability was observed, and in vivo evaluation of repaired tendons using these biomaterials resulted in better organization of the collagen fibers and cell alignment. Therefore, the addition of fibrin into the scaffold has the potential for the treatment of Achilles tendon injuries, which are a frequent problem in orthopedic surgery [55]. In addition, fibrin coated on the microporous walls of macroporous poly(L-lactide) (PLLA) three-dimensional scaffold showed improved scaffold adhesion properties, therefore promoting tissue regeneration [56].

Fibrin is a promising biopolymer that could be used for tissue engineering applications. However, commercially available fibrin glues are costly and are associated with the risk of disease transmission. Whereas single-donor fibrin glues can have a batch-to-batch variation, therefore, the development of fibrinogen and thrombin using a recombinant protein scheme may help solve these problems [57]. Another aspect that requires further investigation related to fibrin includes its preparation from PRP. Platelets are considered a massive pool of growth factors. So, their specific roles in tissue repair are not well understood [58, 59]. Therefore, more investigation is needed to explain the possible advantageous features of platelets to clinical practice.

11.3 Polysaccharide-based Natural Biomaterials

Natural polymers simulate the natural microenvironment of tissue and demonstrate biocompatibility with mechanical properties and the ability to be loaded with growth factors [60]. Polysaccharide-based biomaterials are natural polymers containing sugar monomers. Polysaccharides are obtained from plant or animal sources and are known to have high bioactivity regarding tissue engineering applications. Chitosan, alginate, hyaluronan, and agarose are some of the promising polysaccharide biomaterials employed for cell scaffold developments. Important notice of the applications of polysaccharide recently is the polysaccharide structure in the human body regarding their fundamental properties such as biocompatibility and biodegradability, which opens interest toward polysaccharides. They are reproducible in nature and have a wide range of physical properties and are simple to fabricate into various forms such as beads, capsules, fibers, and films.

11.3.1 Chitosan

Chitosan is a linear polysaccharide polymer derived from partial deacetylation of chitin, which forms structural components in the exoskeleton (shell) of arthropods or in the cell walls of fungi and yeast [61]. Chitosan [poly-(β-1/4)-2-amino-2-deoxy-D-glucopyranose] is a copolymer made of D-glucosamine and N-acetyl-D-glucosamine bonds and β bonds, in which glucosamine is the primary repeating unit in the structure. It is a derivative of the alkaline deacetylation of chitin, and the glucosamine content is named depending on the degree of deacetylation [61–64]. The molecular weight may diverge from 300 to over 1000 kD, with deacetylation between 30 and 95% depending on the procedure of origin and preparation. Chitosan has been the feasible version of the chitin polymer because it is readily soluble in dilute organic acids and therefore possesses greater availability to be incorporated in chemical reactions [65–67]. Chitosan has proved to be a potential biomaterial for porous scaffold synthesis due to its excellent properties such as biodegradability, biocompatibility, cell proliferation, and antimicrobial activities [68, 69].

Chitosan-based materials have been widely applied for tissue engineering applications in the last few decades. The potential of chitosan in tissue engineering as biomaterials are bone and cartilage tissue engineering, blood vessel graft, corneal regeneration, skin regeneration, and periodontal tissue engineering [70, 71]. Yuan and coworkers demonstrated the novel cross-linking chitosan hydrogel with high porosity and seeded MSCs showed the specific histological properties of natural cartilage and the genes for type II collagen were expressed in the same manner as in the natural one [72]. In addition, chitosan had been used to create a chitosan skin-regenerating template, and the results showed the cell-chitosan constructs (primary human dermal fibroblast) within interconnected porous chitosan scaffold, which could be potential for application as a dermal substitute [73]. Another concept in tissue engineering is for vascular graft applications. Wang and coworkers had investigated the co-electrospinning of poly(ε-caprolactone) (PCL)/carboxymethyl chitosan and the PCL/chitosan in the antithrombotic and antibacterial aspects. The materials demonstrated the inner layer of the graft could inhibit thrombosis efficiently, and the outer layer had a particular antibacterial effect according to the addition of chitosan. Moreover, the inner layer of the graft could promote endothelialization and could clinically be applied for vascular grafts in the future [74].

As a low-cost natural polymer with good biodegradability and biocompatibility, as well as having nontoxic, mucoadhesive, and antimicrobial properties, chitosan is a good material for biomedical and biopharmaceutical applications. Mixing chitosan with other natural or synthetic polymers could effectively control their biodegradation rate, enhance their bioactivity, and increase their mechanical properties. The scaffolds also promote the proliferation and differentiation of cells and accelerate tissue regeneration, as well as angiogenesis in different animal models, which could subsequently be used in human clinical trials [75].

11.3.2 Alginate

Alginate or alginic acids are natural biopolymers composed of non-repeating unbranched exopolysaccharides derived from certain seaweeds, bacteria, and brown algae [76]. Alginate is highly biocompatible, hydrophilic, relatively cost-effective, and biodegradable under normal physiological conditions [76]. Fischer and Dörfel identified the L-guluronate residue, D-mannuronate was noticed as the main component of alginate [77]. Alginates are presented with block copolymers, in which the ratio of guluronate to mannuronate varies depending on the natural source. Alginate is currently known to be a whole family of linear copolymers containing blocks of (1,4)-linked β-D-mannuronate (M) and α-L-guluronate (G) residues. The blocks are composed of consecutive G residues (GGGGGG), consecutive M residues (MMMMMM), and alternating M and G residues (GMGMGM) [78]. Alginates extracted from different sources can have alterations in M and G contents, including the length of each block, and more than 200 different alginates are currently being commercially manufactured [79].

There are huge clinical needs for alginate-based biomaterials for tissue engineering and drug delivery. Alginate has been utilized in a wound dressing application. Alginate-based wound dressings such as hydrogels, sponges, and electrospun mats, are promising materials for wound healing, which exhibit many advantages, including hemostatic capability and gel-forming ability [80, 81]. Alginate was found to own many important elements required in a wound dressing such as water absorptivity, optimal water vapor transmission rate, and mild antibacterial activities combined with biocompatibility and biodegradability. There has been a suggestion of certain alginate dressings (e.g. Kaltostat®) that can improve wound healing by activating monocytes to produce pro-inflammatory cytokines such as interleukin-6 and tumor necrosis factor-α [82]. Alginate has become one of the promising biomaterials, as the US Food and Drug Administration (FDA) has approved it for diverse applications in regenerative medicine and nutrition supplements [83].

11.3.3 Agarose

Agarose is a marine algal polysaccharide obtained by the isolation of red algae and seaweed [84] and comprises the repeating unit of 3,6-anhydrous-L-galactose and D-galactose [85]. Due to its physical properties, agarose dissolved into aqueous solutions and forms a gel with a three-dimensional net and porous structure with aqueous content [85]. Agarose gel appears as apyrogenic, colorless, transparent gel, and viscoelastic at temperatures above 45 °C. Since it can form into a hydrogel, it provides a three-dimensional architecture for cells to reside and grow. Moreover, the physical structure of the gels can be tunable by adjusting the agarose concentration, which results in the desired porosity [86]. Agarose is regarded as a choice for numerous biocompatibility tests such as cytotoxicity and subcutaneous implants [87], as well as a substrate for cell growth and microencapsulation ([88, 89]). Therefore, agarose has already been explored as biomaterials in a wide range of preclinical applications, such as three-dimensional tissue growth.

Agarose hydrogels have been employed with different types of cells for generating a range of applications such as bone and cartilage [90, 91]. Evaluation of agarose gels as a biomaterial for chondrogenesis of human bone marrow-derived mesenchymal stem cells (hBM-MSCs) exhibited chondrogenesis of hBM-MSCs in the agarose culture with TGF-β3 with the expression of the chondrogenic markers, including collagen type II and aggrecan [91]. Agarose/silk cell scaffolds have shown to be adjustable in mechanical properties with different gelation degrees. Moreover, the scaffolds were tested for bone tissue engineering application with the presentation of alkaline phosphatase activity and mineralization [92]. A promising application of agarose was also demonstrated in in vivo model. 1.5% agarose subcutaneously implanted in rats induced wound healing process via angiogenesis and fibrous tissue formation abundant with neutrophils. This formation increased until the 12th week and significantly decreased around 16 weeks [85].

11.4 Summary

Organic feedstocks are the carbon-containing raw materials that can be obtained from nature and have been processed for industrial uses. In the medical industry, organic feedstocks are usually extracted from either plants, animals, and sometimes human donors, as suggested in tissue engineering applications. According to the tissue engineering principle, combining cells with natural biomaterial scaffolds as "home for cells" and/or growth factors serves as a promising strategy for constructing substituted tissues for clinical applications. This chapter updated the materials commonly used as cell scaffolds, mainly on protein-based materials, including silk, collagen, decellularized skin, fibrin, and polysaccharide-based materials, including chitosan, alginate, and agarose.

Artificial tissue and organ have been improved by remarkable advances in tissue engineering in the past decades, but still need better biocompatibility and biofunctionality. The challenges remain not only on how to optimize the biomaterials but also push their applications to clinical trials. Organic feedstocks have been shown to be effectively polymerized to obtain gel-like or matrix-like structures compatible with the cells in vitro but need more in vivo studies and preclinical evaluations. They also have advantages in terms of many sources found in nature or harvesting from agriculture. However, the critical point is how to harvest or extract them without losing the biomechanical properties in a sterile condition. Besides, they usually form gel-like or matrix-like structures and be used as polymer sheets, so they have failed to fabricate into a bigger and complex organ structure. To make them into a complex organ, researchers usually coat these materials with different types of metal or plastic and use them as a backbone to support a 3D structure.

The current approach focuses on the biocompatibility of bioscaffolds and targeted tissue regeneration ability. Functionalized biomaterials are under investigation to improve the strength after polymerization by additional chemical bonding between molecules and to incorporate the growth factor molecules to induce better cell attachment, migration, and differentiation. Finally, the next generation of engineered tissues, such as 3D-printed bioscaffold with live cells, is recently developed via 3D printers. They provide promising opportunities to achieve the complex 3D structure of tissues or organs from MRI or CT scan data, which will improve the manufacturing process and overcome current challenges. In conclusion, organic feedstock is an excellent natural source to harvest and produce as biomaterial for tissue engineering applications and may combine with advanced fabrication technology such as 3D printing. This scheme will result in the production of efficient replacements for diseased or damaged tissues and organs, therefore improving the treatment outcome and quality of life for patients.

References

1. Langer, R. and Vacanti, J.P. (1993). Tissue engineering. *Science. American Association for the Advancement of Science* 260 (5110): 920–926.
2. Akter, F. (2016). Principles of tissue engineering. In: *Tissue Engineering Made Easy* (ed. F. Akter), 3–16. London, UK: Academic Press.
3. Ollitrault, D., Legendre, F., Drougard, C. et al. (2015). BMP-2, hypoxia, and COL1A1/HtrA1 siRNAs favor neo-cartilage hyaline matrix formation in chondrocytes. *Tissue Engineering – Part C: Methods* 21 (2): 133–147.
4. Ikada, Y. (2006). Challenges in tissue engineering. *Journal of the Royal Society Interface* 3: 389–601.
5. Sengupta, D. and Heilshorn, S.C. (2010). Protein-engineered biomaterials: highly tunable tissue engineering scaffolds. *Tissue Engineering – Part B: Reviews* 16 (3): 285–293.
6. Suzuki, S., Rayner, C.L., and Chirila, T.V. (2019). Silk fibroin/sericin native blends as potential biomaterial templates. *Advances in Tissue Engineering & Regenerative Medicine: Open Access* 5 (1): 11–19.
7. Hakimi, O., Knight, D.P., Vollrath, F., and Vadgama, P. (2007). Spider and mulberry silkworm silks as compatible biomaterials. *Composites Part B: Engineering* 38 (3): 324–337.
8. Selvi, F., Çakarer, S., Can, T. et al. (2016). Effects of different suture materials on tissue healing. *Journal of Istanbul University Faculty of Dentistry* 50 (1): 35.
9. Yaltirik, M., Dedeoglu, K., Bilgic, B. et al. (2003). Comparison of four different suture materials in soft tissues of rats. *Oral Diseases* 9 (6): 284–286.
10. Vepari, C. and Kaplan, D.L. (2007). Silk as a biomaterial. *Progress in Polymer Science (Oxford)* 32 (8–9): 991–1007.
11. Mandal, B.B., Priya, A.S., and Kundu, S.C. (2009). Novel silk sericin/gelatin 3-D scaffolds and 2-D films: fabrication and characterization for potential tissue engineering applications. *Acta Biomaterialia* 5 (8): 3007–3020.
12. Parenteau-Bareil, R., Gauvin, R., and Berthod, F. (2010). Collagen-based biomaterials for tissue engineering applications. *Materials* 3 (3): 1863–1887.
13. Dong, C. and Lv, Y. (2016). Application of collagen scaffold in tissue engineering: recent advances and new perspectives. *Polymers* 8 (42): 1–20.
14. Boyle, J. (2008). Molecular biology of the cell. In: *Biochemistry and Molecular Biology Education*, 5the (ed. B. Alberts, A. Johnson, J. Lewis, et al.). New York: Garland Science.
15. Purcel, G., Meliţă, D., Andronescu, E., and Grumezescu, A.M. (2016). Collagen-based nanobiomaterials: challenges in soft tissue engineering. In: *Nanobiomaterials in Soft Tissue Engineering: Applications of Nanobiomaterials* (ed. A.M. Grumezescu), 173–200. Oxford, UK: William Andrew.

16. Prockop, D.J. and Kivirikko, K.I. (1995). Collagens: molecular biology, diseases, and potentials for therapy. *Annual Review of Biochemistry* 64: 403–434.
17. Van Der Rest, M. and Garrone, R. (1991). Collagen family of proteins. *The FASEB Journal* 5 (13): 2814–2823.
18. Aszódi, A., Legate, K.R., Nakchbandi, I., and Fässler, R. (2006). What mouse mutants teach us about extracellular matrix function. *Annual Review of Cell and Developmental Biology* 22: 591–621.
19. Allan, B., Ruan, R., Landao-Bassonga, E. et al. (2021). Collagen membrane for guided bone regeneration in dental and orthopedic applications. *Tissue Engineering Part A* 27 (5–6): 372–381. 10.1089/ten.TEA.2020.0140. Epub 2020 Sep 10. PMID:32741266.
20. Atiyeh, B. and Ibrahim, A. (2007). Nonsurgical management of hypertrophic scars: evidence-based therapies, standard practices, and emerging methods: an update. *Aesthetic Plastic Surgery* 31 (5): 468–492.
21. Fan, X., Liang, Y., Cui, Y. et al. (2020). Development of tilapia collagen and chitosan composite hydrogels for nanobody delivery. *Colloids and Surfaces B: Biointerfaces* 195: 1–9.
22. Awad, H.A., Boivin, G.P., Dressler, M.R. et al. (2003). Repair of patellar tendon injuries using a cell-collagen composite. *Journal of Orthopaedic Research* 21 (3): 420–431.
23. Juncosa-Melvin, N., Matlin, K.S., Holdcraft, R.W. et al. (2007). Mechanical stimulation increases collagen type I and collagen type III gene expression of stem cell-collagen sponge constructs for patellar tendon repair. *Tissue Engineering* 13: 1219–1226.
24. Juncosa-Melvin, N., Shearn, J.T., Boivin, G.P. et al. (2006). Effects of mechanical stimulation on the biomechanics and histology of stem cell-collagen sponge constructs for rabbit patellar tendon repair. *Tissue Engineering* 12 (8): 2291–2300.
25. Al-Munajjed, A.A., Plunkett, N.A., Gleeson, J.P. et al. (2009). Development of a biomimetic collagen-hydroxyapatite scaffold for bone tissue engineering using a SBF immersion technique. *Journal of Biomedical Materials Research – Part B Applied Biomaterials* 90 B (2): 584–591.
26. Barua, E., Deoghare, A.B., Deb, P., and Lala, S.D. (2018). Naturally derived biomaterials for development of composite bone scaffold: a review. *IOP Conference Series: Materials Science and Engineering* 377: 8.
27. McBane, J.E., Vulesevic, B., Padavan, D.T. et al. (2013). Evaluation of a collagen-chitosan hydrogel for potential use as a pro-angiogenic site for islet transplantation. *PLoS One* 8 (10): e77538.
28. Mohammadi, F., Samani, S.M., Tanideh, N., and Ahmadi, F. (2018). Hybrid scaffolds of hyaluronic acid and collagen loaded with prednisolone: an interesting system for osteoarthritis. *Advanced Pharmaceutical Bulletin* 8 (1): 11–19.
29. Ahmed, T.A.E., Dare, E.V., and Hincke, M. (2008). Fibrin: a versatile scaffold for tissue engineering applications. *Tissue Engineering – Part B: Reviews* 14 (2): 199–215.
30. Vig, K., Chaudhari, A., Tripathi, S. et al. (2017). Advances in skin regeneration using tissue engineering. *International Journal of Molecular Sciences* 18 (4): 789.
31. Baumann, L.S. and Kerdel, F. (1999). The treatment of bovine collagen allergy with cyclosporin. *Dermatologic Surgery* 25 (3): 247–249.
32. Wang, T., Lew, J., Premkumar, J. et al. (2017). Production of recombinant collagen: state of the art and challenges. *Engineering Biology* 1 (1): 18–23.
33. Blanpain, C. and Fuchs, E. (2009). Epidermal homeostasis: a balancing act of stem cells in the skin. *Nature Reviews Molecular Cell Biology* 10 (3): 207–217.
34. Clark, R.A.F. (2014). Wound repair: Basic Biology to Tissue Engineering. In: *Principles of Tissue Engineering* (ed. R. Lanza, R. Langer and J. Vacanti), 1595–1617. London, UK: Academic Press.
35. Fosnot, J., Kovach III, S.J., and Serletti, J.M. (2011). Acellular dermal matrix: general principles for the plastic surgeon. *Aesthetic Surgery Journal* 31 (7 Suppl): 5S–12S.
36. Gentile, P., Sterodimas, A., Pizzicannella, J. et al. (2020). Systematic review: allogenic use of stromal vascular fraction (SVF) and decellularized extracellular matrices (ECM) as advanced therapy medicinal products (ATMP) in tissue regeneration. *International Journal of Molecular Sciences* 21 (14): 1–14.
37. Przekora, A. (2020). A concise review on tissue engineered artificial skin grafts for chronic wound treatment: can we reconstruct functional skin tissue in vitro? *Cells* 9 (7): 1622–1651.
38. Rahmani Del Bakhshayesh, A., Annabi, N., Khalilov, R. et al. (2018). Recent advances on biomedical applications of scaffolds in wound healing and dermal tissue engineering. *Artificial Cells, Nanomedicine and Biotechnology* 46 (4): 691–705.
39. Urciuolo, F., Casale, C., Imparato, G., and Netti, P.A. (2019). Bioengineered skin substitutes: the role of extracellular matrix and vascularization in the healing of deep wounds. *Journal of Clinical Medicine* 8 (12): 2083.

40. Bello, Y.M., Falabella, A.F., and Eaglstein, W.H. (2001). Tissue-engineered skin: current status in wound healing. *American Journal of Clinical Dermatology* 2 (5): 305–313.
41. MacNeil, S. (2007). Progress and opportunities for tissue-engineered skin. *Nature* 445: 874–880.
42. Metcalfe, A.D. and Ferguson, M.W.J. (2007). Tissue engineering of replacement skin: the crossroads of biomaterials, wound healing, embryonic development, stem cells and regeneration. *Journal of the Royal Society Interface* 4 (14): 413–437.
43. Cen, L., Liu, W., Cui, L. et al. (2008). Collagen tissue engineering: development of novel biomaterials and applications. *Pediatric Research* 63 (5): 492–496.
44. Rössner, E., Smith, M.D., Petschke, B. et al. (2011). Epiflex® A new decellularised human skin tissue transplant: manufacture and properties. *Cell and Tissue Banking* 12 (3): 209–217.
45. Yildirimer, L., Thanh, N.T.K., and Seifalian, A.M. (2012). Skin regeneration scaffolds: a multimodal bottom-up approach. *Trends in Biotechnology* 30 (12): 638–648.
46. Kaur, A., Midha, S., Giri, S., and Mohanty, S. (2019). Functional skin grafts: where biomaterials meet stem cells. *Stem Cells International* 2019 (1286054): 1–20.
47. Sproul, E., Nandi, S., and Brown, A. (2018). Fibrin biomaterials for tissue regeneration and repair. In: *Peptides and Proteins as Biomaterials for Tissue Regeneration and Repair* (ed. M.A. Barbosa and M.C.L. Martins), 151–173. London, UK: Woodhead Publishing.
48. Mescher, A.L. (2018). *Junqueira's Basic Histology: Text and Atlas*, 15ee. London, UK: McGraw-Hill Education.
49. Garg, A.K. (2018). *Autologous Blood Concentrates*, 224. Batavia, IL: Quintessence Publishing.
50. Dohan, D.M., Choukroun, J., Diss, A. et al. (2006). Platelet-rich fibrin (PRF): a second-generation platelet concentrate. Part II: platelet-related biologic features. *Oral Surgery, Oral Medicine, Oral Pathology, Oral Radiology and Endodontology* 101 (3): 45–50.
51. Chen, F.-M., Zhang, J., Zhang, M. et al. (2010). A review on endogenous regenerative technology in periodontal regenerative medicine. *Biomaterials* 31 (31): 7892–7927.
52. Spotnitz, W.D. (2014). Fibrin sealant: the only approved hemostat, sealant, and adhesive – a laboratory and clinical perspective. *ISRN Surgery* 2014: 1–28.
53. Kohno, H., Nagasue, N., Chang, Y.C. et al. (1992). Comparison of topical hemostatic agents in elective hepatic resection: a clinical prospective randomized trial. *World Journal of Surgery* 16 (5): 966–970.
54. Noun, R., Elias, D., Balladur, P. et al. (1996). Fibrin glue effectiveness and tolerance after elective liver resection: a randomized trial. *Hepato-Gastroenterology* 43 (7): 221–224.
55. González-Quevedo, D., Díaz-Ramos, M., Chato-Astrain, J. et al. (2020). Improving the regenerative microenvironment during tendon healing by using nanostructured fibrin/agarose-based hydrogels in a rat Achilles tendon injury model. *The Bone and Joint Journal* 102-B (8): 1095–1106.
56. Gamboa-Martínez, T.C., Gómez Ribelles, J.L., and Gallego Ferrer, G. (2011). Fibrin coating on poly (l-lactide) scaffolds for tissue engineering. *Journal of Bioactive and Compatible Polymers* 26 (5): 464–477.
57. Noori, A., Ashrafi, S.J., Vaez-Ghaemi, R. et al. (2017). A review of fibrin and fibrin composites for bone tissue engineering. *International Journal of Nanomedicine* 12: 4937–4961.
58. Van Den Dolder, J., Mooren, R., Vloon, A.P.G. et al. (2006). Platelet-rich plasma: quantification of growth factor levels and the effect on growth and differentiation of rat bone marrow cells. *Tissue Engineering* 12 (11): 3067–3073.
59. Shiu, H.T., Goss, B., Lutton, C. et al. (2014). Formation of blood clot on biomaterial implants influences bone healing. *Tissue Engineering – Part B: Reviews* 20 (6): 697–712.
60. Turco, G., Marsich, E., Bellomo, F. et al. (2009). Alginate/hydroxyapatite biocomposite for bone ingrowth: a trabecular structure with high and isotropic connectivity. *Biomacromolecules* 10 (6): 1575–1583.
61. Rinaudo, M. (2006). Chitin and chitosan: properties and applications. *Progress in Polymer Science (Oxford)* 31 (7): 603–632.
62. Kim, I.Y., Seo, S.J., Moon, H.S. et al. (2008). Chitosan and its derivatives for tissue engineering applications. *Biotechnology Advances* 26 (1): 1–21.
63. Lodhi, G., Kim, Y.S., Hwang, J.W. et al. (2014). Chitooligosaccharide and its derivatives: preparation and biological applications. *BioMed Research International* 2014: 1–13.
64. Shi, C., Zhu, Y., Ran, X. et al. (2006). Therapeutic potential of chitosan and its derivatives in regenerative medicine. *The Journal of Surgical Research* 133 (2): 185–192.

65. Khor, E. and Lim, L.Y. (2003). Implantable applications of chitin and chitosan. *Biomaterials* 24 (13): 2339–2349.
66. Ravi Kumar, M.N.V. (2000). A review of chitin and chitosan applications. *Reactive and Functional Polymers* 46 (1): 1–27.
67. Rudall, K.M. (2007). Chitin and its association with other molecules. *Journal of Polymer Science Part C: Polymer Symposia* 28 (1): 83–102.
68. Islam, M.M., Shahruzzaman, M., Biswas, S. et al. (2020). Chitosan based bioactive materials in tissue engineering applications – a review. *Bioactive Materials* 5 (1): 164–183.
69. Jiang, T., James, R., Kumbar, S.G., and Laurencin, C.T. (2014). Chitosan as a biomaterial: structure, properties, and applications in tissue engineering and drug delivery. In: *Natural and Synthetic Biomedical Polymers* (ed. S.G. Kumbar, C.T. Laurencin and M. Deng), 91–113. Burlington, MA: Elsevier Science.
70. Saravanan, S., Leena, R.S., and Selvamurugan, N. (2016). Chitosan based biocomposite scaffolds for bone tissue engineering. *International Journal of Biological Macromolecules* 93: 1354–1365.
71. Balagangadharan, K., Dhivya, S., and Selvamurugan, N. (2017). Chitosan based nanofibers in bone tissue engineering. *International Journal of Biological Macromolecules* 104: 1372–1382.
72. Yuan, D., Chen, Z., Lin, T. et al. (2015). Cartilage tissue engineering using combination of chitosan hydrogel and mesenchymal stem cells. *Journal of Chemistry* 2015: 1–6.
73. Hilmi, A.B.M., Halim, A.S., Hassan, A. et al. (2013). in vitro characterization of a chitosan skin regenerating template as a scaffold for cells cultivation. *SpringerPlus* 2 (1): 1–9.
74. Wang, Y., He, C., Feng, Y. et al. (2020). A chitosan modified asymmetric small-diameter vascular graft with anti-thrombotic and anti-bacterial functions for vascular tissue engineering. *Journal of Materials Chemistry B* 8 (3): 568–577.
75. Aguilar, A., Zein, N., Harmouch, E. et al. (2019). Application of chitosan in bone and dental engineering. *Molecules* 24 (26): 3009–3026.
76. Lee, K.Y. and Mooney, D.J. (2012). Alginate: properties and biomedical applications. *Progress in Polymer Science (Oxford)* 37 (1): 106–126.
77. Moradali, M.F., Ghods, S., and Rhem, B.H.A. (2018). Alginate Biosynthesis and Biotechnological Production. In: *Alginates and Their Biomedical Applications*, 1–25. Singapore: Springer Nature Singapore Pte Ltd.
78. Gurikov, P. and Smirnova, I. (2018). Non-conventional methods for gelation of alginate. *Gels* 4 (1): 1–14.
79. Tønnesen, H.H. and Karlsen, J. (2002). Alginate in drug delivery systems. *Drug Development and Industrial Pharmacy* 28 (6): 621–630.
80. Hooper, S.J., Percival, S.L., Hill, K.E. et al. (2012). The visualisation and speed of kill of wound isolates on a silver alginate dressing. *International Wound Journal* 9 (6): 633–642.
81. Li, X., Chen, S., Zhang, B. et al. (2012). In situ injectable nano-composite hydrogel composed of curcumin, N,O-carboxymethyl chitosan and oxidized alginate for wound healing application. *International Journal of Pharmaceutics* 437 (1–2): 110–119.
82. Balakrishnan, B., Mohanty, M., Umashankar, P.R., and Jayakrishnan, A. (2005). Evaluation of an in situ forming hydrogel wound dressing based on oxidized alginate and gelatin. *Biomaterials* 26 (32): 6335–6342.
83. Sun, J. and Tan, H. (2013). Alginate-based biomaterials for regenerative medicine applications. *Materials* 6 (4): 1285–1309.
84. Lee, K.Y. and Mooney, D.J. (2001). Hydrogels for tissue engineering. *Chemical Reviews* 101 (7): 1869–1879.
85. Varoni, E., Tschon, M., Palazzo, B. et al. (2012). Agarose gel as biomaterial or scaffold for implantation surgery: characterization, histological and histomorphometric study on soft tissue response. *Connective Tissue Research* 53 (6): 548–554.
86. Narayanan, J., Xiong, J.Y., and Liu, X.Y. (2006). Determination of agarose gel pore size: absorbance measurements vis a vis other techniques. *Journal of Physics: Conference Series* 28 (1): 83–86.
87. Scarano, A., Carinci, F., and Piattelli, A. (2009). Lip augmentation with a new filler (agarose gel): a 3-year follow-up study. *Oral Surgery, Oral Medicine, Oral Pathology, Oral Radiology and Endodontology* 108 (2): e11–e15.
88. Imani, R., Emami, S.H., Moshtagh, P.R. et al. (2012). Preparation and characterization of agarose-gelatin blend hydrogels as a cell encapsulation matrix: an in-vitro study. *Journal of Macromolecular Science, Part B: Physics* 51 (8): 1606–1616.

89. Gasperini, L., Mano, J.F., and Reis, R.L. (2014). Natural polymers for the microencapsulation of cells. *Journal of the Royal Society Interface* 11 (100): 20140817.
90. Finger, A.R., Sargent, C.Y., Dulaney, K.O. et al. (2007). Differential effects on messenger ribonucleic acid expression by bone marrow-derived human mesenchymal stem cells seeded in agarose constructs due to ramped and steady applications of cyclic hydrostatic pressure. *Tissue Engineering* 13 (6): 1151–1158.
91. Huang, C.Y.C., Reuben, P.M., D'Ppolito, G. et al. (2004). Chondrogenesis of human bone marrow-derived mesenchymal stem cells in agarose culture. *Anatomical Record – Part A Discoveries in Molecular, Cellular, and Evolutionary Biology* 278 (1): 428–436.
92. Lu, Y., Zhang, S., Liu, X. et al. (2017). Silk/agarose scaffolds with tunable properties: via SDS assisted rapid gelation. *RSC Advances* 7 (35): 21740–21748.

12

Green Synthesis of Bio-based Metal–Organic Frameworks

Emile R. Engel[1], Bernardo Castro-Dominguez[2] and Janet L. Scott[1]

[1]*Department of Chemistry, University of Bath, Bath, UK*
[2]*Department of Chemical Engineering, University of Bath, Bath, UK*

12.1 Introduction

Coordination polymers (CPs) are hybrid inorganic–organic assemblies of metal centres linked to organic ligands, effectively forming extended structures. Metal–organic frameworks (MOFs) are a subset of CPs, forming two-dimensional (2D) or three-dimensional (3D) architectures containing potential voids [1].

Metal biomolecule frameworks (MBioFs) have been proposed as a sub-class of MOFs, where the ligands are naturally occurring organic molecules found in living organisms, such as amino acids, nucleobases, or sugars [2]. On the other hand, the term 'BioMOFs' was designated for MOFs relevant to biological and medical applications [3, 4]. While BioMOFs do not explicitly adhere to the principles of green chemistry, they frequently comprise benign components (based on the toxicology of the metal ion and bioactive ligands) to ensure biocompatibility.

This chapter refers to bio-based MOFs as those derived from reactants obtainable from biomass. The term 'bio-based' is usually associated with renewable sources of carbon. MOFs are typically synthesised by the reaction of one or more metal salts with one or more organic ligands; therefore, it is primarily the organic component that determines whether the MOF is bio-based or not.

High-Performance Materials from Bio-based Feedstocks, First Edition. Edited by Andrew J. Hunt, Nontipa Supanchaiyamat, Kaewta Jetsrisuparb and Jesper T.N. Knijnenburg.

Although the inorganic component may not be relevant to the bio-based nature of a MOF, if the intention is to manufacture these materials sustainably at large scales, it is evidently important to select metal ions for sustainability and safety. Thus, in this chapter, we mostly avoid MOFs based on toxic or very hazardous metal ions and devote some content to a discussion of metal ion considerations.

The lack of a standard nomenclature is a major challenge in carrying out broad assessments of MOF materials. Here, we have referred to MOFs according to the names attributed in the literature. While most researchers have accepted the existence of this problem and found ways to work around it, as research into bio-based MOFs expands, it may yet be worthwhile to develop a standard nomenclature system.

Within green chemistry, inorganic synthesis in general, and MOF synthesis in particular, have been relatively neglected. There are, in fact, only a few MOFs that are bio-based *and* for which green synthesis has been demonstrated.

12.2 Green Synthesis of MOFs

Early MOF syntheses were typically carried out under solvothermal conditions, relying on high temperatures, high pressures, and hazardous solvents. Improved sustainable synthesis of MOFs, from the perspective of green chemistry, has been addressed in a number of excellent reviews [5–7]. In one, a useful overview outlined the criteria for assessing the sustainability of MOF production, explicitly applying the 12 principles of green chemistry to industrial-scale MOF production (Figure 12.1) [6]. The authors noted that cost is the main barrier to industrial uptake, but growing consideration is being devoted to environmental impact and safety. Preferable industrial conditions for sustainable synthesis include: open reactors at ambient pressure and temperatures below reflux; organic ligands from renewable sources; solvent-free conditions or water as solvent, and non-corrosive, safe metal salts [7, 8].

Conditions for post-synthetic treatment and activation are frequently overlooked when designing sustainable synthetic methodologies for MOF manufacture [8]. At times, a procedure may be reported as sustainable, but upon close inspection, it involves post-treatment with unfavourable solvents or high temperatures to improve crystallinity or for activation. Even so, it is not always clear how post-treatments improve crystallinity; for instance, it is hypothesised that these treatments facilitate preferential washing to remove amorphous material.

12.2.1 Solvent-Free and Low-Solvent Synthesis

Extensive research has been conducted into the mechanochemical synthesis of MOFs as an alternative to typical solution-based methods, such as solvothermal synthesis. Mechanochemistry involves the intimate mixing of reagents by techniques based on grinding and milling. Typically, reactants are combined in a milling device to carry out neat or liquid-assisted grinding (LAG), where a small, sub-stoichiometric amount of solvent is added. Among the well-researched archetypal MOFs, MOF-5 [9, 10], Hong Kong University of Science and Technology (HKUST)-1 [11, 12], zinc imidazolate framework (ZIF)-8 [13], and UiO-66 [14, 15] are some that have been synthesised by milling or grinding-type methodologies.

By LAG, it has been possible to isolate previously unknown compounds that are apparently more difficult or impossible to prepare by conventional solution-based synthesis [16]. Among these are 3D and 2D Zn(II) fumarate CPs, which were synthesised directly from ZnO to demonstrate the possibility of using metal oxides as starting reagents [17]. As part of the same

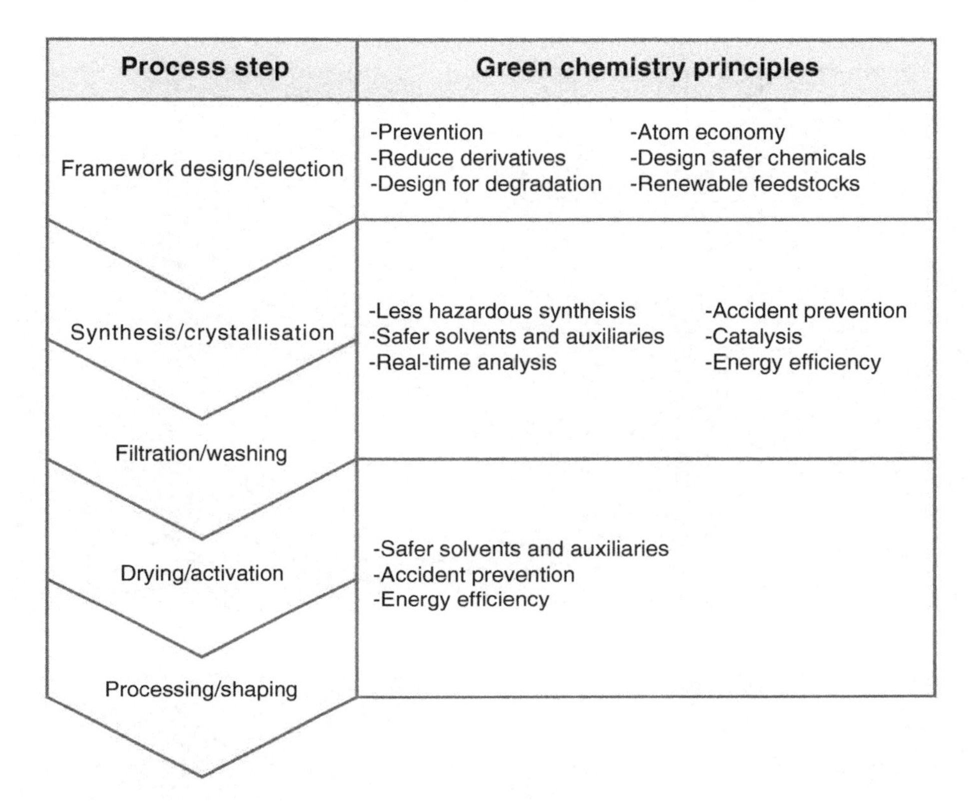

Figure 12.1 *Suggested application of the principles of green chemistry to the process of MOF synthesis at an industrial scale. Source: With permission from the Royal Society of Chemistry. Julien et al. [6].*

study, MOFs were prepared by introducing dipyridyl pillaring ligands. In conventional MOF synthesis, the poor solubility of metal oxides makes them unfavourable as a metal source. Instead, salts such as nitrates, chlorides, and acetates have been more frequently employed. LAG has also been successfully employed to convert ZnO to ZIFs [18].

Metal oxides have also been successfully employed as MOF reagents in methods taking inspiration from mineral neogenesis [19, 20]. A technique called 'accelerated aging' typically uses metal oxides as the MOF metal source under reaction conditions of high humidity and moderate temperatures of up to 45 °C [21]. As a proof-of-concept, accelerated aging was used to prepare a variety of CPs and MOFs from oxalic acid and metal oxides [20]. Surprisingly, differences in metal oxide reactivity under accelerated aging conditions allows for the separation of metal oxide mixtures under mild, solvent-free conditions [20].

The first continuous, scalable, mechanochemical synthesis of MOFs was demonstrated via single- and twin-screw extrusion [22]. Among the first MOFs synthesised by this method was Al(fumarate)(OH), which had not been previously prepared by mechanochemistry. The Zr(IV)-based MOF UiO-66-NH_2 was successfully synthesised by twin-screw extrusion at a rate of 14 kg $hour^{-1}$, although catalytic amounts of MeOH were required (Figure 12.2) [23]. In the same study, at a smaller scale, UiO-66-NH_2 as well as Zr(IV)-based MOF-801 and MOF-804 were prepared by solvent-free or water-assisted planetary milling.

Figure 12.2 *(a) Synthesis of the Zr(IV) MOF UiO-66-NH_2 by twin-screw extrusion. (b) Scanning electron microscopy (SEM) of UiO-66-NH_2 obtained from twin-screw extrusion (left) and planetary milling (right). Source: Reprinted with permission. Copyright (2018) American Chemical Society. Karadeniz et al. [23].*

Dry gel conversion is another low-solvent technique where dry powdered substrates are suspended above an organic solvent or water, within a sealed reactor. The reactor is heated to generate solvent vapour, which, upon contact with the substrates, promotes conversion to the products. This method has been used successfully to prepare a variety of well-known MOFs including Materials Institute Lavoisier (MIL)-100 [24] and UiO-66 [25]. However; the high autogenous pressures associated with this technique are a drawback for scaling up.

12.2.2 Green Solvents

Water is the ideal green solvent and some MOFs previously synthesised in hazardous organic solvents have since been synthesised in water. Ethanol is second to water among conventional solvents considered green or sustainable, although its flammability is a major concern. Acetic acid is a reasonable option where water and ethanol are not feasible [7].

Early ZIF-8 synthesis relied on methanol and 25% aqueous ammonia as solvents [26] or dimethylformamide (DMF) at 140 °C [27]. Later, it was shown that ZIF-8 nanocrystals could be obtained by combining $Zn(NO_3)_2{\cdot}6H_2O$ and 2-methylimidazole in water at room temperature, although a large excess of ligand was required.[18] The synthesis of Al(fumarate)(OH), first commercially produced by BASF as Basolite A520, in an aqueous medium at temperatures well below 100 °C, was a landmark achievement [28]. The synthesis of HKUST-1 has been achieved in water/ethanol at room temperature by using $Cu(OH)_2$ as a metal source [29]. Nanocrystalline HKUST-1 was obtained after five minutes with 63% conversion of $Cu(OH)_2$; however, the total conversion of $Cu(OH)_2$ required 10 hours. A remarkable total surface area of 1700 $m^2\ g^{-1}$ was recorded.

Ionic liquids (ILs) are salts with melting points below 100 °C and are considered an alternative to conventional solvents employed in the solvothermal synthesis of MOFs due to their low vapour pressures. Reactions can be carried out at elevated temperatures without generating the high autogenous pressures associated with conventional solvothermal synthesis. Deep eutectic solvents (DESs) are a special class of ILs where the components are Lewis or Brønsted acids and bases. Synthesis in ILs and DESs has been called 'ionothermal synthesis'.

A number of ZIFs have been synthesised using ILs. Among the first IL-synthesised ZIFs were two previously reported frameworks and two new frameworks discovered at the time [30]. Reactions were carried out at 130 or 150 °C for 48 hours. By microwave synthesis at 140 °C in IL medium, ZIF-8 was synthesised in 60 minutes [31]. UiO-66 has been prepared using the ILs' 1-hexyl-,1-octyl-, and 3-methylimidazolium chloride with short reaction times (<30 minutes) and excellent yields [32]. This was a great improvement over DMF synthesis, which takes 24 hours under solvothermal conditions [33] and up to 120 hours at room temperature [32]. Moreover, gaseous CO_2 has been shown to promote the crystallisation of MOFs from IL reaction mixtures at room temperature; for instance, in HKUST-1, this method yielded nano-sized MOF particles [34].

Although ILs have shown to be excellent for the synthesis of MOFs, their key characteristic, low vapour pressures, is not sufficient to render ILs 'environmentally friendly'. ILs are water-soluble and may yet enter the environment despite their negligible vapour pressures. In fact, a number of ILs commonly employed in green synthesis research may pose serious environmental toxicity hazards [35–37]. In response, there has been some progress in developing ILs of reduced toxicity and improved biodegradability [38–40]. Nevertheless, the environmental and safety risks associated with many ILs, combined with their high energy requirements, when used for MOF synthesis, require close scrutiny. Indeed, these drawbacks undermine their potential deployment as green solvents in industrial settings.

Supercritical CO_2 has been widely exploited for activation of MOFs, as an alternative to conventional solvent exchange and high-temperature evacuation [26]. It is well established that the activation method can influence the structure of the resultant guest-free framework [28]. Supercritical drying is advantageous for preserving porosity and crystallinity in fragile frameworks. To a lesser extent, supercritical CO_2 is another attractive green solvent that has been employed in MOF synthesis [26]. There are limited examples of supercritical CO_2 being used as a reaction medium by itself or combined with reduced amounts of organic solvent. Researchers have had more success in combining supercritical CO_2 with ILs [41]. Mixing ILs with supercritical CO_2 broadens the range of metal salts and organic ligands that can be dissolved. For instance, under typical conditions of supercritical CO_2 mixed with ILs' emimBF_4 or emimBr at 65 °C,

HKUST-1, ZIF-8 and MIL-100 were prepared in less than 10 hours [42]. The reaction times are a major improvement over pure IL synthesis. Again, it is important to point out that although these are novel approaches, and despite progress towards green ILs [38–40], many ILs are potentially dangerous to the environment in general, and aquatic environments in particular [35, 36].

12.2.3 Sonochemical Synthesis

In the context of green synthesis, the main advantage of sonochemical methods is the potential for improved reaction rates (reducing energy requirements), yields, and selectivity [43]. Achievements in MOF synthesis by sonochemical methods have been limited. MOF-5 has been prepared via sonochemical synthesis at reaction temperatures of 129–164 °C, although the method relied on the hazardous solvent 1-methyl-2-pyrrolidone [44]. Mg-MOF-74 was synthesised in only one hour via sonochemical synthesis using DMF as solvent and triethylamine as ligand deprotonating agent [45]. For ZIF-8, scalable sonochemical synthesis was demonstrated, where 85% yield and excellent textural properties were achieved with a reaction time of only two hours [46]. DMF was used as the solvent and pH was modified with NaOH and a minimum amount of triethylamine.

12.2.4 Electrochemical Synthesis

Electrochemical synthesis, or electrosynthesis, of MOFs usually involves an elemental metal anode as the metal ion source and an electrolyte medium in which the ligand is dissolved. The electrochemical approach involves localised synthesis, which allows for the direct deposition of thin films without the need for surface pre-treatment. Other advantages include relatively short reaction times, relatively mild conditions, and direct kinetic control by, for example, manipulation of current density [47, 48]. Researchers at BASF pioneered an electrochemical synthesis of MOFs by preparing HKUST-1, using the anodic dissolution method, at 12–19 V and 1.3 A with a reaction time of 150 minutes [49, 50]. The ligand was dissolved in methanol and Cu plate anodes served as the metal ion source. MOF-5 has also been prepared by the electrochemical synthesis in ILs and DMF. A Zn anode, as a metal ion source, and Ti cathode were used to produce unusual flower-like crystallites of MOF-5 [51]. Additionally, several other archetypal MOFs, including ZIF-8 and MIL-100(Al), have been prepared by the electrochemical synthesis in relatively short reaction times [48].

12.3 Bio-based Ligands

This section focuses on single ligand bio-based MOFs, where bio-based means the ligands are naturally occurring or obtainable from biomass. However, it is not strictly limited and includes selected mixed ligand examples, incorporating both a bio-based and non-bio-based ligand. Unfortunately, many ligands that are naturally occurring or obtainable from biomass are not currently produced as bio-based chemicals for practical and economic reasons; although this will inevitably improve as the development of bio-based economies progresses.

12.3.1 Amino Acids

Aspartic acid and glutamic acid have terminal carboxylate moieties, making them well suited to MOF construction. Materials of the Institute of porous materials from Paris (MIP)-202(Zr), a Zr(IV) aspartate MOF, is one of few hydrolytically stable amino-acid MOFs, which is stable in boiling water across a broad pH range and displays excellent proton conductivity [52]. Gas sorption in MIP-202 was tested for the separation of CO_2/CH_4 and CO_2/N_2 mixtures, displaying an ultra-high selectivity and excellent recyclability, which are promising features for future industrial applications [53]. The water-stable Zn(II) glutamate MOF ZnGlu can be synthesised rapidly at room temperature simply by mixing $ZnSO_4$, glutamic acid, and NaOH in water [54]. ZnGlu showed efficient catalytic features for CO_2 cycloaddition reactions, which can be exploited towards CO_2 fixation. A Cu(II) glutamate MOF was investigated for adsorption of ciprofloxacin, as a model compound for pharmaceuticals as emerging contaminants. At 25 °C and pH 4, ciprofloxacin adsorption amounted to 61.35 mg g^{-1} of MOF [55].

Phenylalanine is an aromatic essential amino acid; however, as a ligand, its structure limits the pursuit of synthesising open framework MOFs. A mixed ligand Zn(II) L-phenylalanine MOF was achieved using bipyridylethene as a pillaring ligand [56]. This MOF is effective in the chiral recognition of other amino acids.

Cysteine is readily oxidised to its dimeric form, cystine, under alkaline conditions. Cysteine has been employed in the preparation of lower-dimensional CPs [57] but 3D MOFs incorporating cysteine are rare. A cysteine-functionalised MOF labelled MIL-101(NH_2)@Au-Cys was prepared by post-synthetic functionalisation via Au nanoparticles [58]. The conjugated MOF showed excellent performance in the enrichment of N-linked glycopeptides. Coordination self-assembly of cysteine has also been reported for lower dimensional CPs [59, 60]. Mg-MBioF is based on cysteine as a ligand [61]. The antioxidant properties of Mg-MBioF were investigated for potential use in therapeutics or the maintenance of antioxidant/oxidant balance as preventative medicine.

Histidine bears a carboxylate moiety at one end and an imidazole group at the other, making it an excellent ligand for MOF construction, which has been surprisingly underexplored. As an approach to imparting chirality to an achiral framework, D-histidine was combined with $Zn(NO_3)$ and methylimidazole to yield a mixed ligand MOF, D-his–ZIF-8, with ZIF-8 architecture [62]. The synthesis was carried out in 85% MeOH in water at room temperature.

A series of homochiral MOFs incorporating amino acids were used as stationary phases for a variety of high-performance liquid chromatography (HPLC) enantiomeric separations [63]. Six MOFs were synthesised; four were mixed ligand MOFs incorporating both an amino acid and non-amino acid ligand, while MOF-1 had only L-tyrosine as ligand and MOF-6 only L-glutamic acid as ligand. MOF-1 and MOF-6 were determined to be the most efficient stationary phases for the separations tested.

12.3.2 Aliphatic Diacids

Fumaric acid is a naturally occurring diacid that was produced from sugars via fungal fermentation before fossil fuel-based production, which still dominates [64]. A wide range of robust MOFs has been synthesised with fumaric acid. A well-known Fe(III) fumarate MOF was originally reported as MIL-88 [65] and later renamed MIL-88A, when an isoreticular series was developed [66]. MIL-88A can be synthesised hydrothermally [67, 68]. MOF-801 (Figure 12.3c) is a Zr(IV) fumarate MOF that has excellent properties as a water sorbent. In a comparison with 23 water-sorbent materials, including 20 MOFs, MOF-801 displayed the best performance based on water stability, sorption/desorption recyclability and regeneration at room temperature [71]. MOF-801 has been prepared by solvent-free mechanochemistry

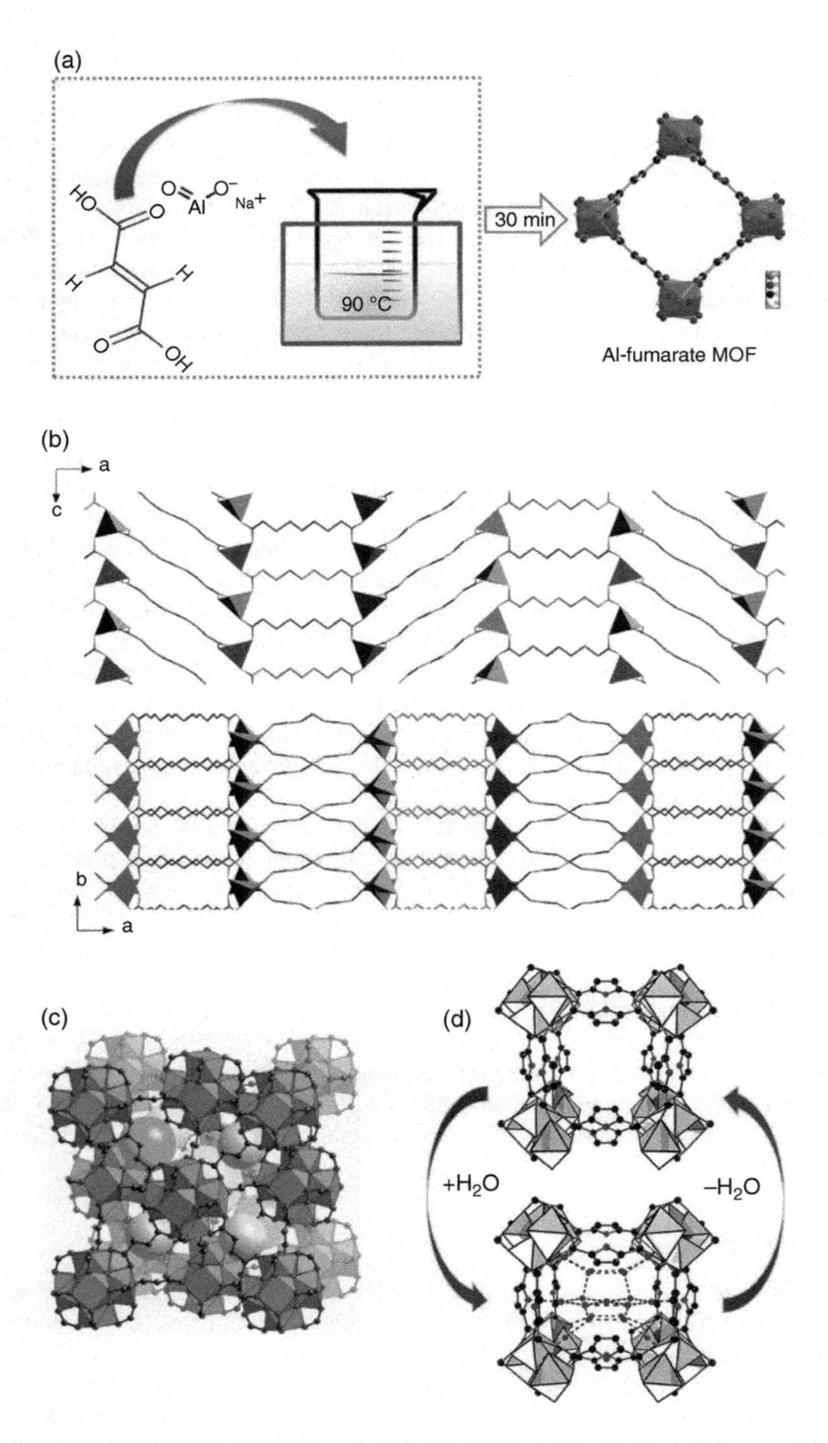

Figure 12.3 *Examples of MOFs with bio-based aliphatic diacids: (a) Synthesis of Al-fumarate MOF 'nano-flakes' in water at 90 °C. Source: Reprinted with permission from Elsevier. Zhou et al. [69]. (b) Zn(II) azelate MOF BioMIL-5, with Zn(II) metal centres represented as polyhedral structures. Source: Reprinted with permission from the Royal Society of Chemistry. Tamames-Tabar et al. [70]. (c) MOF-801 with Zr(IV) metal centres represented as polyhedral shapes and voids represented as spheres. Source: Reprinted with permission. Copyright (2014) American Chemical Society. Furukawa et al. [71]. (d) Hydrated and anhydrous versions of MIL-160. Source: Reprinted with permission from John Wiley and Sons. Wahiduzzaman et al. [72].*

using a planetary mill [23]. Al(fumarate)(OH) is a remarkable bio-based MOF with good water sorption and relevant to heat transfer applications. In one of the most important advances in the green scalable synthesis of MOFs, ton-scale production of Al(fumarate)OH was achieved in an aqueous medium at a space–time yield of more than 3600 kg m^{-3} day^{-1} [28]. The specific surface area and crystallinity of the product obtained from aqueous synthesis were comparable to those obtained from the synthesis in DMF. Al(fumarate)(OH) has also been synthesised by twin-screw extrusion [22] as well as by a facile synthesis from water at 90 °C to generate nano-flakes (Figure 12.3a) [69].

Glutaric acid is a highly water-soluble linear aliphatic dicarboxylic acid. BioMIL-2 is a Ca glutarate MOF obtained by the conventional solvothermal synthesis in DMF [73]. Both a 3D hydrated version and a 2D anhydrous version of BioMIL-2 are known to exist. The anhydrous phase was reported as the first Ca MOF with permanent porosity.

Azelaic acid is a naturally occurring dicarboxylic acid with antibacterial and anti-inflammatory properties. Industrially, it is mainly synthesised by the ozonolysis of oleic acid [74]. BioMIL-5 (Figure 12.3b) is synthesised from a Zn(II) salt and azelaic acid by a hydrothermal method and is active against *Staphilococcus* bacteria [70].

Itaconic acid, or 2-methylenesuccinic acid, is still considered a niche commodity chemical but is readily available in large quantities from industrial fermentation [75]. It has been underutilised as a bio-based ligand for MOF synthesis. A rare luminescent 2D Ca MOF, with an open framework structure, has been reported with itaconic acid as a ligand [76].

2,5-Furandicarboxylic acid (FDC) is a leading bio-based commodity chemical and a useful alternative to terephthalic acid because of their similar molecular structures. Non-toxic Zn(II) FDC MOFs have been investigated as potential drug delivery materials and 5-fluorouracil has frequently been tested as a model drug compound for these systems [77–79]. MIL-160 (Figure 12.3d) is a well-studied MOF based on Al and FDC. Similar to a predecessor based on fumaric acid, MIL-160 has been identified as a potential material for sorption-based heat transfer applications [80]. Synthesis under greener, scalable reaction conditions has been demonstrated [81]. Within a series of five Mg MOFs based on dicarboxylate ligands, CPM-203, incorporating FDC, exhibited the highest CO_2 uptake and good CO_2/CH_4 selectivity [82].

12.3.3 Cyclodextrins

The native CDs are cyclic oligosaccharides comprising 6 (α-CD), 7 (β-CD), and 8 (γ-CD, Figure 12.4a) glucosidic units, and are produced enzymatically from starch. Structurally, CDs resemble truncated cones. The inner cavity of a CD molecule is hydrophobic, while the upper and lower rims are hydrophilic. CDs have been explored extensively in drug delivery, hygiene, and household product applications.

The first MOFs incorporating CDs as ligands (CD-MOFs) were reported in 2010. They were based on γ-CD with Na, K (Figure 12.4b–d), Rb and Cs as metal centres [83]. Since then, a variety of novel α, β, and γ-CD MOF structures have been reported, including with Fe(III) as a metal centre [84]. CD-MOFs have been investigated for a variety of potential applications including molecular separations [85, 86], drug delivery [84, 87], and biomedicine [88].

While the physical properties of CD-MOFs make them attractive for certain applications, their poor hydrolytic stability has been identified as a severe limitation. Post-synthetic modification with cholesterol is one approach that has been used to improve aqueous stability [89].

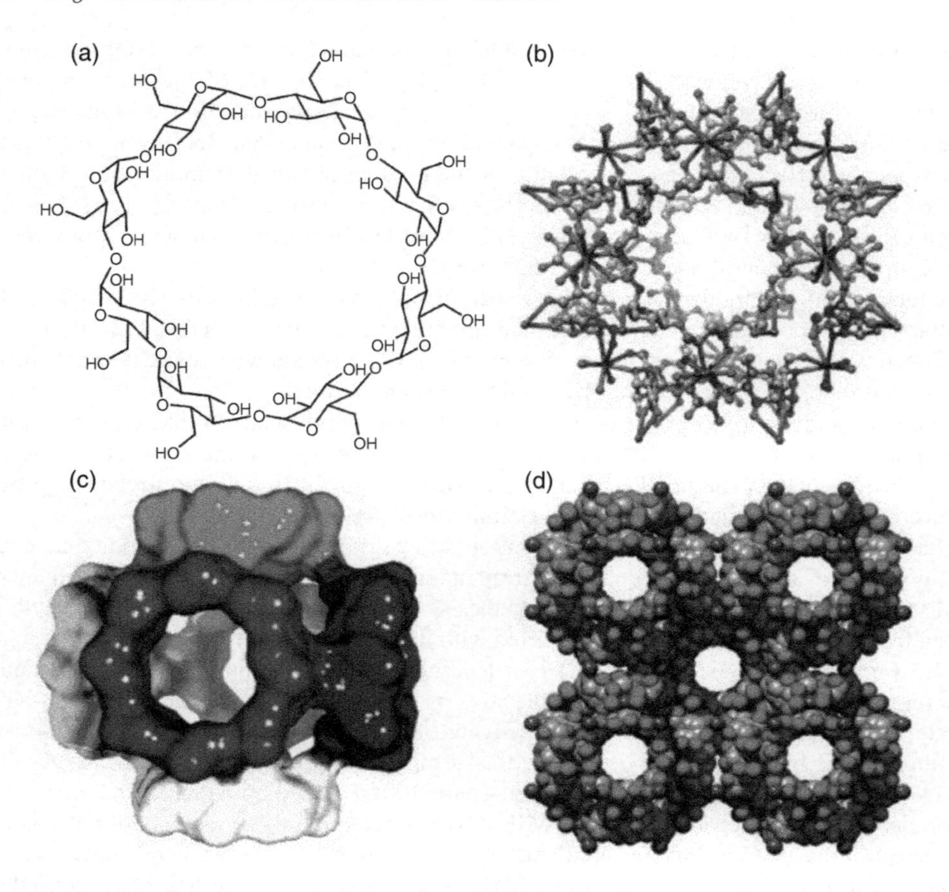

Figure 12.4 *(a) Molecular structure of g-CD. (b) Single crystal X-ray structure of CD-MOF-1 incorporating K and γ-CD: ball-and-stick model of the cubic repeating motif; (c) Cuboidal orientation of the six γ-CD tori, illustrating the 1.7 nm-sized pore at the centre of each* $(\gamma\text{-CD})_6$*-repeating motif, where* K^+ *ions have been omitted for clarity. (d) Space-filling representation of the extended solid-state structure, with body-centred cubic packing arrangement. Source: Reprinted with permission from John Wiley and Sons. Smaldone et al. [83].*

12.3.4 Other

Nicotinic acid, also known as niacin, is a form of Vitamin B3 that was used to prepare BioMIL-1, reported as the first therapeutically active MOF [90]. BioMIL-1 was originally prepared from an Fe(III) salt by solvothermal synthesis; although crystal structure analysis of the MOF revealed mixed valence metal clusters. Within the unit cell, there are three Fe(III) and two Fe(II) sites. The proposed therapeutic delivery of Vitamin B3 via BioMIL-1 is based on the degradation of the MOF under physiological conditions. Elsewhere nicotinic acid MOFs incorporating Co(II) and Ni(II) were prepared by high temperature (180 °C) hydrothermal synthesis [91]. These MOFs were explored for the therapeutic delivery of NO.

Lactic acid has been used as a MOF ligand. The first Ca L-lactate MOFs, MOF-1201, and MOF-1203 were explored as degradable carriers for the agricultural fumigant cis-1,3-dichloropropene [92]. As a demonstration of 'upcycling' polylactic acid waste, MOF-1201 and MOF-1203, as well as a

Table 12.1 *Examples of bio-based MOFs for which green or near-green synthesis has been achieved.*

MOF	Ligand	Metal	Synthesis	Reference
MIP-202	Aspartic acid	Zr(IV)	Reflux in water at ambient pressure	[95]
ZnGlu	Glutamic acid	Zn(II)	Mixing in aqueous alkaline solution	[54]
Al(fumarate)OH	Fumaric acid	Al(III)	Aqueous solution synthesis	[28, 69]
			Twin screw extrusion	[22]
MOF-801	Fumaric acid	Zr(IV)	Twin screw extrusion	[22]
MIL-88A	Fumaric acid	Fe(III)	Hydrothermal synthesis	[2, 6, 11, 14, 23, 26, 54, 57, 67, 68, 87, 94, 96–104]
BioMIL-5	Azelaic acid	Zn(II)	Hydrothermal synthesis	[70]
Poly[aqua-itaconatocalcium(II)]	Itaconic acid	Ca	Crystal growth by gel diffusion	[76]
MIL-160	2,5-Furandicarboxylic acid	Al(III)	Reflux in aqueous solution at ambient pressure	[80]
CD-MOFs	α-CD, β-CD, γ-CD	Na, K, Rb, Cs, Fe	Solution synthesis under ambient conditions	[81, 82]

Zn L-lactate MOF, $[[Zn_2(bdc)(L\text{-lactate})(dmf)]\cdot dmf]_n$, where bdc = 1,4-benzenedicarboxylate, dmf = *N,N*-dimethylformamide (ZnBLD) have been synthesised directly from polylactic acid drinking cups [93].

Nucleobases have frequently been combined with carboxylates or dicarboxylates to prepare open framework MOFs [72]. Bio-MOF-1 is a Zn(II) framework with adenine and biphenyldicarboxylic acid as ligands [83]. As an anionic framework, it has been shown to be a good host for cationic drug molecules. Remarkable biomimetic photoreduction of CO_2 has been demonstrated using AD-MOF-1 and AD-MOF-2, which are mixed ligand Co(II) MOFs incorporating adenine [94].

12.3.5 Exemplars: Bio-based MOFs Obtainable via Green Synthesis

There are a limited number of MOFs that are both bio-based and for which green or near-green synthesis has been demonstrated. Examples of such MOFs are highlighted in Table 12.1.

12.4 Metal Ion Considerations

Green MOF development requires non-toxic, economic, and earth-abundant metal ions. Currently, metal nitrates and chlorides are used in most synthetic protocols; even though they are considered highly toxic, oxidative, and present a risk of explosions [7]. Moreover, the production of these metal salts often generates metal salt waste, which is a major source of pollution. To circumvent these metal salts, scientists and industries have adopted the use of greener alternatives, for example, BASF used sulphate precursors for the production of already mentioned Basolite A520. Certainly, sulphates and acetates represent better alternatives, but they have processability challenges such as poor solubility and hydrolysis. Other precursor materials such as carbonates, hydroxides and oxides, often accessible in minerals, have greener traits as they can react to generate water and CO_2 as by-products. Challenges are related to their lack of solubility and alternative synthetic requirements, such as mechanochemistry.

The selection of environmentally benign metal ions should not only consider MOF synthesis, but also their disposal after MOF decomposition. In fact, toxicology data explored for biological-focused MOFs suggest that metal ions are indeed an environmental concern. To minimise metal ion toxicity, alkaline earth metals should replace transition metals. According to their toxicological profile, Ca, Mg, Mn, Fe, Ti, and Zr are the most suitable materials as they have an oral LD50 of $>1\,g\,kg^{-1}$. Certainly, further studies are required to understand the toxicity and lifecycles of MOFs [6].

12.4.1 Calcium

As a precursor for green MOFs, $CaCO_3$ is excellent since it is commonly found in minerals and it produces only water and carbon dioxide as by-products when reacted with acids. Ca MOFs are important due to their low cost and biocompatible metal centres. Naturally, Ca MOFs have been relevant in biological applications. Shi et al. demonstrated that the application of Ca MOFs promotes bone mineralisation with no signs of toxicity [105]. Indeed, Ca MOFs can be regarded as supplements, while their structure facilitates the delivery of drugs. Examples of drugs delivered by Ca MOFs include curcumin [106] and flurbiprofen [107]. The applicability of Ca MOFs is not limited to biomedical applications. As mentioned earlier, MOF-1201 and MOF-1203 were used for encapsulation and slow release of the agricultural fumigant cis-1,3-dichloropropene [92]. In another instance, a Ca sulfonyldibenzoic acid MOF was used to adsorb CO_2 from a CO_2/N_2 gas mixture [108] and for the uptake of Xe [109]. The sorption capabilities of a non-toxic Ca fumarate MOF (CaFu) were further utilised for the removal of fluoride from brick tea. CaFu showed >80% fluoride removal in only five minutes at room temperature [100]. Ca MOFs have also shown fluorescent properties; Kojima et al. [101] synthesised a MOF using Ca acetylacetonate and 1,4-naphthalenedicarboxylic acid that showed unique fluorescence at 395 nm and room temperature. This fluorescence was attributed to a ligand-to-metal charge transfer transition.

12.4.2 Magnesium

Mg offers abundance, biocompatibility, and flexibility due to its good mix of covalency and ionicity in Mg–O bonds [110]. In fact, Mg^{2+} is considered the most popular alkaline earth metal since it has similar structural chemistry to Zn^{2+}. Studies related to different synthetic methodologies [45] and functionalisation [111] have been reported across the literature. Mg MOFs have been applied to gas adsorption, photoluminescence, and catalysis. [112] For example, Mg-MOF-74, with formula Mg_2(2,5-dihydroxyterephthalate)$(H_2O)_2 \cdot 8H_2O$ and 1-D hexagonal channels of ⊠1.1 nm in diameter, displayed excellent gas sorption behaviour. In another case, Mg_2(pyrazole-3,5-dicarboxylate)$_2(H_2O)_4 \cdot H_2O$ was used for selective H_2/N_2 sorption, achieving 0.56 wt.% at 77 K [112]. This Mg MOF also displayed fluorescence at 480 nm and catalytic properties for aldol condensation reactions of aromatic aldehydes with ketones. Indeed, Mg MOFs have exhibited a variety of properties; for example, Ramaswamy et al. [113] studied a layered Mg-carboxyphosphonate framework (PCMOF [proton-conducting metal organic framework]10), which had a high proton conductivity ($3.55 \times 10^{-2}\,S\,cm^{-1}$) and water stability attributed to Mg phosphonate bonding; while Zhai et al. [114] studied the performance of a Mg(H_2EBTC)(DMF)$_2$ which displayed simultaneous ligand-based fluorescence and phosphorescence.

12.4.3 Manganese

Mn MOFs have been shown to have low toxicity and green traits and are often studied along with Mg MOFs, as in the case of MOF-74. Mn is ubiquitous for living organisms and its various oxidation states are necessary for many biological functions; even so, little is known about the toxicity of Mn at elevated exposures [115]. Mn MOFs have shown excellent catalytic and bio-based features. For instance, Lang et al. [102] synthesised Mn MOFs using Mn acetate and p-phthalic acid and applied these as gas electrode catalysts for lithium–oxygen batteries, displaying coral-like structures and good oxidation capabilities. In drug delivery, MOFs have typically been used to encapsulate pharmaceutical ingredients to control the release of drugs but Andre et al. [116] utilised the pharmaceutical nalidixic acid as a linker in an Mn MOF. In this manner, the water solubility of nalidixic acid was enhanced, enabling antimicrobial activity, while Mn ions displayed an adequate cytotoxicity. Mn MOFs have also been used in non-biological applications. Wang et al. [117] synthesised a layered structure Mn(2,3,5,6-tetrafluoroterephthalic acid)(4,4′-bipyridine)(H_2O)$_2$ as an electrode material for supercapacitors, displaying high specific capacitance levels and good cycling stability. In a different study, an Mn_2(2,5-dioxido-1,4-benzenedicarboxylate) was used as a material applied for CO_2 electrodes in Li–CO_2 batteries, displaying a remarkable discharge capacity and a low charge potential retention [104].

12.4.4 Iron

Fe MOFs have been shown to be relevant in the area of environmental remediation. Liu et al. [118] listed various Fe MOFs applied for contaminant degradation (e.g. various Fe-MIL series, Fe_2O_3@Zn-MOF-74, and Fe–Co PBAs [Prussian blue analogue]) and contaminant adsorption (e.g. Fe-BTC, various Fe-MIL series). In particular, Fe-MIL structures have been extensively studied not only for adsorption and catalysis, but also for bio-based applications. MIL-88A with fumarate as linker (Figure 12.5); MIL-53, MIL-88B, and MIL-101 with a

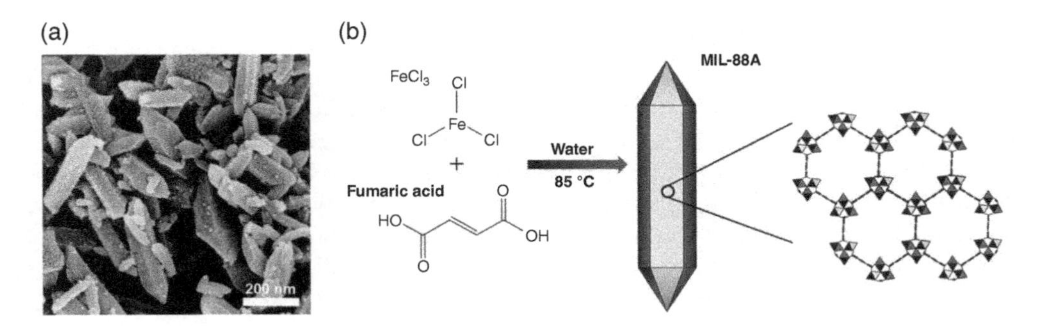

Figure 12.5 (a) Scanning electron micrograph of MIL-88A crystallites and (b) schematic depiction of the synthesis of MIL-88A under relatively mild conditions with only water as a solvent, according to Lin et al. [68] MIL-88A was originally prepared by Serre et al. [65]. Source: Reprinted with permission from the Royal Society of Chemistry. Lin et al. [68].

terephthalate as linker are some of the most important MOFs from this series; nonetheless MIL-100 with 1,3,5-benzene tricarboxylate as linker is considered the most popular MOF. [119] Fe-MIL-100 has low toxicity, safe dosage intake (80 μg ml^{-1}) [120], and excellent biocompatibility; it has been applied as a drug carrier [121], biocompatible oral nanocarriers [122], and others. Fe has also been combined with active pharmaceutical ingredients as linkers for drug delivery, as in the case of BioMIL-1, which contains nicotinic acid and has been described earlier in this text [90]. Beyond biocompatibility and excellent adsorption features, Fe MOFs have shown to promote electrical conductivity. Sun et al. measured the electrical conductivity of twenty different MOFs; Fe ions led to a five orders of magnitude higher electrical conductivity and smaller charge activation energies [123]. Leveraging the high-energy valance electrons and mixed valency of Fe can be used as a strategy for improving electrical conductivity.

12.4.5 Titanium

Ti MOFs have been considered for applications such as photocatalysis, CO_2 reduction, and pollutant degradation due to their optical and photoredox properties [124]. In particular, Ti MOFs are excellent candidates to replace TiO_2, a commercial reagent used in different industries, but classified as a carcinogenic substance. The first Ti MOF reported was MIL-91, which utilises a phosphonate ligand [125]. However, the strategy of combining the high valance of Ti with carboxylate linkers to generate rigid frameworks was introduced in 2009, through the synthesis of MIL-125, which showed excellent H_2 sorption properties [126]. This followed a systematic approach, based on the Ti(IV) isopropoxide clusters $(Ti_3O)(iPrO)_8]^{2+}$ and $[(Ti_4O_2)(iPrO)_6]^{6+}$ and carboxylate linkers to prepare a range of Ti MOFs. In 2013, a heterometallic approach using bimetallic clusters was used by Cui et al. [127] where a Ti-metallosalan linker was initially synthesised and used as a starting reagent to form a Cd Ti oxo MOF from biphenyl-4,4′-dicarboxylic acid and a Cd salt. Indeed, Ti MOFs have shown a variety of developments, especially due to their optical features; for instance, Nguyen et al. synthesised MOF-901 [128] and MOF-902 [129] with an exceptional optical response and low band gap energies at 2.65 and 2.50 eV, respectively. MOF-901/902 were used as photocatalysts to synthesise PolyMMA under visible light irradiation.

12.4.6 Zirconium

Due to their outstanding stability, properties and functions, Zr MOFs, are some of the most promising MOF materials for practical applications. Zr(IV) has an oxidation state with high charge density and bond polarisation, generating strong coordination bonds with carboxylates. In fact, Zr MOFs are considered stable in solvents and acidic aqueous solutions; in particular, UiO-66 has a 12-coordinated $Zr_6(\mu_3\text{-}O)_4(\mu_3\text{-}OH)_4(CO_2)_{12}$ cluster with unprecedented hydrothermal stability beyond most reported MOFs. Zr MOFs have been synthesised differently; Bai et al. [130] depicted the four most relevant (modulated synthesis, isoreticular expansion, topology-guided design, and post-synthetic modification) as well as the types of Zr clusters used across the literature (Zr_6O_8, Zr_8O_6, ZrO_6, ZrO_7, and ZrO_8). Due to the excellent stability and low toxicity of Zr MOFs, they have been used for separation, drug delivery, electrochemistry, [131] fluorescent sensing, and others; even so, catalysis has been a field particularly benefited by Zr MOFs.

The aforementioned metal ions have the lowest reported toxicity for biological systems; nevertheless, the 'green' credentials of MOFs are heavily correlated to the application of the material. For example, consider a hypothetical MOF with unforeseen CO_2 sorption and catalytic capabilities. This MOF can facilitate carbon capture and storage while transforming it into useful chemicals, thus providing significant sustainability gains, even if its starting metal precursor exhibits some toxicity if consumed. Therefore, beyond toxicity, 'earth abundance' of the metal precursor is a good alternative proxy for identifying green sustainable metal ions for the development of MOFs. The most earth-abundant metal ions considered for MOF development are: Al (the most abundant) > Fe (third most abundant) > Ca (fifth most abundant) > Mg (seventh most abundant) > Ti (eighth most abundant) > Zn > Zr. From this list, the properties of Al– and Zn MOFs need to be highlighted.

12.4.7 Aluminium

Al MOFs are stable and highly porous structures that can be synthesised using water as a solvent. The structural characterisation of Al MOFs has been difficult as they are often produced as microcrystalline materials [132]. Even so, the charge of the metal ion and small ionic radius give Al MOFs excellent chemical stability, specially under hydrothermal conditions. In fact, Samokhvalov dedicated a review to the applicability of Al MOFs for sorption in solutions [133]. Férey and coworkers have extensively analysed the properties of Al MOFs and contributed with the development of the MIL-series utilising Al, such as MIL-69 [134], MIL-53 [135], MIL-96 [136], and others. From this series, MIL-53-FA has been commercialised as Basolite A520 and used for natural gas storage. Indeed, the sorption capabilities and stability of Al MOFs is evident in other structures. For example, Feng et al. synthesised three Al MOF structures (BIT-72, 73, and 74) using terephthalic acid ions decorated with monohydroxyl, monomethyl, and dimethyl groups, respectively [103]. These structures displayed good CO_2/N_2 selectivity and high stability when exposed to boiling water and acidic conditions.

12.4.8 Zinc

In 2000, Yaghi and coworkers used N,N′-dimethylformamide, 1,4-benzenedicarboxylate to form MOF-2, MOF-3, and MOF-5 [137]. MOF-5, in particular, was a structure that inspired contemporary MOF research. Bridging two metal ions with four carboxylic groups is known as 'paddle-wheel' and it has been exploited to form mixed ligand MOFs [138]. Certainly, Zn MOFs have good stability; nonetheless, those formed with nitrogen-containing ligands display an even superior stability than those formed with carboxylic acid linkers [139]. Zn MOFs have been studied for a variety of applications, including gas separation. For example, UPC-12 showed a selective affinity towards CO_2 due to the formation of hydrogen bonds between the gas molecules and COOH groups [140]. Similarly, Zn MOF with 1D rectangular channels, decorated with imidazole and methyl groups, displayed a high selectivity towards CO_2 at low pressures. This selectivity is attributed to the synergistic effects of imidazole, methyl functional groups, and the contracted pores of the MOF [141]. Beyond their gas sorption capabilities, Zn MOFs have been studied for biomedical applications, as they also display reduced toxicity. In fact, in a review, Bahrani et al. [139] outlined the use of Zn MOFs for different bio-based applications. Some examples of these applications include the use of

Zn MOFs as drug carriers or nano-capsules (e.g. ZIF-8 [142], Bio-MOF-1 [143], IFMC-1 [144], CMC/MOF-5/GO [98], and Zn-TBDA [145]), as magnetic resonance imaging (MRI) agents (RMOF-3 and ZIF-90 as contrast agents) [146], and as chemo/biosensors (e.g. Bio-MOF-1) [143].

12.5 Challenges for Further Development Towards Applications

The implementation of clean and sustainable methodologies for the development of MOFs has been slow, especially when compared with the synthesis of other materials. An initial step towards the implementation of green chemistry principles in MOFs includes identifying sustainable precursor materials and lab synthetic paths. Nonetheless, to accelerate the commercialisation and deployment of industrially relevant MOFs, challenges related to stability, scalability and cost, market adoption, and competitiveness need further development.

12.5.1 Stability Issues

The wide range of applications that MOFs can offer are typically limited by issues related to their chemical, thermal, hydrothermal, and mechanical stability. Stability is widely defined as the resistance of the MOF structure to degradation [96]. Efforts towards improving their stability have been significantly increased to enable the expansion of MOF applications in different areas. According to the Web of Science, the scientific literature on 'MOF stability' has increased by 48% every year since 2011. Stability studies have focused on improving relevant functional stability properties to match the envisaged MOF application. For example, thermal stability studies are crucial for MOFs applied to gas separation, ion exchange, and high-temperature catalytic processes [95, 147]; mechanical stability for industrial processes where MOFs are compacted into pellets and are required to withstand high pressures [148]; hydrous and anhydrous proton transport stabilities for adsorption of corrosive gases and drug delivery applications [97].

12.5.1.1 Chemical Stability

Although MOFs can be tailored for chemical stability, exceeding the performance of well-known stable structures has represented a challenge [99]. Issues related to chemical stability begin with testing protocols, which have been inconsistent across the literature. Powder X-ray diffraction (PXRD) is often conducted to determine any structural changes of the sample before and after its exposure to aqueous solutions. Although this test is performed across the literature, no standardisation has been implemented to determine the soaking time in the aqueous solution, temperature, or degree of PXRD change. As a complementary technique, gas sorption tests have been conducted to estimate the MOF surface area and measure the degree of pores collapsing and structural amorphisation.

The chemical stability of MOFs in the presence of aqueous solutions or vapour has been extensive since water can cause hydrolysis at the metal-linker nodes. Studies of chemical stability have been limited to hydroxide ions and poorly coordinating acidic and alkaline aqueous solutions. Howarth et al. [96] presented a list of MOFs and their stability at different pH ranges; the most stable include UiO-66-NO_2 (pH range 0–14), Ni_3(BTP) (pH range 2–14), and PCN-426-Cr_{III} (pH range 0–12). Water stability has shown to increase in different cases; for example, the stability of ZIF-8 relies on water exclusion from the pores, which occurs due to the hydrophobic nature and small size of the pores. MOFs presenting metal-node shielding, using positioned hydrophobic linkers, and high bond strengths, also have good water stability. In fact, in azolate MOFs, high pK_a values for nitrogen deprotonation are associated with greater hydrolytic stability of the MOFs. Water-permeable MOFs such as $Ni_3(BTP)_2$ are stable due to their high metal–nitrogen bond strength, while MIL compounds have trivalent metals bonded with strong oxygen anion-terminated linkers.

12.5.1.2 Thermal Stability

Thermally stable MOFs are required for industrial applications. This stability is defined as the ability to avoid irreversible changes in its chemical and physical structures upon heating. The thermal stability of MOFs can occur via amorphisation, melting, node-cluster dehydration, linker dehydrogenation, or graphitisation, all of which result from node–linker bond breakage and linker combustion. These changes are often determined through thermogravimetric analysis (TGA) and in situ PXRD to monitor phase transformations and changes in crystallinity. Secondary tests include CO_2 or cryogenic N_2 sorption and desorption after thermal cycling and differential scanning calorimetry (DSC).

Although some MOFs have shown high thermal stability (e.g. Zr-UiO-66 is stable at temperatures higher than 500 °C); various strategies have been implemented to increase the strength of the node-linker bond and thus enhance the thermal stability of MOFs, including:

1. Using equivalent oxyanion-terminated linkers with higher-valency metal centres. Metal ions in their most stable oxidation state display higher thermal stability [149]. This strategy often results in lower crystal size and crystallinity, complicating the MOF structural characterisation.
2. Changing the composition of linker pendant groups. Makal et al. [150] developed NbO MOFs with various pendant alkoxy groups. In situ synchrotron PXRD revealed that shorter pendant alkoxy groups displayed enhanced thermal stabilities.
3. Framework catenation and densification. In this strategy, by increasing the crystal density increases, more stable polymorphs are attained, enhancing the thermal stability of the material. Unfortunately, densification can reduce desired traits such as large pore volumes.
4. Changes in MOF node structure under thermal dehydration.

12.5.1.3 Hydrothermal Stability

Atomic layer deposition is used for thin-film manufacturing; it requires hydrothermally stable components, as they are exposed to steam at high temperatures. In MOFs, hydrothermal stability strongly depends on the node–linker bond strength. Among the green metal ions previously mentioned, Al^{3+} Fe^{3+} and Zr^{3+} form strong bonds with carboxylates, thus displaying high hydrothermal stability; in contrast, Zn ions are often highly unstable. Other factors that affect the hydrothermal stability include (i) the geometry of the secondary building units, as in the case of MIL-140, which is based on a ZrO-chain subunit; (ii) intermolecular or intramolecular forces (e.g. hydrogen bonding or π stacking); and (iii) the presence of hydrophobic functional groups which hinder water sorption.

12.5.1.4 Mechanical Stability

One of the main advantages of MOFs is their large pores, which, in turn, weakens their mechanical structure. Mechanical stability in MOFs is defined as the ability to avoid structural changes when subject to mechanical loadings, which can cause phase changes, partial collapse of pores, and/or amorphisation. The mechanical stability of MOFs has been assessed by treating the sample using a ball mill or hydraulic pellet press and measuring any changes in crystallinity. Nanoindentation measurements can give information related to the hardness and Young's modulus of the structures. The presence of defects on MOFs affects their mechanical stability; for instance, it was shown that an ideal Hf-UiO-66 has similar shear and bulk moduli to zeolites [151]. Nonetheless, most mechanical instabilities are caused by the removal of pore-filling solvents from the pores of MOFs causing changes in crystallinity. Solvent removal causes capillary force-driven destruction of the structure, which can be avoided by (i) utilising liquid CO_2, for solvent exchange, followed by the removal of CO_2 at supercritical conditions; or (ii) attaching a hydrocarbon or fluorocarbon chain to the nodes to avoid hydrogen-bonding sites between them and the solvents. In general, high coordination metal nodes, strong node-linker bonds, short and rigid linkers, and dense structures with minimised porosities result in enhanced mechanically stable MOFs.

12.5.2 Scalability and Cost

The successful deployment of processes capable of producing MOFs on a large scale with encouraging economic and environmental features relies on the development of continuous production methods; an approach applied to green synthetic methodologies. Compared to batch manufacturing, continuous operations offer higher space–time yields, effectively reducing capital and operating costs. In an attempt to produce a continuous modus operandi, near-critical solvent conditions were utilised in the manufacture of Al-MIL-53, where a space–time yield of $1300\,kg\,m^{-3}\,day^{-1}$ was achieved [152]. Other solvothermal approaches include microwave heating and sonication, which have been shown to reduce energy consumption. Furthermore, the combination of flow reactors with spray drying has shown promising results for shaping MOFs into spherical structures [153]. Nonetheless, the most promising scale-up methodology is a solid-state flow mechanochemical process (reactive extrusion), which has been able to produce HKUST-1, ZIF-8, and Al(fumarate)(OH)) in multi-$kg\,hour^{-1}$ quantities, with little or no solvent [22].

MOF manufacturing costs are dictated by their technological application. For instance, vehicles utilising natural gas require a maximum MOF cost of $\$10\,kg^{-1}$, while hydrogen storage costs cannot exceed $\$18\,kg^{-1}$, overall. DeSantis et al. [154] assessed the economic performance of Ni-MOF-74, Mg-MOF-74, MOF-5, and HKUST-1 manufactured using a solvothermal process, mechanochemical LAG, and an aqueous process. Each manufacturing methodology was comparable to the actual steps of synthesis (e.g. preparation of reactants, synthesis, filtration, and drying) and all costs associated with the metal salt, linker, solvent, precipitation, initial filtration, wash and filter 1, wash and filter 2, rotary drying and activation, particle compaction, and contingency manufacturing cost were considered. This work proposed that solvent utilisation is the biggest drawback for solvothermal MOF manufacturing. LAG and the aqueous process showed a better economic performance, which, if optimised, could generate MOFs under $\$10\,kg^{-1}$.

12.5.3 Competing Alternative Materials

The competitive advantage of MOFs against other materials depends on their specific application. Alternative porous materials range from well-established materials such as zeolites and activated carbons to novel ones such as covalent organic frameworks, porous organic polymers, and porous molecular solids. Slater et al. compared various porous structures according to their functionality [155]. Classifying porous materials according to their structure and function characteristics (e.g. crystallinity, porosity, stability, diversity, processing, and designability), it is possible to distinguish that MOFs have a large structural diversity, but reduced temperature stability when compared to zeolites. Additionally, their processability has limitations such as their lack of solubility, although many recent examples of composites and films have enhanced this feature. Alternatively, the degree of long-range order and the relative strength of the bonds between the atomic or molecular building blocks can also be used to classify the functionality of porous materials. In this manner, MOFs are considered crystalline structures with uniform pores, but unstable to water; whereas porous polymer networks are amorphous but robust. The unique functional drivers of MOFs include their structural and chemical control towards a diverse range of materials.

An understanding of structure–property relationships in MOFs has enabled the modular design of functional porous materials; even though attaining the desired range of traits still represents a challenge (e.g. crystallinity provides order but leads to brittleness). Compared to alternative porous materials, the modular designability of MOFs is superior. A similar trait is found in porous organic cages, where their strong noncovalent intermolecular interactions can be used to control structure and composition. Compared to other materials, the main advantages and disadvantages of MOFs are summarised in Table 12.2 [155].

Table 12.2 *Comparison between MOFs and alternative materials.*

Property	MOFs	Alternative materials
Porosity and surface area	MOFs have high surface areas and pore volumes but low pore stability.	Amorphous polymers have high surface areas and high stability but are not applicable for CO_2 capture or molecular separations.
Selectivity	MOFs are selective in molecular separations and their functionality can be controlled.	Zeolites are selective in molecular separations but their functionality cannot be controlled. Amorphous polymers are selective in molecular separations but their design remains empirical.
Diffusion kinetics	Pore sizes can be tailored to reduce meso- and micropore resistance to diffusion.	Polymers have good diffusion kinetics but permeabilities can decrease over time.
Processability	MOFs have poor solution processability and when shaped as mixed matrix membranes. Non-selective channels are formed at phase interfaces between the organic polymer and filler.	Porous polymers have excellent processability into mixed matrix membranes due to their high compatibility with other polymers.
Stability	MOFs have reduced hydrothermal stability.	Porous organic polymers and porous molecular crystals can have excellent hydrolytic stability.

Source: Adapted from Slater and Cooper [155].

12.6 Conclusion

While early approaches to MOF synthesis were dominated by solvothermal methods, significant research efforts are now devoted to developing green synthesis of MOFs. With consideration for safety and environmental risks, the preferable conditions for industrial implementation of MOF synthesis include open reactors at ambient pressure, temperatures below reflux, and solvent-free conditions or water as solvent. The cost has been identified as the main barrier to industrial uptake. While many synthetic techniques, including sonochemical and electrochemical synthesis, have been explored, water-based solution synthesis and low- or solvent-free continuous synthesis by reactive extrusion are the most promising.

With respect to reactants, the metal salts used should be non-corrosive and safe, and organic ligands should be obtainable from renewable resources (bio-based ligands). Earth-abundant metals should be used where possible. Al- and Fe-based MOFs have been intensively researched for well over a decade and exhibit excellent properties as MOFs based on earth-abundant metals. Bio-based ligands that have been successfully incorporated into MOFs include amino acids, certain aliphatic diacids, CDs, and nucleobases. There remains plenty of scope for the exploration of new bio-based ligands. The costs and availability, in large quantities, of these bio-based ligands will be an important factor affecting the prospects for sustainable industrial manufacture of bio-based MOFs.

The modular design, structural diversity, and long-range order of MOFs are major advantages compared with existing alternative materials. However, the competitive advantage of MOFs against materials such as zeolites and organic polymers is highly dependent on the envisioned application. Molecular separations are a promising application that has been researched for many years. A more recent, rapidly growing area of focus is biomedical applications, where benign bio-based MOFs are particularly interesting. Niche applications where MOFs have a potential competitive advantage, such as in functional composite materials, will continue to emerge. Overcoming important weaknesses, such as poor chemical stability, high cost, and synthesis challenges, will strengthen the case for implementing MOFs, in general, and bio-based MOFs, in particular, ahead of competing for alternative materials.

References

1. Batten, S.R., Champness, N.R., Chen, X.-M. et al. (2013) Terminology of metal–organic frameworks and coordination polymers (IUPAC recommendations 2013), *Pure and Applied Chemistry* 85(8): 1715–1724. doi: https://doi.org/10.1351/PAC-REC-12-11-20.
2. Imaz, I. Rubio-Martínez, M., An, J. et al. (2011) Metal-biomolecule frameworks (MBioFs), *Chemical Communications* 47(26): 7287–7302. doi: https://doi.org/10.1039/c1cc11202c.
3. McKinlay, A.C., Morris, R.E., Horcajada, P. et al. (2010) BioMOFs: metal-organic frameworks for biological and medical applications, *Angewandte Chemie International Edition* 49(36): 6260–6266. doi: https://doi.org/10.1002/anie.201000048.
4. Rojas, S., Devic, T. and Horcajada, P. (2017) Metal organic frameworks based on bioactive components, *Journal of Materials Chemistry B* 5(14): 2560–2573. doi: https://doi.org/10.1039/C6TB03217F.
5. Chen, J., Shen, K. and Li, Y. (2017) Greening the processes of metal–organic framework synthesis and their use in sustainable catalysis, *ChemSusChem* 10(16): 3165–3187. doi: https://doi.org/10.1002/cssc.201700748.
6. Julien, P.A., Mottillo, C. and Friščić, T. (2017) Metal–organic frameworks meet scalable and sustainable synthesis, *Green Chemistry* 19(12): 2729–2747. doi: https://doi.org/10.1039/C7GC01078H.
7. Reinsch, H. (2016) 'Green' synthesis of metal-organic frameworks, *European Journal of Inorganic Chemistry* 2016(27): 4290–4299. doi: https://doi.org/10.1002/ejic.201600286.

8. Reinsch, H. and Stock, N. (2017) Synthesis of MOFs: a personal view on rationalisation, application and exploration, *Dalton Transactions* 46(26): 8339–8349. doi: https://doi.org/10.1039/c7dt01115f.
9. Lv, D. Chen, Y., Li, Y. et al. (2017) Efficient mechanochemical synthesis of MOF-5 for linear alkanes adsorption, *Journal of Chemical and Engineering Data* 62(7): 2030–2036. doi: https://doi.org/10.1021/acs.jced.7b00049.
10. Prochowicz, D., K. Sokołowski, Justyniak, I. et al. (2015) A mechanochemical strategy for IRMOF assembly based on pre-designed oxo-zinc precursors, *Chemical Communications* 51(19): 4032–4035. doi: https://doi.org/10.1039/c4cc09917f.
11. Klimakow, M., Klobes, P., Thünemann, A. F. et al. (2010) Mechanochemical synthesis of metal-organic frameworks: a fast and facile approach toward quantitative yields and high specific surface areas, *Chemistry of Materials* 22(18): 5216–5221. doi: https://doi.org/10.1021/cm1012119.
12. Pichon, A. and James, S. L. (2008) An array-based study of reactivity under solvent-free mechanochemical conditions – insights and trends, *CrystEngComm* 10(12): 1839–1847. doi: https://doi.org/10.1039/b810857a.
13. Beldon, P.J., Fábián, L., Stein, R.S. et al. (2010) Rapid room-temperature synthesis of zeolitic imidazolate frameworks by using mechanochemistry, *Angewandte Chemie International Edition* 49(50): 9640–9643. doi: https://doi.org/10.1002/anie.201005547.
14. Huang, Y.H., Lo, W.-S., Kuo, Y.-W. et al. (2017) Green and rapid synthesis of zirconium metal-organic frameworks via mechanochemistry: UiO-66 analog nanocrystals obtained in one hundred seconds, *Chemical Communications* 53(43): 5818–5821. doi: https://doi.org/10.1039/c7cc03105j.
15. Užarević, K., Wang, T. C., Moon, S.-Y. et al. (2016) Mechanochemical and solvent-free assembly of zirconium-based metal-organic frameworks, *Chemical Communications* 52(10): 2133–2136. doi: https://doi.org/10.1039/c5cc08972g.
16. Friščić, T., Halasz, I., Štrukil, V. et al. (2012) Clean and efficient synthesis using mechanochemistry: coordination polymers, metal-organic frameworks and metallodrugs, *Croatica Chemica Acta* 85(3): 367–378. doi: https://doi.org/10.5562/cca2014.
17. Friščić, T. and Fábián, L. (2009) Mechanochemical conversion of a metal oxide into coordination polymers and porous frameworks using liquid-assisted grinding (LAG), *CrystEngComm* 11(5): 743–745. doi: https://doi.org/10.1039/b822934c.
18. Pan, Y., Liu, Y., Zeng, G. et al. (2011) Rapid synthesis of zeolitic imidazolate framework-8 (ZIF-8) nanocrystals in an aqueous system, *Chemical Communications* 47(7): 2071–2073. doi: https://doi.org/10.1039/c0cc05002d.
19. Mottillo, C., Lu, Y., Pham, M.-H. et al. (2013) Mineral neogenesis as an inspiration for mild, solvent-free synthesis of bulk microporous metal-organic frameworks from metal (Zn, Co) oxides, *Green Chemistry* 15(8): 2121–2131. doi: https://doi.org/10.1039/c3gc40520f.
20. Qi, F., Stein, R.S. and Friščić, T. (2014) Mimicking mineral neogenesis for the clean synthesis of metal-organic materials from mineral feedstocks: coordination polymers, MOFs and metal oxide separation, *Green Chemistry* 16(1): 121–132. doi: https://doi.org/10.1039/c3gc41370e.
21. Cliffe, M.J., Mottillo, C., Stein, R. S. et al. (2012) Accelerated aging: a low energy, solvent-free alternative to solvothermal and mechanochemical synthesis of metal-organic materials, *Chemical Science* 3(8): 2495–2500. doi: https://doi.org/10.1039/c2sc20344h.
22. Crawford, D., Casaban, J., Haydon, R. et al. (2015) Synthesis by extrusion: continuous, large-scale preparation of MOFs using little or no solvent, *Chemical Science* 6(3): 1645–1649. doi: https://doi.org/10.1039/c4sc03217a.
23. Karadeniz, B., Howarth, A. J., Stolar, T. et al. (2018) Benign by design: green and scalable synthesis of zirconium UiO-metal-organic frameworks by water-assisted mechanochemistry, *ACS Sustainable Chemistry and Engineering* 6(11): 15841–15849. doi: https://doi.org/10.1021/acssuschemeng.8b04458.
24. Ahmed, I. Jeon, J., Khan, N. A. et al. (2012) Synthesis of a metal-organic framework, iron-benezenetricarboxylate, from dry gels in the absence of acid and salt, *Crystal Growth and Design* 12(12): 5878–5881. doi: https://doi.org/10.1021/cg3014317.
25. Lu, N. Zhou, F., Jia, H. et al. (2017) Dry-gel conversion synthesis of Zr-based metal-organic frameworks, *Industrial and Engineering Chemistry Research* 56(48): 14155–14163. doi: https://doi.org/10.1021/acs.iecr.7b04010.

26. Huang, X.-C., Lin, Y.-Y., Zhang, J.-P. et al. (2006) Ligand-directed strategy for zeolite-type metal-organic frameworks: zinc(II) imidazolates with unusual zeolitic topologies., *Angewandte Chemie International Edition* 45(10): 1557–1559. doi: https://doi.org/10.1002/anie.200503778.
27. Park, K. S. Ni, Z., Côté, A.P. et al. (2006) Exceptional chemical and thermal stability of zeolitic imidazolate frameworks, *Proceedings of the National Academy of Sciences of the United States of America* 103(27): 10186–10191. doi: https://doi.org/10.1073/pnas.0602439103.
28. Gaab, M. Trukhan, N., Maurer, S. et al. (2012) The progression of Al-based metal-organic frameworks – from academic research to industrial production and applications, *Microporous and Mesoporous Materials* 157: 131–136. doi: https://doi.org/10.1016/j.micromeso.2011.08.016.
29. Majano, G. and Pérez-Ramírez, J. (2013) Scalable room-temperature conversion of copper(II) hydroxide into HKUST-1 ($Cu_3(btc)_2$), *Advanced Materials* 25(7): 1052–1057. doi: https://doi.org/10.1002/adma.201203664.
30. Martins, G.A.V., Byrne, P.J., Allan, P. et al. (2010) The use of ionic liquids in the synthesis of zinc imidazolate frameworks, *Dalton Transactions* 39(7): 1758–1762. doi: https://doi.org/10.1039/b917348j.
31. Yang, L. and Lu, H. (2012) Microwave-assisted ionothermal synthesis and characterization of zeolitic imidazolate framework-8, *Chinese Journal of Chemistry* 30(5): 1040–1044. doi: https://doi.org/10.1002/cjoc.201100595.
32. Sang, X., Zhang, J., Xiang, J. et al. (2017) Ionic liquid accelerates the crystallization of Zr-based metal-organic frameworks, *Nature Communications* 8(1). doi: https://doi.org/10.1038/s41467-017-00226-y.
33. Cavka, J. H., Jakobsen, S., Olsbye, U. et al. (2008) A new zirconium inorganic building brick forming metal organic frameworks with exceptional stability, *Journal of the American Chemical Society* 130(42): 13850–13851. doi: https://doi.org/10.1021/ja8057953.
34. Liu, C., Zhang, B., Zhang, J. et al. (2015) Gas promotes the crystallization of nano-sized metal-organic frameworks in ionic liquid, *Chemical Communications* 51(57): 11445–11448. doi: https://doi.org/10.1039/c5cc02503f.
35. Pretti, C., Chiappe, C., Pieraccini, D. et al. (2006) Acute toxicity of ionic liquids to the zebrafish (Danio rerio), *Green Chemistry* 8(3): 238–240. doi: https://doi.org/10.1039/b511554j.
36. Thuy Pham, T.P., Cho, C.W. and Yun, Y.S. (2010) Environmental fate and toxicity of ionic liquids: a review, *Water Research* 44(2): 352–372. doi: https://doi.org/10.1016/j.watres.2009.09.030.
37. Zhao, D., Liao, Y. and Zhang, Z.D. (2007) Toxicity of ionic liquids, *Clean – Soil, Air, Water* 35(1): 42–48. doi: https://doi.org/10.1002/clen.200600015.
38. Carter, E.B., Culver, S.L., Fox, P.A. et al. (2004) Sweet success: ionic liquids derived from non-nutritive sweeteners, *Chemical Communications* 4(6): 630–631. doi: https://doi.org/10.1039/b313068a.
39. Tao, G. H. He, L., Sun, N. et al. (2005) New generation ionic liquids: cations derived from amino acids, *Chemical Communications* (28): 3562–3564. doi: https://doi.org/10.1039/b504256a.
40. Tao, G.-H. He, L., Liu, W. et al. (2006) Preparation, characterization and application of amino acid-based green ionic liquids, *Green Chemistry* 8(7): 639–646. doi: https://doi.org/10.1039/b600813e.
41. Zhang, B., Zhang, J. and Han, B. (2016) Assembling metal–organic frameworks in ionic liquids and supercritical CO_2, *Chemistry Asian Journal* 11(19): 2610–2619. doi: https://doi.org/10.1002/asia.201600323.
42. López-Periago, A., López-Domínguez, P., Barrio, J.P. et al. (2016) Binary supercritical CO_2 solvent mixtures for the synthesis of 3D metal-organic frameworks, *Microporous and Mesoporous Materials* 234: 155–161. doi: https://doi.org/10.1016/j.micromeso.2016.07.014.
43. Cintas, P. and Luche, J.-L. (1999) The sonochemical approach, *Green Chemistry* 1(3): 115–125. doi: https://doi.org/10.1039/a900593e.
44. Son, W.J., Kim, J., Kim, J. et al. (2008) Sonochemical synthesis of MOF-5, *Chemical Communications* (47): 6336–6338. doi: https://doi.org/10.1039/b814740j.
45. Yang, D.A., Cho, H.-Y., Kim, J. et al. (2012) CO_2 capture and conversion using Mg-MOF-74 prepared by a sonochemical method, *Energy and Environmental Science* 5(4): 6465–6473. doi: https://doi.org/10.1039/c1ee02234b.
46. Cho, H.Y., Kim, J., Kim, S.-N. et al. (2013) High yield 1-L scale synthesis of ZIF-8 via a sonochemical route, *Microporous and Mesoporous Materials* 169: 180–184. doi: https://doi.org/10.1016/j.micromeso.2012.11.012.
47. Al-Kutubi, H., Gascon, J., Sudhölter, E.J.R. et al. (2015) Electrosynthesis of metal–organic frameworks: challenges and opportunities, *ChemElectroChem* 2: 462–474. doi: https://doi.org/10.1002/celc.201402429.

48. Martinez Joaristi, A., Juan-Alcañiz, J., Serra-Crespo, P. et al. (2012) Electrochemical synthesis of some archetypical Zn^{2+}, Cu^{2+}, and Al^{3+} metal organic frameworks, *Crystal Growth and Design* 12(7): 3489–3498. doi: https://doi.org/10.1021/cg300552w.
49. Mueller, U., Schubert, M., Teich, F. et al. (2006) Metal-organic frameworks – prospective industrial applications, *Journal of Materials Chemistry* 16(7): 626–636. doi: https://doi.org/10.1039/b511962f.
50. Mueller, U., Puetter, H., Hesse, M. et al. (2012) Method for electrochemical production of crystalline porous metal organic skeleton material. US Patent 8,163,949 B2, filed 10 January 2011 and issued 24 April 2012.
51. Yang, H.M. Song, X. L., Yang, T. L. et al. (2014) Electrochemical synthesis of flower shaped morphology MOFs in an ionic liquid system and their electrocatalytic application to the hydrogen evolution reaction, *RSC Advances* 4(30): 15720–15726. doi: https://doi.org/10.1039/c3ra47744d.
52. Wang, S., Wahiduzzaman, M., Davis, L. et al. (2018) A robust zirconium amino acid metal-organic framework for proton conduction, *Nature Communications* 9(1): 1–8. doi: https://doi.org/10.1038/s41467-018-07414-4.
53. Lv, D., Chen, J., Yang, K. et al. (2019) Ultrahigh CO_2/CH_4 and CO_2/N_2 adsorption selectivities on a cost-effectively l-aspartic acid based metal-organic framework, *Chemical Engineering Journal* 375, 122074. doi: https://doi.org/10.1016/j.cej.2019.122074.
54. Kathalikkattil, A.C., Roshan, R., Tharun, J. et al. (2016) A sustainable protocol for the facile synthesis of zinc-glutamate MOF: an efficient catalyst for room temperature CO_2 fixation reactions under wet conditions, *Chemical Communications* 52(2): 280–283. doi: https://doi.org/10.1039/c5cc07781h.
55. Olawale, M.D., Tella, A.C., Obaleye, J.A. et al. (2020) Synthesis, characterization and crystal structure of a copper-glutamate metal organic framework (MOF) and its adsorptive removal of ciprofloxacin drug from aqueous solution, *New Journal of Chemistry* 44(10): 3961–3969. doi: https://doi.org/10.1039/d0nj00515k.
56. Ma, X., Zhang, Y., Gao, Y. et al. (2020) Revelation of the chiral recognition of alanine and leucine in an l-phenylalanine-based metal-organic framework, *Chemical Communications* 56(7): 1034–1037. doi: https://doi.org/10.1039/c9cc05912a.
57. Li, C., Deng, K., Tang, Z. et al. (2010). Twisted metal-amino acid nanobelts: chirality transcription. *Journal of the American Chemical Society* 132 (13): 8202–8209.
58. Ma, W., Xu, L., Li, X. et al. (2017) Cysteine-functionalized metal-organic framework: facile synthesis and high efficient enrichment of N-linked glycopeptides in cell lysate, *ACS Applied Materials and Interfaces* 9(23): 19562–19568. doi: https://doi.org/10.1021/acsami.7b02853.
59. Liu, K., Yuan, C., Zou, Q. et al. (2017) Self-assembled zinc/cystine-based chloroplast mimics capable of photoenzymatic reactions for sustainable fuel synthesis, *Angewandte Chemie International Edition* 56(27): 7876–7880. doi: https://doi.org/10.1002/anie.201704678.
60. Zou, Q. and Yan, X. (2018) Amino acid coordinated self-assembly, *Chemistry – A European Journal* 24(4): 755–761. doi: https://doi.org/10.1002/chem.201704032.
61. Wiśniewski, M., Bieniek, A., Roszek, K. et al. (2018) Cystine-based MBioF for maintaining the antioxidant-oxidant balance in airway diseases, *ACS Medicinal Chemistry Letters* 9(12): 1280–1284. doi: https://doi.org/10.1021/acsmedchemlett.8b00468.
62. Zhao, J., Li, H., Han, Y. et al. (2015) Chirality from substitution: enantiomer separation via a modified metal-organic framework, *Journal of Materials Chemistry A* 3(23): 12145–12148. doi: https://doi.org/10.1039/c5ta00998g.
63. Zhang, J.-H., Nong, R.-Y., Xie, S.-M. et al. (2017) Homochiral metal-organic frameworks based on amino acid ligands for HPLC separation of enantiomers, *Electrophoresis* 38(19): 2513–2520. doi: https://doi.org/10.1002/elps.201700122.
64. Yang, S. T., Zhang, K., Zhang, B. et al. (2011) Fumaric acid. In *Comprehensive Biotechnology* (2nde), Elsevier B.V. (ed. M. Moo-Young): 163–177. Amsterdam, The Netherlands. doi: https://doi.org/10.1016/B978-0-08-088504-9.00456-6.
65. Serre, C., Millange, F., Surblé, S. et al. (2004) A route to the synthesis of trivalent transition-metal porous carboxylates with trimeric secondary building units, *Angewandte Chemie International Edition* 43(46): 6286–6289. doi: https://doi.org/10.1002/anie.200454250.
66. Surblé, S., Serre, C., Mellot-Draznieks, C. et al. (2006) A new isoreticular class of metal-organic-frameworks with the MIL-88 topology, *Chemical Communications* (3): 284–286. doi: https://doi.org/10.1039/b512169h.

67. Horcajada, P., Chalati, T., Serre, C. et al. (2010) Porous metal-organic-framework nanoscale carriers as a potential platform for drug delivery and imaging, *Nature Materials* 9(2): 172–178. doi: https://doi.org/10.1038/nmat2608.
68. Lin, K.-Y.A., Chang, H.A. and Hsu, C.J. (2015) Iron-based metal organic framework, MIL-88A, as a heterogeneous persulfate catalyst for decolorization of rhodamine B in water, *RSC Advances* 5(41): 32520–32530. doi: https://doi.org/10.1039/c5ra01447f.
69. Zhou, L., Zhang, X. and Chen, Y. (2017) Facile synthesis of Al-fumarate metal–organic framework nanoflakes and their highly selective adsorption of volatile organic compounds, *Materials Letters* 197: 224–227. doi: https://doi.org/10.1016/j.matlet.2017.01.120.
70. Tamames-Tabar, C., Imbuluzqueta, E., Guillou, N. et al. (2015) A Zn azelate MOF: combining antibacterial effect, *CrystEngComm* 17(2): 456–462. doi: https://doi.org/10.1039/c4ce00885e.
71. Furukawa, H., Gándara, F., Zhang, Y.-B. et al. (2014) Water adsorption in porous metal-organic frameworks and related materials, *Journal of the American Chemical Society* 136(11): 4369–4381. doi: https://doi.org/10.1021/ja500330a.
72. Wahiduzzaman, M., Lenzen, D., Maurin, G. et al. (2018) Rietveld refinement of MIL-160 and its structural flexibility upon H_2O and N_2 adsorption, *European Journal of Inorganic Chemistry* 2018(32): 3626–3632. doi: https://doi.org/10.1002/ejic.201800323.
73. Miller, S.R., Horcajada, P. and Serre, C. (2011) Small chemical causes drastic structural effects: the case of calcium glutarate, *CrystEngComm* 13(6): 1894–1898. doi: https://doi.org/10.1039/c0ce00450b.
74. Benessere, V., Cucciolito, M.E., De Santis, A. et al. (2015) Sustainable process for production of azelaic acid through oxidative cleavage of oleic acid, *Journal of the American Oil Chemists' Society* 92(11–12): 1701–1707. doi: https://doi.org/10.1007/s11746-015-2727-z.
75. Cunha da Cruz, J., Machado de Castro, A. and Camporese Sérvulo, E.F. (2018) World market and biotechnological production of itaconic acid, *3 Biotech* 8: 138. doi: https://doi.org/10.1007/s13205-018-1151-0.
76. Nair, R.M., Sudarsanakumar, M.R., Suma, S. et al. (2016) Crystal structure and characterization of a novel luminescent 2D metal-organic framework, poly[aquaitaconatocalcium(II)] possessing an open framework structure with hydrophobic channels, *Journal of Molecular Structure* 1105: 316–321. doi: https://doi.org/10.1016/j.molstruc.2015.10.035.
77. Ma, D.Y., Xie, J., Zhu, Z. et al. (2017) Drug delivery and selective CO_2 adsorption of a bio-based porous zinc-organic framework from 2,5-furandicarboxylate ligand, *Inorganic Chemistry Communications* 86: 128–132. doi: https://doi.org/10.1016/j.inoche.2017.10.007.
78. Wang, J., Ma, D., Liao, W. et al. (2017) A hydrostable anionic zinc-organic framework carrier with a bcu topology for drug delivery, *CrystEngComm* 19(35): 5244–5250. doi: https://doi.org/10.1039/c7ce01238a.
79. Zhang, Y. and Wang, J. (2018) A water-stable and biofriendly 2D anionic zinc(II) metal–organic framework for drug delivery, *Inorganica Chimica Acta* 477: 8–14. doi: https://doi.org/10.1016/j.ica.2018.01.015.
80. Cadiau, A., Lee, J.S., Borges, D.D. et al. (2015) Design of hydrophilic metal organic framework water adsorbents for heat reallocation, *Advanced Materials* 27(32): 4775–4780. doi: https://doi.org/10.1002/adma.201502418.
81. Permyakova, A., Skrylnyk, O., Courbon, E. et al. (2017) Synthesis optimization, shaping, and heat reallocation evaluation of the hydrophilic metal–organic framework MIL-160(Al), *ChemSusChem* 10(7): 1419–1426. doi: https://doi.org/10.1002/cssc.201700164.
82. Zhai, Q.G., Bu, X., Zhao, X. et al. (2016) Advancing magnesium-organic porous materials through new magnesium cluster chemistry, *Crystal Growth and Design* 16(3): 1261–1267. doi: https://doi.org/10.1021/acs.cgd.5b01297.
83. Smaldone, R.A., Forgan, R.S., Furukawa, H. et al. (2010) Metal-organic frameworks from edible natural products, *Angewandte Chemie International Edition*, 49: 8630–8634. doi: https://doi.org/10.1002/anie.201002343.
84. Abuçafy, M.P., Caetano, B.L., Chiari-Andréo, B.G. et al. (2018) Supramolecular cyclodextrin-based metal-organic frameworks as efficient carrier for anti-inflammatory drugs, *European Journal of Pharmaceutics and Biopharmaceutics* 127: 112–119. doi: https://doi.org/10.1016/j.ejpb.2018.02.009.
85. Hartlieb, K.J., Holcroft, J.M., Moghadam, P.Z. et al. (2016) CD-MOF: a versatile separation medium, *Journal of the American Chemical Society* 138(7): 2292–2301. doi: https://doi.org/10.1021/jacs.5b12860.

86. Holcroft, J.M., Hartlieb, K.J., Moghadam, P.Z. et al. (2015) Carbohydrate-mediated purification of petrochemicals, *Journal of the American Chemical Society* 137(17): 5706–5719. doi: https://doi.org/10.1021/ja511878b.
87. Li, X., Guo, T., Lachmanski, L. et al. (2017) Cyclodextrin-based metal-organic frameworks particles as efficient carriers for lansoprazole: study of morphology and chemical composition of individual particles, *International Journal of Pharmaceutics* 531(2): 424–432. doi: https://doi.org/10.1016/j.ijpharm.2017.05.056.
88. Han, Y., Liu, W., Huang, J. et al. (2018) Cyclodextrin-based metal-organic frameworks (CD-MOFs) in pharmaceutics and biomedicine, *Pharmaceutics* 10(4): 271. doi: https://doi.org/10.3390/pharmaceutics10040271.
89. Singh, V., Guo, T., Xu, H. et al. (2017) Moisture resistant and biofriendly CD-MOF nanoparticles obtained via cholesterol shielding, *Chemical Communications* 53(66): 9246–9249. doi: https://doi.org/10.1039/c7cc03471g.
90. Miller, S.R., Heurtaux, D., Baati, T. et al. (2010) Biodegradable therapeutic MOFs for the delivery of bioactive molecules, *Chemical Communications* 46(25): 4526. doi: https://doi.org/10.1039/c001181a.
91. Pinto, R.V., Antunes, F., Pires J. et al. (2017) Vitamin B3 metal-organic frameworks as potential delivery vehicles for therapeutic nitric oxide, *Acta Biomaterialia* 51: 66–74. doi: https://doi.org/10.1016/j.actbio.2017.01.039.
92. Yang, J., Trickett, C.A., Alahmadi, S.B. et al. (2017) Calcium l-lactate frameworks as naturally degradable carriers for pesticides, *Journal of the American Chemical Society* 139(24): 8118–8121. doi: https://doi.org/10.1021/jacs.7b04542.
93. Slater, B., Wong, S.-O., Duckworth, A. et al. (2019) Upcycling a plastic cup: one-pot synthesis of lactate containing metal organic frameworks from polylactic acid, *Chemical Communications* 55(51): 7319–7322. doi: https://doi.org/10.1039/c9cc02861g.
94. Li, N., Liu, J., Liu, J.-J. et al. (2019) Adenine components in biomimetic metal–organic frameworks for efficient CO_2 photoconversion, *Angewandte Chemie International Edition* 131(16): 5280–5285. doi: https://doi.org/10.1002/ange.201814729.
95. Wang, H.L., Yeh, H., Chen Y.-C. et al. (2018) Thermal stability of metal-organic frameworks and encapsulation of CuO nanocrystals for highly active catalysis, *ACS Applied Materials and Interfaces* 10(11): 9332–9341. doi: https://doi.org/10.1021/acsami.7b17389.
96. Howarth, A.J., Liu, Y., Li, P. et al. (2016) Chemical, thermal and mechanical stabilities of metal-organic frameworks, *Nature Reviews Materials* 1: 15018. doi: https://doi.org/10.1038/natrevmats.2015.18.
97. Huxford, R.C., Della Rocca, J. and Lin, W. (2010) Metal-organic frameworks as potential drug carriers, *Current Opinion in Chemical Biology* 14(2): 262–268. doi: https://doi.org/10.1016/j.cbpa.2009.12.012.
98. Karimzadeh, Z., Javanbakht, S. and Namazi, H. (2019) Carboxymethylcellulose/MOF-5/graphene oxide bio-nanocomposite as antibacterial drug nanocarrier agent, *BioImpacts* 9(1): 5–13. doi: https://doi.org/10.15171/bi.2019.02.
99. Katz, M.J., Brown, Z.J., Colón, Y.J. et al. (2013) A facile synthesis of UiO-66, UiO-67 and their derivatives, *Chemical Communications* 49(82): 9449–9451. doi: https://doi.org/10.1039/c3cc46105j.
100. Ke, F., Peng, C., Zhang, T. et al. (2018) Fumarate-based metal-organic frameworks as a new platform for highly selective removal of fluoride from brick tea, *Scientific Reports* 8: 939. doi: https://doi.org/10.1038/s41598-018-19277-2.
101. Kojima, D., Sanada, T., Wada, N. et al. (2018) Synthesis, structure, and fluorescence properties of a calcium-based metal-organic framework, *RSC Advances* 8(55): 31588–31593. doi: https://doi.org/10.1039/c8ra06043f.
102. Lang, X., Dong, C., Cai, K. et al. (2020) Highly active cluster structure manganese-based metal organic frameworks (Mn-MOFs) like corals in the sea synthesized by facile solvothermal method as gas electrode catalyst for lithium–oxygen batteries, *International Journal of Energy Research* 44(2): 1256–1263. doi: https://doi.org/10.1002/er.4954.
103. Li, H., Feng, X., Ma, D. et al. (2018) Stable aluminum metal-organic frameworks (Al-MOFs) for balanced CO_2 and water selectivity, *ACS Applied Materials and Interfaces* 10(4): 3160–3163. doi: https://doi.org/10.1021/acsami.7b17026.
104. Li, S., Dong, Y., Zhou, J. et al. (2018) Carbon dioxide in the cage: manganese metal-organic frameworks for high performance CO_2 electrodes in Li–CO_2 batteries, *Energy and Environmental Science* 11(5): 1318–1325. doi: https://doi.org/10.1039/c8ee00415c.

105. Shi, F.N., Almeida, J.C., Helguero, L.A. et al. (2015) Calcium phosphonate frameworks for treating bone tissue disorders, *Inorganic Chemistry* 54(20): 9929–9935. doi: https://doi.org/10.1021/acs.inorgchem.5b01634.
106. George, P., Das, R.K. and Chowdhury, P. (2019) Facile microwave synthesis of Ca-BDC metal organic framework for adsorption and controlled release of curcumin, *Microporous and Mesoporous Materials* 281: 161–171. doi: https://doi.org/10.1016/j.micromeso.2019.02.028.
107. Al Haydar, M., Abid, H.R., Sunderland, B. et al. (2017) Metal organic frameworks as a drug delivery system for flurbiprofen, *Drug Design, Development and Therapy* 11: 2685–2695. doi: https://doi.org/10.2147/DDDT.S145716.
108. Plonka, A.M., Banerjee, D., Woerner, W.R. et al. (2013) Mechanism of carbon dioxide adsorption in a highly selective coordination network supported by direct structural evidence, *Angewandte Chemie International Edition* 52(6): 1692–1695. doi: https://doi.org/10.1002/anie.201207808.
109. Banerjee, D., Simon, C.M., Plonka, A.M. et al. (2016) Metal-organic framework with optimally selective xenon adsorption and separation, *Nature Communications* 7: 11831. doi: https://doi.org/10.1038/ncomms11831.
110. Xu, J., Yu, Y., Li, G. et al. (2016) A porous magnesium metal-organic framework showing selective adsorption and separation of nitrile guest molecules, *RSC Advances* 6(106): 104451–104455. doi: https://doi.org/10.1039/c6ra16784e.
111. Su, X., Bromberg, L., Martis, V. et al. (2017) Postsynthetic functionalization of Mg-MOF-74 with tetraethylenepentamine: structural characterization and enhanced CO_2 adsorption, *ACS Applied Materials and Interfaces* 9(12): 11299–11306. doi: https://doi.org/10.1021/acsami.7b02471.
112. Saha, D., Maity, T., Das, S. et al. (2013) A magnesium-based multifunctional metal-organic framework: synthesis, thermally induced structural variation, selective gas adsorption, photoluminescence and heterogeneous catalytic study, *Dalton Transactions* 42(38): 13912–13922. doi: https://doi.org/10.1039/c3dt51509e.
113. Ramaswamy, P., Wong, N.E., Gelfand, B.S. et al. (2015) A water stable magnesium MOF that conducts protons over 10^{-2} S cm^{-1}, *Journal of the American Chemical Society* 137(24): 7640–7643. doi: https://doi.org/10.1021/jacs.5b04399.
114. Zhai, L., Zhang, W.-W., Zuo, J.-L. et al. (2016) Simultaneous observation of ligand-based fluorescence and phosphorescence within a magnesium-based CP/MOF at room temperature, *Dalton Transactions* 45(30): 11935–11938. doi: https://doi.org/10.1039/c6dt02359b.
115. Reaney, S.H., Kwik-Uribe, C.L. and Smith, D.R. (2002) Manganese oxidation state and its implications for toxicity, *Chemical Research in Toxicology* 15(9): 1119–1126. doi: https://doi.org/10.1021/tx025525e.
116. André, V., Ferreira da Silva, A. R., Fernandes, A. et al. (2019) Mg- and Mn-MOFs boost the antibiotic activity of nalidixic acid, *ACS Applied Bio Materials* 2(6): 2347–2354. doi: https://doi.org/10.1021/acsabm.9b00046.
117. Wang, X., Liu, X., Rong, H. et al. (2017) Layered manganese-based metal-organic framework as a high capacity electrode material for supercapacitors, *RSC Advances* 7(47): 29611–29617. doi: https://doi.org/10.1039/c7ra04374k.
118. Liu, X., Zhou, Y., Zhang, J. et al. (2017) Iron containing metal-organic frameworks: structure, synthesis, and applications in environmental remediation, *ACS Applied Materials and Interfaces* 9(24): 20255–20275. doi: https://doi.org/10.1021/acsami.7b02563.
119. Zhong, G., Liu, D. and Zhang, J. (2018) Applications of porous metal-organic framework MIL-100(M) (M = Cr, Fe, Sc, Al, V), *Crystal Growth and Design* 18(12): 7730–7744. doi: https://doi.org/10.1021/acs.cgd.8b01353.
120. Chen, G., Leng, X., Luo, J. et al. (2019) in vitro toxicity study of a porous iron(III) metal-organic framework, *Molecules* 24(7). doi: https://doi.org/10.3390/molecules24071211.
121. Wyszogrodzka, G., Dorożyński, P., Gil, B. et al. (2018) Iron-based metal-organic frameworks as a theranostic carrier for local tuberculosis therapy, *Pharmaceutical Research* 35(7). doi: https://doi.org/10.1007/s11095-018-2425-2.
122. Hidalgo, T., Giménez-Marqués, M., Bellido, E. et al. (2017) Chitosan-coated mesoporous MIL-100(Fe) nanoparticles as improved bio-compatible oral nanocarriers, *Scientific Reports* 7: 1–14. doi: https://doi.org/10.1038/srep43099.

123. Sun, L., Hendon, C.H., Park, S.S. et al. (2017) Is iron unique in promoting electrical conductivity in MOFs?, *Chemical Science* 8(6): 4450–4457. doi: https://doi.org/10.1039/c7sc00647k.
124. Nguyen, H.L. (2017) The chemistry of titanium-based metal-organic frameworks, *New Journal of Chemistry* 41(23): 14030–14043. doi: https://doi.org/10.1039/c7nj03153j.
125. Serre, C., Groves, J.A., Lightfoot, P. et al. (2006) Synthesis, structure and properties of related microporous N,N′-piperazinebismethylenephosphonates of aluminum and titanium, *Chemistry of Materials* 18(6): 1451–1457. doi: https://doi.org/10.1021/cm052149l.
126. Dan-Hardi, M., Serre, C., Frot, T. et al. (2009) A new photoactive crystalline highly porous titanium(IV) dicarboxylate, *Journal of the American Chemical Society* 131(31): 10857–10859. doi: https://doi.org/10.1021/ja903726m.
127. Xuan, W., Ye, C., Zhang, M. et al. (2013) A chiral porous metallosalan-organic framework containing titanium-oxo clusters for enantioselective catalytic sulfoxidation, *Chemical Science* 4(8): 3154–3159. doi: https://doi.org/10.1039/c3sc50487e.
128. Nguyen, H.L., Gándara, F., Furukawa, H. et al. (2016) A titanium-organic framework as an exemplar of combining the chemistry of metal- and covalent-organic frameworks, *Journal of the American Chemical Society* 138(13): 4330–4333. doi: https://doi.org/10.1021/jacs.6b01233.
129. Nguyen, H.L., Vu, T.T., Le, D. et al. (2017) A titanium-organic framework: engineering of the band-gap energy for photocatalytic property enhancement, *ACS Catalysis* 7(1): 338–342. doi: https://doi.org/10.1021/acscatal.6b02642.
130. Bai, Y., Dou, Y., Xie, L.-H. et al. (2016) Zr-based metal-organic frameworks: design, synthesis, structure, and applications, *Chemical Society Reviews* 45(8): 2327–2367. doi: https://doi.org/10.1039/c5cs00837a.
131. Xia, W., Mahmood, A., Zou, R. et al. (2015) Metal-organic frameworks and their derived nanostructures for electrochemical energy storage and conversion, *Energy and Environmental Science* 8(7): 1837–1866. doi: https://doi.org/10.1039/c5ee00762c.
132. Stock, N. (2014) Metal-organic frameworks: aluminium-based frameworks. In *Encyclopedia of Inorganic and Bioinorganic Chemistry*, 1–16. John Wiley & Sons: Hoboken, NJ (ed. R.A. Scott). doi: https://doi.org/10.1002/9781119951438.eibc2197.
133. Samokhvalov, A. (2018) Aluminum metal–organic frameworks for sorption in solution: a review, *Coordination Chemistry Reviews* 374: 236–253. doi: https://doi.org/10.1016/j.ccr.2018.06.011.
134. Loiseau, T., Mellot-Draznieks, C., Muguerra, H. et al. (2005) Hydrothermal synthesis and crystal structure of a new three-dimensional aluminum-organic framework MIL-69 with 2,6-naphthalenedicarboxylate (ndc), Al(OH)(ndc)·H_2O, *Comptes Rendus Chimie* 8: 765–772. doi: https://doi.org/10.1016/j.crci.2004.10.011.
135. Loiseau, T., Volkringer, C., Haouas, M. et al. (2015) Crystal chemistry of aluminium carboxylates: from molecular species towards porous infinite three-dimensional networks, *Comptes Rendus Chimie* 18(12): 1350–1369. doi: https://doi.org/10.1016/j.crci.2015.08.006.
136. Loiseau, T., Lecroq, L., Volkringer, C. et al. (2006) MIL-96, a porous aluminum trimesate 3D structure constructed from a hexagonal network of 18-membered rings and μ3-oxo-centered trinuclear units, *Journal of the American Chemical Society* 128(31): 10223–10230. doi: https://doi.org/10.1021/ja0621086.
137. Eddaoudi, M., Li, H. and Yaghi, O.M. (2000) Highly porous and stable metal-organic frameworks: structure design and sorption properties, *Journal of the American Chemical Society* 122(7): 1391–1397. doi: https://doi.org/10.1021/ja9933386.
138. Vagin, S.I., Ott, A.K. and Rieger, B. (2007) Paddle-wheel zinc carboxylate clusters as building units for metal-organic frameworks, *Chemie Ingenieur Technik* 79(6): 767–780. doi: https://doi.org/10.1002/cite.200700062.
139. Bahrani, S., Hashemi, S.A., Mousavi, S.M. et al. (2019) Zinc-based metal–organic frameworks as nontoxic and biodegradable platforms for biomedical applications: review study, *Drug Metabolism Reviews* 51(3): 356–377. doi: https://doi.org/10.1080/03602532.2019.1632887.
140. Wang, R., Liu, X., Qi, D. et al. (2015) A Zn metal-organic framework with high stability and sorption selectivity for CO_2, *Inorganic Chemistry* 54(22): 10587–10592. doi: https://doi.org/10.1021/acs.inorgchem.5b01232.
141. Lv, X., Li, L., Tang, S. et al. (2014) High CO_2/N_2 and CO_2/CH_4 selectivity in a chiral metal–organic framework with contracted pores and multiple functionalities, *Chemical Communications* 50(52): 6886–6889. doi: https://doi.org/10.1039/c4cc00334a.

142. Wang, H., Tong, L., Jiawei, L. et al. (2019) One-pot synthesis of poly(ethylene glycol) modified zeolitic imidazolate framework-8 nanoparticles: size control, surface modification and drug encapsulation, *Colloids and Surfaces A: Physicochemical and Engineering Aspects* 568: 224–230. doi: https://doi.org/10.1016/j.colsurfa.2019.02.025.
143. An, J., Geib, S.J. and Rosi, N.L. (2009) Cation-triggered drug release from a porous zinc-adeninate metal-organic framework, *Journal of the American Chemical Society* 131(24): 8376–8377. doi: https://doi.org/10.1021/ja902972w.
144. Qin, J.S., Du, D.-Y., Li, W.-L. et al. (2012) N-rich zeolite-like metal-organic framework with sodalite topology: high CO_2 uptake, selective gas adsorption and efficient drug delivery, *Chemical Science* 3(6): 2114–2118. doi: https://doi.org/10.1039/c2sc00017b.
145. Dong, K., Zhang, Y., Zhang, L. et al. (2019) Facile preparation of metal–organic frameworks-based hydrophobic anticancer drug delivery nanoplatform for targeted and enhanced cancer treatment, *Talanta* 194: 703–708. doi: https://doi.org/10.1016/j.talanta.2018.10.101.
146. Yang, X., Tang, Q., Jiang, Y. et al. (2019) Nanoscale ATP-responsive zeolitic imidazole framework-90 as a general platform for cytosolic protein delivery and genome editing, *Journal of the American Chemical Society* 141(9): 3782–3786. doi: https://doi.org/10.1021/jacs.8b11996.
147. Zhao, X., Bu, X., Wu, T. et al. (2013) Selective anion exchange with nanogated isoreticular positive metal-organic frameworks, *Nature Communications* 4: 2344. doi: https://doi.org/10.1038/ncomms3344.
148. Redfern, L.R. and Farha, O.K. (2019) Mechanical properties of metal-organic frameworks, *Chemical Science* 10(46): 10666–10679. doi: https://doi.org/10.1039/c9sc04249k.
149. Mouchaham, G., Wang, S. and Serre, C. (2018) The stability of metal–organic frameworks. In *Metal-Organic Frameworks*, Weinheim, Germany Wiley-VCH (ed. H. Garcia and S. Navalón). 1–28. doi: doi:https://doi.org/10.1002/9783527809097.ch1.
150. Makal, T.A., Wang, X. and Zhou, H.C. (2013) Tuning the moisture and thermal stability of metal-organic frameworks through incorporation of pendant hydrophobic groups, *Crystal Growth and Design* 13(11): 4760–4768. doi: https://doi.org/10.1021/cg4009224.
151. Wu, H., Yildirim, T. and Zhou, W. (2013) Exceptional mechanical stability of highly porous zirconium metal-organic framework UiO-66 and its important implications, *Journal of Physical Chemistry Letters* 4(6): 925–930. doi: https://doi.org/10.1021/jz4002345.
152. Bayliss, P.A., Ibarra, I.A., Pérez, E. et al. (2014) Synthesis of metal-organic frameworks by continuous flow, *Green Chemistry*, 16(8): 3796–3802. doi: https://doi.org/10.1039/c4gc00313f.
153. Rubio-Martinez, M., Avci-Camur, C., Thornton, A.W. et al. (2017) New synthetic routes towards MOF production at scale, *Chemical Society Reviews* 46(11): 3453–3480. doi: https://doi.org/10.1039/c7cs00109f.
154. DeSantis, D., Mason, J.A., James, B.D. et al. (2017) Techno-economic analysis of metal-organic frameworks for hydrogen and natural gas storage, *Energy and Fuels* 31(2): 2024–2032. doi: https://doi.org/10.1021/acs.energyfuels.6b02510.
155. Slater, A.G. and Cooper, A.I. (2015) Function-led design of new porous materials, *Science* 348(6238): aaa8075. doi: https://doi.org/10.1126/science.aaa8075.

13

Geopolymers Based on Biomass Ash and Bio-based Additives for Construction Industry

Prinya Chindaprasirt[1,2], Ubolluk Rattanasak[3] and Patcharapol Posi[4]

[1] Sustainable Infrastructure Research and Development Center, Department of Civil Engineering, Khon Kaen University, Khon Kaen, Thailand
[2] Academy of Science, Royal Society of Thailand, Bangkok, Thailand
[3] Department of Chemistry, Burapha University, Chonburi, Thailand
[4] Department of Civil Engineering, Rajamangala University of Technology Isan Khon Kaen Campus, Khon Kaen, Thailand

13.1 Introduction

Portland cement is the most used construction material. However, its production emits a large amount of greenhouse gases. As the production of Portland cement involves the calcination of calcareous compounds and the use of high energy in firing and processing, it is not an environmentally friendly material. The production of a ton of Portland cement generates almost a ton of carbon dioxide (CO_2). In 2017, the world cement production was 4.1 billion tons, which represents the third-largest source of CO_2 emissions [1], approximately 7 wt% of the total global CO_2 output [2]. In the cement production process, CO_2 is released from two sources. The first is from the chemical reaction from the calcination process by the decomposition of calcium carbonate ($CaCO_3$) in limestone that results in calcium oxide (CaO) and CO_2. The second source is from the energy required for the heating and processing of cement. The calcination of raw materials for the cement production requires high temperatures of over 1200 °C to form a cement clinker [1]. It has been reported that 61.7 million tons of clinker production

High-Performance Materials from Bio-based Feedstocks, First Edition. Edited by Andrew J. Hunt, Nontipa Supanchaiyamat, Kaewta Jetsrisuparb and Jesper T.N. Knijnenburg.

consume around 11.0 million tons of coal which is equivalent to 0.5 million m^3 oil or 1.8 million m^3 of natural gas [3]. It is therefore imperative that an alternative binder or green cement is required to reduce or help in the reduction of Portland cement usage.

One of the promising alternative binders to Portland cement is geopolymer, which is made from starting materials rich in silica (SiO_2) and alumina (Al_2O_3) that are activated with alkali solutions. The properties of geopolymer are comparable to those of Portland cement products. The geopolymer binder provides good binding capacity and the manufacturing process is similar to that of the normal Portland cement and concrete but with different starting materials and reaction pathways. The Portland cement paste is made from cement and water, whereas the geopolymer paste is prepared from a starting material of a fine amorphous powder and alkali solution. Preferred starting materials for geopolymer are industrial by-products such as coal fly ash and ground granulated blast furnace slag (GGBFS) due to their availability, high strength and good performances achieved. In addition, a large amount of other ashes such as from the energy harvesting of agricultural waste are receiving a lot of attention as they also contain a high amount of silica and thus could be used as a starting material or additive in the geopolymer system. In this respect, the enhanced properties of a geopolymer made with biomass are therefore required to produce sufficiently strong and durable composites with a reasonably long service life [4–6].

13.2 Pozzolan and Agricultural Waste Ash

Pozzolanic material or pozzolan is defined by the ASTM C 618–19 [7] as a siliceous and/or aluminous material which will chemically react with calcium hydroxide ($Ca(OH)_2$) in the presence of moisture at room temperature and form compounds possessing cementitious properties. Pozzolans can be natural products or by-products from various manufacturing processes such as fly ash, volcanic ash, GGBFS, calcined shale, calcined kaolin, diatomaceous earth, silica fume, rice husk ash, and bagasse ash.

Fly ash is a by-product of the combustion of pulverized coal in thermal power plants. At present, because of its availability and good reactivity, coal fly ash is extensively used worldwide as pozzolan as a partial replacement for more expensive Portland cement in making blended fly ash/Portland cement. The cost of fly ash mainly comes from handling and transportation, which typically represent about half of the cost for cement. Blended Portland pozzolana cements with up to 60% pozzolans have now been successfully used in structural applications [8], which not only reduces the cost of the material but also enhances durability. The other coal ashes, viz. fly ashes from other small power plants with fluidized bed-burning technology using low burning temperature, and bottom ashes with larger particle sizes are much less used due to their low pozzolanic activity indices [9]. Metakaolin and GGBFS also have high pozzolanic reactivity and have been demonstrated for their suitability in making concrete [10].

In recent years, the use of agricultural ashes produced from biomass waste has been increasing due to concerns regarding environmental and waste disposal problems. The use of these agricultural ashes as pozzolans is shown to be beneficial due to their high reactivity and availability and is expected to play an important role in the construction industries. Biomass ashes are available in substantial quantity and also cost much less than Portland cement.

The utilization of pozzolan results in a reduction in temperature rise and improvement in durability and strength enhancement of concrete [11]. Pozzolans improve the properties of cement-based materials due to the refinement of the pore structure and porosity, which has a significant effect on the properties of paste [4]. In addition, fly ash with mainly spherical particles can physically disperse cement flocs and improve the workability of concrete. Ground bottom ash with similar chemical composition to those of fly ash can also be used as supplementary cementitious materials [12, 13]. It is agreed that the increase in the fineness of pozzolan generally improves the properties of the binder and concrete [4, 14]. A fine pozzolan increases the mortar strength compared to that made from the original as-received coarser fly ash and reduces the average pore size of the paste compared to the use of a coarse pozzolan. In addition, the application of the proper amount of pozzolan improves the resistance of concrete against harmful solutions [15–17].

The SiO_2, Al_2O_3, and CaO contents of pozzolan play an important role in the reaction and subsequent properties of the binder. The ternary diagram of CaO–SiO_2–Al_2O_3 of some pozzolans is shown in Figure 13.1. Fly ash contains a high content of SiO_2, Al_2O_3, and CaO (for high calcium content fly ash), which are important ingredients for pozzolan and as starting materials for geopolymer. The fly ash particles are mostly spherical with a smooth surface and their fineness is similar to that of Portland cement. It has also been shown that fly ashes, both class F (low calcium content) and class C (high calcium content), are suitable source materials for making good-quality geopolymer [18, 19]. Silica fume, rice husk ash, and rice husk bark ash have a high SiO_2 content exceeding 70%, whereas GGBFS and Portland cement have a high CaO content. Metakaolin and class F fly ash are rich in SiO_2 and Al_2O_3 with low-calcium compound content at less than 10%. So, they are usually classified as aluminosilicate materials. For class C fly ash, the CaO content is generally more than 15%. Some lignite fly ash contains a calcium content of up to 40% [20].

The pozzolans from agricultural waste are presently receiving more attention since their uses generally improve the properties of the blended cement concrete and also reduce environmental problems. Palm oil fuel ash (POFA), rice husk ash, rice husk bark ash, and bagasse ash are promising agricultural pozzolans. The annual outputs of biomass ashes are considerable. In 2017, Thailand produced 1.3 million tons of rice husk ash, 2.4 million tons of POFA, and 5.6 million

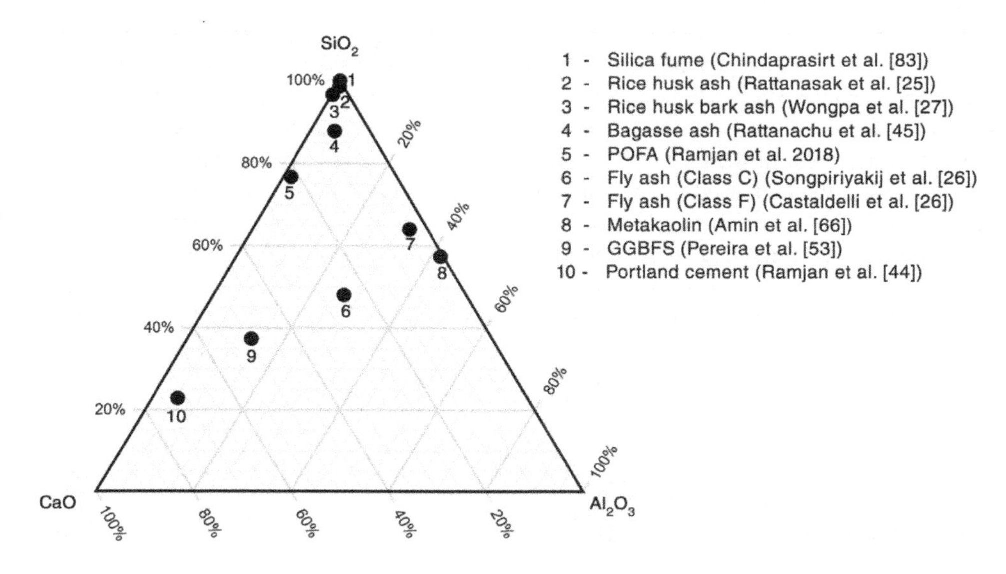

Figure 13.1 Ternary diagram of CaO–SiO_2–Al_2O_3 of selected pozzolans.

tons of bagasse ash [21]. At present, these ashes are not commercially used and large quantities have to be disposed of in landfills resulting in environmental problems. Research on the performances of these agricultural pozzolans as supplementary cementitious materials for partial replacement of Portland cement and starting materials for geopolymer is therefore needed.

13.3 Geopolymer

In addition to pozzolana cement, geopolymer is an alternative environmentally friendly cement for which various ashes with high SiO_2 and/or Al_2O_3 content can be used as starting materials. The ashes in fine powder form with high surface area and increased reactivity are activated with alkali solutions (e.g. sodium hydroxide (NaOH) and sodium silicate (Na_2SiO_3)) and geopolymerization is increasingly obtained with additional heat curing [5]. Geopolymer was first discovered when various calcined clays, predominately calcined kaolinite (metakaolin), were activated with an alkaline solution to produce hardened ceramic-like products at a low temperature of 90–115 °C. In the geopolymerization process, amorphous phases of SiO_2 and Al_2O_3 in starting materials leach in the presence of a high alkali solution forming an inorganic polymer chain [22]. Geopolymer is classified as a low-temperature ceramic or inorganic polymer as it is prepared without the conventional burning of calcareous material for making cement. The use of geopolymer materials could thus reduce the emission of additional CO_2 as they can be made from waste ashes including coal fly ash and industrial wastes. Geopolymer also provides good mechanical properties and excellent acid and fire-resistant properties [23]. Since geopolymer provides less energy consumption, low production costs, high early strength, and fast setting, these properties make geopolymer suitable for applications in many fields of industry, such as civil engineering, automotive and aerospace industries, non-ferrous foundries and metallurgy, plastics industries, waste management, art and decoration, and the retrofit of buildings.

Coal fly ash is widely used as a source material for making geopolymer as it contains a sufficient amount of reactive SiO_2 and Al_2O_3. Since coal fly ash is an industrial by-product, the utilization of such material to produce value-added products is of considerable commercial interest. Nowadays, agricultural ashes are also used as optional source material for geopolymer precursors. Various agricultural ashes such as rice husk ash, rice husk bark ash, POFA, and bagasse ash have been studied and shown to be suitable source materials for making geopolymer [24–27]. These bio-based geopolymers not only reduce the waste by reusing industrial and/or agricultural wastes but also reduce the use of virgin materials and greenhouse gas emissions from the increased use of Portland cement. Geopolymers are, therefore, alternative binders that could compliment the use of Portland cement. As a consequence, geopolymer technology has attracted a great deal of attention worldwide in the past 30 years.

The geopolymerization reaction is a complicated one. In summary, it is a result of a chemical reaction between various aluminosilicate oxides with an alkali solution such as NaOH and KOH (and some heat curing for the acceleration of reaction) yielding an inorganic polymer Si–O–Al–O bonds with general formula as shown in Eq. (13.1) [28, 29]:

$$M_n\left[-(-Si-O_2)_z-O-Al-O\right]_n(wH_2O) \tag{13.1}$$

where – indicates a bond,

z is 1, 2, or 3, and

n is the degree of polymerization.

Geopolymer is the synthetic analogue of natural zeolite and for an enhanced reaction, some heat curing is used in the synthesis. However, the reaction times are substantially faster than that of comparable zeolite and amorphous to semi-crystalline products with some zeolitic properties are obtained [30]. The chemistry of geopolymerization is, therefore, close to that of zeolites [28]. However, the products are different in composition and structure. The synthesis of zeolite involves nucleation in solution and crystal growth with resulting zeolite usually with low mechanical strength. Whereas the synthesis of geopolymer involves the leaching of Si^{4+} and Al^{3+} ions followed by condensation and formation of amorphous to semi-crystalline gel resulting in geopolymer with high mechanical strength [31].

The exact mechanism of geopolymer setting and hardening is not yet fully understood. Most proposed mechanisms consist of dissolution, transportation, or orientation, as well as a reprecipitation step. Hua and van Deventer [29] divided the geopolymerization mechanism into two main steps:

1. Dissolution; the vitreous component of starting material (aluminosilicate glass) in contact with the alkali solution is dissolved forming the series of complex ionic species. The reaction occurs at the surface of the starting material due to the presence of a glassy phase [32].
2. Polymerization; the small molecules presented in the dissolution can join each other and form a large molecule that precipitates in the form of a gel, in which some degree of short-range structural order can be identified [33].

The crystal structure in the source material also plays an important role in determining the dissolution products. Another factor that significantly affects the phase development of geopolymer products is the order of mixing [22]. Silica and alumina are supplied in both soluble and insoluble components. Normally, the order of mixing of geopolymer paste is as follows:

Step 1. NaOH and sodium silicate solutions are prepared in advance. The preparation of NaOH solution from NaOH flakes is usually done one day prior to the mixing.

Step 2. Source materials (Al–Si material) are prepared in fine powder form. Grinding and/or sieving are required for materials that are not in fine powder form.

Step 3. The solid powder is mixed with alkali solutions to obtain a consistent geopolymer paste.

Step 4. The geopolymer paste is cast. Heat curing is usually required to speed up the geopolymerization reaction. For mix with high calcium content, curing at ambient conditions is possible.

Any materials that contain a significant amount of silicon (Si) and aluminum (Al) in an amorphous phase can be used as source material for making geopolymer [34]. Therefore, many industrial waste materials such as coal fly ash, rice husk ash, GGBFS, and silica fume (particularly if used as an alternative to sodium silicate which normally forms part of the reactant solution) can be used [35]. However, Davidovits has suggested that the oxide molar ratios in the geopolymer matrix should be within the following ranges: $Na_2O/SiO_2 = 0.2–0.48$, $SiO_2/Al_2O_3 = 3.3–4.5$, $H_2O/Na_2O = 10–25$, and $Na_2O/Al_2O_3 = 0.8–1.2$ [5, 28].

Currently, pulverized coal fly ashes are widely used since they are available in large quantities in many parts of the world and, most importantly, their use results in geopolymer with satisfactory high strength [20]. In addition, there are concerted efforts in the utilization of industrial and agricultural waste ashes for making geopolymer [25–27].

13.4 Combustion of Biomass

The combustion of biomass material consists of the removal of free moisture and the burning of organic matter resulting in ashes as a by-product. For example, the burning of rice husk at various temperatures with adequate oxygen supply has been reported by Damer [36]. When rice husk is heated to 105 °C, free water or moisture is first removed. At 200 °C, the color of the rice husk changes from yellow to brown indicating the onset of combustion. The organic matter then starts to burn and is completely burnt at 400 °C with resulting ash as a by-product [36]. However, with insufficient oxygen, a large amount of carbon and other organic matter could still be left unburnt and the resulting ash is black in color with a high loss on ignition (LOI) [24]. With sufficient oxygen and adequate burning duration, the complete combustion of rice husk is obtained at 450–500 °C resulting in white or light brown ash [36]. For normal burning conditions, a higher temperature of around 700–750 °C and a longer burning duration are needed to obtain complete combustion [37]. For rice husk, open burning and controlled burning are possible. For most of the biomasses, boiler burning and fluidized bed burning are now employed.

13.4.1 Open Field Burning

Rice husk is normally disposed of by burning in open fields. The quality of ash is variable depending on the size of the heap and supply of air. The temperature and burning conditions vary considerably within the heap. The exterior of the heap will burn freely, whereas the interior will burn under a limitation of air supply. As burning continues, the temperature of the external layer of the heap is lower than that of the interior part owing to the air circulation around the heap. The combustion temperature is normally in the range of 550–800 °C for small heaps of less than 20 kg [24]. For practical open field burning, a fire on a perforated plate raised slightly above the ground is a good practice to supply air. However, for a large heap of husk, the temperature could be higher at 1000 °C or slightly more. This open field-burning method can result in dirt and pollution problems when it is windy. Various covers have been applied during the combustion; however, the air supply is reduced resulting in more unburnt material with high LOI.

13.4.2 Controlled Burning

Low-cost incinerators have been made from a 200 l oil drum with wire gauze, and from clay bricks with a perforated bucket. The former was developed by the Pakistan Council of Scientific and Industrial Research and the ferrocement drum with wire gauze basket by the Asian Institute of Technology for the controlled burning of rice husk. These incinerators are attractive as they are simple to construct and low-cost with the burning temperature controlled by the amount of rice husk [38]. Under controlled burning conditions, the volatile organic matter in the rice husk is removed which is consisting of cellulose and lignin. In addition, the residual ash is mainly amorphous silica with a microporous structure [39].

13.4.3 Boiler Burning

Normally, the biomass waste is burnt as a fuel source in the mill and the ash is named according to the source material, such as POFA, bagasse ash, and rice husk bark ash. The temperature in the boiler burning varies from around 800–1200 °C depending on the mill-burning kiln condition and process. The ash obtained is normally black in color with a reasonably high LOI and/or carbon content. Rice husk ash burnt this way is specifically called "boiler ash". Generally, the boiler ash is burnt quickly and ash with high LOI is usually obtained.

13.4.4 Fluidized Bed Burning

Fluidized bed burning is also used quite extensively used for biomass waste. The burning is done within a hot bed of sand or inert particles. A mixture of fuel and inert particles is suspended by an upward flow of combustion air. This system allows oxygen to reach the biomass fuel more readily and the efficiency of the combustion process is increased. The heat transfer of this process is increased for low operating temperatures (750–900 °C) and leads to a reduced amount of nitrogen oxide (NO_x) produced [40]. This is an environmental benefit for the combustion of high nitrogen-content biomass fuels. Generally, limestone is added into the bed to capture sulfur dioxide (SO_2) during the combustion; therefore, the SO_2 emission is small. However, the resultant ash contains a high CaO and $CaSO_4$ content that limit the use as pozzolan in a cement mixture but could be used in the geopolymer system.

13.5 Properties and Utilization of Biomass Ashes

Biomass is a significant source of renewable energy for agricultural countries. In 2017, a total of 120 million tons of biomass was generated in Thailand comprising 1.1 million tons of fuelwood, 6.9 million tons of rice husk, 28.1 million tons of bagasse, and 83.9 million tons of other agricultural waste [21]. These wastes are abundant in agricultural countries like China, India, and other Asian countries. Biomass has been mainly used as an alternative energy resource for industrial plants and biomass power plants owing to their availability and low cost. Ashes are the by-product of the combustion process and comprise approximately 15–20 wt% of the original biomass. Various types of utilizations and discards for the ashes are summarized as follows:

Landfill: Since biomass ashes do not contain hazardous matters, there is no ban on disposal at the landfill. In addition, the low bulk density can pose handling and transportation problems for industrial reuse. A large amount of biomass ash is therefore still not being used and has to be disposed of at the landfill.

Soil amendment: Biomass ashes are often used as a soil conditioner to make the soil less cohesive and suitable for growing crops and watering [41]. The biomass ashes (especially rice husk ash) are also used in combination with lime for soil stabilization [41].

Cement and concrete fillers: Ground biomass ashes are used to partially replace cement in a concrete mixture. With the addition of a proper amount of ash, the strength of concrete could be increased due to the pozzolanic effect and the filler effect. Depending on the chemical composition and burning condition of biomass, the optimum cement replacement level is usually not more than 20% [42].

Geopolymer precursor: Biomass ashes have the potential to be used as a geopolymer precursor. However, as these ashes are predominantly rich in silica, they are used either as a silica source or in conjunction with other pozzolans containing alumina for making geopolymer [24, 26]. This is discussed in more detail in the next section.

Rice husk ash is the most common agricultural waste ash and has been researched quite extensively [15, 43–45]. Other ashes such as rice husk bark ash, POFA, and bagasse ash also possess pozzolanic properties and can therefore be used as good pozzolan for making pozzolanic cement or as primary silica source material for making geopolymer. Figure 13.2 shows the morphology of the agricultural waste ashes and fly ash for comparison. The fly ash particles from the combustion of pulverized coal (conventional coal fly ash) are mostly spherical and its surface is smooth due to the high burning temperature.

The shapes of the biomass ashes are very irregular with a porous surface originating from the cellular structure of the agricultural product. The intensity of the surface porosity depends on the type of agricultural waste and burning conditions. For example, rice husk ash prepared using proper burning conditions with a temperature usually below 700 °C exhibits a very porous surface from the remnants of the cellular structure [24, 37]. Burning at a higher temperature results in the collapse of cellular structure and reduction in porosity. Rice husk bark ash is derived from the burning of rice husk and wood bark and the ash shows a less porous structure on the particle surface. This is mainly due to the presence of wood bark which is denser than rice husk. POFA also shows less porosity as it is derived from the fruit bunch containing both skin and kernel. The surface of bagasse ash is also quite porous due to the original cellular structure.

Typical chemical compositions of biomass ashes are shown in Table 13.1. Rice husk ash contains over 80 wt% silica with a small amount of impurities such as K_2O, Na_2O, and Fe_2O_3 [25, 48, 49]. The commercial use of rice husk ash is mainly in the silica extraction process [61, 62] and as a supplementary cementitious material to partially replace Portland cement [16, 63]. It has also been used to make lightweight aggregate (LWA) with the incorporation of boric acid (H_3BO_3) to increase product stability [64].

Bagasse ash is a waste from the sugar industry produced by burning bagasse for heat and energy in the sugar-processing mill. Normal burning usually results in an incomplete burning condition with an inadequate supply of air or too short burning time leaving a high carbon content in the ash with a high LOI. The chemical composition and physical properties of bagasse ash are dependent on the burning condition, viz. burning temperature, burning duration, and air or oxygen availability. Silica is the main composition of bagasse ash (~70–80 wt%) similar to rice husk ash with a small amount of Al_2O_3 [45, 65, 66]. The specific gravity of bagasse ash obtained from normal burning is approximately 2.2 [15]. To increase its reactivity, grinding is needed to reduce the particle sizes and increase the surface area [11, 15, 43] similar to other biomass ashes. Bagasse ash has been used to synthesize zeolite using hydrothermal treatment process for various applications including adsorption of heavy metals [51], dyes [55], and phenolic compounds [56]. However, thus far, only a small amount of bagasse ash is used, and the remaining amount is treated as waste.

POFA is black in color due to the high amount of unburnt carbon left from the combustion. Generally, the as-received POFA particles are quite coarse (particle size 60–180 μm) and have a porous surface with high porosity [57, 58]. Most of the POFA is disposed of in landfills. For the utilization of this ash, an initial process involving the removal of moisture by heating at 105 °C and sieving out foreign particles, unburnt kernels, and fibers is recommended. Similarly, the POFA needs to be ground to increase the fineness and surface area. The ground POFA-specific gravity is usually in the range of 1.9–2.3. Similar to other biomass ashes, POFA contains a high amount of SiO_2 and also substantially high calcium (CaO) and potassium (K_2O) contents [44, 53, 54], making it suitable for use in concrete [44] and geopolymer mixtures. In a number of situations, where the loss of ignition is very high, it is recommended that the POFA is burnt again with a temperature of around 500 °C for an hour to remove the unburnt carbon [67].

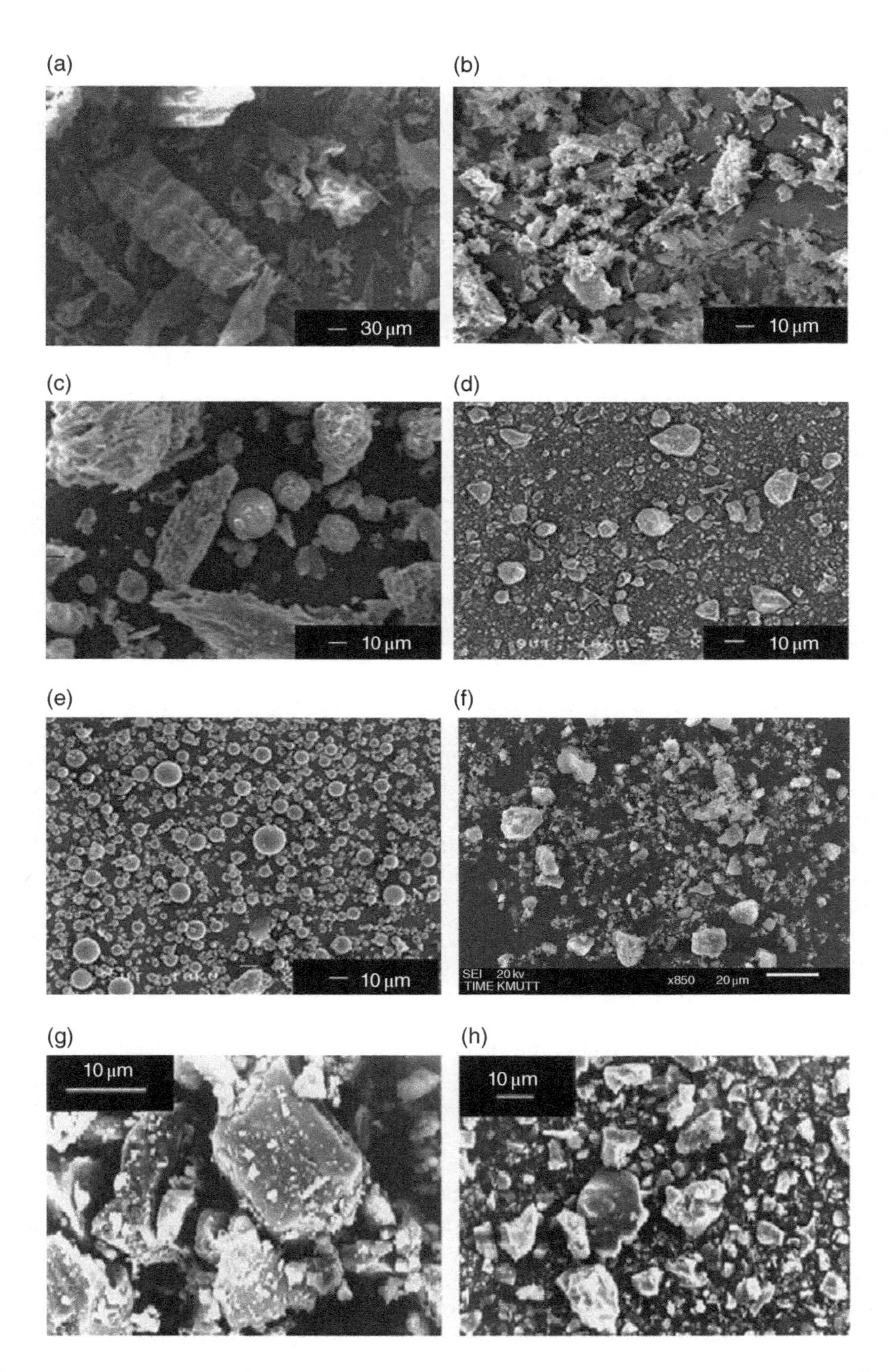

Figure 13.2 *Morphology of fly ash, cement, and various agricultural pozzolans. (a) Original rice husk ash, (b) Ground rice husk ash, (c) Original palm oil ash, (d) Ground palm oil ash, (e) Fine fly ash, (f) Bagasse ash, (g) Portland cement, (h) Rice husk bark ash. Source: (a–e) From Chindaprasirt et al. [46]; (f) from Rerkpiboon et al. [15]; (g–h) from Chindaprasirt et al. [47]. Reproduced with permission from Elsevier.*

Table 13.1 *Typical chemical compositions of rice husk ash, palm oil fuel ash (POFA) and bagasse ash.*

Composition (wt%)	Rice husk ash	Bagasse ash	Palm oil fuel ash
SiO_2	85–95	70–80	45–60
Al_2O_3	0.5–0.8	3–7	1–7
CaO	0.7–3	2–4	5–14
Fe_2O_3	0.2–1	2–6	1–6
MgO	0.4–2	1–5	3–7
K_2O	2–4	0.4–5	0.1–20
Loss on ignition (LOI)	0.1–5	0.1–8	2–8
References	Aukkadet et al. [15]; Montakarntiwong et al. [43]; Ramjan et al. [44]; Rattanachu et al. [45]; Rattanasak et al. [25]; Tippayasam et al. [48]; Vassilev et al. [41]; Wen et al. [49]	Castaldelli et al. [50]; Cordeiro et al. [14]; Oliveira et al. [51]; Pereira et al. [53]; Ramjan et al. [44]; Runyut et al. [53]; Salami et al. [54]; Shah et al. [55]; Shah et al. [56]	Sata et al. [57]; Tangchirapat et al. [58]; Castaldelli et al. [50]; Ranjbar et al. [59]; Elbasir et al. [60]; Runyut et al. [53]; Salami et al. [54]

The XRD patterns of typical biomass ashes are shown in Figure 13.3. These patterns are used to identify the mineral content in the form of crystalline and amorphous phases, which are used to gauge the reactivity or ability of biomass ash to take part in the reaction. The broad hump and its height represent the amount of amorphous phases, while the sharp peaks represent the crystalline phases in the materials. The amount of amorphous phases indicate the reactivity and pozzolanic properties of these materials [68]. Bagasse ash has a high content of SiO_2 in the form of quartz and cristobalite. Quartz is usually from the sand particles attached to the sugarcane during harvesting [14]. Cristobalite is a crystal morphology of silica resulting from the high combustion temperature of bagasse. The amorphous phases in biomass ashes come from the crystallization loss at 300–750 °C of the origin resulting in the amorphization [41]. At burning temperatures above 750 °C, the amorphous silica transforms into crystalline silica. Amorphous mineral phases are required when ash is used as the reactant in the chemical reaction, otherwise temperature curing is needed to activate the reaction [5].

Biomass ashes currently offer the opportunities to be utilized as source materials for making construction products. These ashes are greatly enriched by various components including active compounds (e.g. lime, anhydrite Ca, and Ca–Mg silicates and aluminosilicates), semi-active compounds (such as portlandite, gypsum, carbonates, mica, and iron oxides), and inert or inactive compounds (such as quartz, mullite, and some char) [41]. The active and semi-active compounds and phases of biomass ashes react and form reactive substances in water or alkaline solution leading to the setting and hardening of the usual binder system. This results from the formation of new silicates, aluminosilicates, sulfates, carbonates, and hydrates which can bind the inert phases in the multicomponent system. Some biomass ashes do not comply with the criteria of BS EN-450 (fly ash for concrete: definition, specifications, and conformity criteria), especially in the contents of alkali, chlorides, sulfates, and LOI. The amount of biomass in the cement and geopolymer mixture should not be too high and is usually up to around 20% in the Portland cement system. High ash content can be added into the cement mixture if

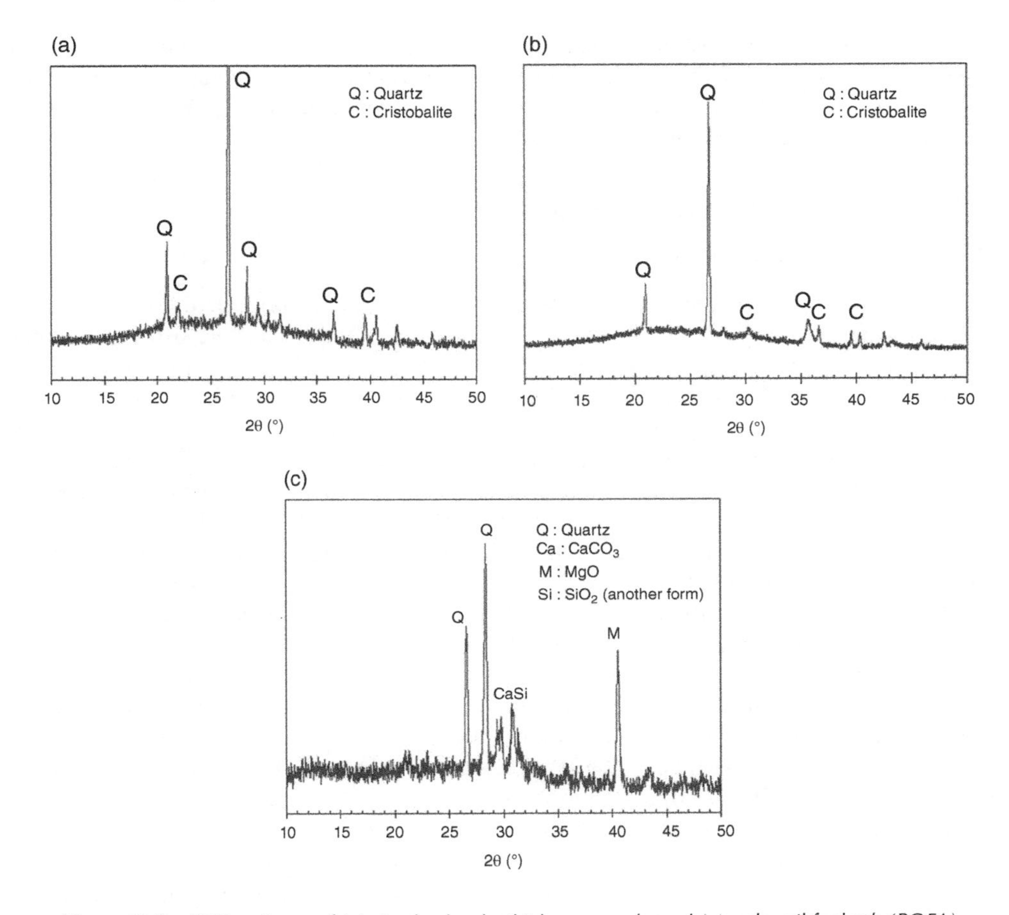

Figure 13.3 *XRD patterns of (a) rice husk ash, (b) bagasse ash, and (c) palm oil fuel ash (POFA).*

it does not alter the cement chemistry and the cement products should have an equal or even better performance than the normal product [41].

13.6 Biomass Ash-based Geopolymer

It has been shown that rice husk ash, bagasse ash, and POFA are good pozzolans and have the potential for use as a starting material or additive in making geopolymer since they contain amorphous (or glassy) phases of SiO_2 that can react with alkali solutions in the geopolymer mixture. However, the reactivity of these ashes needs to be enhanced by grinding them into fine particles with an increase in surface area. The SiO_2 and Al_2O_3 contents in the ash are also a significant factor to obtain the optimum mix design. Most biomass ashes contain a high amount of SiO_2 but low Al_2O_3 content and thus hinder its sole use as a starting material in making geopolymer. Therefore, an external source of amorphous or reactive Al_2O_3 is required to provide a proper SiO_2/Al_2O_3 ratio for the formation of a Si–O–Al polymeric chain in the matrix.

The most used biomass ash in the construction field is rice husk ash. There is therefore a relatively large amount of literature on this ash compared to other agricultural ashes.

The information on rice husk ash is therefore quite comprehensive and could be used as a possible trend for other ashes. Although, this is not always true as the chemical compositions of various ashes are not the same and care should be taken to extend or apply the knowledge on similar but different agricultural ashes.

13.6.1 Rice Husk Ash-based Geopolymer

13.6.1.1 Rice Husk Ash Geopolymer

Rice husk ash is mainly composed of silica which accounts for more than 70% and could reach 90%. The silica exists in two forms, viz. amorphous and crystalline silica. A number of researchers have studied the use of rice husk ash as a source material for the synthesis of geopolymer [24–27, 49, 69]. Rice husk ash is obtained from the burning of rice husk as a fuel source in the biomass power plant. The properties of this power plant-burnt rice husk ash are quite variable depending on the type of furnace and burning conditions. Admittedly, the laboratory-burnt rice husk ash is more uniform and provides the desired amorphous phase and low carbon content. The whitish-grey color ash with low LOI is obtained with a burning temperature around 700 °C [24]. The control open field burning is also obtained with the burning of a small 20 kg heap of rice husk on a perforated steel plate raised slightly above the ground to allow the flow of air. The resulting grey rice husk ash contains a low LOI of only 3.25% with a specific gravity of 1.97. The grey color results from the almost complete combustion with an adequate air supply. Both ashes obtained from biomass power plants and from the laboratory are ground into small particles before utilization. This leads to the collapse of the cellular structure of the rice husk ash and an increase in specific gravity. As a consequence, the ash contains very little Al_2O_3, and an external source of Al_2O_3 is required. The use of a high SiO_2/Al_2O_3 ratio in the mixture results in a geopolymer with high elasticity compared to that of a normal geopolymer. In addition, it is recommended that a stabilizing agent such as boric acid together with a slightly higher curing temperature is needed to make the geopolymer stable when immersed in water or exposed to a humid environment [25].

13.6.1.2 Rice Husk Ash Geopolymer with External Alumina Source

Rice husk ash is not solely used as a starting material in geopolymer synthesis owing to the high SiO_2 content. In this case, SiO_2 gel is generated when alkali solutions and heat are applied during the synthesis process. The resulting SiO_2 gel causes the expansion and water absorption of the composite leading to instability of the material. An external source of Al_2O_3 is therefore required for use in conjunction with rice husk ash. This includes coal fly ash [24, 26, 27, 69], sludge [49], metakaolin [12, 48], and aluminum hydroxide [25]. The typical chemical compositions of Al_2O_3 external source materials are tabulated in Table 13.2.

Coal fly ash is mostly preferred owing to its high alumina content, amorphous phase, and wide availability. Up to 60% of rice husk ash by weight is mixed with fly ash in the synthesis. The heat-curing temperature in the range of 50–115°C with dry oven condition or steam curing is used to accelerate the reaction depending on the mix proportions. However, the use of a high percentage of rice husk ash in the geopolymer mixture results in a significant reduction in compressive strength. The increase in SiO_2/Al_2O_3 molar ratio to 16 has been shown to reduce the compressive strength by more than half of the control geopolymer [24]. When using high-calcium fly ash [26, 27], heat curing is not needed as the reaction is enhanced

Table 13.2 Typical chemical compositions of external-source materials.

Composition (wt%)	High-calcium fly ash	Low-calcium fly ash	Metakaolin	Ground granulated blast furnace slag
SiO_2	25–40	50–60	45–55	30–40
Al_2O_3	13–25	25–30	30–45	10–15
CaO	15–30	3–5	0–0.5	35–45
Fe_2O_3	10–15	5–12	1–7	0.3–2
MgO	1–2	~1	0.1–2	5–8
K_2O	1–3	1–2	0.3–2	0.4–1
Loss on ignition (LOI)	0.1–4	1–7	1–14	0.1–6
References	Chindaprasirt et al. [19]; Kroehong et al. [70]; Ramjan et al. [44]; Runyut et al. [53]; Salami et al. [54]; Songpiriyakij et al. [26]; Wongpa et al. [27]	Alvarez-Ayuso et al. [18]; Castaldelli et al. [50]; Ranjbar et al. [59]	Amin et al. [65]; Bashar et al., [71]; Chindaprasirt [12]; Poon et al. [10]; Tchakoute et al. [72]; Tippayasam et al. [48]; van Jaarsveld et al. [73]	Arulrajah et al. [66]; Castaldelli et al. [74]; Pereira et al. [52]

by the presence of calcium. From the reaction between alkali solution, calcium, silica, and alumina compounds in fly ash and silica from rice husk ash, the resulting matrix consists of calcium silicate hydrate (CSH) that coexists with calcium aluminosilicate hydrate (CASH) and sodium aluminosilicate hydrate (NASH) [75]. This alkali-activated reaction results in the reaction of $Ca(OH)_2$ with amorphous SiO_2 from rice husk ash and fly ash to form additional CSH. This reaction differs from the normal geopolymerization reaction since the pathway and main product are different. Without heat curing, the Si–O–Al bond in the geopolymer product is not easily formed and the strength is thus low [76]. For the use of rice husk ash with sludge, the strength is rather low and additional treatment of rice husk is recommended. For example, rice husk ash has been modified with 2 M $AlCl_3$ solution, air-dried, pulverized, and sieved through a 0.8 mm sieve opening before mixing with sludge [49]. As the sludge contains quite variable and diverse compounds, the compressive strength of this material is rather low at 5.0 MPa. This material can be used for pavements or housing blocks where high strength is generally not required.

A very high volume of rice husk ash with a weight percentage of 90.0–97.5 in solid proportion has also been used to prepare a geopolymer [25]. In this work, $Al(OH)_3$ powder at 2.75–10.0% by weight of rice husk ash is applied to the mixture as the external Al^{3+} ion source. Since the mixture contains a very high SiO_2 content originating from the rice husk ash, the hardened geopolymer is not stable and disintegrates when immersed in water. To overcome this problem, powdered boric acid (H_3BO_3) is incorporated into the mixture followed by oven curing at 115 °C for 48 hours. There is no sign of expansion after heat curing and dissolution in water during the boiling test. Since there is a non-bridging oxygen in the geopolymer structure with a high silica content, the addition of H_3BO_3 provides the oxygen to bridge leading to a more stable matrix structure and more solidification [77]. A geopolymer matrix with a reasonably high compressive strength of 15–20 MPa is obtained [25].

13.6.1.3 Properties of Rice Husk Ash-based Geopolymer

13.6.1.3.1 Fly Ash/Rice Husk Ash Geopolymer

When high-calcium fly ash is used as an external source of alumina, up to 60% by weight of rice husk ash can be mixed with this fly ash with heat curing at 60 °C for 24 hours. The increase in rice husk ash content (i.e. increase in SiO_2/Al_2O_3 ratio) beyond this level results in the instability of the geopolymer after 90 days due to moisture absorption resulting in very low strength as shown in Figure 13.4 [26]. The optimum mix of this rice husk ash/fly ash geopolymer concrete gives the compressive strength up to 72.0 MPa at 90 days. Owing to the high calcium content in fly ash, the products in this geopolymeric system are CSH coexisting with NASH and CASH compounds. The compressive strength of this system also increases with time, mainly due to the increase in strength of the CSH compound with time, which is highly desirable. The increase in NaOH solution concentration to 18 M together with the use of Na_2SiO_3 solution increases the compressive strength of the mixture. It has been reported that the type and concentration of alkali solutions affect the dissolution of the source material. The leaching of Al^{3+} and Si^{4+} ions is normally high with a high concentration of NaOH [22].

Rice husk ash also improves the thermal properties of the fly ash/rice husk ash geopolymer. The fire test at 1100 °C with a Bunsen burner on a 10 mm geopolymer plate has been performed [38]. The temperature of the opposite side from the frame of a geopolymer is measured every two minutes for 30 minutes and shows that the temperature also increases with time but at a slower rate. The use of rice husk ash reduces the temperature at the surface. The temperature significantly drops with the increase in rice husk ash content from 40 to 60% by weight with the lowest temperature of 180 °C detected. When rice husk ash is used at not more than 50%, the geopolymer shows good thermal stability and test results render the possibility for the use of this material as a heat insulator. However, cracks are observed at the surface due to the expansion of the sodium silicate compound.

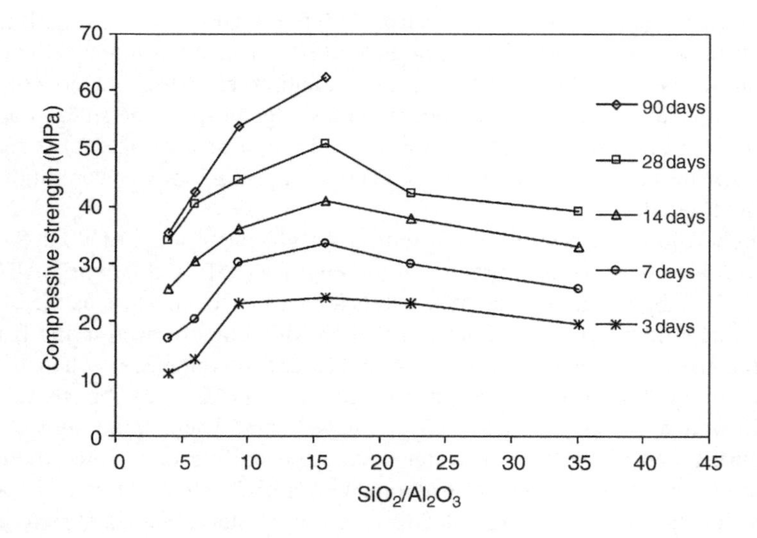

Figure 13.4 Compressive strength of rice husk-ash/fly-ash geopolymer with SiO_2/Al_2O_3 ratio. Source: Songpiriyakij et al. [26]. Reproduced with permission from Elsevier.

13.6.1.3.2 Rice Husk Ash/Metakaolin Geopolymer

Metakaolin is a good source of aluminosilicate as its main chemical compounds are SiO_2 and Al_2O_3 in amorphous phases. Metakaolin is prepared by the calcination of kaolin clay at 500–600 °C to make it suitable as a source material for geopolymer production [12, 73]. This calcination temperature results in a satisfactory high-strength metakaolin geopolymer. The use of a higher calcination temperature leads to metakaolin with low reactivity due to the increased crystalline product in the form of mullite. A number of researchers have studied the use of metakaolin combined with rice husk ash in geopolymer. Two types of rice husk ashes, viz. a grey rice husk ash (laboratory-prepared ash) and a black rice husk ash (biomass power plant ash) are used for most of the studies. It has been reported that a maximum of 60% of grey rice husk ash can be mixed with metakaolin to obtain a good rice husk ash/metakaolin geopolymer. A higher dosage of more than 60% rice husk ash results in excessive expansion during heat curing due to the low strength obtained with the high silica-to-alumina ratio in the mixture [12]. For the comparison of grey rice husk ash and black rice husk ash in the metakaolin geopolymer, the study was done using mortar with metakaolin/rice husk ash ratios of 100/0, 80/20, 60/40, 40/60, and 20/80 with 10 M NaOH, sodium hydroxide-to-sodium silicate ratio of 1, liquid-to-solid binder ratio of 0.7, and curing at 60 °C for 24 hours. The compressive strengths of geopolymer mortars using black rice husk ash were consistently and substantially higher than those using grey rice husk ash. In addition, no swelling or expansion of composites has been observed during heat curing. This is mainly due to the presence of both amorphous and crystalline phases in the black rice husk ash. It has been shown that some formation of the semicrystalline or crystalline phases in the geopolymer matrix during the geopolymerization and heat curing leads to a desirable mechanical performance [78].

13.6.1.3.3 Rice Husk Ash/$Al(OH)_3$ Geopolymer

Additional alumina can also be achieved with the incorporation of $Al(OH)_3$. The study by Rattanasak et al. [25] showed that the use of high volume rice husk ash in making geopolymer with the aid of $Al(OH)_3$ is applicable but the problem of the stability of the product is of concern. The stability is maintained with the use of H_3BO_3 and heat curing at 115 °C for 48 hours and a mortar compressive strength of around 20.0 MPa is achieved. With the use of $Al(OH)_3$, the increase in Al content improves the Si/Al ratio and leads to a geopolymer with increased strength. The geopolymer sample also shows high elasticity and does not crack under the maximum load but continues to deform with the continuation of the loading. For durability, the compressive strengths of the geopolymer samples immersed in a 3% H_2SO_4 solution (representing the severe acid condition) and immersed in 5% $MgSO_4$ solution (representing the salt condition) increase approximately 20–30% with the increase in the $Al(OH)_3$ content.

The deterioration of rice husk ash/$Al(OH)_3$ geopolymer immersed in $MgSO_4$ solution involves the reaction of the hydroxyl ions in geopolymer with magnesium ions in 5% $MgSO_4$ solution resulting in magnesium hydroxide ($Mg(OH)_2$) [25]. This action involves the migration of hydroxide ions to the surface of the samples and the formation of insoluble brucite on the surface of the sample. In addition, the unit weight of immersed samples reduces with the increase in porosity. In the presence of high silica content, this magnesium hydroxide can slowly react with silica gel to form a hydrated magnesium silicate compound with no binding capacity. The immersion in $MgSO_4$ solution, therefore, resulted in the deterioration of materials and the loss of compressive strength [25].

Similarly, the strength loss occurs after immersion in H_2SO_4 solution but to a lesser degree [25]. Owing to the excess SiO_2 in rice husk ash geopolymer, sodium silicate (Na_2SiO_3) can be formed. When this product comes into contact with the acid solution (i.e. H_2SO_4), silicic acid (H_4SiO_4) is formed leading to mass loss as presented in Eqs. (13.2) and (13.3).

$$SiO_2(s) + 2NaOH(aq) \rightarrow Na_2O \cdot SiO_2(s) + H_2O(l) \tag{13.2}$$

$$Na_2O \cdot SiO_2(s) + 2H_2SO_4(aq) + H_2O(l) \rightarrow H_4SiO_4(aq) + Na_2SO_4(aq) \tag{13.3}$$

The immersion in H_2SO_4 solution also therefore resulted in the deterioration of materials and the loss in compressive strength but to a lesser extent compared to the immersion in $MgSO_4$ solution [25].

13.6.2 Bagasse Ash-based Geopolymer

13.6.2.1 Bagasse Ash Geopolymer

Bagasse ash is the waste from sugarcane processing and bioethanol plants. The by-product bagasse is used as a biofuel for steam generator use in the evaporation and drying processes. The burning of bagasse results in heat and energy and bagasse ash is the residue of the burning. Similar to rice husk ash, bagasse ash also contains approximately 70% silica. However, some bagasse ash can contain high CaO owing to the burning condition and dumping method. A number of researchers have reported that some bagasse ash is sometimes obtained from the settling lagoon of a sugarcane-processing plant. This ash is mixed with washed water from the sugarcane processing which is disposed of into the lagoon [74]. The washed water normally contains some lime ($Ca(OH)_2$) which is used to precipitate the sugarcane juice before its transfer to an evaporation drum. This collected bagasse ash thus has a high calcium content of around 15% and also a high LOI of around 30%. The SiO_2 content of this ash is reported to be at 31% and the Al_2O_3 content at 7% [74]. This Al_2O_3 content is sufficient to use bagasse ash as a starting material for geopolymer. However, the use of sole bagasse ash as source material is not recommended for a similar reason to that of rice husk ash.

To improve the properties of geopolymer, it is suggested that the bagasse ash undergoes a simple treatment or removal of foreign matters and contaminants of bagasse ash [52]. For example, 10 kg bagasse is burnt for 12–14 hours and the ash is sieved through a mesh with 2.38 mm opening to remove the unburnt matter before the ash is ground into a fine powder before the mixing step. This is important when the obtained bagasse ash is contaminated or mixed with undesirable substances. The obtained geopolymer shows a significant improvement in the compressive strength compared to the control mix [52].

13.6.2.2 Bagasse Ash Geopolymer with External Alumina Source

A number of researchers have investigated the use of various external alumina sources in combination with bagasse ash. Using only bagasse ash in a geopolymer mixture provides a low compressive strength of only 5.0 MPa, even with heat curing at 80 °C for 24 hours [48]. The external alumina materials employed in conjunction with bagasse ash include blast furnace slag [52, 74], fly ash [50], kaolin [48, 65], and spent coffee grounds (SCGs) [67]. Kaolin and fly ash contain both reactive SiO_2 and Al_2O_3 compounds, whereas blast furnace slag also contains high calcium content in addition to silica and alumina [52]. The maximum use of up to 50% of bagasse ash has been reported for making bagasse ash geopolymer with satisfactory strength [52].

13.6.2.3 Properties of Bagasse Ash-based Geopolymer

13.6.2.3.1 Fly Ash/Bagasse Ash Geopolymer

Low-calcium fly ash is used in conjunction with bagasse ash at a fly ash/bagasse ash ratio of 50/50 to 25/75 by weight [50]. Since this ash is retrieved from a lagoon, it has to be oven-dried at 105 °C and ground into fine particles. However, the mixture has a problem in setting and hardening as it contains a high LOI. So, re-calcination at 650 °C for two hours is required to improve its properties. The bagasse ash geopolymers with KOH and K_2SiO_3 cured at 20 and 65 °C for three days are reported to have good properties. The SiO_2/K_2O molar ratio affects the compressive strength of the geopolymer. The presence of potassium cation (K^+) supports the stability of the gel structure of the geopolymer and the mix with SiO_2/K_2O ratio of 0.75 provides a good compressive strength of 25.0–31.0 MPa at three days. When cured at low temperature (20 °C or room temperature), the geopolymer mixture with low-calcium fly ash as an external source material can set and give a reasonable strength of 7.0 MPa at three days. The use of bagasse ash as a sole starting material with potassium alkali, however, does not set or harden. It can be seen that heat curing is also a significant factor to improve the strength of a bagasse ash geopolymer. In addition, the use of high-volume bagasse ash in geopolymer results in a highly porous matrix, which absorbs the moisture in a humid environment and deteriorates with the formation of cracks. It is, therefore, important to reduce the LOI of bagasse ash, incorporate additional starting material with an active alumina content up to 20%, and use temperature curing to enhance the properties of bagasse ash geopolymer.

13.6.2.3.2 Metakaolin/Bagasse Ash Geopolymer

Even though metakaolin is a reactive material, a high volume of bagasse ash of more than 50% by weight is not recommended in the metakaolin/bagasse ash geopolymer system. The maximum strength at seven days of the metakaolin/bagasse ash geopolymer was 15.3 MPa when a metakaolin/bagasse ratio of 50/50 and heat curing at 60 °C for 24 hours were applied [65]. Kaolin can also be used instead of metakaolin. However, the early strength of kaolin bagasse ash geopolymer is low at 7 MPa. The strength is developed with increased curing time and reaches 11.0 MPa at 91 days [48]. The microstructural study with scanning electron microscope (SEM) shows that the heterogeneous geopolymer matrix can be clearly observed. Cracks and loose packing of particles have been clearly identified which are the main reasons for the low composite strength. The increase in fine aggregate in this system also causes a reduction in the geopolymer strength. When increasing the sand/source material ratio from 1/1 to 2/1 and 3/1, the strength reduces from 13.4 to 9.7 MPa and 6.4 MPa, respectively. The incorporation of sand into the geopolymer paste increases the heterogeneity of the matrix and thus reduces the strength of the geopolymer.

13.6.2.3.3 Slag/Bagasse Ash Geopolymer

Apart from fly ash, blast furnace slag is an alternative source material to be used in conjunction with bagasse ash. The slag contains a high amorphous phase and high calcium content that can react with the alkaline solution. In this system, CSH compound is also formed in addition to the aluminosilicate compound, and the strength of slag/bagasse ash geopolymer cured at 65 °C for three days is relatively high at 43.0 MPa. When the geopolymer is cured at room temperature, the strength increases with a longer curing time and reaches 70.0 MPa with a slag/bagasse ash ratio of 60/40. At high bagasse ash content, the matrix becomes more porous than the low bagasse ash matrix [74]. In this system, the pozzolanic reaction between SiO_2 and Al_2O_3 with $Ca(OH)_2$ in the geopolymer matrix proceeds over time. When using a high bagasse ash content of 50% with temperature curing at 65 °C for seven hours, the strength at 90 days is significantly reduced to 26.0 MPa [52]. Because the bagasse ash has a high porosity, the use of a high amount of bagasse ash in the mixture results in a mixture that is harsh and sometimes results in segregation due to the fact that the bagasse ash absorbs liquid in the mixture. To offset this, an additional alkali solution is usually required to maintain the workability or flow of the mixture.

13.6.3 Palm Oil Fuel Ash-based Geopolymer

13.6.3.1 Palm Oil Fuel Ash Geopolymer

POFA is a waste product from the steam and power generation of palm oil-processing mills. All parts of palm including frond, fiber, shell, and empty fruit bunch are used as a fuel source in the combustion process. The chemical compositions of POFA from different processing mills are quite variable depending on the geological conditions, fertilizer used, soil chemistry, and cultivation practices [79]. The wastes from different parts of the palm affect the chemical composition of the ash due to the different nutrient uptake of the palm trees. Therefore, the composition of the ash can vary in the same mill with different collection times. A number of large particles and unburnt material can be found in the ash owing to incomplete combustion. Therefore, sieving and grinding are commonly applied to reduce the POFA particle size and increase uniformity before utilization [17]. A POFA with a high degree of unburnt material needs to be burnt again. Elbasir et al. [60] used reburning at 500 °C for 90 minutes to remove unburnt carbon followed by grinding into small particle sizes.

13.6.3.2 Palm Oil Fuel Ash with External Alumina Source

The chemical compositions of POFA are similar to other biomass ashes with a low content of Al_2O_3 (less than 5%) and reasonably high calcium content of up to 10%. However, with some Al_2O_3 and reasonably high calcium content, it can be used as a sole starting material for making geopolymer mixtures. Depending on the amount of Al_2O_3, in the case of low Al_2O_3, the main binding compound is in the form of CSH due to the reaction between the calcium compound (~5–10%) and alkali solutions [53, 54, 60]. Coal fly ash [59, 70] and metakaolin [71] are good additional starting materials to synthesize a geopolymer with good mechanical properties.

13.6.3.3 Properties of Palm Oil Fuel Ash-based Geopolymer

13.6.3.3.1 Fly Ash/Palm Oil Fuel Ash Geopolymer

When low-calcium fly ash is used with POFA in the geopolymer mixture, heat curing (for example, at 65 °C for 24 hours) is needed [59], while for high-calcium fly ash, the curing can be done at room temperature [70]. In the latter case, CSH is a significant product that accelerates the setting and hardening with a resulting increase in compressive strength. Similar to other biomass ashes, the use of a high POFA content results in high SiO_2/Al_2O_3 ratio and leads to a low compressive-strength geopolymer [59] due to the lower amount of $Al(OH)_4^{4-}$ species available for geopolymerization reaction. The presence of Al species enables a higher rate of geopolymerization between aluminate and silicate ions resulting in a higher initial compressive strength.

13.6.3.3.2 Metakaolin/Palm Oil Fuel Ash Geopolymer

A small amount of metakaolin can be used in conjunction with POFA in making geopolymer. Metakaolin at 10% by weight with a high-concentration NaOH of 14 M and steel fiber is successfully used in making the high-volume POFA geopolymer concrete [71] with a maximum compressive strength of 30.0 MPa. Addition of water and/or alkali solution is/are needed to control the flow and workability of the mixture. The incorporation of steel fiber has little effect on the compressive strength of the geopolymer. However, the fiber enhances the bond resistance and delays crack propagation in the geopolymer matrix.

Apart from the steel fiber and polymer fiber, oil palm fiber has been used in high-calcium fly ash geopolymer to improve the flexural behavior of the composite [70]. The empty bunches are washed and boiled in water for two hours. Fibers are then dried and cut before the treatment with 2% NaOH solution for 24 hours. The fiber is mixed with fly ash at 1–3% by weight. It is found that the oil palm fiber significantly improves the flexural strength and toughness of the geopolymer. The high fiber content in the mixture transfers the brittle failure of the composite to a ductile failure.

13.6.3.3.3 Palm Oil Fuel Ash Geopolymer

POFA has been used successfully as a sole starting material in making an alkali-activated composite [53, 54, 60]. Without any external Al_2O_3, the aluminosilicate chain of Al–O–Si has difficulty in forming in the composite. With little CASH and NASH formed, the main products of the system are CSH and $Ca(OH)_2$ which are mainly responsible for the strength gain. Pretreatment of POFA with calcination at 500 °C for 90 minutes and grinding into small particle sizes have been performed in the case using only POFA as the sole source material. To improve the strength, highly concentrated NaOH (10–14 M) and a high amount of sodium silicate solution are preferred to obtain a good strength. In this system, the sodium silicate solution provides the Na^+ ions that facilitate the formation of sodium-silicate-rich gels, leading to a finer pore structure and influences the mechanical property by increasing the compressive strength of materials. The Na-binding gel and calcium-based gels of CSH and CASH are extensively detected in this composite [53]. The compressive strength of POFA composites increases with an increase in curing time due to the further development of the CSH compound. In addition, the increase in fineness of POFA particles contributes to a denser and more compact morphology of composites leading to high strength gain [60].

To improve the flexural strength, polyvinyl alcohol (PVA) fiber can be mixed into the geopolymer mixture [54]. The composites have a good resistance to the sulfate environment. However, the high concentration of NaOH results in adverse effects due to the rapid loss of alkalinity in the first week after exposure to 5% Na_2SO_4 and $MgSO_4$ solutions. A white substance is deposited at the composite surface, which was identified from the

XRD pattern as magnesium hydroxide ($Mg(OH)_2$) and calcium sulfate ($CaSO_4$) [80]. These composites are the product of the reaction between Mg^{2+}, SO_4^{2-}, and hydroxyl ion (OH^-) as shown in the following equations:

$$Mg^{2+}(aq)+SO_4^{2-}(aq)+Ca(OH)_2(s)+H_2O(l)\rightarrow Mg(OH)_2(s)+CaSO_4{\cdot}2H_2O(s) \tag{13.4}$$

The CSH compound in the composite can be attacked quite rapidly by $MgSO_4$ resulting in the formation of silica gel and loss in strength. The equation is described as follows:

$$\begin{aligned}&3Mg^{2+}(aq)+3SO_4^{2-}(aq)+3CaO{\cdot}SiO_2(s)+9H_2O(l)\rightarrow 3Mg(OH)_2(s)\\&+3[CaSO_4{\cdot}2H_2O](s)+SiO_2(s)\end{aligned} \tag{13.5}$$

The deterioration of samples exposed to a $MgSO_4$ solution is more severe than that exposed to a Na_2SO_4 solution. The reaction between Mg^{2+} and $Ca(OH)_2$ leads to a high void ratio and results in a low compressive strength of composites [80].

13.6.4 Other Biomass-based Geopolymers

Apart from the conventional biomass ashes, SCG has been studied as the source material of geopolymer synthesis [66]. SCG is an organic waste containing a high nitrogen content and therefore it is usually used as a fertilizer. Since the SCG waste increases annually, the utilization of SCG is widely studied. Due to the high moisture content in the SCG, an initial drying step is required. The main oxide compositions of SCG measured by XRF are 34% CaO, 43% K_2O, and 16% TiO_2, with little or no SiO_2 and Al_2O_3 detected [66]. Therefore, bagasse ash and blast furnace slag are selected to mix with 70% SCG using 8 M NaOH for making a geopolymer mixture. However, the composites give a low compressive strength of only 0.2–0.3 MPa. Materials with this compressive strength value can be developed for use in the road subgrade materials as the required compressive strength was only 0.3 MPa [66]. A slightly higher strength (1.5 MPa) is obtained with a medium temperature curing at 50 °C and a longer curing time of 90 days.

13.6.5 Use of Biomass in Making Sodium Silicate Solution and Other Products

13.6.5.1 Use of Biomass in Making Sodium Silicate Solution

A sodium silicate solution is an alkali solution that is usually preferred in making geopolymer. This solution can be prepared from a variety of biomass ashes. The common biomass used for this purpose is rice husk ash due to its availability, high silica content, and high reactivity. The sodium silicate solution is obtained from boiling a mixture of rice husk ash and NaOH solution at 90 °C for one hour [81]. The resulting sodium silicate solution is then used to prepare the fly ash geopolymer mortar. The strength of this geopolymer is similar to the geopolymer prepared from a commercial sodium silicate solution. In addition, SiO_2 can also be extracted from bagasse ash to produce a sodium silicate solution or water glass for use in a geopolymer mixture [72]. These data therefore confirm that most biomass ashes are suitable as raw materials for making a sodium silicate solution.

13.6.5.2 Use of Biomass Ash in Making Lightweight Aggregate

Rice husk ash was also used to make LWA using alkali-activation technology [64]. In this process, rice husk ash is mixed with 10 M NaOH solution and powdered boric acid, which is then heat-cured at 115 °C for 48 hours to obtain a hardened sodium silicate paste. The hardened paste is then crushed and heated at 550 °C for three hours to form the LWA with a density of 0.20–0.40 g cm^{-3} [64]. This LWA was suggested as usable for lightweight concrete panels, thermal insulation, noise absorption, and fire protection.

13.6.6 Fire Resistance of Bio-based Geopolymer

Regarding the thermal behavior of geopolymer, the use of high-volume POFA in the geopolymer matrix results in thermal instability when the composite is subjected to high temperature [70]. The strength of a geopolymer prepared from 50% POFA and 50% fly ash is significantly reduced from 20.0 MPa when subjected to a temperature of 300 °C to only 5.0 MPa at 1000 °C. At high POFA contents, the mixture contains increased organic matter and additional water is required to control the flow of the geopolymer mixture. For the performance of the geopolymer under high temperature, initially water is released during exposure to the high temperature. The thermogravimetric analysis also shows a high weight loss at 100 and 400 °C indicating water evaporation and decomposition of $Ca(OH)_2$, respectively [82]. In addition, a high-volume POFA geopolymer shows the bubble and foam-like structure when the composite is heated to 500 °C owing to the high SiO_2 content [59]. Therefore, using more than 50% of POFA is not recommended when the geopolymer is subjected to high temperatures.

Rice husk ash is used to improve the fire resistance of recycled aggregate high-calcium fly ash geopolymer concrete [83]. The addition of the ground rice husk ash effectively improved the geopolymer strength. The 28-days' compressive strengths of geopolymer are reasonably high at 36.0–38.1 MPa due to the improved microstructure and dense matrix. The incorporation of rice husk ash increases the silica content in the system and with the high calcium content, additional CSH is formed that coexists with NASH and CASH gels [75]. However, the inclusion of SiO_2-rich materials also had an adverse effect on the post-fire residual strength of geopolymer concretes, primarily due to the reduced porosity. It has been reported that using 20% rice husk ash blended with fly ash presents a good fire resistance similar to that of the original fly ash geopolymer. When 1000 °C fire was applied on the geopolymer surface with 1 cm thickness, the temperature at the opposite side of this rice husk ash geopolymer was 350 °C at 15 minutes [38]. When the rice husk ash content was increased to 40%, the temperature on the opposite side was reduced to 250 °C, presenting a good fire resistance of the rice husk ash geopolymer. However, cracks are formed due to the expansion of sodium silicate in the matrix.

13.7 Conclusion

Biomass ashes, viz. rice husk ash, POFA, bagasse ash, and rice husk bark ash are potential source materials for geopolymer synthesis for the building industry. With proper burning conditions (i.e. burning at temperatures of up to 700 °C with an adequate supply of air or oxygen), the obtained ash contains amorphous silica and low carbon or LOI value. These burning conditions are also obtainable from a simple open field burning. For the biomass ash obtained from the burning as a fuel source in a steam generator in a mill, the burning conditions are variable and the ashes are also variable and could contain large amounts of unburnt carbon and some re-calcination or reburning may be required.

Biomass ashes contain a high amount of amorphous SiO_2 which is one of the important compounds for making a geopolymer. It is, however, required to have other starting materials with some reactive alumina compound so that the mixture contains a usable ratio of SiO_2/Al_2O_3 to form a geopolymer binder under activation with high alkali solution. The other starting materials for an additional source of alumina mixture are fly ash, metakaolin, and GGBFS, and it has been shown that $Al(OH)_3$ solution could also be used as the alumina source to make a satisfactory geopolymer. The strength and durability of a biomass-based geopolymer are comparable to those of other geopolymer or Portland cement products. An increase in the silica content in the mixture is beneficial and enhances the durability of the geopolymer in both acid and sulfate conditions. For a larger increase in the silica content, the strength of the geopolymer is reduced with a change from a brittle material to a more elastic one.

The problems encountered with the use of biomass ash are as follows:

1. Transportation: the ash is bulky and is therefore difficult to transport which involves extra cost.
2. The burning conditions are not uniform, resulting in inhomogeneous biomass ashes. Extra effort is needed to make the feedstock for the starting material using biomass ash more uniform.

Despite these challenges, the use of biomass ashes is of much interest as it is a by-product and available in many parts of the world. The good cementitious products as well as the utilization of waste or by-products are attractive in the reduction of the greenhouse gas produced in the construction industry. The supplementary use of the biomass ash geopolymer should thus be increased in order to ensure the sustainability of the construction industry. Apart from the use of biomass-based geopolymer for building materials, biomass ashes can also be used to synthesize sodium silicate, LWA, and fire-resistant materials. The diverse applications should lead to the sustainable use of biomass ashes in construction materials.

References

1. Andrew, R.M. (2018). Global CO_2 emissions from cement production, 1928–2017. *Earth System Science Data* 10: 2213–2239. https://doi.org/10.5281/zenodo.831455.
2. Malhotra, V.M. (2002). Introduction: sustainable development and concrete technology. *ACI Concrete International* 24 (7): 22.
3. Austin, G.T. (1984). *Shreve's Chemical Process Industries*, 5th edition (International edition)e. Singapore: McGraw-Hill Book Company.
4. Chindaprasirt, P., Homwuttiwong, S., and Sirivivatnanon, V. (2004). Influence of fly ash fineness on strength, drying shrinkage and sulfate resistance of blended cement mortar. *Cement and Concrete Research* 34: 1087–1092. https://doi.org/10.1016/j.cemconres.2003.11.021.
5. Davidovits, J. (1991). Geopolymer: inorganic polymeric new materials. *Journal of Thermal Analysis* 37: 1633–1659. https://doi.org/10.1007/BF01912193.
6. Swamy, R.N. (1986). *Concrete Technology and Design vol. 3: Cement Replacement Materials*. London: Surrey University Press.
7. ASTM 618-19 (2019). *Standard Specification for Coal Fly Ash and Raw or Calcined Natural Pozzolan for Use in Concrete*. Pennsylvania: ASTM International.
8. Duràn-Herrera, A., Juàrez, C.A., Valdez, P., and Bentz, D.P. (2011). Evaluation of sustainable high-volume fly ash concretes. *Cement and Concrete Composites* 33: 39–45. https://doi.org/10.1016/j.cemconcomp.2010.09.020.

9. Chindaprasirt, P., Jaturapitakkul, C., Chalee, W., and Rattanasak, U. (2009). Comparative study on characteristic of fly ash and bottom ash geopolymers. *Waste Management* 29 (2): 539–543. https://doi.org/10.1016/j.wasman.2008.06.023.
10. Poon, C.S., Azhar, S., Anson, M., and Wong, Y.N. (2003). Performance of metakaolin concrete at elevated temperatures. *Construction and Building Materials* 25 (10): 83–89. https://doi.org/10.1016/S0958-9465(01)00061-0.
11. Somna, R., Jaturapitakkul, C., and Made, A.M. (2012). Effect of ground fly ash and ground bagasse ash on the durability of recycled aggregate concrete. *Cement and Concrete Composite* 34 (7): 848–854. https://doi.org/10.1016/j.cemconcomp.2012.03.003.
12. Chindaprasirt, P. (2009). *A Development of Rice Husk Ash-based Geopolymer*. Bangkok: National Science and Technology Development Agency.
13. Jaturapitakkul, C. and Cheerarot, R. (2003). Development of bottom ash as pozzolanic material. *Journal of Materials in Civil Engineering* 15 (1): 48–53. https://doi.org/10.1061/(ASCE)0899-1561(2003)15:1(48).
14. Cordeiro, G.C., Toledo Filho, R.D., Tavares, L.M., and Fairbairn, E.D.M.R. (2009). Ultrafine grinding of sugar cane bagasse ash for application as pozzolanic admixture in concrete. *Cement and Concrete Research* 39 (2): 110–115. https://doi.org/10.1016/j.cemconres.2008.11.005.
15. Rerkpiboon, A., Tangchirapat, W., and Jaturapitakkul, C. (2015). Strength, chloride resistance, and expansion of concretes containing ground bagasse ash. *Construction and Building Materials* 101: 983–989. https://doi.org/10.1016/j.conbuildmat.2015.10.140.
16. Chindaprasirt, P. and Rukzon, S. (2015). Strength and chloride resistance of the blended Portland cement mortar containing rice husk ash and ground river sand. *Materials and Structures* 48 (11): 3771–3777. https://doi.org/10.1016/S1674-4799(09)60083-2.
17. Kroehong, W., Damrongwiriyanupap, N., Sinsiri, T., and Jaturapitakkul, C. (2016). The effect of palm oil fuel ash as a supplementary cementitious material on chloride penetration and microstructure of blended cement paste. *Arabian Journal for Science and Engineering* 41 (12): 4799–4808. https://doi.org/10.1007/s13369-016-2143-1.
18. Alvarez-Ayuso, E., Querol, X., Plana, F. et al. (2008). Environmental, physical and structural characterisation of geopolymer matrixs synthesized from coal (co-)combustion fly ashes. *Journal of Hazardous Materials* 154: 175–183. https://doi.org/10.1016/j.jhazmat.2007.10.008.
19. Chindaprasirt, P., Chareerat, T., and Sirivivananon, V. (2007). Workability and strength of coarse high calcium fly ash geopolymer. *Cement and Concrete Composites* 29 (3): 224–229. https://doi.org/10.1016/j.cemconcomp.2006.11.002.
20. Chindaprasirt, P. and Rattanasak, U. (2016). Improvement of durability of cement pipe with high calcium fly ash geopolymer covering. *Construction and Building Materials* 112: 956–961. https://doi.org/10.1016/j.conbuildmat.2016.03.023.
21. Department of Alternative Energy Development and Efficiency, Thailand (2017). *Thailand Alternative Energy Situation*. Bangkok: Ministry of Energy.
22. Rattanasak, U. and Chindaprasirt, P. (2009). Influence of NaOH solution on the synthesis of fly ash geopolymer. *Minerals Engineering* 22 (12): 1073–1078. https://doi.org/10.1016/j.mineng.2009.03.022.
23. Hardjito, D., Wallah, S.E., Sumajouw, D.M.J., and Rangan, B.V. (2004). On the development of fly ash–based geopolymer concrete. *ACI Material Journal* 101 (6): 467–472.
24. Detphan, S. and Chindaprasirt, P. (2009). Preparation of fly ash and rice husk ash geopolymer. *International Journal of Minerals, Metallurgy and Materials* 16 (6): 720–726. https://doi.org/10.1016/S1674-4799(10)60019-2.
25. Rattanasak, U., Chindaprasirt, P., and Suwanvitaya, P. (2010). Development of high volume rice husk ash alumino-silicate composite. *International Journal of Minerals, Metallurgy and Materials* 17 (5): 654–659. https://doi.org/10.1007/s12613-010-0370-0.
26. Songpiriyakij, S., Kubprasit, T., Jaturapitakkul, C., and Chindaprasirt, P. (2010). Compressive strength and degree of reaction of biomass- and fly ash-based geopolymer. *Construction and Building Materials* 24 (3): 236–240. https://doi.org/10.1016/j.conbuildmat.2009.09.002.
27. Wongpa, J., Kiattikomol, K., Jaturapitakkul, C., and Chindaprasirt, P. (2010). Compressive strength, modulus of elasticity, and water permeability of inorganic polymer concrete. *Materials and Design* 31 (10): 4748–4754. https://doi.org/10.1016/j.matdes.2010.05.012.

28. Davidovits, J. (1999). Chemistry of geopolymeric systems, terminology. Geopolymer 1999 International Conference, Saint-Quentin, France (30 June–2 July 1999), 9–40.
29. Hua, X., van Deventer, J.S.J. (1999). The geopolymerisation of natural alumino-silicates. Proceedings: Second International Conference on Geopolymere 1999, Saint-Quentin, France (30 June–2 July 1999), 43–63.
30. van Jaarsveld, J.G.S., van Deventer, J.S.J., and Lorenzen, L. (1998). Factors affecting the immobilisation of metals in geopolymerised fly ash. *Metallurgical and Materials Transactions B* 29B: 283–291. https://doi.org/10.1007/s11663-998-0032-z.
31. Comrie, D.C. and Kriven, W.M. (2003). Composite cold ceramic geopolymer in a refractory application. *Advances in Ceramic Matrix Composites IX* 153: 211–225. https://doi.org/10.1002/9781118406892.ch14.
32. Inada, M., Tsujimoto, H., Eguchi, Y. et al. (2005). Microwave-assisted zeolite synthesis from coal fly ash in hydrothermal process. *Fuel* 84: 1482–1486. https://doi.org/10.1016/j.fuel.2005.02.002.
33. Palomo, A., Alonso, S., Fernandez-Jimenez, A. et al. (2004). Alkaline activation of fly ashes A 29Si NMR study of the reaction products. *Journal of the American Ceramic Society* 87: 1141–1145. https://doi.org/10.1111/j.1551-2916.2004.01141.x.
34. Gourley, J.T. (2003). Geopolymers; opportunities for environmentally friendly construction materials. Proceedings: Materials 2003 Conference, Adaptive Materials for a Modern Society, Sydney, New South Wales, Australia (1–3 October 2003).
35. Vladimir, Z. (2004). High effective silica fume alkali activator. *Bulletin of Material Science* 27 (2): 179–182. https://doi.org/10.1007/BF02708502.
36. Damer, S.A. (1976) Rice hull ash as pozzolanic material. M Eng. Thesis no. 953. Bangkok: Asian Institute of Technology.
37. Chindaprasirt, P., Kanchanda, P., Sathonsaowaphak, A., and Cao, H.T. (2007). Sulfate resistance of blended cements containing fly ash and rice husk ash. *Construction and Building Materials* 21 (6): 1356–1361. https://doi.org/10.1016/j.conbuildmat.2005.10.005.
38. Rattanasak, U. and Chindaprasirt, P. (2009). *Rice Husk Ash in Concrete*. Ubon Ratchathani, Thailand: Science and Engineering Publishing.
39. Singh, B. (2018). Rice husk ash. In: *Waste and Supplementary Cementitious Materials in Concrete* (ed. R. Siddique and P. Cachim), 417–460. Oxford, UK: Woodhead Publishing https://doi.org/10.1016/B978-0-08-102156-9.00013-4.
40. Basu, P. (1999). Combustion of coal in circulating fluidized-bed boilers: a review. *Chemical Engineering Science* 54: 5547–5557. https://doi.org/10.1016/S0009-2509(99)00285-7.
41. Vassilev, S.V., Baxter, D., Andersen, L.K., and Vassileva, C.G. (2013). An overview of the composition and application of biomass ash: part 2. Potential utilisation, technological and ecological advantages and challenges. *Fuel* 105: 19–39. https://doi.org/10.1016/j.fuel.2012.10.001.
42. Awang, H.B. and Al-Mulali, M.Z. (2018). The inclusion of palm oil fuel ash biomass waste in concrete: a literature review. In: *Palm Oil* (ed. V. Waisundara), 117–132. London, UK: IntechOpen https://doi.org/10.5772/intechopen.76632.
43. Montakarntiwong, K., Chusilp, N., Tangchirapat, W., and Jaturapitakkul, C. (2013). Strength and heat evolution of concretes containing bagasse ash from thermal power plants in sugar industry. *Materials and Design* 49: 414–420. https://doi.org/10.1016/j.matdes.2013.01.031.
44. Ramjan, S., Tangchirapat, W., and Jaturapitakkul, C. (2018). Effects of binary and ternary blended cements made from palm oil fuel ash and rice husk ash on alkali-silica reaction of mortar. *Arabian Journal for Science and Engineering* 43 (4): 1941–1954. https://doi.org/10.1007/s13369-017-2843-1.
45. Rattanachu, P., Tangchirapat, W., and Jaturapitakkul, C. (2019). Water permeability and sulfate resistance of eco-friendly high-strength concrete composed of ground bagasse ash and recycled concrete aggregate. *Journal of Materials in Civil Engineering* 31 (6): 1–8. https://doi.org/10.1061/(ASCE)MT.1943-5533.0002740.
46. Chindaprasirt, P., Rukzon, S., and Sirivivatnanon, V. (2008). Resistance to chloride penetration of blended Portland cement mortar containing palm oil fuel ash, rice husk ash and fly ash. *Construction and Building Materials* 22 (5): 932–938. https://doi.org/10.1016/j.conbuildmat.2006.12.001.
47. Chindaprasirt, P., Homwuttiwong, S., and Jaturapitakkul, C. (2007). Strength and water permeability of concrete containing palm oil fuel ash and rice husk-bark ash. *Construction and Building Materials* 21 (7): 1492–1499. https://doi.org/10.1016/j.conbuildmat.2006.06.015.

48. Tippayasam, C., Kaewpapasson, P., Thavorniti, P. et al. (2014). Effect of Thai kaolin on properties of agricultural ash blended geopolymers. *Construction and Building Materials* 51: 455–459. https://doi.org/10.1016/j.conbuildmat.2013.11.079.
49. Wen, N., Zhao, Y., Yu, Z., and Liu, M. (2019). A sludge and modified rice husk ash-based geopolymer: synthesis and characterization analysis. *Journal of Cleaner Production* 226: 805–814. https://doi.org/10.1016/j.jclepro.2019.04.045.
50. Castaldelli, V.N., Moraes, J.C.B., Akasaki, J.L. et al. (2016). Study of the binary system fly ash/sugarcane bagasse ash (FA/SCBA) in SiO_2/K_2O alkali-activated binders. *Fuel* 174: 307–316. https://doi.org/10.1016/j.fuel.2016.02.020.
51. Oliveira, J.A., Cunha, F.A., and Ruotolo, L.A.M. (2019). Synthesis of zeolite from sugarcane bagasse fly ash and its application as a low-cost absorbent to remove heavy metals. *Journal of Cleaner Production* 229: 956–963. https://doi.org/10.1016/j.jclepro.2019.05.069.
52. Pereira, A., Akasaki, J.L., Melges, J.L.P. et al. (2015). Mechanical and durability properties of alkali-activated mortar based on sugarcane bagasse and blast furnace slag. *Ceramics International* 41: 13012–13024. https://doi.org/10.1016/j.ceramint.2015.07.001.
53. Runyut, D.A., Robert, S., Ismail, I. et al. (2018). Microstructure and mechanical characterization of alkali-activated palm oil fuel ash. *Journal of Materials in Civil Engineering* 30 (7): 04018119. https://doi.org/10.1061/(ASCE)MT.1943-5533.0002303.
54. Salami, B.A., Johari, M.A.M., Ahmad, Z.A., and Maslehuddin, M. (2017). Durability performance of palm oil fuel ash-based engineering alkaline-activated cementitious composite (POFA-EACC) mortar in sulfate environment. *Construction and Building Materials* 131: 229–244. https://doi.org/10.1016/j.conbuildmat.2016.11.048.
55. Shah, B.A., Shah, A.V., Mistry, C.B. et al. (2011). Surface modified bagasse fly ash zeolites for removal of reactive black-5. *Journal of Dispersion Science and Technology* 32 (9): 1247–1255. https://doi.org/10.1080/01932691.2010.505550.
56. Shah, B., Tailor, R., and Shah, A. (2012). Zeolitic bagasse fly ash as a low-cost sorbent for the sequestration of p-nitrophenol: equilibrium, kinetics, and column studies. *Environmental Science and Pollution Research* 19 (4): 1171–1186. https://doi.org/10.1007/s11356-011-0638-6.
57. Sata, V., Jaturapitakkul, C., and Kiattikomol, K. (2004). Utilization of palm oil fuel ash in high-strength concrete. *Journal of Materials in Civil Engineering* 16: 623–628. https://doi.org/10.1061/(ASCE)0899-1561(2004)16:6(623).
58. Tangchirapat, W., Jaturapitakkul, C., and Chindaprasirt, P. (2009). Use of palm oil fuel ash as a supplementary cementitious material for producing high-strength concrete. *Construction and Building Materials* 23 (7): 2641–2646. https://doi.org/10.1016/j.conbuildmat.2009.01.008.
59. Ranjbar, N., Mahrali, M., Alengaram, J. et al. (2014). Compressive strength and microstructural analysis of fly ash/palm oil fuel ash based geopolymer mortar under elevated temperatures. *Construction and Building Materials* 65: 144–121. https://doi.org/10.1016/j.conbuildmat.2014.04.064.
60. Elbasir, O.M.M., Jahari, M.A.M., and Ahmad, Z.A. (2019). Effect of finesse of palm oil fuel ash on compressive strength and microstructure of alkaline activated mortar. *European Journal of Environmental and Civil Engineering* 23 (2): 136–152. https://doi.org/10.1080/19648189.2016.1271362.
61. Costa, J.A.S. and Paranhos, C.M. (2018). Systematic evaluation of amorphous silica production from rice husk ashes. *Journal of Cleaner Production* 192: 688–697. https://doi.org/10.1016/j.jclepro.2018.05.028.
62. Kalapathy, U., Proctor, A., and Shutz, J. (2000). A simple method for production of pure silica from rice hull ash. *Bioresource Technology* 73 (3): 257–262. https://doi.org/10.1016/S0960-8524(99)00127-3.
63. Padhi, R.S., Patra, R.K., Mukharjee, B.B., and Dey, T. (2018). Influence of incorporation of rice husk ash and coarse recycled concrete aggregates on properties of concrete. *Construction and Building Materials* 173: 289–297. https://doi.org/10.1016/j.conbuildmat.2018.03.270.
64. Chindaprasirt, P., Jaturapitakkul, C., and Rattanasak, U. (2009). Influence of fineness of rice husk ash and additives on the properties of lightweight aggregate. *Fuel* 88 (1): 158–162. https://doi.org/10.1016/j.fuel.2008.07.024.
65. Amin, N., Faisal, M., Muhammad, K., and Gul, S. (2016). Synthesis and characterization of geopolymer from bagasse bottom ash, waste of sugar industries and naturally available china clay. *Journal of Cleaner Production* 129: 491–495. https://doi.org/10.1016/j.jclepro.2016.04.024.

66. Arulrajah, A., Kua, T.A., Suksiripattanapong, C. et al. (2017). Compressive strength and microstructural properties of spent coffee grounds-bagasse ash based geopolymers with slag supplement. *Journal of Cleaner Production* 162: 1491–1501. https://doi.org/10.1016/j.jclepro.2017.06.171.
67. Chandara, C., Sakai, E., Azizli, K.M. et al. (2010). The effect of unburned carbon in palm oil fuel ash on fluidity of cement pastes containing superplasticizer. *Construction and Building Materials* 24 (9): 1590–1593. https://doi.org/10.1016/j.conbuildmat.2010.02.036.
68. Chindaprasirt, P. and Rattanasak, U. (2018). Fire-resistant geopolymer bricks synthesized from high-calcium fly ash with outdoor heat exposure. *Clean Technologies and Environmental Policy* 20 (5): 1097–1103. https://doi.org/10.1007/s10098-018-1532-4.
69. Zhu, H., Liang, G., Xu, J. et al. (2019). Influence of rice husk ash on the waterproof properties of ultrafine fly ash based geopolymer. *Construction and Building Materials* 208: 394–401. https://doi.org/10.1016/j.conbuildmat.2019.03.035.
70. Kroehong, W., Jaturapitakkul, C., Pothisiri, T., and Chindaprasirt, P. (2018). Effect of oil palm fiber content on the physical and mechanical properties and microstructure of high-calcium fly ash geopolymer paste. *Arabian Journal for Science and Engineering* 43 (10): 5215–5224. https://doi.org/10.1007/s13369-017-3059-0.
71. Bashar, I.I., Alengaram, U.J., Jumaat, M.Z. et al. (2016). Engineering properties and fracture behaviour of high volume palm oil fuel ash based fibre reinforced geopolymer concrete. *Construction and Building Materials* 111: 286–297. https://doi.org/10.1016/j.conbuildmat.2016.02.022.
72. Tchakoute, H.K., Ruscher, C.H., Hinch, M. et al. (2017). Utilization f sodium waterglass from sugar cane bagasse ash as a new alternative hardener for producing metakaolin-based geopolymer cement. *Chemie der Erde* 77: 257–266. https://doi.org/10.1016/j.chemer.2017.04.003.
73. van Jaarsveld, J.G.S., van Deventer, J.S.J., and Lukey, G.C. (2002). The effect of composition and temperature on the properties of fly ash- and kaolinite-based geopolymers. *Chemical Engineering Journal* 89 (1–3): 36–73. https://doi.org/10.1016/S1385-8947(02)00025-6.
74. Castaldelli, V.N., Akasaki, J.L., Melges, J.L.P. et al. (2013). Use of slag/sugar cane bagasse ash (SCBA) blends in the production of alkali-activated materials. *Materials* 6: 3108–3127. https://10.3390/ma6083108.
75. Somna, K., Jaturapitakkul, C., Kajitvichyanukul, P., and Chindaprasirt, P. (2011). NaOH-activated ground fly ash geopolymer cured at ambient temperature. *Fuel* 90 (6): 2118–2124. https://doi.org/10.1016/j.fuel.2011.01.018.
76. Chindaprasirt, P. and Rattanasak, U. (2017). Characterization of the high-calcium fly ash geopolymer mortar with hot-weather curing systems for sustainable application. *Advanced Powder Technology* 28 (9): 2317–2324. https://doi.org/10.1016/j.apt.2017.06.013.
77. Kim, K.D., Choi, K.Y, Yang, J.W. (2005) Formation of spherical hollow silica particles from sodium silicate solution by ultrasonic spray pyrolysis method. *Colloids and Surfaces A* 254(1–3): 193–198. https://doi.org/10.1016/j.colsurfa.2004.12.009.
78. Hussain, M., Varley, R.J., Cheng, Y.B. et al. (2005). Synthesis and thermal bahavior of inorganic-organic hybrid geopolymer composites. *Journal of Applied Polymer Science* 96: 112–121. https://doi.org/10.1002/app.21413.
79. Ooi, Z.X., Ismail, H., Bakar, A.A., and Teoh, Y.P. (2014). A review on recycling ash derived from *Elaeis guineensis* by-product. *Bioresources* 9 (4): 7926–7940.
80. Chindaprasirt, P., Jenjirapanya, J., and Rattanasak, U. (2014). Characterizations of FBC/PCC fly ash geopolymeric composites. *Construction and Building Materials* 66 (1): 72–78. https://doi.org/10.1016/j.conbuildmat.2014.05.067.
81. Chindaprasirt, P., Rattanasak, U., Sittichot, J., Songpiriyakij, S. (2009) Thai petty patent no. 4836: production of sodium silicate solution from rice husk ash. Bangkok: DIP Thailand.
82. Chindaprasirt, P., Paisitsrisawat, P., and Rattanasak, U. (2014). Strength and resistance to sulfate and sulfuric acid of ground fluidized bed combustion fly ash–silica fume alkali-activated composite. *Advanced Powder Technology* 25 (3): 1087–1093. https://doi.org/10.1016/j.apt.2014.02.007.
83. Nuaklong, P., Jongvivatsakul, P., Pothisiri, T. et al. (2020). Influence of rice husk ash on mechanical properties and fire resistance of recycled aggregate high-calcium fly ash geopolymer concrete. *Journal of Cleaner Production* 252 (10): 119797. https://doi.org/10.1016/j.jclepro.2019.119797.

14

The Role of Bio-based Excipients in the Formulation of Lipophilic Nutraceuticals

Alexandra Teleki[1], Christos Tsekou[2] and Alan Connolly[2]

[1]*Science for Life Laboratory, Department of Pharmacy, Uppsala University, Uppsala, Sweden*
[2]*DSM Nutritional Products Ltd., Nutrition R&D Center Forms and Application, Basel, Switzerland*

14.1 Introduction

Nutraceuticals play an essential role in human health and well-being. They have been defined as food or part of food that provides medical or health benefits including the prevention and/or treatment of a disease [1]. A healthy, well-balanced diet typically provides all necessary micronutrients such as vitamins, minerals, and trace elements. However, different population groups such as infants, pregnant and/or lactating women, and the elderly, can have specific vitamin requirements. These groups run a higher risk of suffering from micronutrient deficiencies and their associated health problems [2]. In many countries, an adequate diet might not be available to the entire population due to lack of availability or as a result of the high costs associated with fresh fruits and vegetables. Dietary supplements and/or food fortification with vitamins and nutraceuticals can contribute to improved health in these populations. Examples include vitamin A and iron supplementation in developing countries, folic acid for pregnant women, and vitamin D for infants, children, and the elderly [2]. Furthermore, in the era of industrialised food production, there is a growing consumer demand for products of high nutritional value [3].

Lipophilic nutraceuticals such as vitamins A, D, E, and K are rarely sold in their pure form due to their sensitivity to light and oxygen. These vitamins are typically encapsulated in a

High-Performance Materials from Bio-based Feedstocks, First Edition. Edited by Andrew J. Hunt, Nontipa Supanchaiyamat, Kaewta Jetsrisuparb and Jesper T.N. Knijnenburg.

protective matrix before being fortified into food products or dietary supplements (i.e. tablets and capsules). The role of the formulation process is to ensure vitamin stability during processing and subsequent storage (i.e. protect against light, humidity, and oxygen), taste and odour masking, controlled release, increased bioavailability, and improved handling. Vitamin formulation most commonly entails microencapsulation technologies to produce microcapsules that entrap particles or droplets of the bioactive ingredient in a protective matrix [4]. Here, biopolymers or other bio-based raw materials play a key role as wall materials, emulsifiers, and/or stabilisers. The most common food-grade biopolymers include maltodextrin, gum acacia, starch, gelatin, and plant proteins. Originally, formulated vitamins were produced with the aid of surface-active agents or surfactants such as Tween and Brij. However, there is growing concern regarding the adverse health risks associated with such synthetic surfactants. This has led to an increased consumer demand for natural (plant-, microbial-, or animal-based) and Generally Recognized As Safe (GRAS) alternatives [5, 6].

This chapter focuses on the use of bio-based raw materials as functional ingredients used in the formulation of lipophilic nutraceuticals intended for food applications. The initial focus of this chapter will be on the role these bio-based raw materials play as processing aids for the generation of stable nano-emulsions, a prerequisite for the delivery of stable lipophilic nutraceuticals [7, 8]. Special focus will be given to current bio-based raw materials being utilised as well as alternative biomaterials currently being explored, e.g. saponins, lecithins, cellulose, and zein protein [9, 10]. Another focus of this chapter will be the generation of colloidal delivery vesicles which are being designed for the formulation of lipophilic nutraceuticals. These techniques have been highlighted for their ability to stabilise sensitive lipophilic nutraceuticals while also increasing bio-accessibility [11, 12]. The final section of this chapter will focus on drying technologies currently employed to deliver dry lipophilic formulated nutraceuticals (e.g. spray drying) as well as exploring promising alternatives such as spray-freeze drying, electrohydrodynamic focusing, fluid bed granulation, and extrusion. The chapter aims to give an overview of the most recent advances made in the field and thus primarily focuses on literature published in the past 10 years. Earlier studies have partially been covered in a previous review article published by the first author for the 100th year anniversary of vitamins [4]. For a more general overview of microencapsulation [3], the interested reader is referred to the following reviews specifically for vitamin A [13], vitamin D [14], and pro-vitamin A (β-carotene) [15] as well as technologies to enhance the dispersibility of bioactive compounds [16].

14.2 Emulsions and the Importance of Bio-based Materials as Emulsifiers

14.2.1 Conventional Micro- and Nanoemulsions

Fat-soluble vitamins are often formulated as oil-in-water (O/W) emulsions to improve their bioavailability when used as food fortificants [7, 8]. Further enhanced functionalities of emulsions include vitamin protection against degradation via oxidative, hydrolytic, or thermal pathways as well as controlled release of the active lipophilic compound. Emulsions are typically classified according to their size as microemulsions, nanoemulsions (<300 nm), or according to their characteristics, i.e. self-emulsifying delivery systems [16]. In emulsions, naturally derived raw materials play a crucial role as emulsifiers as they adsorb at the O/W interface, thereby reducing the interfacial tension. The resulting emulsions are often further processed into dry powders by drying technologies (see Section 14.4).

The most commonly used emulsifiers in vitamin formulations for food applications include starches (specially modified with octenyl succinic anhydride (OSA) [17]), gum acacia, and proteins (especially whey protein). The chemical modification of starch with OSA is achieved via a standard esterification reaction, resulting in an amphiphilic character. Due to the hydrophobic and steric attributes of the OSA-starch, it is used as an effective stabiliser of O/W emulsions [18]. Gum acacia is an amphiphilic polysaccharide derived from acacia trees and is commonly used in beverage emulsions. It has a hydrophobic protein fraction covalently linked to the hydrophilic polysaccharide structure, thus rendering the molecule active at the O/W interface. Whey protein isolate (WPI) contains a mixture of globular proteins, primarily β-lactoglobulin. WPI can be used to form O/W emulsions with a small droplet size at low concentrations, especially when compared to gum acacia. This can be attributed to the low molecular weight of WPI, its high surface activity, and the thin interfacial protein layer formed in comparison to the previously discussed polysaccharide [19]. Droplet stabilisation by gum acacia primarily occurs by a steric hindrance to flocculation. Gum acacia increases the viscosity of the aqueous phase which can hinder droplet break-up in emulsions. Compared to polysaccharides, emulsions stabilised by proteins are more sensitive to environmental factors such as pH, salt concentration, and temperature [8].

Lipophilic nutraceuticals including vitamins are commonly formulated into **nano-emulsions** [8]. This is important for improving the stability and bio-accessibility of the emulsified nutraceutical while also minimising turbidity in beverage applications [20]. Furthermore, the improved oral bioavailability of nano-emulsions compared to conventional emulsions (>1 μm) has been reported [21]. This can be correlated to the enhanced dispersibility as the surface area increases in nanoemulsions compared to conventional emulsions [10]. In order to achieve droplet sizes <100 nm, small non-ionic surfactants such as Tween 20 or Tween 80 are typically used [8]. However, current research efforts are focused on reducing the use of these surfactants and replacing them with bio-based materials. An emerging group of research is focused on the natural surfactant **Quillaja saponins** (e.g. Q-Naturale®), which is isolated from the bark of the *Quillaja saponaria* Molina tree [22, 23]. *Quillaja* saponins are triterpenoid saponins consisting of a hydrophobic quillaic acid backbone with hydrophilic sugar moieties, thus giving rise to the amphiphilic nature of the molecule. The hydrophobic backbone is either a steroid or a triterpene [22]. *Quillaja* saponins have been used as natural surfactants in the formation of pro-vitamin A [24], vitamin D3 [25], and vitamin E-based emulsions [9, 23, 26]. Yang and McClements [23] demonstrated that *Quillaja* saponin was more effective at stabilising vitamin E emulsions with increased concentrations compared to Tween 80. However, the viscosity of the oil and aqueous phases had to be carefully tuned to achieve small droplet sizes (<250 nm).

Lecithins are a class of naturally derived surfactants based on phospholipids that consist of glycerol, two fatty acids, and choline phosphate. The chain length, position, and degree of unsaturation of the fatty acids can vary. Phosphatidylcholine is an example of a commonly used lecithin. However, naturally derived lecithins are typically composed of a complex mixture of several lipid structures. Ozturk et al. [26] prepared vitamin E nanoemulsions using soy lecithin or *Quillaja* saponin as surfactants. The authors showed that lower *Quillaja* saponin concentrations (0.5%) were required compared to lecithin (2%) to stabilise emulsions with a small droplet size (<130 nm). While lecithin nanoemulsions were only stable at pH 7–8 and with low salt concentrations (emulsions unstable >100 mM NaCl), *Quillaja* saponin emulsions were stable over a wider pH range (pH 3–8) and with higher salt concentrations (unstable >500 mM NaCl). Reichert et al. [5] studied the use of natural cosurfactants such as sodium caseinate, pea protein, rapeseed lecithin, and egg lecithin together with *Quillaja* saponins on

O/W emulsion characteristics. Synergistic emulsifying properties were obtained for *Quillaja* saponin-lecithin systems when both surfactants were used at low concentrations. *Quillaja* saponin and lecithins have also been used as stabilisers in solid lipid nanoparticles containing vitamin A, pro-vitamin A, and Omega-3 fish oils [27].

Lv et al. [9] recently carried out a comparative study of vitamin E emulsions stabilised by gum acacia, *Quillaja* saponin or WPI. Significantly smaller droplet sizes were obtained in *Quillaja* saponin and WPI emulsions compared to gum acacia (Figure 14.1). *Quillaja* saponin

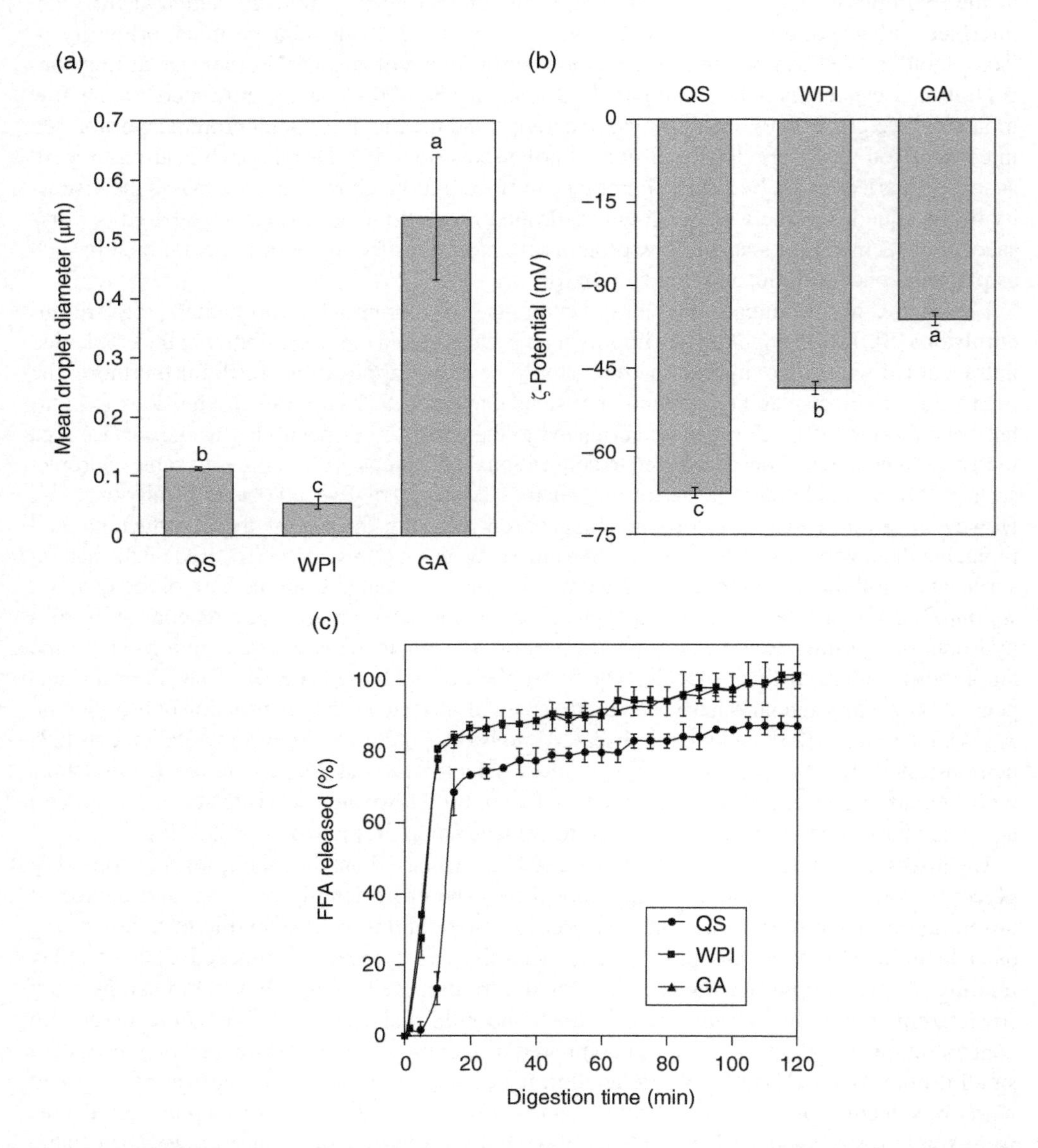

Figure 14.1 *Mean droplet diameter ($d_{3,2}$) (a) and zeta potential at pH 7 (b) of vitamin E emulsions stabilised with Quillaja saponin, whey protein isolate (WPI) or gum acacia. Free fatty acid released during in vitro digestion simulating the small intestine (c). Source: Reproduced with permission from [9].*

and WPI can both reduce interfacial surface tension more effectively compared to gum acacia [28]. The molecular weight of gum acacia is greater than that of both *Quillaja* saponin and WPI emulsifiers resulting in slower adsorption onto the freshly formed oil droplet surface during emulsification [28]. Saponin and gum acacia emulsions were resistant to droplet aggregation over a wide pH range (2–8). In contrast, WPI-stabilised emulsions were highly sensitive to pH as it affects the electrostatic stabilisation of the droplets, especially around the isoelectric point of the protein (pH 5). Interestingly, lipid digestion was slower in saponin emulsions compared to gum acacia and WPI-based emulsions. This was attributed to the strong interfacial film formed by the saponin on the oil droplet surface which inhibited bile salt and enzyme adsorption [9].

14.2.2 Pickering-Stabilised Emulsions

Emulsions stabilised by particles, so-called Pickering emulsions, are also gaining attention as alternative delivery systems [29, 30]. In these emulsions, solid particles adsorb at the O/W interface (rather than surfactants as in conventional emulsions). The main advantage of particle-stabilised emulsions is their high stability against coalescence. This occurs because the energy required to remove a particle from the interface is considerably higher than that needed to desorb surfactants [31]. The use of Pickering emulsions as delivery vehicles in food applications has primarily been hindered by the challenge to find effective GRAS stabilisers. Another challenge is the size of Pickering emulsions (typically 1–10 μm) that might affect the appearance and stability of the emulsion in food products. Silica, although not a bio-based raw material, is the most extensively studied Pickering emulsion stabiliser and it often has to be modified in order to control its wetting properties and render it partially hydrophobic. Alison et al. [32, 33] used silica nanoparticles non-covalently modified with chitosan to stabilise O/W emulsions and studied the influence of pH on their structural properties. In that system, the chitosan concentration also governed emulsion structure and stability as both Pickering and network stabilisation mechanisms were involved [32, 33]. However, the use of silica or modified silica particles may not be in line with all consumer demands for clean-label, naturally derived food additives [34]. Other **naturally derived particles** include starch granules, chitin nanocrystals, cellulose microparticles, zein, and soy protein particles, flavonoid particles, and protein microgel particles [30]. The reader is referred to several recent reviews on Pickering emulsions for food applications for more detailed accounts on such emulsion stabilisers [35–38].

Despite this large interest in Pickering emulsions for food applications, only a few studies have demonstrated the use of these systems as delivery vehicles for vitamins. Recently, Mitbumrung et al. [39] encapsulated vitamin D3 in Pickering emulsions stabilised by nanofibrillated **cellulose** (NFC) extracted from mangosteen rind (*Garcinia mangostana*). NFC isolated from this rind has a diameter of approximately 60 nm and can readily be dispersed as fibres in water. The vitamin D3 emulsions stabilised by NFC had relatively large droplet sizes (10–15 μm), but exhibited good stability over a range of pH (3–7) and ionic strength (<500 mM NaCl). Hedjazi and Razavi [40] investigated the impact of cellulose nanocrystals from different sources on the stability of Pickering emulsions. The authors concluded that cellulose nanocrystals derived from cotton exhibited better stability compared to those produced using bacteria. This was primarily attributed to the smaller particle size of the former.

Pickering emulsions prepared from **protein nanoparticles** such as zein and gliadin (a wheat protein) are often highly unstable under certain conditions due to the pH sensitivity of such proteins. This can be overcome by the formation of protein–biopolymer complexes (analogous to the silica–chitosan studies described previously) that stabilise emulsions by synergistic Pickering and network mechanisms. Fu et al. [41] prepared a wheat gluten–xanthan gum electrostatic complex as Pickering stabilisers of pro-vitamin A emulsions. Xanthan gum hindered the creaming of the emulsions, which was crucial for the emulsion stability. The improved stability of the Pickering emulsions was demonstrated with respect to temperature, pH, and salt levels. These emulsions were also effective at inhibiting chemical degradation of pro-vitamin A. Ma et al. [42] prepared Pickering pro-vitamin A emulsions using gliadin nanoparticle–gum acacia complexes. Oil droplets of approximately 6 μm in size were formed in the presence of gum acacia (compared to 9.4 μm with only gliadin) with enhanced resistance to changes in pH, ionic strength, and temperature.

14.3 Novel Formulation Technologies: Colloidal Delivery Vesicles

14.3.1 Microgels

A formulation technology gaining increased interest is the use of soft solids (microgels) to entrap lipophilic vitamins or nutraceuticals in a cross-linked polymer matrix [43–45]. Microgels are thus defined as a colloidal dispersion (or suspension) of gel-like particles [46]. These gel-like particles are kinetically stable structures held together by strong covalent bonding [42]. In microgels, emulsion droplets are stabilised by an emulsifier in combination with a gelling agent. This creates a soft solid shell around several emulsion droplets which are then incorporated into a continuous gel matrix. These gel-like particles can also be used as stabilisers for Pickering emulsions [11]. A wide range of food biopolymers can be used for microgel fabrication including casein, starch, and whey protein.

Zhang et al. [47] prepared a carboxymethyl starch/β-cyclodextrin microgel loaded with ascorbic acid, where the release rate of ascorbic acid was governed by the swelling degree of the microgel. The swelling of microgel particles can be triggered by changes in pH, temperature, or the presence of ions [45]. Encapsulated substances are also released due to the breakdown of the microgel structure which is triggered by external factors such as enzyme activity or high shear forces. Mun et al. [48] prepared rice starch hydrogels loaded with a pro-vitamin A emulsion (stabilised by WPI; Figure 14.2). The authors showed that the hydrogel formulation exhibited higher bio-accessibility of pro-vitamin A compared to conventional emulsions. The hydrogel matrix protects the emulsion droplets from aggregation during the simulated oral and gastric phases of digestion. Thus, the emulsion droplets carrying pro-vitamin A remained evenly distributed throughout the system while the pure emulsion droplets aggregated together. In a follow-up study [49], the authors also demonstrated that the lipid digestion rate could be further modulated by the choice of starch used to produce the hydrogels (rice vs. mung bean starch). Starches with a higher amylose content generate gels with a more compact and rigid structure, which positively affects gel strength [50]. Stronger gels, in turn, exhibited better protection against matrix erosion by amylase digestion. The lipid digestion rate and pro-vitamin A bio-accessibility can be further controlled by the addition of indigestible polysaccharides (i.e. methylcellulose) to the microgels [12]. Chen et al. [51] encapsulated vitamin E in WPI gels and evaluated the effect of emulsifiers such as glycerol monostearate, Tween 20, OSA-modified starch, and gum

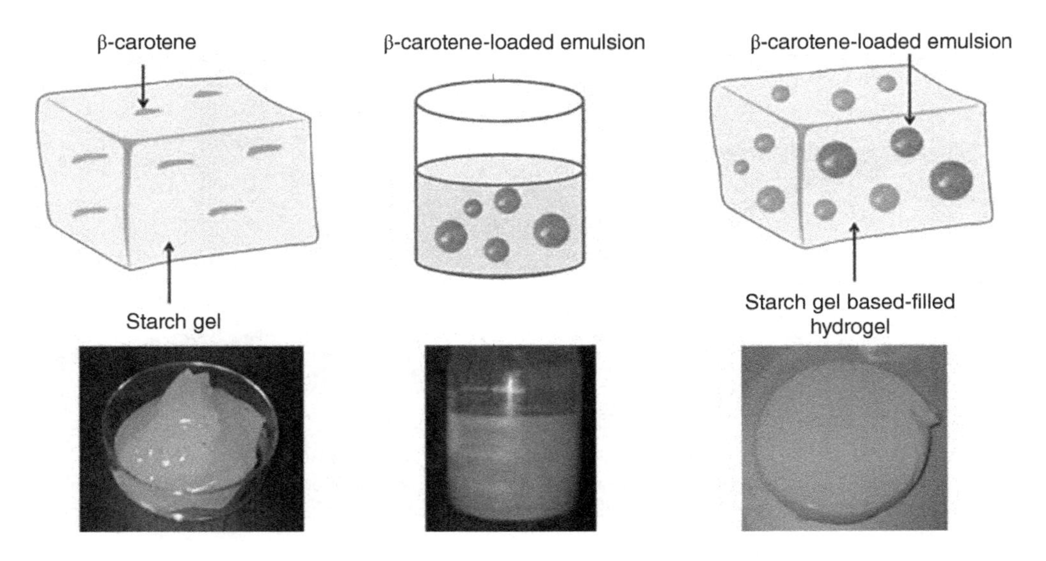

Figure 14.2 Schematic representation and visual appearance of starch hydrogel, oil-in-water (O/W) pro-vitamin A emulsion and filled hydrogel. Source: Reproduced with permission from [48].

acacia on the structure of the gels. The biopolymer emulsifiers (OSA-starch and gum acacia) strengthened the gel network by forming complexes with WPI, which, in turn, improved the storage stability of vitamin E.

14.3.2 Nanoprecipitation

Colloidal delivery vesicles can be prepared by the nanoprecipitation of proteins. Zein, the major storage protein in corn, has previously been highlighted for food and nutritional applications [52]. **Zein** is a low-cost GRAS co-product of corn biodiesel and starch production which is soluble in 70–90% aqueous ethanol mixtures [53]. When these mixtures are dispersed in water under shear, zein nanoparticles are formed (so-called solvent-induced nanoprecipitation). Sensitive hydrophobic ingredients such as vitamins can be encapsulated in these zein nanoparticles resulting in increased protection against oxidation and hydrolysis. Pan et al. [54] compared the oxidative stability of retinol encapsulated in either zein colloidal particles or an O/W emulsion. The authors reported enhanced barrier property of zein colloidal particles against peroxyl radical-induced oxidation but oxygen permeation could not be limited in either of the delivery systems (Figure 14.3). Digestion of zein nanoparticles is limited in the stomach, while complete digestion occurs in the intestine [54]. Studies have also suggested that zein exhibits mucoadhesive properties, prolonging its residence time in the gastrointestinal tract and thus making it attractive for controlled release applications [55].

Zang et al. co-encapsulated resveratrol and α-tocopherol in zein nanoparticles resulting in 67 and 97% encapsulation efficiency, respectively [56]. Higher loading of active ingredients in zein nanoparticles can be achieved by flash nanoprecipitation. In this process, amphiphilic molecules such as casein self-assemble around the hydrophobic (zein) core. The organic and

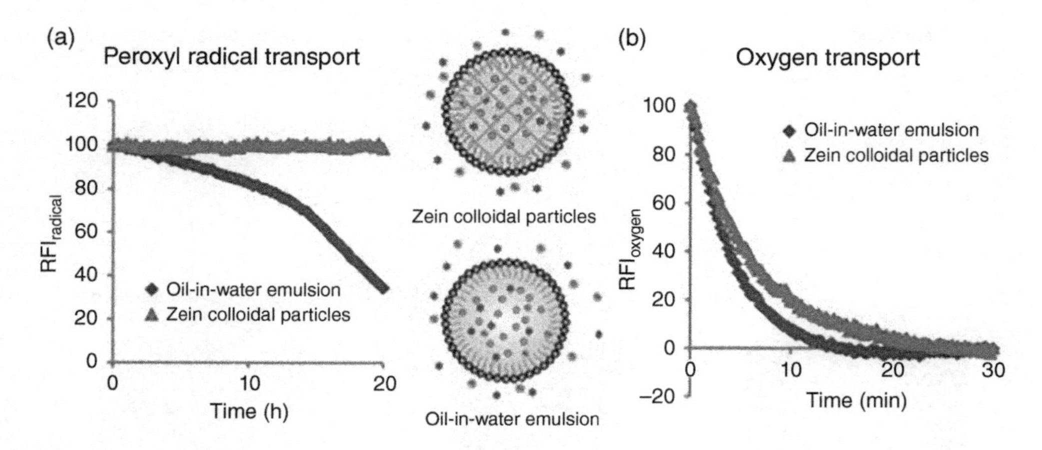

Figure 14.3 Permeation of peroxyl radicals (a) and oxygen (b) from the aqueous phase to the core of zein colloidal particles and O/W emulsion. Source: Reproduced with permission from [54].

aqueous solutions are rapidly mixed within a confined geometry to ensure homogeneous supersaturation prior to nucleation and growth of zein nanoparticles. This process was used to form particles with a vitamin E acetate core (45%) encapsulated by zein and casein [57].

14.3.3 Liposomes

Liposomes have been widely used for medical and pharmaceutical applications and have also gained interest in the food industry for the controlled delivery of functional ingredients such as proteins, enzymes, vitamins, and flavours [58]. Liposomes are spherical vesicles (Figure 14.4), typically a few 100 nm in size, composed of bio-based phospholipids (polar lipids) in a bilayer structure that enables the co-encapsulation of both hydrophilic and hydrophobic bioactives [59] such as vitamins E and C [60]. Liposomes are considered natural, biodegradable, and non-toxic delivery vehicles. The most commonly used phospholipid in liposomes is phosphatidylcholine, also referred to as lecithin which has already been discussed here (Section 14.2.1). Encapsulation efficiency of bioactives in liposomes is governed by the lipid bilayer stability and structure [61]. Incorporation of cholesterol or stearic acid can

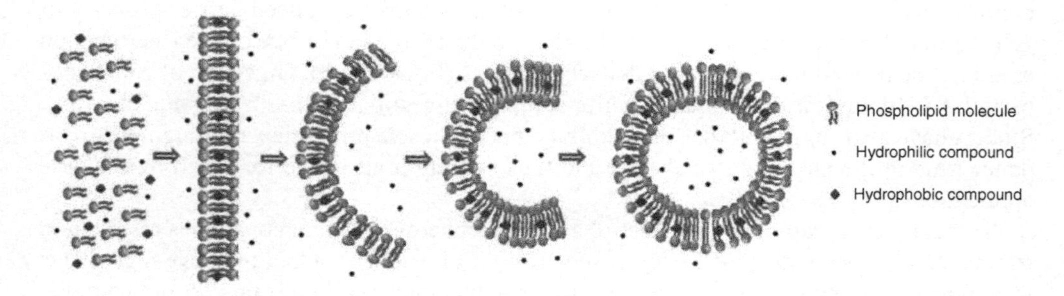

Figure 14.4 Schematic of phospholipids, a phospholipid bilayer, and formation of liposome structure encapsulating hydrophilic and hydrophobic compounds. Source: Reproduced with permission from [58].

enhance bilayer stability [62]. For example, in a liposomal formulation containing vitamins E and C, the bilayer stability and rigidity are increased by the incorporation of stearic acid. The resulting liposomal formulation exhibited a protective effect on the antioxidant activity of the heat-labile vitamin C, even after pasteurisation [63].

Chaves et al. [64] produced multi-lamellar liposomes co-encapsulating two hydrophobic compounds: curcumin and vitamin D3. The liposome dispersion was stabilised by the addition of a mixture of xanthan gum and guar gum and showed good stability upon storage for more than 40 days. The polysaccharides act as thickening agents and can thereby delay the destabilisation of the phospholipid vehicles [65]. Liposomes have also been encapsulated in alginate or alginate-pectin microparticles. However, this did not improve the vitamin encapsulation efficiency or retention compared to the free liposome formulation control [66].

14.3.4 Complex Coacervation

Coacervation is a promising microencapsulation technology that has been used in the food and pharmaceutical industries for the encapsulation of lipophilic active ingredients [67]. This formulation technology can result in encapsulation efficiencies of up to 99%, even with high loading of lipophilic active ingredients. There are two types of coacervation processes, namely simple and complex coacervation. During simple coacervation, a single polymer is involved and the process is characterised by the separation of the liquid phase through the addition of an electrolyte [68]. Meanwhile, during complex coacervation, two or more oppositely charged polymers in an aqueous solution are involved [69]. Microencapsulation of an active ingredient by coacervation occurs by the phase separation of one or more hydrocolloids from the initial solution and the subsequent deposition of the newly formed coacervate phase around the active ingredient that is suspended or emulsified in the same reaction medium (Figure 14.5).

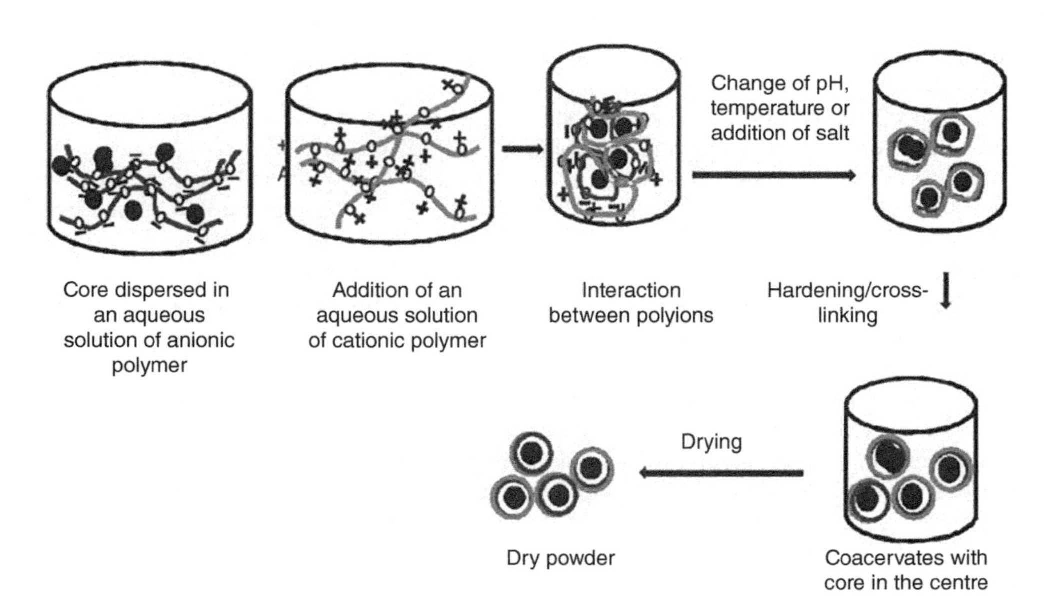

Figure 14.5 Overview of the complex coacervation process. Source: Reproduced with permission from [67].

The resulting shell can then be cross-linked by enzymatic or chemical cross-linking mechanisms to provide a more stable structure [70]. The interior structure of the resulting coacervates can be mononuclear (comprised of one oil droplet within the core) or multinuclear (comprised of several discrete oil droplets within the core). Studies have shown that a multinuclear internal structure can result in coacervates with an improved release profile while also significantly reducing the cost during industrial manufacturing [71, 72]. Bio-based materials such as gelatin and gum acacia are traditionally used to generate complex coacervates, where gelatin acts as the positive electrolyte and gum acacia acts as the negative electrolyte [73]. However, due to ethnic and religious reasons, current research is focused on identifying natural bio-based alternatives to gelatin. Plant-derived proteins are increasingly viewed as a suitable alternative to gelatin but certain limitations need to be overcome before they can be commercially applied [74]. As a result of EU law, all novel protein sources must be assessed for their potential to cause an allergenic response in humans prior to being launched onto the food market (European Commission (EC) regulation No. 258/97 and European Union (EU) recommendation 97/618 EC; http://eur-lex.europa.eu/). Also, as previously mentioned, emulsions stabilised by proteins are more sensitive to environmental factors such as pH, salt concentration, and temperature [8].

Complex coacervation produces a rigid, non-porous structure that offers a good barrier to oxygen. As a result, it has been widely used in order to improve the stability of ingredients that are sensitive to oxidation [75]. For example, pro-vitamin A coacervates formed using gum tragacanth and casein resulted in significantly higher stability after three months' storage compared to pro-vitamin A in oil [76]. Yuan et al. [77] demonstrated that the combination of soy protein isolate (SPI) with chitosan resulted in complex coacervates with improved oxidative stability when compared to simple coacervates generated with SPI alone. In the same study, it was found that enzymatic crosslinking with transglutaminase further improved oxidative stability, most likely due to the increased mechanical strength of the coacervate wall. Huang et al. [78] also demonstrated that enzymatic cross-linking increased the mechanical wall strength for coacervates generated using SPI-chitosan.

Complex coacervation has also been shown to protect sensitive lipophilic compounds during gastrointestinal digestion. Timilsena et al. [70] encapsulated chia seed oil in a matrix comprised of chia seed protein isolate with chia seed gum and demonstrated that 60% of the oil was hydrolysed during *in vitro* digestion, whereas almost all the non-encapsulated oil was hydrolysed, suggesting improved bio-accessibility as a result of encapsulation within the coacervate. Ilyasoglu and El [79] formed eicosapentaenoic acid (EPA) and docosahexaenoic acid (DHA) coacervates using sodium caseinate and gum acacia. The results demonstrated that 56.2% EPA and 47.4% DHA reached the large intestine without digestion. This would suggest that the combination of coacervating agents used is important in order to ensure maximal bio-accessibility when targeted release in the small intestine is required.

Co-encapsulation of actives is also possible using complex coacervation. Co-encapsulation can be used to deliver multiple active ingredients in a singular vessel while also improving the stability of sensitive active ingredients. Wang et al. [80] produced a gelatin and hexametaphosphate coacervate containing tuna oil, curcumin, coenzyme Q10 as well the vitamins A, E, D_3, and K_2. FTIR analysis of the resulting microcapsules indicated that no oxidation of the actives occurred, which suggests that there was no negative interaction between the individual components. Eratte et al. [81] co-encapsulated probiotic bacteria with tuna oil in a WPI and gum acacia matrix. The resulting coacervate demonstrated significantly higher viability of the probiotic bacteria while also greatly improving the oxidative stability of the tuna oil, suggesting a synergistic effect. Tuna oil, being high in Omega-3 oils, may have exhibited a protective effect and maintained the viability of the bacteria during processing.

14.3.5 Complexation

Cyclodextrins (CDs) are natural bio-based cyclic oligosaccharides derived from starch with six (α), seven (β), or eight (γ) glucose residues linked by α(1–4) glycosidic bonds [82]. The steric arrangement of these glucose units in the CD molecule results in a hollow truncated cone with a hydrophilic outer surface, which makes CDs water-soluble, and a hydrophobic internal cavity, which enables CDs to form inclusion complexes with various hydrophobic guest molecules. Thus, the complexation of lipophilic bioactive ingredients in CDs has been researched for decades, although their application in food products remains limited due to their high cost and the limited loading capacity of active compounds achievable. Kimura and Ooya [83] compared the solubilisation of α-tocopherol in hydroxypropyl (HP)-βCDs and hyperbranched polyglycerol-modified β-CD and found complexation efficacy significantly higher in the latter.

14.4 Key Drying Technologies Employed During Formulation

The majority of industrially produced vitamin and lipophilic nutraceutical formulations are marketed as dry powders due to their enhanced chemical and microbiological stability (compared to, e.g., emulsions) as well as their ease of handling, transportation, and storage. Bio-based raw materials play an important role as excipients during drying processes as they constitute the wall material of the dry product powder which infers the desired physical and chemical properties on the product.

14.4.1 Spray Drying

Spray drying is the most widely used drying technology and is associated with large-scale, continuous operation, and low production costs [84]. During spray drying, emulsions or suspensions are atomised at elevated temperature and pressure via a nozzle into a heated chamber. Evaporation of water and particle formation occur simultaneously under controlled conditions resulting in shriveled particle morphology, typical for spray-dried powders (Figure 14.6c). Emulsion properties such as droplet size and viscosity as well as the choice of stabilisers determine the resulting particle characteristics. In addition, the viscosity of the emulsion must not be too high in order to facilitate the atomisation process while also ensuring spherical particles are obtained [86]. The emulsion must be physically stable with submicron oil droplet size in order to maximise encapsulation efficiency. Encapsulation efficiency is one of the most important parameters during spray drying as it directly affects the properties of the final product. Low encapsulation efficiency results in high surface oil on the final dried particles which can result in poor product characteristics, i.e. stability and flowability [87]. Numerous factors influence encapsulation efficiency including emulsion droplet size and viscosity. The amorphous fraction and molecular weight of matrix materials can also influence encapsulation efficiency [88].

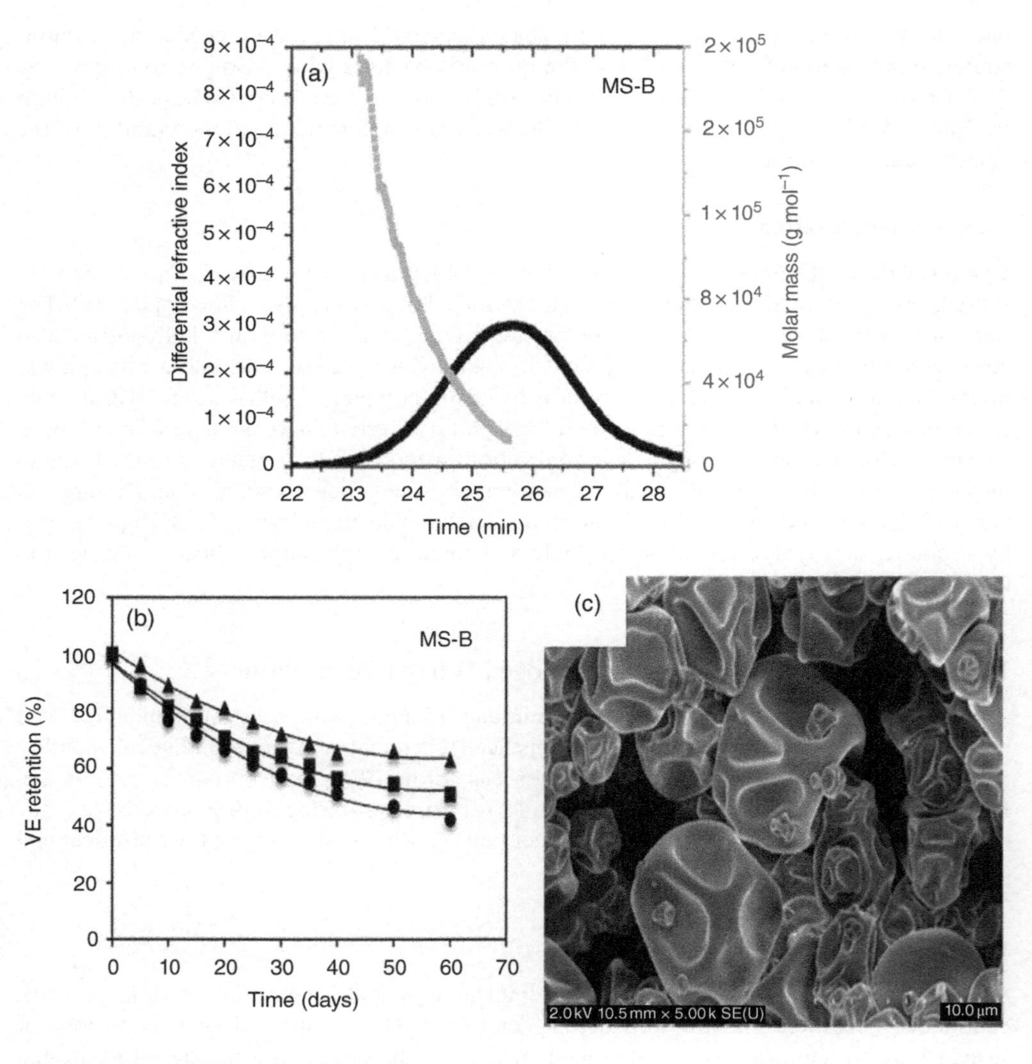

Figure 14.6 Weight-average molecular weight and differential refractive index as a function of elution time for an OSA-modified food starch (a). Vitamin E retention in spray-dried capsules with OSA-modified starch at different storage temperatures (triangles: 4 °C, squares: 20 °C, and circles: 35 °C) as a function of time (b). Scanning electron microscope image of vitamin E capsule morphology after 10 days of storage at room temperature and 73% RH. The scale bar is 10 μm (c). Source: Reproduced with permission from [85].

Recent studies have applied systematic experimental designs to better understand the interplay between process conditions and material properties used during spray drying. For example, Corrêa-Filho et al. [89] used response surface methodology to study the effect of wall material (gum acacia) concentration and drying inlet temperature on product particle properties (i.e. active content, encapsulation efficiency and drying yield) of encapsulated pro-vitamin A. Goncalves et al. [90] also studied the effect of gum acacia concentration on spray-dried vitamin A microcapsules and identified a minimal gum acacia concentration necessary for vitamin A stabilisation. Jafari et al. [91] applied a Taguchi design to study the effect of inlet air temperature during spray drying and different combinations of carrier agents (maltodextrin, gum acacia, modified starch, and whey protein concentrate) on the physicochemical characteristics of vitamin D3-enriched whey protein powder. An increase in maltodextrin concentration in the vitamin D3-containing feed solution resulted in smoother and more spherical particles.

To increase the physicochemical stability of the encapsulated actives, a mixture of matrix materials is often used [92]. Maltodextrin, OSA-modified starches [17], and gum acacia are well known bio-based wall materials used for particle formation during spray drying. Combinations of polysaccharides and sugars are also frequently applied due to their low viscosities at high concentrations coupled with their high aqueous solubility. Maltodextrins are manufactured enzymatically or by acid hydrolysis and their properties are characterised by their dextrose equivalent (DE) [93]. Maltodextrins are commonly used in combination with other surface-active carbohydrates, such as OSA-maltodextrin to obtain stable emulsions [94]. OSA-modified starches also contain a fraction of these simple sugars that act as fillers and can enhance the stability of encapsulated actives (e.g. pro-vitamin A) after spray drying [95]. Hategekirnana et al. [85] studied the effect of different OSA-modified starches on the properties of spray-dried vitamin E. Low molecular weight OSA-starches were the most effective at generating stable vitamin E particles (Figure 14.6). The OSA-modified starch had a monomodal molecular weight distribution with an average of $4.3 \times 10^4\,g\,mol^{-1}$ (Figure 14.6a) and the highest vitamin E retention was obtained from emulsions spray-dried with the lowest molecular weight OSA-modified starch. The degradation of vitamin E as a function of storage temperature followed the Weibull model (Figure 14.6b). Proteins such as WPI are also widely used during spray drying. However, they are sensitive to denaturation and aggregation encountered during the drying process, which can alter the resulting product if not adequately controlled [6]. In order to avoid this, Nesterenko et al. [96] modified SPI with fatty acid chains via acylation to render the macromolecule more amphiphilic. This increased amphiphilic nature enhanced the encapsulation efficiency of vitamin E.

14.4.2 Spray-Freeze Drying

During spray drying, high inlet temperatures can result in thermal degradation of heat-sensitive materials or loss of volatile active ingredients. As a result, alternative technologies are required when the drying of such material is required. Freeze drying is an alternative to spray drying and is typically employed for these heat-sensitive compounds. However, the long drying times as well as batch operation result in high operational costs limiting their use to certain sectors. A recent development in drying technologies is spray-freeze drying. This combines both the spray-drying and freeze-drying processes in a single unit operation which exploits the advantages of both technologies [97, 98]. Spray-freeze drying consists of three steps: (i) atomisation of droplets, (ii) solidification of droplets in contact with cold fluid, and (iii) sublimation at low temperature and pressure conditions [99]. The process has gained interest for processing volatiles, biological, and food supplements.

Parthasarathi and Anandharamakrishnan [100] prepared vitamin E microcapsules by spray drying, freeze drying, and spray-freeze drying using WPI as a carrier agent. Figure 14.7 shows the product particle morphology obtained from the different drying processes. Freeze-dried

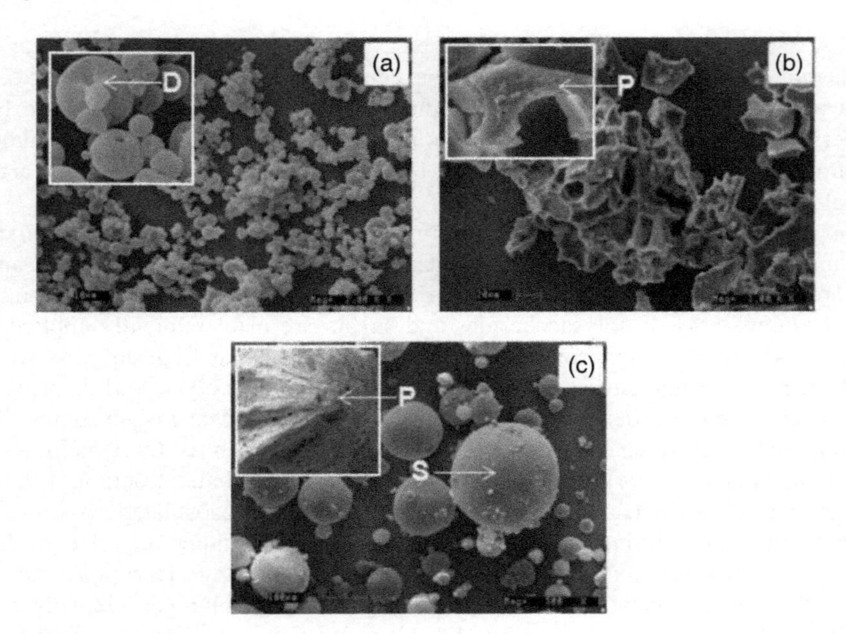

Figure 14.7 Scanning electron microscope images of vitamin E microcapsules produced by spray drying (a), freeze drying (b), and spray-freeze drying (c). Source: Reproduced with permission from [100].

and spray-freeze-dried particles had a porous structure that was formed during sublimation of ice crystals which resulted in higher dissolution rates compared to spray-dried microcapsules. *In vivo* data indicated that spray-freeze-dried microcapsules exhibited slightly higher vitamin E bioavailability upon oral administration in Wistar rats. This demonstrates that spray-freeze drying is a suitable alternative to conventional drying technologies and could be used to control the release of bioactive ingredients [98].

14.4.3 Electrohydrodynamic Processing

Electrohydrodynamic processes such as electrospraying or electrospinning are alternative technologies to spray drying which can circumvent the use of high temperatures during formulation [101]. In these technologies, a high-voltage electrostatic field charges the surface of a biopolymer solution, forming a liquid jet of fine droplets that are then directed towards a grounded collector. The jet/droplets are subjected to electrostatic forces on their way to the collector and solidify by solvent evaporation and cooling resulting in fibre or particle formation. The encapsulation of a wide range of different biocompounds has been demonstrated in electrospun/electrosprayed structures, i.e. Omega-3 fatty acids and vitamins A and E. Biopolymers used in these processes include amaranth, pullulan, zein, whey, and soy proteins [101]. The thermal stability and photosensitivity of folic acid were improved by encapsulation in electrospun amaranth protein isolate–pullulan fibres [102]. Pérez-Masiá et al. [103] compared electrospraying and spray drying for the formulation of a lipophilic active ingredient using whey protein concentrate as a matrix material. Higher encapsulation efficiencies and thermal stability of the active ingredient were obtained using electrospraying. Celebioglu and Uyar [104] produced inclusion complexes of vitamin E and HP-β-CD. The resulting nanofibres could rapidly dissolve in water and release the encapsulated vitamin E.

14.4.4 Fluid Bed Drying

Fluid bed drying (FBD) has shown promise as an alternative technology for lipid encapsulation [105, 106]. FBD combines encapsulation, drying and agglomeration into a single process step resulting in granules with good flow and dissolution characteristics. The position of the spray nozzle is important and can be altered depending on the starting material. Top spray, bottom spray, and tangential spray are most commonly applied. During FBD, an emulsion of the sensitively active ingredient is sprayed onto a fluidised bed of 'seed' particles. The emulsion droplets then adhere to the seed particles and are dried by a jet of warm fluidised air. Choice of seed particle is important as it directly affects the morphology of the final granules. Spherical seed particles will result in granules with a smoother outer surface, while coarser seed particles will result in granules with a less regular outer surface [106]. The size of the resulting granules can be controlled by adjusting the spraying time and increasing/decreasing the atomising pressure of the spray nozzle.

So far, no studies have reported the FBD of vitamins or pro-vitamins. However, Benelli et al. [105] compared the efficiency of spray drying with FBD of phytochemical extracts. The biobased excipients gum acacia and OSA-starch were used as emulsifiers and the respective emulsions were then dried using both technologies. Fluidised bed granules exhibited better retention of the active compounds while also resulting in better flow properties. A follow-up study compared the efficiency of FBD using an emulsion generated with gum acacia or whey protein concentrate. Again, high retention of the volatile markers was observed in both cases compared to spray drying [106]. In a similar study, Pellicer et al. [107] compared different drying techniques and their effect on the stability of volatile flavour compounds. Strawberry flavours were encapsulated using OSA-starch and maltodextrin with or without β-cyclodextrin. Results indicate that for all compounds, FBD resulted in the lowest retention of flavour volatiles when compared to spray and freeze drying. However, the addition of β-cyclodextrin significantly improved volatile retention over five months' storage at 4 and 25 °C. This effect was associated with the structure of the β-cyclodextrin which binds the volatile ingredients within its core and thereby increasing retention demonstrating that the choice of material used to form the wall material is important in maximising volatile retention.

14.4.5 Extrusion

Although technically not a drying technology, extrusion is carried out at low moisture contents which can negate the need for a drying step. Extrusion is a relatively simple, scalable, solvent-free process, that includes hot-melt extrusion, hot-melt granulation or melt extrusion, and wet granulation [108, 109]. Melt extrusion is an all-in-one process where the matrix materials, liquid oils, and water can be injected directly into the extruder. By applying sufficient shear forces within the extruder, an emulsion can be formed. The wet emulsified molten mass is then forced through a die under pressure and allowed to cool and harden for further downstream processing by pelletising or milling. Carbohydrates are commonly used during melt extrusion as they enter a rubbery state once sufficient temperature or/and plasticisation has been applied and then return to the glassy state upon cooling [110]. This dense glassy state minimises the free volume and molecular mobility providing a suitable environment for volatile flavour retention and extended shelf-life stability of sensitive materials [111]. Bio-based raw materials such as starch and maltodextrin are typically used due to their low hygroscopicity as well as their neutral taste and aroma [112]. Depending on the characteristics of the desired product, proteins derived from milk, cereals, and animals may also be utilised [113, 114].

The Durarome® process was the first commercially available line of encapsulated flavours introduced by Firminech S.A., which took advantage of these characteristics [115]. The resulting glassy products exhibited minimal volatile loss over time. However, volatile loss significantly increased when the product was exposed to higher relative humidity as a result of the sample reverting to a rubbery state. Tackenberg et al. [116] encapsulated orange terpenes in a glassy carbohydrate matrix. Orange terpene content remained unchanged after 12 weeks' storage. Chang et al. [117] encapsulated vitamin E in a matrix of OSA starch and lecithin and retained 93% of the vitamin E while minimal losses were observed after four weeks' storage at 25 and 30 °C.

14.5 Conclusions and Future Perspectives

Bio-based raw materials are important ingredients used during the formulation of lipophilic nutraceuticals for application in fortified foods, beverages, and dietary supplements. Current research has largely focused on well-known bio-based, GRAS raw materials such as modified starches, gum acacia, and plant proteins that are food-approved. Key properties of the formulation, such as stability, dissolution rate and bioavailability, are governed by these biomaterials. Emulsion technologies coupled with spray drying are the most prevalent for the production of dry lipophilic nutraceutical-containing powders. Despite their industrial use for decades, the intricate interplay between these lipophilic nutraceuticals, the formulation ingredients, and the process parameters are not widely understood outside of certain industries. Recent studies have proposed the use of systematic experimental design and statistical data evaluation to advance our understanding across different length scales in these systems. Comparative studies across different formulation technologies (i.e. using different drying technologies) are important to study the role of biopolymers in different processes. Such efforts are also key to tackle the inherent variability of physicochemical properties of bio-based raw materials. *Quillaja* saponins are a promising new group of natural surfactants with the desired properties to replace synthetic surfactants currently used as formulation aids. New interfacial structures with novel functionalities are also sought in order to develop synergistic systems that combine the properties of different formulation aids. For example, emulsion microgels can provide unprecedented lipophilic nutraceutical microencapsulation properties. However, novel characterisation techniques are required to study these complex systems at the nanoscale (e.g. small-angle X-ray scattering, atomic force microscopy, and positron annihilation spectroscopy [118–121]. By increasing our knowledge of these properties, we can then optimise the formulation process to ensure the delivery and release of the active ingredient in a controlled manner. The performance of the formulation then needs to be verified using a combination of *in vitro* and *in vivo* assays to ensure they protect the active ingredient from gastric and/or intestinal digestion in order to improve their bioavailability.

References

1. Chauhan, B., Kumar, G., Kalam, N., and Ansari, S.H. (2013). Current concepts and prospects of herbal nutraceutical: a review. *Journal of Advanced Pharmaceutical Technology & Research* 4 (1): 4–8.
2. Rautiainen, S., Manson, J.E., Lichtenstein, A.H., and Sesso, H.D. (2016). Dietary supplements and disease prevention – a global overview. *Nature Reviews Endocrinology* 12 (7): 407–420.
3. Correa-Filho, L.C., Moldao-Martins, M., and Alves, V.D. (2019). Advances in the application of microcapsules as carriers of functional compounds for food products. *Applied Sciences* 9 (3): 571.

4. Teleki, A., Hitzfeld, A., and Eggersdorfer, M. (2013). 100 years of vitamins: the science of formulation is the key to functionality. *Kona Powder and Particle Journal* 30: 144–163.
5. Reichert, C.L., Salminen, H., Bonisch, G.B. et al. (2018). Concentration effect of *Quillaja saponin* – Co-surfactant mixtures on emulsifying properties. *Journal of Colloid and Interface Science* 519: 71–80.
6. Silva, E.K., Azevedo, V.M., Cunha, R.L. et al. (2016). Ultrasound-assisted encapsulation of annatto seed oil: whey protein isolate versus modified starch. *Food Hydrocolloids* 56: 71–83.
7. McClements, D.J. (2018). Enhanced delivery of lipophilic bioactives using emulsions: a review of major factors affecting vitamin, nutraceutical, and lipid bioaccessibility. *Food & Function* 9 (1): 22–41.
8. Ozturk, B. (2017). Nanoemulsions for food fortification with lipophilic vitamins: production challenges, stability, and bioavailability. *European Journal of Lipid Science and Technology* 119 (7): 1500539.
9. Lv, S.S., Zhang, Y.H., Tan, H.Y. et al. (2019). Vitamin E encapsulation within oil-in-water emulsions: impact of emulsifier type on physicochemical stability and bioaccessibility. *Journal of Agricultural and Food Chemistry* 67 (5): 1521–1529.
10. Odriozola-Serrano, I., Oms-Oliu, G., and Martín-Belloso, O. (2014). Nanoemulsion-based delivery systems to improve functionality of lipophilic components. *Frontiers in Nutrition* 1: 24.
11. Destribats, M., Wolfs, M., Pinaud, F. et al. (2013). Pickering emulsions stabilized by soft microgels: influence of the emulsification process on particle interfacial organization and emulsion properties. *Langmuir* 29 (40): 12367–12374.
12. Mun, S., Park, S., Kim, Y.R., and McClements, D.J. (2016). Influence of methylcellulose on attributes of beta-carotene fortified starch-based filled hydrogels: optical, rheological, structural, digestibility, and bioaccessibility properties. *Food Research International* 87: 18–24.
13. Gonçalves, A., Estevinho, B.N., and Rocha, F. (2016). Microencapsulation of vitamin A: a review. *Trends in Food Science & Technology* 51: 76–87.
14. Maurya, V.K., Bashir, K., and Aggarwal, M. (2020). Vitamin D microencapsulation and fortification: trends and technologies. *The Journal of Steroid Biochemistry and Molecular Biology* 196: 105489.
15. Mao, L., Wang, D., Liu, F., and Gao, Y. (2018). Emulsion design for the delivery of β-carotene in complex food systems. *Critical Reviews in Food Science and Nutrition* 58 (5): 770–784.
16. Recharla, N., Riaz, M., Ko, S., and Park, S. (2017). Novel technologies to enhance solubility of food-derived bioactive compounds: a review. *Journal of Functional Foods* 39: 63–73.
17. Sweedman, M.C., Tizzotti, M.J., Schäfer, C., and Gilbert, R.G. (2013). Structure and physicochemical properties of octenyl succinic anhydride modified starches: a review. *Carbohydrate Polymers* 92 (1): 905–920.
18. Agama-Acevedo, E. and Bello-Perez, L.A. (2017). Starch as an emulsions stability: the case of octenyl succinic anhydride (OSA) starch. *Current Opinion in Food Science* 13: 78–83.
19. Ozturk, B., Argin, S., Ozilgen, M., and McClements, D.J. (2015). Formation and stabilization of nanoemulsion-based vitamin E delivery systems using natural biopolymers: whey protein isolate and gum arabic. *Food Chemistry* 188: 256–263.
20. Chen, C.C. and Wagner, G. (2004). Vitamin E nanoparticle for beverage applications. *Chemical Engineering Research and Design* 82 (A11): 1432–1437.
21. Parthasarathi, S., Muthukumar, S.P., and Anandharamakrishnan, C. (2016). The influence of droplet size on the stability, in vivo digestion, and oral bioavailability of vitamin E emulsions. *Food & Function* 7 (5): 2294–2302.
22. Reichert, C.L., Salminen, H., and Weiss, J. (2019). *Quillaja saponin* characteristics and functional properties. *Annual Review of Food Science and Technology* 10: 43–73.
23. Yang, Y. and McClements, D.J. (2013). Encapsulation of vitamin E in edible emulsions fabricated using a natural surfactant. *Food Hydrocolloids* 30 (2): 712–720.
24. Luo, X., Zhou, Y.Y., Bai, L. et al. (2017). Fabrication of beta-carotene nanoemulsion-based delivery systems using dual-channel microfluidization: physical and chemical stability. *Journal of Colloid and Interface Science* 490: 328–335.
25. Ozturk, B., Argin, S., Ozilgen, M., and McClements, D.J. (2015). Nanoemulsion delivery systems for oil-soluble vitamins: influence of carrier oil type on lipid digestion and vitamin D-3 bioaccessibility. *Food Chemistry* 187: 499–506.
26. Ozturk, B., Argin, S., Ozilgen, M., and McClements, D.J. (2014). Formation and stabilization of nanoemulsion-based vitamin E delivery systems using natural surfactants: *Quillaja saponin* and lecithin. *Journal of Food Engineering* 142: 57–63.

27. Salminen, H., Gommel, C., Leuenberger, B.H., and Weiss, J. (2016). Influence of encapsulated functional lipids on crystal structure and chemical stability in solid lipid nanoparticles: towards bioactive-based design of delivery systems. *Food Chemistry* 190: 928–937.
28. Bai, L., Huan, S., Gu, J., and McClements, D.J. (2016). Fabrication of oil-in-water nanoemulsions by dual-channel microfluidization using natural emulsifiers: saponins, phospholipids, proteins, and polysaccharides. *Food Hydrocolloids* 61: 703–711.
29. Chevalier, Y. and Bolzinger, M.A. (2013). Emulsions stabilized with solid nanoparticles: Pickering emulsions. *Colloids and Surfaces A:Physicochemical and Engineering Aspects* 439: 23–34.
30. Yang, Y., Fang, Z., Chen, X. et al. (2017). An overview of Pickering emulsions: solid-particle materials, classification, morphology, and applications. *Frontiers in Pharmacology* 8: 287–287.
31. Sarkar, A., Murray, B., Holmes, M. et al. (2016). in vitro digestion of Pickering emulsions stabilized by soft whey protein microgel particles: influence of thermal treatment. *Soft Matter* 12 (15): 3558–3569.
32. Alison, L., Demirörs, A.F., Tervoort, E. et al. (2018). Emulsions stabilized by chitosan-modified silica nanoparticles: pH control of structure–property relations. *Langmuir* 34 (21): 6147–6160.
33. Alison, L., Ruhs, P.A., Tervoort, E. et al. (2016). Pickering and network stabilization of biocompatible emulsions using chitosan-modified silica nanoparticles. *Langmuir* 32 (50): 13446–13457.
34. Linke, C. and Drusch, S. (2018). Pickering emulsions in foods – opportunities and limitations. *Critical Reviews in Food Science and Nutrition* 58 (12): 1971–1985.
35. Berton-Carabin, C.C. and Schroen, K. (2015). Pickering emulsions for food applications: background, trends, and challenges. *Annual Review of Food Science and Technology* 6: 263–297.
36. Murray, B.S. (2019). Pickering emulsions for food and drinks. *Current Opinion in Food Science* 27: 57–63.
37. Tavernier, I., Wijaya, W., Van der Meeren, P. et al. (2016). Food-grade particles for emulsion stabilization. *Trends in Food Science & Technology* 50: 159–174.
38. Xiao, J., Li, Y.Q., and Huang, Q.R. (2016). Recent advances on food-grade particles stabilized Pickering emulsions: fabrication, characterization and research trends. *Trends in Food Science & Technology* 55: 48–60.
39. Mitbumrung, W., Suphantharika, M., McClements, D.J., and Winuprasith, T. (2019). Encapsulation of vitamin D3 in Pickering emulsion stabilized by nanofibrillated mangosteen cellulose: effect of environmental stresses. *Journal of Food Science* 84 (11): 3213–3221.
40. Hedjazi, S. and Razavi, S.H. (2018). A comparison of canthaxanthine Pickering emulsions, stabilized with cellulose nanocrystals of different origins. *International Journal of Biological Macromolecules* 106: 489–497.
41. Fu, D.W., Deng, S.M., McClements, D.J. et al. (2019). Encapsulation of beta-carotene in wheat gluten nanoparticle-xanthan gum-stabilized Pickering emulsions: enhancement of carotenoid stability and bioaccessibility. *Food Hydrocolloids* 89: 80–89.
42. Ma, L., Zou, L., McClements, D.J., and Liu, W. (2020). One-step preparation of high internal phase emulsions using natural edible Pickering stabilizers: gliadin nanoparticles/gum Arabic. *Food Hydrocolloids* 100: 105381.
43. Dickinson, E. (2015). Microgels –an alternative colloidal ingredient for stabilization of food emulsions. *Trends in Food Science & Technology* 43 (2): 178–188.
44. Lu, Y., Mao, L.K., Hou, Z.Q. et al. (2019). Development of emulsion gels for the delivery of functional food ingredients: from structure to functionality. *Food Engineering Reviews* 11 (4): 245–258.
45. Torres, O., Murray, B., and Sarkar, A. (2016). Emulsion microgel particles: novel encapsulation strategy for lipophilic molecules. *Trends in Food Science & Technology* 55: 98–108.
46. Pelton, R. and Hoare, T. (2011). Microgels and their synthesis: an introduction. In: *Microgel Suspensions: Fundamentals and Applications* (ed. A. Fernandez-Nieves, H.M. Wyss, J. Mattsson and D.A. Weitz), 1–32. Weinheim, Germany: Wiley-VCH Verlag GmbH & Co. KGaA, Boschstr. 12, 69469.
47. Zhang, B., Li, H., Li, X. et al. (2015). Preparation, characterization, and in vitro release of carboxymethyl starch/β-cyclodextrin microgel–ascorbic acid inclusion complexes. *RSC Advances* 5 (76): 61815–61820.
48. Mun, S., Kim, Y.-R., and McClements, D.J. (2015). Control of β-carotene bioaccessibility using starch-based filled hydrogels. *Food Chemistry* 173: 454–461.
49. Mun, S., Kim, Y.-R., Shin, M., and McClements, D.J. (2015). Control of lipid digestion and nutraceutical bioaccessibility using starch-based filled hydrogels: influence of starch and surfactant type. *Food Hydrocolloids* 44: 380–389.

50. Tangsrianugul, N., Suphantharika, M., and McClements, D.J. (2015). Simulated gastrointestinal fate of lipids encapsulated in starch hydrogels: impact of normal and high amylose corn starch. *Food Research International* 78: 79–87.
51. Chen, H., Mao, L., Hou, Z. et al. (2020). Roles of additional emulsifiers in the structures of emulsion gels and stability of vitamin E. *Food Hydrocolloids* 99: 105372.
52. Kasaai, M.R. (2018). Zein and zein -based nano-materials for food and nutrition applications: a review. *Trends in Food Science & Technology* 79: 184–197.
53. Patel, A.R. and Velikov, K.P. (2014). Zein as a source of functional colloidal nano- and microstructures. *Current Opinion in Colloid & Interface Science* 19 (5): 450–458.
54. Pan, Y., Tikekar, R.V., Wang, M.S. et al. (2015). Effect of barrier properties of zein colloidal particles and oil-in-water emulsions on oxidative stability of encapsulated bioactive compounds. *Food Hydrocolloids* 43: 82–90.
55. Cheng, C.J., Ferruzzi, M., and Jones, O.G. (2019). Fate of lutein-containing zein nanoparticles following simulated gastric and intestinal digestion. *Food Hydrocolloids* 87: 229–236.
56. Zhang, F., Khan, M.A., Cheng, H., and Liang, L. (2019). Co-encapsulation of α-tocopherol and resveratrol within zein nanoparticles: impact on antioxidant activity and stability. *Journal of Food Engineering* 247: 9–18.
57. Weissmueller, N.T., Lu, H.D., Hurley, A., and Prud'homme, R.K. (2016). Nanocarriers from GRAS zein proteins to encapsulate hydrophobic actives. *Biomacromolecules* 17 (11): 3828–3837.
58. Emami, S., Azadmard-Damirchi, S., Peighambardoust, S.H. et al. (2016). Liposomes as carrier vehicles for functional compounds in food sector. *Journal of Experimental Nanoscience* 11 (9): 737–759.
59. Mazur, F., Bally, M., Städler, B., and Chandrawati, R. (2017). Liposomes and lipid bilayers in biosensors. *Advances in Colloid and Interface Science* 249: 88–99.
60. Tsai, W.C. and Rizvi, S.S.H. (2017). Simultaneous microencapsulation of hydrophilic and lipophilic bioactives in liposomes produced by an ecofriendly supercritical fluid process. *Food Research International* 99 (Pt 1): 256–262.
61. Keller, B.C. (2001). Liposomes in nutrition. *Trends in Food Science & Technology* 12 (1): 25–31.
62. Kaddah, S., Khreich, N., Kaddah, F. et al. (2018). Cholesterol modulates the liposome membrane fluidity and permeability for a hydrophilic molecule. *Food and Chemical Toxicology* 113: 40–48.
63. Marsanasco, M., Marquez, A.L., Wagner, J.R. et al. (2011). Liposomes as vehicles for vitamins E and C: an alternative to fortify orange juice and offer vitamin C protection after heat treatment. *Food Research International* 44 (9): 3039–3046.
64. Chaves, M.A., Oseliero Filho, P.L., Jange, C.G. et al. (2018). Structural characterization of multilamellar liposomes coencapsulating curcumin and vitamin D3. *Colloids and Surfaces A: Physicochemical and Engineering Aspects* 549: 112–121.
65. Toniazzo, T., Berbel, I.F., Cho, S. et al. (2014). β-carotene-loaded liposome dispersions stabilized with xanthan and guar gums: physico-chemical stability and feasibility of application in yogurt. *LWT – Food Science and Technology* 59 (2, Part 2): 1265–1273.
66. Ota, A., Istenič, K., Skrt, M. et al. (2018). Encapsulation of pantothenic acid into liposomes and into alginate or alginate–pectin microparticles loaded with liposomes. *Journal of Food Engineering* 229: 21–31.
67. Timilsena, Y.P., Akanbi, T.O., Khalid, N. et al. (2019). Complex coacervation: principles, mechanisms and applications in microencapsulation. *International Journal of Biological Macromolecules* 121: 1276–1286.
68. Vasiliu, S., Popa, M., and Rinaudo, M. (2005). Polyelectrolyte capsules made of two biocompatible natural polymers. *European Polymer Journal* 41 (5): 923–932.
69. Timilsena, Y.P., Wang, B., Adhikari, R., and Adhikari, B. (2017). Advances in microencapsulation of polyunsaturated fatty acids (PUFAs)-rich plant oils using complex coacervation: a review. *Food Hydrocolloids* 69: 369–381.
70. Timilsena, Y.P., Wang, B., Adhikari, R., and Adhikari, B. (2016). Preparation and characterization of chia seed protein isolate-chia seed gum complex coacervates. *Food Hydrocolloids* 52: 554–563.
71. Dong, Z.-J., Xia, S.-Q., Hua, S. et al. (2008). Optimization of cross-linking parameters during production of transglutaminase-hardened spherical multinuclear microcapsules by complex coacervation. *Colloids and Surfaces B: Biointerfaces* 63 (1): 41–47.

72. Wang, B., Adhikari, B., and Barrow, C.J. (2014). Optimisation of the microencapsulation of tuna oil in gelatin-sodium hexametaphosphate using complex coacervation. *Food Chemistry* 158: 358–365.
73. Lemetter, C.Y.G., Meeuse, F.M., and Zuidam, N.J. (2009). Control of the morphology and the size of complex coacervate microcapsules during scale-up. *AIChE Journal* 55 (6): 1487–1496.
74. Warnakulasuriya, S.N. and Nickerson, M.T. (2018). Review on plant protein–polysaccharide complex coacervation, and the functionality and applicability of formed complexes. *Journal of the Science of Food and Agriculture* 98 (15): 5559–5571.
75. Eratte, D., Wang, B., Dowling, K. et al. (2014). Complex coacervation with whey protein isolate and gum arabic for the microencapsulation of omega-3 rich tuna oil. *Food & Function* 5 (11): 2743–2750.
76. Jain, A., Thakur, D., Ghoshal, G. et al. (2016). Characterization of microcapsulated β-carotene formed by complex coacervation using casein and gum tragacanth. *International Journal of Biological Macromolecules* 87: 101–113.
77. Yuan, Y., Kong, Z.-Y., Sun, Y.-E. et al. (2017). Complex coacervation of soy protein with chitosan: constructing antioxidant microcapsule for algal oil delivery. *LWT – Food Science and Technology* 75: 171–179.
78. Huang, G.Q., Xiao, J.X., Qiu, H.W., and Yang, J. (2014). Cross-linking of soybean protein isolate-chitosan coacervate with transglutaminase utilizing capsanthin as the model core. *Journal of Microencapsulation* 31 (7): 708–715.
79. Ilyasoglu, H. and El, S.N. (2014). Nanoencapsulation of EPA/DHA with sodium caseinate–gum arabic complex and its usage in the enrichment of fruit juice. *LWT – Food Science and Technology* 56 (2): 461–468.
80. Wang, B., Vongsvivut, J., Adhikari, B., and Barrow, C.J. (2015). Microencapsulation of tuna oil fortified with the multiple lipophilic ingredients vitamins A, D3, E, K2, curcumin and coenzyme Q10. *Journal of Functional Foods* 19: 893–901.
81. Eratte, D., McKnight, S., Gengenbach, T.R. et al. (2015). Co-encapsulation and characterisation of omega-3 fatty acids and probiotic bacteria in whey protein isolate–gum Arabic complex coacervates. *Journal of Functional Foods* 19: 882–892.
82. Lopez-Nicolas, J.M., Rodriguez-Bonilla, P., and Garcia-Carmona, F. (2014). Cyclodextrins and antioxidants. *Critical Reviews in Food Science and Nutrition* 54 (2): 251–276.
83. Kimura, M. and Ooya, T. (2016). Enhanced solubilization of alpha-tocopherol by hyperbranched polyglycerol-modified beta-cyclodextin. *Journal of Drug Delivery Science and Technology* 35: 30–33.
84. Reineccius, G.A. (2004). The spray drying of food flavors. *Drying Technology* 22 (6): 1289–1324.
85. Hategekirnana, J., Masamba, K.G., Ma, J.G., and Zhong, F. (2015). Encapsulation of vitamin E: effect of physicochemical properties of wall material on retention and stability. *Carbohydrate Polymers* 124: 172–179.
86. Drusch, S., Serfert, Y., Van Den Heuvel, A., and Schwarz, K. (2006). Physicochemical characterization and oxidative stability of fish oil encapsulated in an amorphous matrix containing trehalose. *Food Research International* 39 (7): 807–815.
87. Pourashouri, P., Shabanpour, B., Razavi, S.H. et al. (2014). Impact of wall materials on physicochemical properties of microencapsulated fish oil by spray drying. *Food and Bioprocess Technology* 7 (8): 2354–2365.
88. Ramakrishnan, Y., Adzahan, N.M., Yusof, Y.A., and Muhammad, K. (2018). Effect of wall materials on the spray drying efficiency, powder properties and stability of bioactive compounds in tamarillo juice microencapsulation. *Powder Technology* 328: 406–414.
89. Corrêa-Filho, L.C., Lourenço, M.M., Moldão-Martins, M., and Alves, V.D. (2019). Microencapsulation of β-carotene by spray drying: effect of wall material concentration and drying inlet temperature. *International Journal of Food Science* 2019: 8914852.
90. Goncalves, A., Estevinho, B.N., and Rocha, F. (2017). Design and characterization of controlled-release vitamin A microparticles prepared by a spray-drying process. *Powder Technology* 305: 411–417.
91. Jafari, S.M., Masoudi, S., and Bahrami, A. (2019). A Taguchi approach production of spray-dried whey powder enriched with nanoencapsulated vitamin D-3. *Drying Technology* 37 (16): 2059–2071.
92. Breternitz, N.R., Fidelis, C.H.d.V., Silva, V.M. et al. (2017). Volatile composition and physicochemical characteristics of mussel (Perna perna) protein hydrolysate microencapsulated with maltodextrin and n-OSA modified starch. *Food and Bioproducts Processing* 105: 12–25.

93. Pycia, K., Gryszkin, A., Berski, W., and Juszczak, L. (2018). The influence of chemically modified potato maltodextrins on stability and rheological properties of model oil-in-water emulsions. *Polymers* 10 (1): 67.
94. Sotelo-Bautista, M., Bello-Perez, L.A., Gonzalez-Soto, R.A. et al. (2020). OSA-maltodextrin as wall material for encapsulation of essential avocado oil by spray drying. *Journal of Dispersion Science and Technology* 41 (2): 235–242.
95. Shaaruddin, S., Mahmood, Z., Ismail, H. et al. (2019). Stability of β-carotene in carrot powder and sugar confection as affected by resistant maltodextrin and octenyl succinate anhydride (OSA) starches. *Journal of Food Science and Technology* 56 (7): 3461–3470.
96. Nesterenko, A., Alric, I., Silvestre, F., and Durrieu, V. (2014). Comparative study of encapsulation of vitamins with native and modified soy protein. *Food Hydrocolloids* 38: 172–179.
97. Dutta, S., Moses, J.A., and Anandharamakrishnan, C. (2018). Modern frontiers and applications of spray-freeze-drying in design of food and biological supplements. *Journal of Food Process Engineering* 41 (8): e12881.
98. Ishwarya, S.P., Anandharamakrishnan, C., and Stapley, A.G.F. (2015). Spray-freeze-drying: a novel process for the drying of foods and bioproducts. *Trends in Food Science & Technology* 41 (2): 161–181.
99. Leuenberger, H. (2002). Spray freeze-drying – the process of choice for low water soluble drugs? *Journal of Nanoparticle Research* 4 (1–2): 111–119.
100. Parthasarathi, S. and Anandharamakrishnan, C. (2016). Enhancement of oral bioavailability of vitamin E by spray-freeze drying of whey protein microcapsules. *Food and Bioproducts Processing* 100: 469–476.
101. Jacobsen, C., Garcia-Moreno, P.J., Mendes, A.C. et al. (2018). Use of electrohydrodynamic processing for encapsulation of sensitive bioactive compounds and applications in food. *Annual Review of Food Science and Technology* 9: 525–549.
102. Aceituno-Medina, M., Mendoza, S., Lagaron, J.M., and López-Rubio, A. (2015). Photoprotection of folic acid upon encapsulation in food-grade amaranth (*Amaranthus hypochondriacus* L.) protein isolate – pullulan electrospun fibres. *LWT – Food Science and Technology* 62 (2): 970–975.
103. Pérez-Masiá, R., Lagaron, J.M., and Lopez-Rubio, A. (2015). Morphology and stability of edible lycopene-containing micro- and nanocapsules produced through electrospraying and spray drying. *Food and Bioprocess Technology* 8 (2): 459–470.
104. Celebioglu, A. and Uyar, T. (2017). Antioxidant vitamin E/cyclodextrin inclusion complex electrospun nanofibres: enhanced water solubility, prolonged shelf life, and photostability of vitamin E. *Journal of Agricultural and Food Chemistry* 65 (26): 5404–5412.
105. Benelli, L., Cortés-Rojas, D.F., Souza, C.R.F., and Oliveira, W.P. (2015). Fluid bed drying and agglomeration of phytopharmaceutical compositions. *Powder Technology* 273: 145–153.
106. Benelli, L. and Oliveira, W.P. (2019). Fluidized bed coating of inert cores with a lipid-based system loaded with a polyphenol-rich *Rosmarinus officinalis* extract. *Food and Bioproducts Processing* 114: 216–226.
107. Pellicer, J.A., Fortea, M.I., Trabal, J. et al. (2019). Stability of microencapsulated strawberry flavour by spray drying, freeze drying and fluid bed. *Powder Technology* 347: 179–185.
108. Kallakunta, V.R., Sarabu, S., Bandari, S. et al. (2019). An update on the contribution of hot-melt extrusion technology to novel drug delivery in the twenty-first century: part I. *Expert Opinion on Drug Delivery* 16 (5): 539–550.
109. Sarabu, S., Bandari, S., Kallakunta, V.R. et al. (2019). An update on the contribution of hot-melt extrusion technology to novel drug delivery in the twenty-first century: part II. *Expert Opinion on Drug Delivery* 16 (6): 567–582.
110. Tackenberg, M.W., Thommes, M., Schuchmann, H.P., and Kleinebudde, P. (2014). Solid state of processed carbohydrate matrices from maltodextrin and sucrose. *Journal of Food Engineering* 129: 30–37.
111. Castro, N., Durrieu, V., Raynaud, C. et al. (2016). Melt extrusion encapsulation of flavors: a review. *Polymer Reviews* 56 (1): 137–186.
112. Castro, N., Durrieu, V., Raynaud, C., and Rouilly, A. (2016). Influence of DE-value on the physicochemical properties of maltodextrin for melt extrusion processes. *Carbohydrate Polymers* 144: 464–473.
113. Masavang, S., Roudaut, G., and Champion, D. (2019). Identification of complex glass transition phenomena by DSC in expanded cereal-based food extrudates: impact of plasticization by water and sucrose. *Journal of Food Engineering* 245: 43–52.

114. Philipp, C., Emin, M.A., Buckow, R. et al. (2018). Pea protein-fortified extruded snacks: linking melt viscosity and glass transition temperature with expansion behaviour. *Journal of Food Engineering* 217: 93–100.
115. Bohn, D.M., Cadwallader, K.R., and Schmidt, S.J. (2005). Use of DSC, DVS-DSC, and DVS-fast GC-FID to evaluate the physicochemical changes that occur in artificial cherry Durarome® upon humidification. *Journal of Food Science* 70 (2): E109–E116.
116. Tackenberg, M.W., Krauss, R., Marmann, A. et al. (2015). Encapsulation of liquids using a counter rotating twin screw extruder. *European Journal of Pharmaceutics and Biopharmaceutics* 89: 9–17.
117. Chang, D.W., Zhang, X.M., and Kim, J.M. (2014). Encapsulation of vitamin E in glassy carbohydrates by extrusion. *Advanced Materials Research* 842: 95–99.
118. Dong, A.W., Pascual-Izarra, C., Pas, S.J. et al. (2009). Positron annihilation lifetime spectroscopy (PALS) as a characterization technique for nanostructured self-assembled amphiphile systems. *Journal of Physical Chemistry B* 113 (1): 84–91.
119. Hughes, D.J., Bönisch, G.B., Zwick, T. et al. (2018). Phase separation in amorphous hydrophobically modified starch–sucrose blends: glass transition, matrix dynamics and phase behavior. *Carbohydrate Polymers* 199: 1–10.
120. Maiorova, L.A., Erokhina, S.I., Pisani, M. et al. (2019). Encapsulation of vitamin B-12 into nanoengineered capsules and soft matter nanosystems for targeted delivery. *Colloids and Surfaces B: Biointerfaces* 182: 110366.
121. Mougin, K., Bruntz, A., Severin, D., and Teleki, A. (2016). Morphological stability of microencapsulated vitamin formulations by AFM imaging. *Food Structure* 9: 1–12.

15

Bio-derived Polymers for Packaging

Pornnapa Kasemsiri[1], Uraiwan Pongsa[2], Manunya Okhawilai[3], Salim Hiziroglu[4], Nawadon Petchwattana[5], Wilaiporn Kraisuwan[5] and Benjatham Sukkaneewat[6]

[1]*Department of Chemical Engineering, Khon Kaen University, Khon Kaen, Thailand*
[2]*Division of Industrial Engineering Technology, Rajamangala University of Technology Rattanakosin Wang Klai Kang Won Campus, Prachuap Khiri Khan, Thailand*
[3]*Metallurgy and Materials Science Research Institute, Chulalongkorn University, Bangkok, Thailand*
[4]*Department of Natural Resource Ecology and Management, Oklahoma State University, Stillwater, OK, USA*
[5]*Department of Chemical Engineering, Srinakharinwirot University, Nakhon Nayok, Thailand*
[6]*Division of Chemistry, Udon Thani Rajabhat University, Udon Thani, Thailand*

15.1 Introduction

Petroleum-based polymers have been extensively used for various packaging applications, such as food and medical, due to their unique properties for final products and low cost [1]. Nowadays, the continuous increase in the consumption of petroleum-based products has created a critical environmental problem. These non-sustainable products require several years to degrade and generatsse toxic decomposition products [2, 3]. Incineration and recycling have been used to eliminate plastic waste. However, these methods are insufficient for solving global environmental pollution [4, 5]. The development of materials from renewable resources has received great attention. The use of bio-based polymers is a promising approach to reduce the critical environmental problems from petroleum-based products consumption [6, 7]. Bio-based polymers can be classified into three groups as summarized in Table 15.1.

High-Performance Materials from Bio-based Feedstocks, First Edition. Edited by Andrew J. Hunt, Nontipa Supanchaiyamat, Kaewta Jetsrisuparb and Jesper T.N. Knijnenburg.

Table 15.1 Classification of bio-based polymers.

Types of bio-based polymer	Examples
1. Extracted or isolated bio-based polymer from biomass	Polysaccharides and proteins: starch, chitosan, gelatin, etc.
2. Produced biopolymer by microorganisms	Bacterial cellulose and pullulan
3. Produced by chemical synthesis	Polylactic acid, polyvinyl alcohol (PVA), and polybutyl succinate

Source: Modified from Inamuddin [8].

The biopolymers can be applied as functional and structural materials for packaging. Several approaches have been used to improve the properties of biopolymers such as modification with functional groups, blending with compatible polymers, and reinforcing with micro-/nano-fillers. This chapter covers the development of biopolymers and their composites for packaging purposes.

15.2 Starch

Recently, starch has attracted much attention from academics and industries to develop biodegradable packaging due to its low cost, biodegradability, and low toxicity. Starch is the major polysaccharide that is produced by various types of plants such as cassava, maize, and wheat. The main components of starch are amylose and amylopectin. Amylose is a linear and long molecule of 1,4-α-D-glucan with few branches whereas the main structure of amylopectin is short chains with a high number of 1,6-α-D-glucan [9]. The ratio of amylopectin/amylose depends on the type of plant. The content of amylopectin shows a high impact on the properties of starch, such as thermal stability, rheology, and processing properties [10]. Jha et al. [11] investigated the amylose/amylopectin ratio on bionanocomposite starch. The authors studied starches with different amylose/amylopectin ratios, i.e. potato starch (20/80), corn starch (28/72), wheat starch (25/75), and high amylose corn starch (70/30). The starches were blended with carboxymethyl cellulose (CMC) and nanoclay. The mechanical properties and water resistance ability of films are depicted in Figure 15.1. The bionanocomposite film from corn starch showed the highest mechanical strength and water resistance due to the formation of strong intermolecular hydrogen bonds between polymer matrix and filler.

Conventional polymer-processing techniques have been applied to process starch for biodegradable packaging applications such as solution casting, internal mixing, extrusion, injection and/or compression molding, and thermoforming [9]. Starch has a good ability to form a continuous matrix during its processing [7, 8]. However, starch still has some limitations due to its low water resistance and poor mechanical properties. To overcome these disadvantages, two approaches, viz. modification structure of starch and blending starch with compatible polymers have been applied in starch processing.

The modification of starch via esterification, etherification, acetylation, hydroxylation, and oxidation has been commonly applied to improve the properties of starch [12]. Biduski et al. [13] studied the modification of sorghum starch by oxidation and lactic acid treatment. The water vapor permeability of the starch film with single modification by acid ($2.41 \pm 0.06\,g\,mm\,m^{-2}\,day^{-1}\,kPa^{-1}$) or oxidation ($2.94 \pm 0.12\,g\,mm\,m^{-2}\,day^{-1}\,kPa^{-1}$) was lower than that of the neat starch film ($3.94 \pm 0.06\,g\,mm\,m^{-2}\,day^{-1}\,kPa^{-1}$), whereas the dual modification with both acid and oxidative treatment showed the highest water vapor permeability

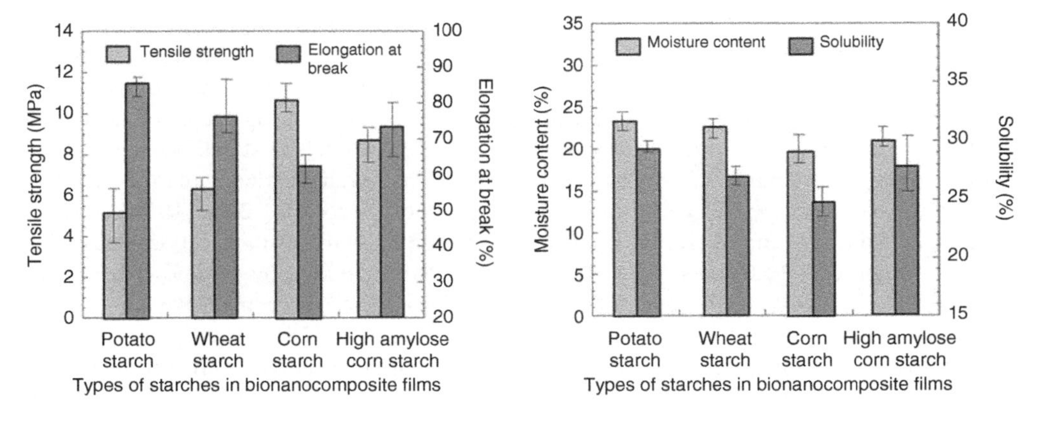

Figure 15.1 *Effect of different amylose/amylopectin ratios on properties of bionanocomposite films. Source: Jha et al. [11].*

($5.48 \pm 0.88\,g\,mm\,m^{-2}\,day^{-1}\,kPa^{-1}$). The dual modification provided a higher carboxyl and carbonyl group content that induced the repulsive force between the starch polymer chains. This phenomenon led to higher water mobility in the starch film. Hence, the insertion of carboxyl and carbonyl groups into starch should be controlled to achieve the appropriate characteristics of starch films with low water vapor permeability. Li et al. [14] studied the inhibition of plasticizer migration from the esterified starch film at various degrees of substitution (DS) of acetate in the range of 1.17–2.61. Diethyl phthalate was used to investigate the plasticizer migration from the packaging film. After microwave heating of the esterified starch film, the highest diethyl phthalate migration was observed for the esterified starch film at DS of 1.17. The migration content decreased to 12.18% when starch was esterified at DS of 1.81. The migration content depended on the three-dimensional structure resulting from the interactions between ester linkages in starch and diethyl phthalate. Wokadala et al. [15] prepared modified starch blended with poly(lactic acid) (PLA), where the high amylose starch was modified by butyl-esterification. The PLA/modified starch had higher tensile strength and elongation at break compared to PLA/unmodified starch.

Blending starch with biodegradable polymers has resulted in improved mechanical and physical properties of starch. Various types of biodegradable polymers such as gelatin [16], chitosan [17, 18], polyvinyl alcohol (PVA) [19], PLA [20], and poly(butylene succinate) (PBS) [21] have been blended with starch to produce packaging materials. Tian et al. [22] studied the effect of PVA/starch ratio on film properties. The tensile strength, elongation at break, and modulus of PVA/starch blend films decreased when the amount of starch was increased. The 50/50 PVA/starch film showed a tensile strength of 9 MPa which was higher than that of a low-density polyethylene package film. Priya et al. [23] reported the effect of citric acid on the properties of PVA/starch blend films. The citric acid acts as a cross-linker between PVA and starch via ester linkages. An increase in citric acid content in the PVA/starch blend films enhanced the film flexibility due to the plasticizing effect of citric acid residue in the polymer matrix. Furthermore, the incorporation of 20 wt% citric acid into the film can inhibit the growth of *Escherichia coli* (*E. coli*) and *Staphylococcus aureus* (*S. aureus*). Ounkaew et al. [19] also prepared starch/PVA-blended films at various citric acid contents

(0–30 wt%). The biodegradability of starch/PVA films was observed by the soil burial method. After 30 days, the weight loss of films ranged from 65.28 to 86.64%, with lower values at increasing citric acid contents. The ester linkages in the film resisted the penetration of moisture resulting in a decreased microbial attack. Zeng et al. [24] investigated starch blended with modified PBS terminated with hydroxyl groups (OH–PBS–OH). The starch was gelatinized with a plasticizer to obtain thermoplastic starch (TPS). The starch blended with modified PBS at 50% decreased the overall water absorption and soluble ratio to 21.2 and 10.5%, respectively. The addition of modified PBS improved the hydrophobicity of starch. Furthermore, the tensile strength of TPS blended with 10% modified PBS increased by 10 times when compared to TPS. The polymer blend of TPS/modified PBS enabled more extensive applications due to the compatibility of PLA and modified starch. Şen et al. [25] studied polyelectrolyte films based on cationic starch and sodium alginate. The positive charges were attached to the hydroxyl groups of the starch backbone by using sodium hydroxide (NaOH) as a catalyst. The inhibition zone diameter of *E. coli* and *S. aureus* increased when the content of cationic starch increased, whereas the hydrophobicity of the polyelectrolyte film increased with increasing sodium alginate content; sodium alginate had a higher hydrophobicity than cationic starch. Ren et al. [26] studied the effect of chitosan concentration on properties of corn starch/chitosan blend films. The flexibility of the starch/chitosan blend films increased with increasing chitosan content up to 41% starch. The intermolecular hydrogen bonding between chitosan and starch enhanced the mechanical properties of the samples. The water resistance of starch/chitosan blend films also decreased when the chitosan concentration in the film increased due to the hydrophobicity of chitosan.

15.3 Chitin/Chitosan

Chitin is one of the most abundant natural polymers after cellulose and can be found in the exoskeleton of insects and shells of crab, shrimp, and crawfish [27]. Chitosan is a linear amino-polysaccharide with glucosamine and N-acetyl glucosamine units and is derived from chitin after deacetylation. Chitosan possesses many outstanding biological properties including biocompatibility, biodegradability, non-toxicity and, particularly, antimicrobial activity, and thus it has been applied in, e.g. medical devices (wound dressing) [28], food, and textile. For decades, food packaging films based on chitosan have been extensively developed due to their bioactivities, particularly antimicrobial ability. This section reviews recent advances in food and preservation packaging of chitin and chitosan in detail. For example, the preservation properties of chitosan films on chilled meat were improved by neutralization using NaOH solution. It was found that the mechanical properties of the chitosan casting film increased while the swelling value, water vapor permeability, and oxygen permeability decreased with the increase of NaOH concentration and neutralization time [29].

Chitosan, however, has some disadvantageous properties such as low mechanical strength, rigid structure, and low thermal stability. Therefore, the preparation of polymer blends containing chitosan and other high-performance materials is a simple and effective choice to improve those mentioned properties of chitosan. Moreover, the blending of natural and synthetic polymers having biocompatibility gradually becomes an innovative approach to improve the cost–performance ratio of the resulting polymer. Due to environmental considerations with respect to petroleum-based polymers, blends of chitosan with naturally derived and synthetic biopolymers through simple solution casting have been developed. As an example, PVA is a synthetic polymer with low cost, non-toxicity, excellent chemical resistance, and high tensile strength used to improve the functional characteristics of chitosan. PVA is water-soluble; thus it can be blended with weak acid-soluble chitosan. Polymer blends of chitosan with PVA showed good compatibility of both biopolymers attributed to the strong interaction of the positively charged chitosan and oppositely charged PVA. These chitosan/PVA blends demonstrated an increased film stretching, i.e. tensile elongation at break. However, the addition of PVA into chitosan films exhibited lower mechanical and antimicrobial properties [30].

Gelatin is a natural water-soluble protein, obtained from the partial hydrolysis of collagen. With a high content of proline, glycine, and hydroxyproline sequences, gelatin films show excellent gas and volatile compound barriers [31]. Chitosan/gelatin films containing natural extracts and plant essential oils (EOs) such as cinnamon, nutmeg, and thyme EO as alternatives to synthetic food additives (owing to their proven wide-spectrum antimicrobial capacity) showed a heterogeneous surface with notable pores on the surface [32]. It was found that these three EOs could inhibit the growth of foodborne bacterial pathogens, with thyme being the most effective EO. Moreover, the chitosan/gelatin films with EOs acted as a UV light barrier that protects against photo-oxidation of organic compounds and degradation of vitamins and pigments [33].

Polymer blends of starches and chitosan have been reported to modify the relatively low moisture barrier behavior of starches due to their strong hydrophilic nature. The different sources of starch to prepare chitosan/starch blends include corn [26, 34], wheat [35–37], and rice [38].

Chitosan/CMC films have been extensively prepared due to their high natural abundance and low cost [26, 39]. However, their hydrophilicities limit the shelf life because the films have low water protection as well as low mechanical properties. Cross-linking of the polymers using a cross-linking agent is an effective method to overcome such drawbacks. The most commonly used cross-linking agent is glutaraldehyde. Greater mechanical strength and low water vapor penetration were obtained from chitosan/CMC using glutaraldehyde as a cross-linking agent [40]. Poly(L-isomer lactic acid, PLLA) is a biopolymer with hydrophobic properties causing contamination from bacteria adsorption. Thus, the development of chitosan/PLLA with an antimicrobial agent is an alternative choice [41, 42].

Solution film casting is a versatile fabrication method to prepare chitosan blends and composites for food packaging. Electrospinning is a straightforward and cost-effective technique to fabricate nonwoven fiber mats having submicron to nanoscale fiber diameter. However, chitosan has been widely reported to pose difficulties in electrospinning due to the charge repulsion of ionic groups. The development of electrospun fibrous mats from chitosan/PVA [32] loaded with antioxidant peptide for food biopackaging has been studied. With an incorporation of chitosan, the tensile strength of fibrous mats sharply increased while the elongation at break of the mats decreased compared to the pristine PVA. Interestingly, the water vapor permeability of the chitosan/PVA mats fabricated by electrospinning technique was 10 times lower than that reported in the literature using film casting [30], which may be due to the highly linked 3D network that decreased the migration of water vapor through the nanofibrous mats. Multifunctional chitosan food packaging films have been studied as well. Bioactive agents have been incorporated into chitosan films, including chitosan/PVA fibrous mats encapsulating fish-purified antioxidant peptide. Uniform fibrous mats without any evidence of bead formation can be prepared [32].

Carotenoids, a member of tetraterpenoids, have various outstanding bioactivities, mainly antioxidant, antimicrobial, anticancer, immunomodulatory, and antidiabetic. New active composite packaging films of chitosan/carotenoids have been developed [43]. Unexpectedly, the extracted chitosan and carotenoids were prepared from a similar source of blue crab shells. The prepared films showed a more heterogeneous and rough structure. Darkening of the films by incorporation of carotenoproteins protects the packaged food films from visible and UV light. The thermal stability of the composite film increased, whereas tensile properties in terms of tensile strength and elongation at break decreased. However, the bioactivities of carotenoproteins in the chitosan film improved.

Chitin-based food packaging films have rarely been investigated because it is difficult to fabricate chitin into a film by solution casting. Chitin nanocrystals incorporated into chitosan-based films have been developed [44, 45]. Such reviews highlight the potential and promising nature of chitosan as an alternative for active food packaging films.

15.4 Cellulose and Its Derivatives

The finding of renewable, biodegradable, and eco-friendly resources to replace petroleum-based materials has attracted much attention due to environmental concerns. Among bio-based materials, cellulose is abundant, cost-effective, biocompatible, and biodegradable, and has therefore received growing attention for food packaging applications [46, 47]. As a novel type of cellulose material, nanocellulose has been extensively considered as a promising sustainable material due to its nanoscale dimensions and excellent properties such as lightweight, high surface area, and good mechanical properties [48]. This section describes the development and applications of nanocellulose and its derivatives for packaging materials.

Cellulose is a natural linear polysaccharide consisting of a long chain of anhydroglucose units linked together through β(1–4)-glycosidic bonds [49]. The interconnected cellulose chains via hydrogen bonding create elementary fibrils that are packed into bundled fibril structures with nanoscale diameters and lengths of up to several micrometers [50]. Generally, a mix of crystalline and amorphous areas is arbitrarily arranged along the length of the cellulose. The crystalline region is formed by the highly ordered packing of cellulose chains and provides high strength and stiffness, whereas the amorphous or disordered region influences the flexibility of the material [48, 51].

Cellulose is generally obtained from the fiber cell walls of plants, some marine animals, fungi, and bacteria [52, 53]. The most important source of cellulose is wood, which can be divided into two types according to their anatomical features: hardwood and softwood. Hardwood has a more complex and rigid structure than softwood, and thus a stronger mechanical treatment is required to achieve an equivalent isolation level [54, 55]. Nowadays, agricultural biomass residues and wastes are gaining attention as renewable sources of cellulose. Rice straw, wheat straw, sugarcane bagasse, rapeseed straw, corn stalk, cassava bagasse, and banana have been used as raw materials to produce cellulose with a less harmful process [56–58]. Zhao et al. [59] investigated the isolation of high-purity cellulose from food waste durian rind, providing a potential application of cellulosic waste resource conversion into biodegradable packaging materials. The durian rind cellulose film exhibited excellent properties, good appearance, and fast biodegradability.

Nanocellulose has been widely applied in industrial applications. There are three types of nanocellulose, i.e. cellulose nanocrystals (CNCs), cellulose nanofibers (CNFs), and bacterial nanocellulose (BNC). These types have a similar chemical composition, but their characteristics (i.e. morphology, particle size, and crystallinity) strongly depend on the source and extraction method.

15.4.1 Cellulose Nanocrystals

CNC, also called microcrystalline cellulose (MCC) or cellulose nanowhisker, is usually extracted from natural cellulose fibers by the acid hydrolysis process [60]. The extracted nanocellulose exists as rod-like or whisker-shaped particles with widely varying dimensions, with a typical diameter of 3–70 nm and length of 100–250 nm. The characteristic properties of CNC depend on the cellulose source and hydrolysis conditions [51]. Generally, sulfuric acid is used for hydrolysis to remove the amorphous cellulose regions so that highly crystalline cellulose nanoparticles remain [61]. The use of sulfuric acid in the hydrolysis process introduces negatively charged sulfate groups on the CNC surface that stabilize the nanocrystal suspensions by electrostatic repulsion [62]. Several authors have studied the potential use of CNCs as reinforcing fillers for developing biodegradable nanocomposites for food packaging. CNCs isolated from coffee or rice husk by sulfuric acid hydrolysis coupled with alkali and bleaching treatments were added into corn starch films to investigate their reinforcing ability [143]. The elastic modulus of TPS films filled with 1 wt% of coffee and rice husk CNCs increased by 121 and 186%, respectively, due to very efficient load transfer through percolated CNCs' network within the polymer matrix. Starch nanocomposite films containing 5–15% CNCs extracted from *Alicante Bouschet* grape pomace by sulfuric acid hydrolysis showed more opaque, greater mechanical properties and decreased permeability compared to neat starch film, making them attractive for food packaging applications [63]. Noorbakhsh-Soltani et al. [64] reported the formation of individual cellulose nanowhiskers (average diameter >10 nm and average length >100 nm) by sulfuric acid hydrolysis of cellulose microcrystals. These nanowhiskers could be used as a nano-additive into gelatin and starch-based nanocomposite films. In addition, CNCs can be derived by mild chemical and enzymatic hydrolysis as an alternative to the harsh reaction condition in strong acid. CNCs obtained by ammonium persulfate hydrolysis exhibited higher crystallinity and higher charge densities due to the presence of carboxylic groups compared to sulfuric acid hydrolysis [65]. The CNC-based coatings showed lower oxygen permeability coefficients than synthetic resins commonly used in food packaging materials. CNCs isolated from rice and oat husks via enzymatic hydrolysis were used to develop PVA-based aerogel with water-absorption capacity for food packaging [66]. Due to the larger pore size of oat CNCs, the aerogel incorporated with oat CNCs showed higher water-absorption capacity compared to other aerogels and might be used as absorbers to decrease water activity and restrict microbial growth in fresh meat packaging.

15.4.2 Cellulose Nanofibers

CNFs consist of long, flexible, entangled fibrils with a typical diameter of 5–50 nm [58] and length of 1–5 μm [57]. CNFs can be extracted by the cleavage of cellulose fibrils in the longitudinal axis through shear and impact forces applied by mechanical processes [52]. Compared to CNC, CNF has a long fibril shape with a higher aspect ratio, relative low crystallinity (CNF contains both crystalline and amorphous regions), and a high abundance of hydroxyl groups that are more easily accessible for surface modification [48, 52, 53]. Turbak et al. [67] first produced nanofibrils from a softwood pulp by high-pressure homogenization (HPH) and micro-fluidization. Even though CNF has a promising potential in various applications, major limitations are the process scalability, expensive industrial equipment, and high energy requirements [57, 58]. Currently, alternative methods for mechanical disintegration such as ultrafine friction grinding, high-intensity ultrasonication, and ball milling have recently acquired a growing interest in academia and industry [50, 58] due to the cost-effectiveness, ease of operation, reliability, and applicability in wet and dry conditions on a wide range of materials. Martins et al. [68] prepared CNFs with a diameter of 10–30 nm from peach palm residue by mechanical defibrillation in an ultrafine grinder and investigated their application in cassava starch films. 306% increase in tensile strength and 26% reduction in water vapor permeability were observed for cassava starch-based films incorporated with 5.37% CNFs compared to the CNF-free film.

Chemical pretreatment can effectively improve the delamination of CNF. This can be achieved via 2,2, 6, 6-tetramethyl-1-piperidine oxoammonium salt (TEMPO)-mediated oxidation or via sulfonation with sodium bisulfate or quaternization [46, 57, 58, 69]. Hai et al. [70] prepared green nanocomposites from chitin and bamboo CNFs for food packaging. The CNFs obtained from bamboo via TEMPO oxidation were 16.4 ± 5.3 nm in width and 1267 ± 623 nm in length as observed by AFM and image analysis techniques. The tensile strength of the nanocomposites containing chitin and bamboo nanofibers increases from 26.3 ± 11.7 to 80.1 ± 18.7 MPa with increasing CNF content from 6 to 90%.

15.4.3 Bacterial Nanocellulose

BNC is high-purity cellulose that is typically synthesized by bacteria in culture media based on sugar sources [29]. BNC is prepared in the form of twisting ribbon-shaped fibrils with a diameter of 20–100 nm and a length of more than 2 μm [48, 58]. Processing times ranging from a few days up to two weeks are required for biosynthesis. *Acetobacter xylinum* (or *Gluconacetobacter xylinus*) is considered one of the most efficient producers because of its ability to produce relatively high BNC yields [47, 52]. The chemical composition of BNC and plant-derived cellulose is quite similar but BNC is free from other polymers such as wax, lignin, hemicelluloses, or pectin; thus, it has been broadly utilized in biomedical applications for its high purity. BNC has a degree of polymerization of 3000–9000 and a distinct crystallinity of 80–90% [58]. Due to its excellent properties, BNC has been used to develop food packaging as well. Wang et al. [71] investigated the properties of PVA/BNC films containing silver nanoparticles. The films showed a high oxygen barrier capacity, enhanced water vapor permeability, and good antimicrobial activities against *E. coli*, indicating that developed films have a great potential in flexible antimicrobial food packaging applications. Biopolymer films fabricated from potato peel powder, BNCs, and curcumin were studied by Xie et al. [72]. By the addition of BNCs in potato peel films, the tensile strength was greatly enhanced, whereas moisture permeability, oxygen permeability, and moisture content decreased. The active biodegradable films can be applied in food packaging to prevent lipid oxidation and prolong the shelf life of fatty food products.

15.4.4 Carboxymethyl Cellulose

The large existence of hydrogen bonding within the cellulose structure makes it rather hard to dissolve in water and most common solvents, which further limits its applicability. One solution to overcoming these problems is the chemical modification of nanocellulose with a careful selection of the appropriate chemical reactive groups to convert native cellulose into water-soluble cellulose derivatives for a large variety of applications. CMC or sodium CMC is a major derivative of cellulose that is primarily obtained by alkalization of cellulose followed by carboxymethylation to link the hydroxyl groups on the cellulose backbone with carboxymethyl groups ($-CH_2-COONa$) [23]. CMC has been recognized as an excellent alternative biomaterial because of its solubility, non-toxicity, biocompatibility, biodegradability, chelating ability, and pH sensitivity [73]. Recently, the use of alternative abundant and renewable low-cost cellulose sources as raw material has received great attention for CMC production. For example, CMC has been prepared from bamboo scrap [74], sugarcane bagasse [75], pineapple peel, and seaweed [76]. Besides, BNC can be considered as a great raw material for CMC synthesis [77] because of its high intrinsic purity and molecular formula that is very similar to that of plant cellulose. Moreover, BNC does not require any chemical treatment for the removal of lignin, hemicelluloses, and pectin prior to cellulose derivatization. CMC has been widely used in numerous industrial fields including food biofilm packaging, drug delivery, medicine, sensors, and as adsorbent for wastewater treatment [29, 74]. The addition of CMC into the polymer matrix can enhance mechanical, thermal, and barrier properties and make material favorable for packaging application. Tavares et al. [78] demonstrated that the mechanical properties and vapor permeability of corn starch films improved upon the addition of 40 wt% CMC. Makwana et al. [79] prepared active films based on agar, CMC, and silver-modified montmorillonite as a biodegradable alternative to the commodity plastic films for food packaging applications. The developed films showed excellent antibacterial activity against both *Bacillus subtilis* and *E. coli* bacteria, and high mechanical and thermal stability. Biodegradable antibacterial PVA-based composite films containing CMCs and TiO_2 with appropriate potential for packaging applications were fabricated by Behnezhad et al. [80].

15.5 Poly(Lactic Acid)

Nowadays, thermoplastics are widely used as a food packaging material due to their low cost, durability, lightweight, and lower processing temperature. The extensive use of thermoplastics, however, results in considerable plastic consumption and major waste accumulation. There have been many attempts to replace nonbiodegradable packaging materials by using natural products such as wood, paper, and leaves. However, these materials are incompatible with the industrial packing and filing processes and their poor gas-barrier properties negatively affect the quality of food products. Thus, the use of biodegradable plastics seems appropriate for both the sustainable environment and industrial-processing points of view.

While there are many commercially available bioplastics in the world market, PLA or polylactide is widely known and used as packaging due to its good mechanical strength, transparency, and good processability with reasonable material cost. PLA is expected to substitute fossil-based polymers such as polystyrene (PS) and poly(ethylene terephthalate) (PET) [81, 82] and as an alternative solution of the plastic waste accumulated in the environment especially for single-use food packaging. Nowadays, the use of PLA in food packaging has already drawn significant attention from our society and industrial section. PLA-based food packaging and containers including bottles, trays, cups, and bags are currently used in various countries. These applications increase the awareness and demands of bioplastics to develop green packaging materials following the Sustainable Development Goals (SDGs).

However, the problems of poor resistance to impact forces have limited the use of PLA in several value-added applications [83]. Thus, the toughness of PLA should be improved before its application especially for films, bottles, and thermoforming and injection-molding products. This challenges researchers and plastic developers to modify PLA with higher toughness. The most effective way is to add toughening agents that can induce the toughening mechanism. This section discusses several properties related to packaging applications of toughened PLA in order to explore the potential of this material for sustainable food packaging applications.

PLA is a biodegradable, compostable, and biocompatible polymer that is classified as thermoplastic aliphatic polyester. PLA is derived from renewable raw materials and can be produced by ring-opening polymerization (ROP) of lactide or the polycondensation of lactic acid [82–84]. The global market of PLA increases by approximately 20–30% $year^{-1}$. Depending on the D : L-lactide composition, PLA can be classified as an amorphous, semicrystalline, or highly crystalline polymer. PLA has a glass transition temperature of 60–65 °C and a melting range of 140–180 °C [85, 86]. The mechanical properties of PLA are between those of PS and PET with heat distortion temperature (HDT) ranging from 55 to 110 °C depending on the presence of crystalline regions [52].

15.5.1 Bio-based Toughening Agents Used in PLA Toughness Improvement

Many groups of additives have been reported as toughening agents in PLA. The main purpose of this modification is to improve the impact resistance and elongation at the break of PLA. Most of the bio-based toughening agents applied to PLA are epoxidized natural rubber (ENR), bio-elastomers, and other biodegradable flexible polymers. Numerous reports indicate that PLA has been toughened with various flexible polymers such as polycaprolactone (PCL), PBS, polyhydroxyalkanoates (PHAs), poly(butylene adipate-co-terephthalate) (PBAT), poly(propylene carbonate) (PPC), and others. Normally, PLA blends are less stiff and tougher than their parent polymer compositions. Some synergistic behaviors were also reported with the incorporation of compatibilizers such as PBS, poly(butylene succinate-co-adipate) (PBSA) copolymer, and PBAT.

15.5.2 Toughening of PLA and Its Properties Related to Packaging Applications

The global concerns on fossil-based plastic materials have drawn much attention to biodegradable packaging materials. Nowadays, PLA is already being applied for short shelf-life food packagings like containers, beverage cups, yoghurt cups, laminated films, and blister packages. In practical use, food packages should possess suitable toughness and impact properties to resist cracking during transportation, handling by consumers, or storage at low temperature, while packaging films need an improvement in good transparency, toughness, and gas barrier properties [83]. For this reason, PLA blended with flexible polyesters have been widely investigated in the past years. Injection-molded specimens [82], blown films [84], and extruded specimens [20] prepared from PLA/PCL blends could increase their elongation at break to 140, 105 (in machine direction) and 70%, respectively, in comparison to neat PLA. Impact strength of PLA/PCL blends revealed a brittleness reduction of PLA [82, 87] while tensile strength and modulus were slightly dropped when using the flexible PCL modifier [82]. The dramatic increase in ductility of PLA was obtained when blended with PBS, and the elongation at break of PLA/PBS blends could reach more than 250% [88, 89]. A higher impact strength of the blends was also reported [8]. PLA/PBS multilayer membranes prepared by Messin et al. [86] provided desirable barrier performance of 30% for O_2, 40% for H_2O, and 70% for CO_2, which were superior to neat PLA membranes. However, PLA/PBS membranes were obtained as translucent materials. In addition, several studies have paid attention to investigate the properties of PLA/polyhydroxybutyrate (PHB) blends. Slightly tougher PLA/PHB blends with an increase in Young's modulus were attained due to the fact that PHB is a semicrystalline polymer and does not clearly show flexible characteristics [90–93]. The miscibility of both polymers seems to be the important factor determining the final properties of

PLA/PHB blends. More specifically, PHB with a low molecular weight provided better miscibility with PLA resulting in more desirable mechanical properties [94–96]. The crystallization of PLA is directly related to the thermomechanical resistance of PLA-based food packaging materials [97, 98] and could be accelerated by the incorporation of PHB [96]. The improvement in toughness and impact properties of those PLA/polyester blends led to their recommendation for application as homeware packaging, food packaging, and extruded films for food product preservation [88, 91].

Apart from mechanical properties, other interesting properties related to food packaging applications were observed in the case of PLA/copolyester blends. Peelable lidding films for food packages were fabricated from PLA/PBAT blends. The proper optical properties (4% of transmission haze and 99% of clarity) and peel strength (8–10 N/15 mm) were achieved with the optimum PLA : PBAT ratio of 80 : 20 [99]. Furthermore, PLA/PBAT antifogging films prepared by Wang et al. [100] using the solvent casting method exhibited the UV-screening properties and appropriate water vapor permeability, which minimized fogging on the film surface and prolonged the freshness of vegetables during the packaging test. As expected, PLA/PBAT and PLA/PBSA blends possessed a higher ductility than neat PLA. Recently, these tougher PLA-based materials have been further developed into antibacterial films for food packaging applications [101, 102]. Super-toughened blown films originating from PLA/PBSA blends were suggested as flexible packaging materials [103]. PLA/PBSA blends with the presence of a chain extender (epoxy-functionalized styrene acrylate (ESA)) showed a large increase in ductility along with a significant improvement in the optical, barrier, and anti-slip properties. ESA modified the interaction between both polymers resulting in better miscibility of the blends. It is noteworthy that the immiscibility between PLA and the above-mentioned polyesters (or copolyester) seems to be the crucial challenge limiting the improvement in properties of the blends [99, 104, 105]. For example, phase separation of PLA/PCL blends caused a reduction in gas barrier properties and a marginal enhancement in thermal stability [97]. However, this problem could be solved by the addition of reactive additives such as dicumyl peroxide [106–108], triphenyl phosphate [109], isocyanate-containing compounds [110, 111], epoxy-containing compounds [112, 113], and third polymers [114] as coupling agents to ameliorate the miscibility of the blends.

Similarly, polymer blends tailored from PLA with natural rubber (NR) and its derivatives, especially ENR, also need their miscibility [115]. The remarkable improvement in mechanical properties of these polymer blends seems to be the main point discussed for several works. The blends had an improved tensile toughness owing to the increased elongation at break of PLA by the elastic nature of the rubbers, while their higher-impact toughness was negligible [116]. In fact, NR and ENR are very efficient impact modifiers for PLA. Super-toughened PLA/NR and PLA/ENR blends having their interfacial reaction via dynamic vulcanization or interfacial compatibilization could absorb high impact energy and provided a higher impact strength when compared to neat PLA [117, 118]. The increase in PLA crystallization by the addition of NR was also reported [115, 119]. However, PLA-based blends modified with a large amount of NR or ENR might sacrifice their transparency after the toughening process [120]. Due to the outstanding impact toughness of PLA/NR (or ENR) blends, the development of packaging materials with high impact resistance derived from these polymer blends should be further investigated. Table 15.2 summarized the mechanical properties of selected biodegradable polymer and PLA blends.

In conclusion, PLA is expected to be used in petroleum-based plastics' substitution due to its biodegradability and renewability. However, PLA has very low impact strength that limits

Table 15.2 Mechanical properties of PLA blends with bio-based toughening agents.

Polymer	Blended composition (wt%)	Tensile strength (MPa)	Elongation at break (%)	Impact strength (J m^{-1})	Reference
PLA	—	67.4 ± 1.46	4.98 ± 0.29	34.2 ± 0.32	[121]
PLA/PBS	10	~57	~16	~14	[122]
PLA/PBAT	10	~28	~60	~60	[113]
PLA/PCL	20	~27	~42	—	[123]
PLA/PHB	15	31 ± 5	140 ± 60	—	[124]
PLA/NR	10	~19	~1.1	~21	[125]
PLA/ethylene-co-vinyl acetate (EVA)	20	37.3 ± 3.3	~12	—	[126]
PLA/polypropylene carbonate (PPC)	30	37.5 ± 2.9	~24	—	[126]
PLA/PHA	10	42.3 ± 3.7	~5.2	—	[126]

its application in situations that require a high level of impact toughness. This motivated researchers and plastic product developers to produce PLA packaging with high toughness. Modifying PLA with toughening agents and plasticizers showed a significant improvement in the elongation at break and impact strength with better processability. These toughened PLA blends have been processed with improved mechanical properties and can be adequately used as packaging such as thermoformed trays, blow-molded bottles, injection-molded products, and films.

15.6 Bio-based Active and Intelligent Agents for Packaging

Common polymer packaging materials have been improved to overcome their shortcomings for the last decade. Nowadays, the main limitation is a new design that needs to meet specific requirements for marketing purposes such as complex and multilayered materials. However, it has been observed that polymer packaging for specific applications such as food can no longer provide sufficient protection and maintain the quality of food products. Therefore, concepts of active and intelligent packaging have attracted substantial interest in extending shelf life or maintaining or monitoring the quality and safety of food [127].

15.6.1 Active Agents

The main purpose of using active agents in polymer packaging is to extend the shelf life and maintain the quality of food. Such active agents have been incorporated into polymer matrices to achieve active packaging. Various approaches have been used such as grafting of active agents on functional groups of polymers and incorporation of active agents inside or on the surface of polymers [19, 128–130]. In the first type, nonmigratory active packaging acts without the migration of active agents from the packaging into foods, whereas in the second type, the actively releasing packaging allows the release of active agents in the atmosphere surrounding the food [127].

Most of the nonmigratory active packaging materials absorb moisture or gas to control the quality of food products. Bovi et al. [131] used fructose as a moisture absorber for the storage of strawberries. A three-layer structure was used to prepare FruitPad, which consisted of polyethylene at the top and bottom whereas the middle layer was cellulose fiber incorporated with different fructose concentrations of 20 and 30%. The highest moisture absorption at 0.94 g of water per g of pad was observed for FruitPad containing 30% fructose at 20 °C and 100% relative humidity. The weight loss of strawberries was 0.94%, which was lower than the acceptable value of 6%. The authors concluded that 30% fructose in FruitPad was optimal for transpiration of strawberries and minimal condensation. Warsiki et al. [132] studied chitosan and potassium permanganate as active packaging for ethylene absorbers. Ethylene is a ripening hormone of fruits, and to delay the ripening of fruit, the amount of ethylene in packaging needs to be controlled. The tomatoes were stored in a chitosan film containing potassium permanganate. It was observed that the incorporation of 7 g of potassium permanganate into the film can absorb ethylene and delay the ripening of tomatoes. The hardness of tomatoes stored in chitosan/potassium permanganate film was higher than the control.

Active release packaging has been widely developed in recent years. Various types of active agents such as EOs, extracted plants and herbs, organic acids, and nanoparticles have been incorporated in food packaging to preserve food products and their quality [127].

EOs can be obtained from many parts of plants such as bud, bark, leaves, and seed. EOs used in packaging for food industries have been classified as GRAS (generally recognized as safe). Various types of EOs have been applied in active packaging, such as EO of thyme [17], basil [133], cinnamon [144], and oregano [128]. Furthermore, the main components of EOs such as thymol and carvacrol were also extracted and used in packaging. Hosseini et al. [134] studied chitosan films containing thyme, clove, and cinnamon EO. The chitosan films containing thyme EOl showed the highest antibacterial activities for Gram-positive bacteria (*S. aureus*) and Gram-negative bacteria (*Salmonella enteritidis*) when compared to chitosan films containing clove and cinnamon EOs at the same content. The tensile strength of the chitosan film increased from 15.93 to 21.35 MPa when 0.5–1.5% of cinnamon EO was added. The cross-linker effect of cinnamon EO with the chitosan matrix reduced the free volume and mobility of polymer chains. The chitosan films incorporated with thyme and clove EO showed a decrease in tensile strength due to the reduced compactness of the film structure. Ketkaew et al. [128] observed the effect of oregano EO (OEO) content on the properties of cassava starch foam composite. The minimum concentration of OEO in starch foam composite against *S. aureus* and *E. coli* was 8 wt%. The OEO consists of hydrophobic compounds such as carvacrol, thymol, and p-cymene that can penetrate the lipids of the bacterial cell membrane, causing the leakage of ions from the cytoplasm which leads to cell death. The water resistance of the starch foam composite was improved with increasing OEO content owing to its hydrophobicity. The soil biodegradation test revealed that the increase in OEO content reduced the degradation rate. The EO might inhibit the growth of spores and fungal mycelia. Trongchuen et al. [130] investigated starch foam composites incorporated with OEO, spent coffee grounds, and extracted spent coffee grounds. The synergistic effect of antibacterial activities was observed for the starch foam composite containing extracted spent coffee grounds, 10 wt% spent coffee grounds, and 8 wt% OEO (SFE-10/8 wt% OEO). The inhibition zones for *S. aureus* and *E. coli* of SFE-10/8 wt% OEO were larger than those of other types of starch foam composites. The spent coffee grounds and extracted spent coffee grounds contain bioactive compounds such as caffeine, caffeic acid, and melanoidins that can assist OEO to attack bacteria cells. Petchwattana and Naknaen [135] studied the antibacterial activities of PBS incorporated with thymol as an active agent. The PBS film containing 10 wt% thymol showed the most effective growth inhibition of *S. aureus* and *E. coli*, and can release active agents over 15 days in different food simulants as an indicator for the shelf life of the film. Alparslan and Baygar [136] prepared active films based on chitosan and orange peel EO. The

1% orange peel EO in film was the minimum concentration against *B. subtilis*, *S. aureus*, *E. coli*, *Pseudomonas aeruginosa*, and *Candida albicans*. The chitosan film containing orange peel EO was used as packaging for shrimps under refrigerated conditions. Based on the microbiological and sensorial analysis, the stored shrimp in the chitosan film containing orange peel EO has a shelf life of up to 15 days, which was longer than the neat chitosan film (10 days). The incorporation of orange peel EO into the chitosan film can preserve and maintain the shelf life of food.

Organic acids have been used in active packaging as antibacterial and cross-linking agents for some types of polymers. Most often, organic acids such as citric acid, malic acid, lactic acid, and tartaric acid have been applied in active food packaging [137] to inhibit the growth of bacteria via reduction of pH of microbial cells and disruption of cell membrane permeability [138]. Ounkaew et al. [19] prepared active films based on starch/PVA enriched with antioxidants from spent coffee grounds and citric acid. The citric acid acted as an antimicrobial agent and cross-linker between the hydroxyl groups of starch/PVA. The minimum concentration of citric acid that inhibits the growth of *S. aureus* and *E. coli* was observed at 30 wt%. Furthermore, the synergistic effect on antibacterial activities was also investigated when the film was incorporated with citric acid and extracted spent coffee ground (PSt-E/30 wt%). The antioxidant release in food simulants showed that PSt-E/30 wt% had the highest 2,2-diphenyl-1-picrylhydrazyl (DPPH) scavenging capacity. This result implied that PSt-E/30 wt% can be applied for the prevention of lipid oxidation and maintaining the quality of foods. The optimal citric acid concentration for mechanical properties was 10 wt%, and the tensile strength of the film increased from 5.63 to 7.44 MPa. The film degradation was the in range of 65.28–86.64% after 30 days of soil burial. Halim et al. [139] studied biodegradable films containing tannic acid to preserve cherry tomatoes (*Solanum lycopersicum* var. *cerasiforme*) and grapes (*Vitis vinifera*). The presence of 15 wt% tannin acid in three types of films, i.e. chitosan, gelatin, and methylcellulose enhanced antibacterial activities against *S. aureus* and *E. coli*. The addition of tannic acid increased the film acidity that blocked the oxygen source of aerobic bacteria and affected bacterial cell lysis. The 14-days preservation study demonstrated that the addition of tannic acid into films can reduce weight loss and browning index of both cherry tomatoes and grapes. This observation indicated that tannic acid improved fruit freshness. An increase in tensile strength was observed in gelatin containing tannic acid (14.7 MPa), whereas the tensile strength of chitosan and methylcellulose containing tannic acid decreased. This phenomenon can be explained by the strong interaction that tannin forms with gelatin but by creating a weak and unstable bond with chitosan and methylcellulose.

Nanoparticles such as ZnO, TiO_2, and AgNPs have been applied in active packaging due to their non-toxicity to human cells [127]. Petchwattana et al. [140] studied the effect of ZnO on PBS films. The 6 wt% ZnO in the film was the minimum concentration that is active against *S. aureus* and *E. coli*. The released zinc ions damaged the bacterial cell membrane resulting in leakage of intercellular substances and destruction of bacterial cells. The release study in food simulant indicated that zinc ions migrated in acetic acid over 15 days. The uniform distribution of ZnO in the polymer matrix improved mechanical properties: the tensile strength of the films increased from 31.24 to 51.93 MPa when 2–10 wt% ZnO was added. Ortega et al. [141] developed active starch films incorporated with AgNPs via green synthesis. Maltose was used to reduce silver nitrate to AgNPs. The starch film containing 71.5 ppm AgNPs was the minimum concentration that can inhibit the growth of *E. coli* and *Salmonella* spp. To achieve antimicrobial activities, the starch film containing 143 ppm AgNPs was selected to test food storage. These active films extended the shelf life of fresh cheese samples for 21 days. The AgNPs could interact with proteins and deoxyribonucleic acid (DNA) via phosphorus and sulfur. Furthermore, it also prevented DNA replication leading to cell death.

15.6.2 Intelligent Packaging

Intelligent packaging has been used in the form of labels printed onto food packaging to monitor product quality. These intelligent tags can be obtained from various types of materials such as electronic labels designed with a built-in battery system and pH indicator based on natural products. Due to the simplicity, low cost, and efficiency, the color change of natural dyes by external stimuli has been widely applied in intelligent packaging [127]. This color change is due to the presence of phenolic or conjugated substances that change their structure when there is a change in pH. Anthocyanin is a pigment that has received attention for use in intelligent packaging. Anthocyanin-rich purple eggplant extract (PEE) or black eggplant extract (BEE) were added into active and intelligent chitosan food packaging films to increase the antioxidant ability of chitosan films. The two extracts were pH-sensitive and color changes in different buffer solutions could be used to monitor milk spoilage. The UV-visible light barrier and mechanical properties of the resulting film were also improved. However, only low contents of PEE and BEE could be well dispersed in the films [142]. Chitosan/anthocyanin-rich purple-fleshed sweet potato extract (PSPE) has also been used to develop antioxidant and intelligent pH-sensing films for their anthocyanin content. However, the film exhibited less film stretching and moisture absorption due to anthocyanin addition while the water vapor barrier and mechanical strength remained unchanged [142].

15.7 Conclusion

Biopolymer packaging is expected to be used as a substitute for petroleum-based plastics due to its biodegradability and renewability. To meet the target properties, biopolymers have been modified with various techniques such as chemical modification, blending with other polymers, and compounding to form composites. In addition, biopolymer packaging has also been developed as active packaging by incorporating intelligent additives. The active and intelligent biopolymer packagings have been applied to monitor and maintain the quality of food products. The development of biopolymer packaging with smart properties has attracted much attention to serving fresh, tasty, and convenient food products with prolonged shelf-life for consumer and food industries.

References

1. Amorim, J., Souza, K., Duarte, C., et al. (2020) Plant and bacterial nanocellulose: production, properties and applications in medicine, food, cosmetics, electronics and engineering. A review, *Environmental Chemistry Letters* 18(3): 851–869. https://doi.org/10.1007/s10311-020-00989-9.
2. Mello, L., Mali, S. (2014) Use of malt bagasse to produce biodegradable baked foams made from cassava starch, *Industrial Crops and Products* 55: 187–193. https://doi.org/10.1016/j.indcrop.2014.02.015.
3. Vercelheze, A., Fakhouri, F., Dall'Antônia, L. et al. (2012) Properties of baked foams based on cassava starch, sugarcane bagasse fibers and montmorillonite, *Carbohydrate Polymers* 87(2): 1302–1310. https://doi.org/10.1016/j.carbpol.2011.09.016.
4. Guarás, M., Alvarez, V., Ludueña, L. (2016) Biodegradable nanocomposites based on starch/polycaprolactone/compatibilizer ternary blends reinforced with natural and organo-modified montmorillonite, *Journal of Applied Polymer Science* 133(44): 44163–44173. https://doi.org/10.1002/app.44163.

5. Pornsuksomboon, K., Holló B., Szécsényi, K. et al. (2016) Properties of baked foams from citric acid modified cassava starch and native cassava starch blends, *Carbohydrate Polymers* 136: 107–112. https://doi.org/10.1016/j.carbpol.2015.09.019.
6. Edhirej, A., Sapuan, S., Jawaid, M., Ismarrubie, Z., et al. (2018) Preparation and characterization of cassava starch/peel composite film, *Polymer Composites* 39(5): 1704–1715. https://doi.org/10.1002/pc.24121.
7. Rodrigues, C., Tofanello, A., Nantes, I. (2015) Biological oxidative mechanisms for degradation of poly (lactic acid) blended with thermoplastic starch, *ACS Sustainable Chemistry and Engineering* 3(11): 2756–2766. https://doi.org/10.1021/acssuschemeng.5b00639.
8. Inamuddin (2016). *Green Polymer Composites Technology: Properties And Applications*. Boca Raton: CRC Press. https://doi.org/10.1201/9781315371184.
9. Ortega, R., Bonilla, J., Talens, P. et al. (2017). Future of Starch-Based Materials in Food Packaging, Starch-Based Materials in Food Packaging: Processing, Characterization and Applications. Elsevier, Oxford, UK. https://doi.org/10.1016/B978-0-12-809439-6.00009-1.
10. Zhu, J., Zhang, S., Zhang, B. et al. (2017) Structural features and thermal property of propionylated starches with different amylose/amylopectin ratio, *International Journal of Biological Macromolecules* 97: 123–130. https://doi.org/10.1016/j.ijbiomac.2017.01.033.
11. Jha, P., Dharmalingam, K., Nishizu, T. et al. (2020) Effect of amylose–amylopectin ratios on physical, mechanical, and thermal properties of starch-based bionanocomposite films incorporated with CMC and nanoclay, *Starch – Staerke* 72(1–2): 1–9. https://doi.org/10.1002/star.201900121.
12. Masina, N., Choonara, Y., Kumar, P. et al. (2017) A review of the chemical modification techniques of starch, *Carbohydrate Polymers* 157: 1226–1236. https://doi.org/10.1016/j.carbpol.2016.09.094.
13. Biduski, B., da Silva F., da Silva W. et al. (2017) Impact of acid and oxidative modifications, single or dual, of sorghum starch on biodegradable films, *Food Chemistry* 214: 53–60. https://doi.org/10.1016/j.foodchem.2016.07.039.
14. Li, X., He, Y., Huang, C. et al. (2016) Inhibition of plasticizer migration from packaging to foods during microwave heating by controlling the esterified starch film structure, *Food Control* 66: 130–136. https://doi.org/10.1016/j.foodcont.2016.01.046.
15. Wokadala, O., Emmambux, N., Ray S. (2014) Inducing PLA/starch compatibility through butyl-etherification of waxy and high amylose starch, *Carbohydrate Polymers* 112: 216–224. https://doi.org/10.1016/j.carbpol.2014.05.095.
16. Acosta, S., Chiralt, A., Santamarina, P. et al. (2016) Antifungal films based on starch-gelatin blend, containing essential oils, *Food Hydrocolloids* 61: 233–240. https://doi.org/10.1016/j.foodhyd.2016.05.008.
17. Talón, E., Trifkovic, K., Vargas, M. et al. (2017) Release of polyphenols from starch-chitosan based films containing thyme extract, *Carbohydrate Polymers* 175: 122–130. https://doi.org/10.1016/j.carbpol.2017.07.067.
18. Valencia-Sullca, C., Vargas, M., Atarés, L. et al. (2018). Thermoplastic cassava starch-chitosan bilayer films containing essential oils. *Food Hydrocolloids* 75: 107–115.
19. Ounkaew, A., Kasemsiri, P., Kamwilaisak, K. et al. (2018) Polyvinyl alcohol (PVA)/starch bioactive packaging film enriched with antioxidants from spent coffee ground and citric acid, *Journal of Polymers and the Environment* 26(9): 3762–3772. https://doi.org/10.1007/s10924-018-1254-z.
20. Ferri, J.M., Fenollar, O., Jorda-Vilaplana, A., et al. (2016) Effect of miscibility on mechanical and thermal properties of poly(lactic acid)/ polycaprolactone blends, *Polymer International* 65(4): 453–463. https://doi.org/10.1002/pi.5079.
21. Correa, A.C., Carmona, V., Simão, J. et al. (2017). Biodegradable blends of urea plasticized thermoplastic starch (UTPS) and poly (ε-caprolactone)(PCL): morphological, rheological, thermal and mechanical properties. *Carbohydrate Polymers* 167: 177–184. https://doi.org/10.1016/j.carbpol.2017.03.051.
22. Tian, H., Yan, J., Rajulu, A. et al. (2017). Fabrication and properties of polyvinyl alcohol/starch blend films: effect of composition and humidity. *International Journal of Biological Macromolecules* 96: 518–523. https://doi.org/10.1016/j.ijbiomac.2016.12.067.
23. Priya, B., Gupta, V., Pathania, D. et al. (2014) Synthesis, characterization and antibacterial activity of biodegradable starch/PVA composite films reinforced with cellulosic fibre, *Carbohydrate Polymers* 109: 171–179. https://doi.org/10.1016/j.carbpol.2014.03.044.

24. Zeng, J., Jiao, L., Li, Y. et al. (2011) Bio-based blends of starch and poly (butylene succinate) with improved miscibility, mechanical properties, and reduced water absorption, *Carbohydrate Polymers* 83(2): 762–768. https://doi.org/10.1016/j.carbpol.2010.08.051.
25. Şen, F., Uzunsoy, İ., Baştürk, E. et al. (2017) Antimicrobial agent-free hybrid cationic starch/sodium alginate polyelectrolyte films for food packaging materials, *Carbohydrate Polymers* 170: 264–270. https://doi.org/10.1016/j.carbpol.2017.04.079.
26. Ren, L., Yan, X., Zhou, J. et al. (2017) Influence of chitosan concentration on mechanical and barrier properties of corn starch/chitosan films, *International Journal of Biological Macromolecules* 105: 1636–1643. https://doi.org/10.1016/j.ijbiomac.2017.02.008.
27. Tokura, S. and Tamura H. (2007). *Chitin and Chitosan*. Amsterdam, The Netherlands: Elsevier. https://doi.org/10.1016/B978-044451967-2/00127-6.
28. Colobatiu, L., Gavan, A., Potarniche, A. et al. (2019) Evaluation of bioactive compounds-loaded chitosan films as a novel and potential diabetic wound dressing material, *Reactive and Functional Polymers* 145: 104369. https://doi.org/10.1016/j.reactfunctpolym.2019.104369.
29. Chen, Y., Cui, G., Dan, N. et al. (2019) Preparation and characterization of dopamine–sodium carboxymethyl cellulose hydrogel, *Microbial Biotechnology* 12(4): 677–687. https://doi.org/10.1111/1751-7915.13401.
30. Bonilla, J., Fortunati, E., Atarés, L. (2014) Physical, structural and antimicrobial properties of poly vinyl alcohol–chitosan biodegradable films, *Food Hydrocolloids* 35: 463–470. https://doi.org/10.1016/j.foodhyd.2013.07.002.
31. Figueroa, K., Andrade, M., Torres, O. (2018) Development of antimicrobial biocomposite films to preserve the quality of bread, *Molecules* 23(1): 1–18. https://doi.org/10.3390/molecules23010212.
32. Haghighi, H., Biard, S., Bigi, F. et al. (2019) Comprehensive characterization of active chitosan-gelatin blend films enriched with different essential oils, *Food Hydrocolloids* 95: 33–42. https://doi.org/10.1016/j.foodhyd.2019.04.019.
33. da Rocha, M., Alemán, A., Romani, V. et al. (2018) Effects of agar films incorporated with fish protein hydrolysate or clove essential oil on flounder (Paralichthys orbignyanus) fillets shelf-life. *Food Hydrocolloids* 81: 351–363. https://doi.org/10.1016/j.foodhyd.2018.03.017.
34. Wang, H., Gong, X., Miao, Y. et al. (2019) Preparation and characterization of multilayer films composed of chitosan, sodium alginate and carboxymethyl chitosan-ZnO nanoparticles, *Food Chemistry* 283: 397–403. https://doi.org/10.1016/j.foodchem.2019.01.022.
35. Bonilla, J., Atarés, L., Vargaset, M.al. (2013) Properties of wheat starch film-forming dispersions and films as affected by chitosan addition, *Journal of Food Engineering* 114(3): 303–312. https://doi.org/10.1016/j.jfoodeng.2012.08.005.
36. Costa, M., Cerqueira, M., Ruiz, H. (2015) Use of wheat bran arabinoxylans in chitosan-based films: Effect on physicochemical properties, *Industrial Crops and Products* 66: 305. https://doi.org/10.1016/j.indcrop.2015.01.003.
37. Dang, C., Yin, Y., and Pu, J. (2017). Preparation and synthesis of water-soluble chitosan derivative incorporated in ultrasonic-assistant wheat straw paper for antibacterial food-packaging. *Nordic Pulp & Paper Research Journal* 32 (4): 606–614. https://doi.org/10.3183/npprj-2017-32-04_p606-614_pu.
38. Suriyatem, R., Auras, R., Rachtanapun, C. et al. (2018) Biodegradable rice starch/carboxymethyl chitosan films with added propolis extract for potential use as active food packaging, *Polymers* 10(9): 10090954. https://doi.org/10.3390/polym10090954.
39. Shahbazi, Y. (2018) Application of carboxymethyl cellulose and chitosan coatings containing Mentha spicata essential oil in fresh strawberries *International Journal of Biological Macromolecules* 112: 264–272. https://doi.org/10.1016/j.ijbiomac.2018.01.186.
40. Valizadeh, S., Naseri, M., Babaei, S. et al. (2019) Development of bioactive composite films from chitosan and carboxymethyl cellulose using glutaraldehyde, cinnamon essential oil and oleic acid, *International Journal of Biological Macromolecules* 134: 604–612. https://doi.org/10.1016/j.ijbiomac.2019.05.071.
41. Andrade-Del Olmo, J. Pérez-Álvarez, L., Hernáez, E., et al. (2019) Antibacterial multilayer of chitosan and (2-carboxyethyl)- β-cyclodextrin onto polylactic acid (PLLA), *Food Hydrocolloids* 88(July 2018): 228–236. https://doi.org/10.1016/j.foodhyd.2018.10.014.
42. Castro, E., Iniguez, F., Samsudin, H. et al. (2016) Poly(lactic acid)-mass production, processing, industrial applications, and end of life, *Advanced Drug Delivery Reviews* 107: 333–366. https://doi.org/10.1016/j.addr.2016.03.010.

43. Hamdi, M., Nasri, R., Li, S., et al. (2019) Bioactive composite films with chitosan and carotenoproteins extract from blue crab shells: Bbiological potential and structural, thermal, and mechanical characterization, *Food Hydrocolloids* 89: 802–812. https://doi.org/10.1016/j.foodhyd.2018.11.062.
44. Sun, J., Du, Y., Ma, J., et al. (2019) Transparent bionanocomposite films based on konjac glucomannan, chitosan, and TEMPO-oxidized chitin nanocrystals with enhanced mechanical and barrier properties, *International Journal of Biological Macromolecules* 138: 866–873. https://doi.org/10.1016/j.ijbiomac.2019.07.170.
45. Wu, C., Sun, J., Zheng, P., et al. (2019) Preparation of an intelligent film based on chitosan/oxidized chitin nanocrystals incorporating black rice bran anthocyanins for seafood spoilage monitoring, *Carbohydrate Polymers* 222: 115006. https://doi.org/10.1016/j.carbpol.2019.115006.
46. Habibi, Y., Lucia, L., and Rojas, O. (2010). Cellulose nanocrystals: chemistry, self-assembly, and applications. *Chemical Reviews* 110 (6): 3479–3500. https://doi.org/10.1021/cr900339w.
47. Salas, C., Nypelö, T., Rodriguez, C. et al. (2014) Nanocellulose properties and applications in colloids and interfaces, *Current Opinion in Colloid & Interface Science* 19(5): 383–96. https://doi.org/10.1016/j.cocis.2014.10.003.
48. Phanthong, P., Reubroycharoen, P., Hao, X. et al. (2018). Nanocellulose: Extraction and application. *Carbon Resources Conversion* 1(1): 32–43. https://doi.org/10.1016/j.crcon.2018.05.004.
49. Jiang, F. and Hsieh, Y. (2013). Chemically and mechanically isolated nanocellulose and their self-assembled structures. *Carbohydrate Polymers* 95 (1): 32–40. https://doi.org/10.1016/j.carbpol.2013.02.022.
50. Piras, C., Fernández, S. and Borggraeve, W. (2019) Ball milling: a green technology for the preparation and functionalisation of nanocellulose derivatives, *Nanoscale Advances* 1(3): 937–47. https://doi.org/10.1039/C8NA00238J.
51. Sharma, A., Thakur, M., Bhattacharya, M. et al. (2019). Commercial application of cellulose nanocomposites – a review. *Biotechnology Reports* 21: e00316. https://doi.org/10.1016/j.btre.2019.e00316.
52. Lin, N. and Dufresne, A. (2014). Nanocellulose in biomedicine: current status and future prospect. *European Polymer Journal* 59: 302–325. https://doi.org/10.1016/j.eurpolymj.2014.07.025.
53. Abitbol, T., Rivkin, A., Cao, Y. et al. (2016). Nanocellulose, a tiny fiber with huge applications. *Current Opinion in Biotechnology* 39: 76–88. http://dx.doi.org/10.1016/j.copbio.2016.01.002.
54. Stelte, W. and Sanadi, A. (2009). Preparation and characterization of cellulose nanofibers from two commercial hardwood and softwood pulps. *Industrial & Engineering Chemistry Research* 48 (24): 11211–11219. https://doi.org/10.1021/ie9011672.
55. He, M., Yang, G., Chen, J. et al. (2018) Production and characterization of cellulose nanofibrils from different chemical and mechanical pulps, *Journal of Wood Chemistry and Technology* 38(2):149–58. https://doi.org/10.1039/b808639g.
56. Chaker, A., Mutjé, P., Vilar, M. et al. (2014). Agriculture crop residues as a source for the production of nanofibrillated cellulose with low energy demand. *Cellulose* 21 (6): 4247–4259. https://doi.org/10.1007/s10570-014-0454-5.
57. Boufi, S. and Chaker, A. (2016). Easy production of cellulose nanofibrils from corn stalk by a conventional high speed blender. *Industrial Crops and Products* 93: 39–47. https://doi.org/10.1016/j.indcrop.2016.05.030.
58. Nechyporchuk, O., Belgacem, M., and Bras, J. (2016). Production of cellulose nanofibrils: a review of recent advances. *Industrial Crops and Products* 93: 2–25. https://doi.org/10.1016/j.indcrop.2016.02.016.
59. Zhao, G., Lyu, X., Lee, J. et al. (2019). Biodegradable and transparent cellulose film prepared eco-friendly from durian rind for packaging application. *Food Packaging and Shelf Life* 21: 100345. https://doi.org/10.1016/j.fpsl.2019.100345.
60. Mu, R., Hong, X., Ni, Y. et al. (2019). Recent trends and applications of cellulose nanocrystals in food industry. *Trends in Food Science & Technology* 93: 136–144. https://doi.org/10.1016/j.tifs.2019.09.013.
61. Wulandari, W., Rochliadi, A., and Arcana, I. (2016). Nanocellulose prepared by acid hydrolysis of isolated cellulose from sugarcane bagasse. *Proceedings of Materials Science and Engineering* 107: 12045.
62. Cheng, S., Zhang, Y., Cha, R. et al. (2016) Water-soluble nanocrystalline cellulose films with highly transparent and oxygen barrier properties, *Nanoscale* 8(2): 973–978. https://doi.org/10.1039/c5nr07647a.

63. de Souza Coelho, C.C., Silva, R.B.S., Carvalho, C.W.P. et al. (2020). Cellulose nanocrystals from grape pomace and their use for the development of starch-based nanocomposite films. *International Journal of Biological Macromolecules* 159: 1048–1061. https://doi.org/10.1016/j.ijbiomac.2020.05.046.
64. Noorbakhsh-Soltani, S.M., Zerafat, M.M., and Sabbaghi, S. (2018). A comparative study of gelatin and starch-based nano-composite films modified by nano-cellulose and chitosan for food packaging applications. *Carbohydrate Polymers* 189: 48–55. https://doi.org/10.1016/j.carbpol.2018.02.012.
65. Mascheroni, E., Rampazzo, R., Ortenzi, M. A. et al. (2016) Comparison of cellulose nanocrystals obtained by sulfuric acid hydrolysis and ammonium persulfate, to be used as coating on flexible food-packaging materials, *Cellulose* 23: 779–793. https://doi.org/10.1007/s10570-015-0853-2.
66. de Oliveira, J.P., Bruni, G.P., Halal, S.L.M. et al. (2019). Cellulose nanocrystals from rice and oat husks and their application in aerogels for food packaging. *International Journal of Biological Macromolecules* 124: 175–184. https://doi.org/10.1016/j.ijbiomac.2018.11.205.
67. Turbak, A., Snyder, F., and Sandberg, K. (1983). Microfibrillated cellulose, a new cellulose product: properties, uses, and commercial potential. *Journal of Applied Polymer Science: Applied Polymer Symposium* 37: 815–827.
68. Martins, M.P., Dagostin, J.L.A., Franco, T.S. et al. (2020). Application of cellulose nanofibrils isolated from an agroindustrial residue of peach palm in cassava starch films. *Food Biophysics* 15: 323–334. https://doi.org/10.1007/s11483-020-09626-y.
69. Kalia, S., Boufi, S. and Celli A. et al.(2014). Kango S. Nanofibrillated cellulose: surface modification and potential applications, *Colloid and Polymer Science* 292(1): 5–31. https://doi.org/10.1007/s00396-013-3112-9.
70. Hai, L.V., Choi, E.S., Zhai, L. et al. (2020). Green nanocomposite made with chitin and bamboo nanofibers and its mechanical, thermal and biodegradable properties for food packaging. *International Journal of Biological Macromolecules* 144: 491–499. https://doi.org/10.1016/j.ijbiomac.2019.12.124.
71. Wang, W., Yu, Z., Alsammarraie, F.K. et al. (2020). Properties and antimicrobial activity of polyvinyl alcohol-modified bacterial nanocellulose packaging films incorporated with silver nanoparticles. *Food Hydrocolloids* 100: 105411. https://doi.org/10.1016/j.foodhyd.2019.105411.
72. Xie, Y., Niu, X., and Yang, J. (2020). Active biodegradable films based on the whole potato peel incorporated with bacterial cellulose and curcumin. *International Journal of Biological Macromolecules* 150: 480–491. https://doi.org/10.1016/j.ijbiomac.2020.01.291.
73. Long, L., Li, F., Shu, M. et al. (2019) Fabrication and application of carboxymethyl cellulose-carbon nanotube aerogels, *Materials* 12(11): 1867. https://doi.org/10.3390/ma14144059.
74. Chen, W. and Wen, Y. (2019). Optimization of reaction conditions for preparing carboxymethyl cellulose from bamboo scraps by response surface methodology. *Proceedings of Materials Science and Engineering* 611: 12041.
75. Ye, H., Xu, S., Wu, S. et al. (2018). Optimization of sodium carboxymethyl cellulose preparation from bagasse by response surface methodology. *Proceedings of Materials Science and Engineering* 381: 12043.
76. Chumee, J. and Khemmakama, P. (2014). Carboxymethyl cellulose from pineapple peel: useful green bioplastic. *Advanced Materials Research* 979: 366–369.
77. Casaburi, A., Montoya, R., Cerrutti, P. et al. (2018). Carboxymethyl cellulose with tailored degree of substitution obtained from bacterial cellulose. *Food Hydrocolloids* 75: 147–156. https://doi.org/10.1016/j.foodhyd.2017.09.002.
78. Tavares, K.M., de Campos, A., and Mitsuyuki, M.C. (2019). Corn and cassava starch with carboxymethyl cellulose films and its mechanical and hydrophobic properties. *Carbohydrate Polymers* 223: 115055. https://doi.org/10.1016/j.carbpol.2019.115055.
79. Makwana, D., Castaño, J., Somani, R.S. et al. (2020). Characterization of aAgar-CMC/Ag-MMT nanocomposite and evaluation of antibacterial and mechanical properties for packaging applications. *Arabian Journal of Chemistry* 13 (1): 3092–3099. https://doi.org/10.1016/j.arabjc.2018.08.017.
80. Behnezhad, M. and Goodarzi1, M. and Baniasadi, H. (2020). Fabrication and characterization of polyvinyl alcohol/carboxymethyl cellulose/titanium dioxide degradable composite films: an RSM study. *Materials Research Express* 6: 125548.
81. Liu, H. and Zhang, J. (2011) Research progress in toughening modification of poly(lactic acid), *Journal of Polymer Science, Part B: Polymer Physics* 49(15): 1051–1083. https://doi.org/10.1002/polb.22283.

82. Urquijo, J., Guerrica-Echevarría, G. and Eguiazábal, J. I. (2015) Melt processed PLA/PCL blends: Eeffect of processing method on phase structure, morphology, and mechanical properties, *Journal of Applied Polymer Science* 132(41): 1–9. https://doi.org/10.1002/app.42641.
83. Armentano, I., Bitinis, N., Fortunati, E. et al. (2013) Multifunctional nanostructured PLA materials for packaging and tissue engineering, *Progress in Polymer Science* 38(10–11): 1720–1747. https://doi.org/10.1016/j.progpolymsci.2013.05.010.
84. Zengwen, C. Pan, H., Chen, Y. et al. (2019) Transform poly (lactic acid) packaging film from brittleness to toughness using traditional industrial equipments, *Polymer* 180(July): 121728. https://doi.org/10.1016/j.polymer.2019.121728.
85. Lertwongpipat, N., Petchwatana, N. and Covavisaruch, S. (2014) Enhancing the flexural and impact properties of bioplastic poly(lactic acid) by melt blending with poly(butylene succinate), *Advanced Materials Research* 931–932: 106–110. https://doi.org/10.4028/www.scientific.net/AMR.931-932.106.
86. Messin, T., Marais, S., Follain, N. et al. (2020) Biodegradable PLA/PBS multinanolayer membrane with enhanced barrier performances, *Journal of Membrane Science* 598: 117777. https://doi.org/10.1016/j.memsci.2019.117777.
87. Ostafinska, A., Fortelny, I., Nevoralova, M. et al. (2015) Synergistic effects in mechanical properties of PLA/PCL blends with optimized composition, processing, and morphology, *RSC Advances* 5(120): 98971–98982. https://doi.org/10.1039/c5ra21178f.
88. Deng, Y. and Thomas, N. L. (2015) Blending poly(butylene succinate) with poly(lactic acid): ductility and phase inversion effects, *European Polymer Journal* 71: 534–546. https://doi.org/10.1016/j.eurpolymj.2015.08.029.
89. Wang, R., Wang, S., Zhang, Y. et al. (2009) Toughening modification of PLLA/PBS blends via in situ compatibilization, *Polymer Engineering & Science* 49(1): 26–33. https://doi.org/10.1002/pen.21210.
90. Armentano, I., Fortunati, E., Burgos, N. et al. (2015a) Bio-based PLA_PHB plasticized blend films: processing and structural characterization, *LWT-Food Science and Technology* 64(2): 980–988. https://doi.org/10.1016/j.lwt.2015.06.032.
91. Arrieta, M. P., Samper, M. D., Aldas, M. et al. (2017) On the use of PLA-PHB blends for sustainable food packaging applications, *Materials* 10(9): 1–26. https://doi.org/10.3390/ma10091008.
92. Yoon, J., Lee, W.S., Kim, K.S. et al. (2000). Effect of poly(ethylene glycol)-block-poly(L-lactide) on the poly[(R)-3-hydroxybutyrate]/poly(L-lactide) blends. *European Polymer Journal* 36: 3–10.
93. Zhang, M. and Thomas, N. L. (2011) Blending polylactic acid with polyhydroxybutyrate: the effect on thermal, mechanical, and biodegradation properties, *Advances in Polymer Technology* 30(2): 67–79. https://doi.org/10.1002/adv.20235.
94. Blümm, E. and Owen, A. J. (1995) Miscibility, crystallization and melting of poly(3-hydroxybutyrate)/poly(L-lactide) blends, *Polymer* 36(21): 4077–4081. https://doi.org/10.1016/0032-3861(95)90987-D.
95. Koyama, N. and Doi, Y. (1997) Miscibility of binary blends of poly[(R)-3-hydroxybutyric acid] and poly[(S)-lactic acid], *Polymer* 38(7): 1589–1593. https://doi.org/10.1016/S0032-3861(96)00685-4.
96. Ohkoshi, I., Abe, H. and Doi, Y. (2000) Miscibility and solid-state structures for blends of poly[(S)-lactide] with atactic poly[(R,S)-3-hydroxybutyrate], *Polymer* 41(15): 5985–5992. https://doi.org/10.1016/S0032-3861(99)00781-8.
97. Cabedo, L., Luis Feijoo J., Pilar Villanueva, M. et al. (2006) Optimization of biodegradable nanocomposites based on aPLA/PCL blends for food packaging applications, *Macromolecular Symposia* 233: 191–197. https://doi.org/10.1002/masy.200690017.
98. Li, H. and Huneault, M. A. (2007) Effect of nucleation and plasticization on the crystallization of poly(lactic acid), *Polymer* 48(23): 6855–6866. https://doi.org/10.1016/j.polymer.2007.09.020.
99. Liewchirakorn, P., Aht-Ong, D. and Chinsirikul, W. (2018) Practical approach in developing desirable peel–seal and clear lidding films based on poly(lactic acid) and poly(butylene adipate-co-terephthalate) blends, *Packaging Technology and Science* 31(5), 296–309. https://doi.org/10.1002/pts.2321.
100. Wang, L. F., Rhim, J. W. and Hong, S. I. (2016) Preparation of poly(lactide)/poly(butylene adipate-co-terephthalate) blend films using a solven casting method and their food packaging application, *LWT-Food Science and Technology* 68: 454–461. https://doi.org/10.1016/j.lwt.2015.12.062.
101. Sharma, S., Jaiswal, A. K., Duffy, B. et al. (2020) Ferulic acid incorporated active films based on poly(lactide) /poly(butylene adipate-co-terephthalate) blend for food packaging, *Food Packaging and Shelf Life* 24: 100491. https://doi.org/10.1016/j.fpsl.2020.100491.

102. Zhang, J., Cao, C., Zheng, S. et al. (2020). Poly (butylene adipate-co-terephthalate)/magnesium oxide/silver ternary composite biofilms for food packaging application. *Food Packaging and Shelf Life* 24: 100487. https://doi.org/10.1016/j.fpsl.2020.100487.
103. Palai, B., Mohanty, S., and Nayak, S.K. (2020). Synergistic effect of polylactic acid (PLA) and poly(butylene succinate-co-adipate) (PBSA) based sustainable, reactive, super toughened eco-composite blown films for flexible packaging applications. *Polymer Testing* 83: 106130. https://doi.org/10.1016/j.polymertesting.2019.106130.
104. Vilay, V., Mariatti, M., Ahmad, Z. et al. (2009) Characterization of the mechanical and thermal properties and morphological behavior of biodegradable poly(L-lactide)/poly(ε-caprolactone) and poly(L-lactide)/poly(butylene succinate-co-L-lactate) polymeric blends, *Journal of Applied Polymer Science* 114(3): 1784–1792. https://doi.org/10.1002/app.30683.
105. Finotti, P.F.M., Costa, L.C., Capote, T.S.O. et al. (2017). Immiscible poly(lactic acid)/poly(ε-caprolactone) for temporary implants: compatibility and cytotoxicity. *Journal of the Mechanical Behavior of Biomedical Materials* 68: 155–162. https://doi.org/10.1016/j.jmbbm.2017.01.050.
106. Ma, P., Cai, X., Zhang, Y. et al. (2014). In-situ compatibilization of poly(lactic acid) and poly(butylene adipate-co-terephthalate) blends by using dicumyl peroxide as a free-radical initiator. *Polymer Degradation and Stability* 102: 145–151. https://doi.org/10.1016/j.polymdegradstab.2014.01.025.
107. Semba, T., Kitagawa, K., Ishiaku, U.S. et al. (2006) The effect of crosslinking on the mechanical properties of polylactic acid/polycaprolactone blends, *Journal of Applied Polymer Science* 101(3): 1816–1825. https://doi.org/10.1002/app.23589.
108. Zhang, X. and Zhang, Y. (2016). Reinforcement effect of poly(butylene succinate) (PBS)-grafted cellulose nanocrystal on toughened PBS/polylactic acid blends. *Carbohydrate Polymers* 140: 374–382. https://doi.org/10.1016/j.carbpol.2015.12.073.
109. Ojijo, V. and Ray, S.S. (2015). Super toughened biodegradable polylactide blends with non-linear copolymer interfacial architecture obtained via facile in-situ reactive compatibilization. *Polymer* 80: 1–17. https://doi.org/10.1016/j.polymer.2015.10.038.
110. Harada, M., Ohya, T., Iida, K., et al. (2007) Increased impact strength of biodegradable poly(lactic acid)/poly(butylene succinate) blend composites by using isocyanate as a reactive processing agent, *Journal of Applied Polymer Science* 106(3): 1813–1820. https://doi.org/10.1002/app.26717.
111. Vannaladsaysy, V., Todo, M., Takayama, T. et al. (2009) Effects of lysine triisocyanate on the mode i fracture behavior of polymer blend of poly (L-lactic acid) and poly (butylene succinate-co-L-lactate), *Journal of Materials Science* 44(11): 3006–3009. https://doi.org/10.1007/s10853-009-3428-5.
112. Kumar, M., Mohanty, S., Nayak, S.K. et al. (2010). Effect of glycidyl methacrylate (GMA) on the thermal, mechanical and morphological property of biodegradable PLA/PBAT blend and its nanocomposites. *Bioresource Technology* 101(21): 8406–8415. https://doi.org/10.1016/j.biortech.2010.05.075.
113. Zhang, N., Wang, Q., Ren, J., et al. (2009) Preparation and properties of biodegradable poly(lactic acid)/poly(butylene adipate-co-terephthalate) blend with glycidyl methacrylate as reactive processing agent, *Journal of Materials Science* 44(1): 250–256. https://doi.org/10.1007/s10853-008-3049-4.
114. Pivsa-Art, S., Thumsorn, S., Pavasupree, S. et al. (2013). Effect of additive on crystallization and mechanical properties of polymer blends of poly(lactic acid) and poly[(butylene succinate)-co-adipate]. *Energy Procedia* 34: 563–571. https://doi.org/10.1016/j.egypro.2013.06.786.
115. Bitinis, N., Verdejo, R., Cassagnau, P. et al. (2011). Structure and properties of polylactide/natural rubber blends. *Materials Chemistry and Physics* 129(3): 823–831. https://doi.org/10.1016/j.matchemphys.2011.05.016.
116. Pongtanayut, K., Thongpin, C., and Santawitee, O. (2013). The effect of rubber on morphology, thermal properties and mechanical properties of PLA/NR and PLA/ENR blends. *Energy Procedia* 34: 888–897. https://doi.org/10.1016/j.egypro.2013.06.826.
117. Ayutthaya, W. D. N. and Poompradub, S. (2014) Thermal and mechanical properties of poly(lactic acid)/natural rubber blend using epoxidized natural rubber and poly(methyl methacrylate) as co-compatibilizers, *Macromolecular Research* 22(7): 686–692. https://doi.org/10.1007/s13233-014-2102-1.
118. Chen, Y., Yuan, D. and Xu, C. (2014) Dynamically vulcanized biobased polylactide/natural rubber blend material with continuous cross-linked rubber phase, *ACS Applied Materials & Interfaces* 6(6): 3811–3816. https://doi.org/10.1021/am5004766.

119. Pattamaprom, C., Chareonsalung, W., Teerawattananon, C., et al. (2016) Improvement in impact resistance of polylactic acid by masticated and compatibilized natural rubber, *Iranian Polymer Journal (English Edition)* 25(2): 169–178. 10.1007/s13726-015-0411-7.82. Zhang, C., Wang, W., Huang, Y. et al. (2013) Thermal, mechanical and rheological properties of toughened by expoxidized natural rubber, *Materials & Design* 45: 198–205. https://doi.org/10.1016/j.matdes.2012.09.024.
120. Yang, W., Weng, Y., Puglia, D. et al. (2020). Poly(lactic acid)/lignin films with enhanced toughness and anti-oxidation performance for active food packaging. *International Journal of Biological Macromolecules* 144: 102–110. https://doi.org/10.1016/j.ijbiomac.2019.12.085.
121. Petchwattana, N., Naknaen, P., Sanetuntikul, J. et al. (2018) Crystallisation behaviour and transparency of poly(lactic acid) nucleated with dimethylbenzylidene sorbitol, *Plastics, Rubber and Composites* 47(4): 147–155. https://doi.org/10.1080/14658011.2018.1447338.
122. Hassan, E., Wei, Y., Jiao, H. et al. (2013) Dynamic mechanical properties and thermal stability of poly(lactic acid) and poly(butylene succinate) blends composites, *Journal of Fiber Bioengineering and Informatics* 6(1): 85–94. https://doi.org/10.3993/jfbi03201308.
123. Wachirahuttapong, S., Thongpin, C. and Sombatsompop, N. (2016) Effect of PCL and compatibility contents on the morphology, crystallization and mechanical properties of PLA/PCL blends, *Energy Procedia* 89: 198–206. https://doi.org/10.1016/j.egypro.2016.05.026.
124. Armentano, I., Fortunati, E., Burgos, N. et al. (2015b) Processing and characterization of plasticized PLA/PHB blends for biodegradable multiphase systems, *Express Polymer Letters* 9(7): 583–596. https://doi.org/10.3144/expresspolymlett.2015.55.
125. Mohammad, N. N. B., Arsad, A., Rahmat, A. R. et al. (2014) Influence of compatibilizer on mechanical properties of polylactic acid/natural rubber blends, *Applied Mechanics and Materials* 554: 81–85. https://doi.org/10.4028/www.scientific.net/AMM.554.81.
126. Diaz, C. and Kim, S. (2015). Film performance of poly(lactic acid) blends for packaging applications. *Journal of Applied Packaging Research* 8: 43–51.
127. Dainelli, D., Gontard, N., Spyropoulos, D. et al. (2008). Active and intelligent food packaging: legal aspects and safety concerns. *Trends in Food Science & Technology* 19: S103–S112. https://doi.org/10.1016/j.tifs.2008.09.011.
128. Ketkaew, S., Kasemsiri, P., Hiziroglu, S. et al. (2018) Effect of oregano essential oil content on properties of green biocomposites based on cassava starch and sugarcane bagasse for Bbioactive packaging, *Journal of Polymers and the Environment* 26(1): 311–318. https://doi.org/10.1007/s10924-017-0957-x.
129. Tornuk, F., Hancer, M., Sagdic, O. et al. (2015). LLDPE based food packaging incorporated with nanoclays grafted with bioactive compounds to extend shelf life of some meat products. *LWT-Food Science and Technology* 64 (2): 540–546. https://doi.org/10.1016/j.lwt.2015.06.030.
130. Trongchuen, K., Ounkaew, A., Kasemsiri, P. et al. (2018) Bioactive starch foam composite enriched with natural antioxidants from spent coffee ground and essential oil, *Starch-Staerke* 70(7–8): 1700238. https://doi.org/10.1002/star.201700238.
131. Bovi, G.G., Caleb, O.J., Klaus, E. et al. (2018). Moisture absorption kinetics of FruitPad for packaging of fresh strawberry. *Journal of Food Engineering* 223: 248–254. https://doi.org/10.1016/j.jfoodeng.2017.10.012.
132. Warsiki, E. (2018) Application of chitosan as biomaterial for active packaging of ethylene absorber, *IOP Conference Series: Earth and Environmental Science* 141: 12036. https://doi.org/10.1088/1755-1315/141/1/012036.
133. Tongnuanchan, P., Benjakul, S., Prodpran, T. et al. (2016) Mechanical, thermal and heat sealing properties of fish skin gelatin film containing palm oil and basil essential oil with different surfactants, *Food Hydrocolloids* 56: 93–107. https://doi.org/10.1016/j.foodhyd.2015.12.005.
134. Hosseini, M. H., Razavi, S. H. and Mousavi, M. A. (2009) Antimicrobial, physical and mechanical properties of chitosan-based films incorporated with thyme, clove and cinnamon essential oils, *Journal of Food Processing and Preservation* 33(6): 727–743. https://doi.org/10.1111/j.1745-4549.2008.00307.x.
135. Petchwattana, N. and Naknaen, P. (2015). Utilization of thymol as an antimicrobial agent for biodegradable poly(butylene succinate). *Materials Chemistry and Physics* 163: 369–375. https://doi.org/10.1016/j.matchemphys.2015.07.052.

136. Alparslan, Y. and Baygar, T. (2017). Effect of chitosan film coating combined with orange peel essential oil on the shelf life of deepwater pink shrimp. *Food and Bioprocess Technology* 10: 842–853. https://doi.org/10.1007/s11947-017-1862-y.
137. Coban, H. B. (2020) Organic acids as antimicrobial food agents: applications and microbial productions, *Bioprocess and Biosystems Engineering* 43(4): 569–591. https://doi.org/10.1007/s00449-019-02256-w.
138. Eswaranandam, S., Hettiarachchy, N. S. and Johnson, M. G. (2004) Antimicrobial activity of citric, lactic, malic, or tartaric acids and nisin-incorporated soy protein film against *Listeria monocytogenes, Escherichia coli O157:H7*, and Salmonella gaminara, *Journal of Food Science* 69(3): FMS79–FMS84. https://doi.org/10.1111/j.1365-2621.2004.tb13375.x.
139. Halim, A.L.A., Kamari, A., and Phillip, E. (2018). Chitosan, gelatin and methylcellulose films incorporated with tannic acid for food packaging. *International Journal of Biological Macromolecules* 120: 1119–1126. https://doi.org/10.1016/j.ijbiomac.2018.08.169.
140. Petchwattana, N., Covavisaruch, S., Wibooranawong, S. et al. (2016). Antimicrobial food packaging prepared from poly(butylene succinate) and zinc oxide. *Measurement* 93: 442–448. https://doi.org/10.1016/j.measurement.2016.07.048.
141. Ortega, F., Giannuzzi, L., Arce, V. B. et al. (2017) Active composite starch films containing green synthetized silver nanoparticles, *Food Hydrocolloids* 70: 152–162. https://doi.org/10.1016/j.foodhyd.2017.03.036.
142. Yong, H., Wang, X., Bai, R., et al. (2019) Development of antioxidant and intelligent pH-sensing packaging films by incorporating purple-fleshed sweet potato extract into chitosan matrix, *Food Hydrocolloids* 90: 216–224. https://doi.org/10.1016/j.foodhyd.2018.12.015.
143. Collazo-Bigliardi S., Ortega-Toro, R., and Chiralt Boix, A. (2018). Isolation and characterisation of microcrystalline cellulose and cellulose nanocrystals from coffee husk and comparative study with rice husk. *Carbohydrate Polymers* 191: 205–215. https://doi.org/10.1016/j.carbpol.2018.03.022.
144. Wang, Y., Li, R., Lu, R. et al. (2019) Preparation of chitosan/corn starch/cinnamaldehyde films for strawberry preservation, *Foods* 8: 423. https://doi.org/10.3390/foods8090423.

16

Recent Developments in Bio-Based Materials for Controlled-Release Fertilizers

Kritapas Laohhasurayotin, Doungporn Yiamsawas and Wiyong Kangwansupamonkon

National Nanotechnology Center, National Science and Technology Development Agency, Pathum Thani, Thailand

16.1 Introduction and Historical Review

16.1.1 Early Fertilizer Development and Its Impact on Environment

Traditional fertilization is one of the earliest known practices in agriculture developed within several thousand years of human civilization. Fertilization is necessary to give sufficient crop yield to supply the matching food quantities against the rising global populations. However, it was not until the early decades of the twentieth century that fertilizer technology had changed forever after Fritz Haber developed the first method to make NH_3 from atmospheric N_2 [1] followed by its industrial-scale synthesis by Carl Bosch [2]. Ever since, chemical fertilizers, together with modern agricultural technologies, have played major roles in producing food crops to support global populations reaching six billion in 2002 [3]. Double global grain demand is projected by 2050 when populations are expected to increase by another 50%. Hence, an even more increased utilization of macronutrient fertilizers such as nitrogen (N), phosphorus (P), and potassium (K) is required for the mass production of food crops. Especially N and P are essential because lack of these nutrients can limit plant growth [4] and agricultural yield [5]. However, only the soluble fraction can be taken up by plants, which is

High-Performance Materials from Bio-based Feedstocks, First Edition. Edited by Andrew J. Hunt, Nontipa Supanchaiyamat, Kaewta Jetsrisuparb and Jesper T.N. Knijnenburg.

often lost by leaching, resulting in contamination of surface and groundwater [6]. The benefits of synthetic fertilizers for the agricultural process have to trade with the environmental impact.

Growing plants on deteriorated agricultural land requires large quantities of fertilizers to supply adequate nutrients, especially after multiple repeated cycles of the same crop in the same field. Most common solid fertilizers exist as water-soluble uncoated or pristine granules. Such materials show limitations due to the fast release and are thus vulnerable to nutrient losses by leaching and volatilization or surface run-off [7]. In their early growth stages, crops cannot absorb all the supplied nutrients, thus leaving unused nutrients to leach out and be collected in the water reservoir, resulting in eutrophication and economic deficits [8]. This overuse of water-soluble fertilizers and pesticides as well as other synthetic compounds can increase pollutions in soil, surface water, and groundwater, incurring health and decontamination costs. As a result, fertilizer usage alters the ecosystem balance and causes several serious consequences, such as eutrophication, drought, and climate change (Figure 16.1).

Nitrogen fertilizers were first obtained by the conversion of NH_3 to HNO_3 which were subsequently converted into ammonium nitrate (NH_4NO_3) and explosives [9]. Combinations of ammonium nitrate with other synthetic sources of P and K were developed in order to increase plant nutrient uptake. It is estimated that only 30% of N and 10% of P applied as water-soluble fertilizers can be taken up by plants, and the rest are lost to the environment due to, e.g. volatilization of N [10] or P accumulated in the soil in plant-unavailable forms [5]. In fact, loss of N can be traced into several paths, mainly due to nitrification, denitrification, leaching, and volatilization [11]. When applied incorrectly, NH_4NO_3 causes the N loss via ammonium volatilization, N_2O emission, and NO_3^- leaching. All these mechanisms can be found in most types of major N fertilizers such as urea, aqua ammonia, sodium nitrate, and ammonium phosphates. While N and P losses have received significant attention, the loss of K is under little consideration because it does not result in eutrophication or accumulation in an unavailable form like P. Significant potassium losses to the environment occur by leaching and surface runoff that may cause a potential economic issue to crop owners [12].

A significant number of technologies has been developed to reduce NPK nutrient losses, such as polymer shell coatings [13], bio-based adsorbents [14], inorganic fertilizers [15, 16], and crop rotation. A slow- or controlled-release approach has been more preferable to traditional fertilizers for many years now since the fertilizer release is more effective and controllable [17]. Figure 16.2 displays nutrient release profiles from different simulated and experimental scenarios that indicate the successful control of nitrogen release in the form of urea with selected coating materials.

16.1.2 Controlled-Release Fertilizer

Unlike conventional fertilizers that readily dissolve in water, modern fertilizers are frequently engineered to have controllable nutrient release properties. Although there is no distinct difference in the definition of slow-release fertilizer (SRF) and controlled-release fertilizer (CRF), SRF and CRF can be identified based on their main nutrient release mechanisms. A slow- or controlled-release fertilizer receives its general concept through the patterns in delaying the plant nutrients to be available for the plant uptake in the soil after application or extending the nutrients' availability at a significantly longer period [19]. The specific mechanisms of controlled fertilizer release may slightly vary from one concept to the other, but the main objective of CRFs is to slow down the rate of nutrient dissolution in water. For example, coating-type CRFs are usually formed in a core-shell form containing a fertilizer core within a coating shell,

(a)

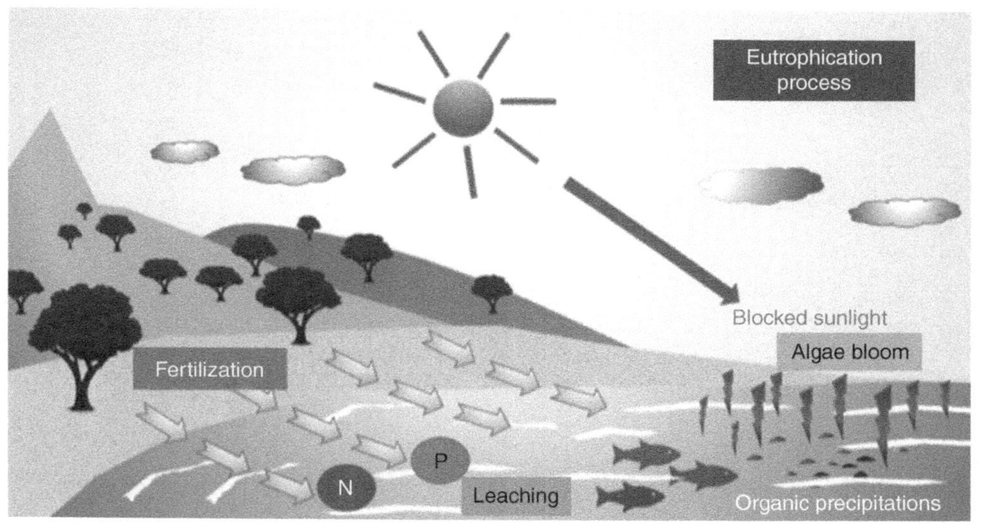

(b)

Figure 16.1 Overfertilization incident causes (a) eutrophication and (b) drought land (Lukas/Adobe Stock).

such as urea submerged within a supergranule or covered by a polymer coating layer [20]. A different type of CRF acts via the slow decomposition of the fertilizer (like urea-formaldehyde compound) to release plant nutrients due to the microbial activities in the soil [17, 21].

Slow-release fertilizers were initially prepared from bio-based solids such as manures and composts (Figure 16.3). Synthetic CRFs, e.g. urea-formaldehyde and methylene urea, were later employed due to their advantages in giving high N content and lower water solubility [22]

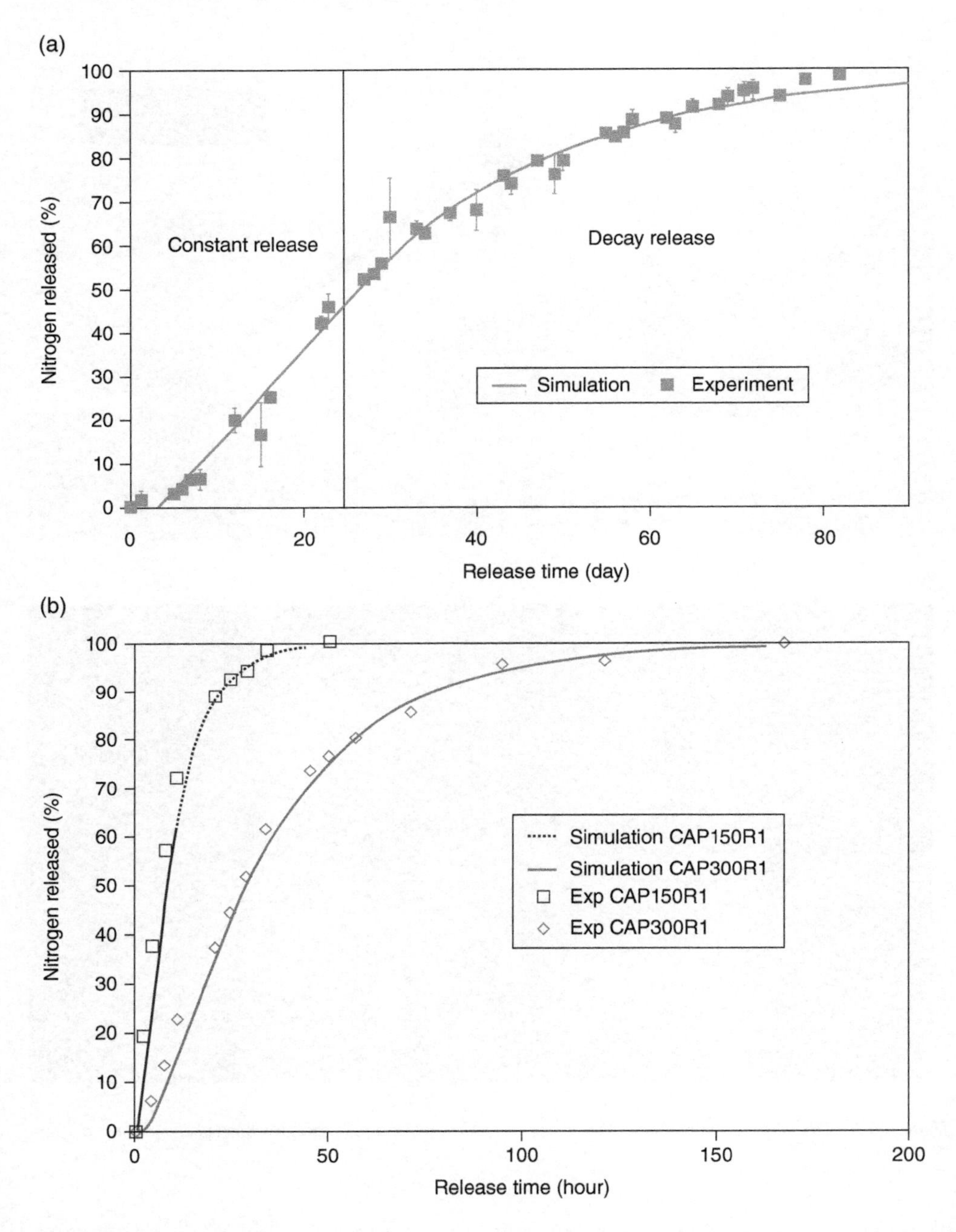

Figure 16.2 *Examples of nitrogen release profiles controllable by the coating materials: (a) comparison study via simulation model vs. experiment and (b) influence of coating thickness on nitrogen release rates. Source: Trinh et al. [18]. Reprinted from Trinh et al., Copyright© (2015), with permission from Elsevier.*

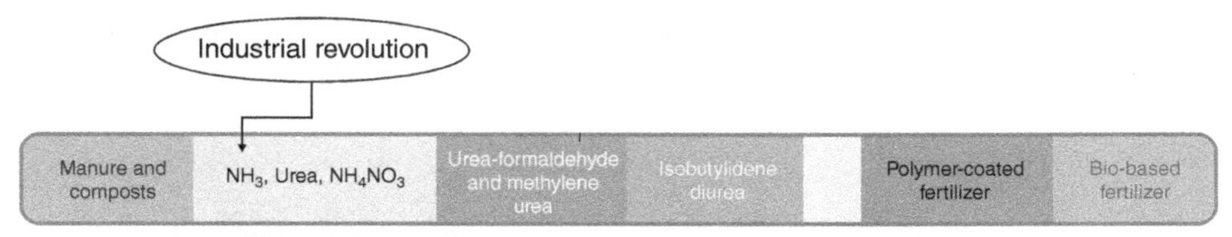

Figure 16.3 *Development of controlled-release fertilizers (CRFs), which emerged from bio-resources to chemical synthetics and more complex bio-based systems.*

compared to the highly water-soluble urea. The decomposition of these two urea compounds still requires microbial activity to help the distribution of stored nutrients. However, a synthetic compound like isobutylidine diurea can offer slow-release features without microbe dependence even at a low temperature.

The coated fertilizers use insoluble materials such as organic polymers (thermoplastics and resins) or inorganic substances (sulfur and other minerals) as a barrier to prevent the water-soluble core from releasing [23]. The release rate can be controlled by nutrient diffusion through a semi-impermeable coating and depends on the composition and thickness of the coating.

Nutrient release profiles from polymer-coated granules are designed to match the nutrient release rate to the maximum nutrient uptake rate by plants, which varies throughout the growing season depending on the crop maturity. Hence, the use of CRFs increases the fertilizer use efficiency and crop yield while the environmental burden is decreased by reducing or eliminating the nutrients applied in excess of what the plants can absorb [24].

Even though polymer coatings have received much attention in fertilizer development, these materials have become more restrictive since petroleum-based synthetic materials such as polyolefins, acrylic resin, glycerol ester, and polysulfone are made from nonrenewable resources [25]. These polymers are often nonbiodegradable and harmful to the environment [26]. Good alternatives to these fossil fuel-derived materials are biomaterials, which are relatively low-cost, abundant, environmentally friendly, and sustainable in terms of supplying resources. Biopolymers can be found in any form from agricultural produce to biomass as wastes and include cellulose, starch, keratin, chitin, and polyamino acids [25]. However, these biomaterials mainly contain hydrophilic groups. Their hydrophobicity must be additionally modified, but can also be obtained through the incorporation of other renewable polymers such as bio-based polyurethane (PU) from locust sawdust [27].

Nitrogen is not the only plant nutrient of concern for agricultural applications. Other macronutrients such as K, P, Ca, and Mg, and micronutrients like Fe, B, Zn, and Mn are also studied in a controlled release manner because excessive application of these elements can eventually lead to environmental pollution as well as soil acidity. In an ideal controlled-release concept, the micronutrient can be continuously released in a precise portion, reducing the nutrients' losses and root damage caused by their high concentrations [28].

16.2 Mechanistic View of Controlled-Release Fertilizer from Bio-based Materials

The sustainable release of plant nutrients from modified fertilizers which excludes natural organic compounds and inorganic low solubility compounds [23] is the main focus of this chapter. A physical barrier to control the release of water-soluble fertilizers can be obtained via coating the fertilizer granules (Section 16.2.1) or by dispersing the fertilizer in a matrix (Section 16.2.2). In addition, other controlled release mechanisms are also briefly discussed in Section 16.2.3.

16.2.1 Coating Type

In coating-type fertilizers, the top surface of the fertilizer is directly coated with a physical barrier that controls the release of the inner core when it encounters a release medium. Most coated fertilizers use an (typically hydrophobic) organic polymer coating such as thermoplastics or resins [29]. In the most frequent representation of polymer-coated CRF, this extended release involves the semipermeable characteristics of coating materials, whose physical or chemical properties are changed in response to external stimuli, e.g. temperature, pH, or moisture, which results in good release efficiency due to controlled water penetration to the fertilizer core. Although the actual release process of this core-shell fertilizer is not clearly understood yet, the release mechanism for coated fertilizers was suggested to proceed via a multistage diffusion model. This model involves the penetration of irrigation water through the coating layer into the inner fertilizer core and results in a partial nutrient dissolution [23, 30, 31] as shown in Figure 16.4.

When the osmotic pressure building within the containment surpasses the threshold membrane resistance, the incident that the coating bursts and the nutrients spontaneously release is referred to as the "failure mechanism", which depends on the stability of the coating layer. This is often observed in frail coatings with (modified) sulfur [32]. Most bio-based materials that are hydrophilic are thus likely to encounter this failure mechanism as it is hard to sustain the release over 30 days [33]. Starch, cellulose, or other bio-resource coatings without hydrophobic modification will readily allow water absorption, penetration, and finally break into pieces and the fertilizer is released spontaneously. Moreover, in addition to the water absorption, the coating layer can be degraded by soil microbes [20]. Therefore, directly obtained natural bio-based polymers are not ideal for use in CRF. However, when the membrane is able to withstand the developing pressure (for example, via hydrophobicity modification), the core fertilizer can be released gradually via diffusion through the polymer coating. The key factors in governing the nutrient release are related to the diffusion/swelling, degradation of the polymer coating, and fracture or dissolution. Bio-based polymers such as alkyd [13] and PU resins [34] including lignin [35] and other water-insoluble materials have been developed to substitute petroleum-based materials. Their controlled release depends on the ambient temperature and moisture as they accelerate the nutrient release rate through the coating. It should be mentioned that besides the traditional coating material that aims to exploit its hydrophobicity to retain the release of water-soluble core, some hydrophilic polymers designed to function

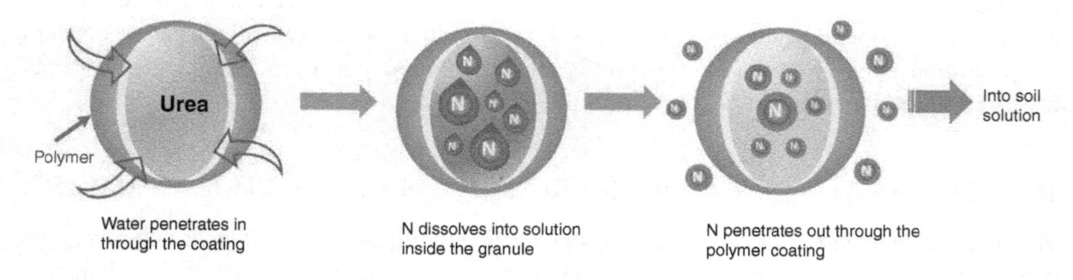

Figure 16.4 Release mechanism of polymer-coated fertilizer granule.

as hydrogel with controllable swelling/deswelling capacity have also been widely studied. This will be discussed later in Section 16.3.1.

The conventional SRF is covered with a protective coating/encapsulating consisting of one or more layers of a water-insoluble, semipermeable material. Typical physical methods for encapsulating fertilizers include:

Spray method: This method is the most common method for coating the fertilizer granules in state of the art. Usually, the polymer is dissolved in a suitable solvent and is sprayed on the granule of fertilizer, which is dried to remove the solvent through evaporation. The treatment is repeated as often as necessary until the desired coating thickness is reached. Typical spray methods for encapsulating fertilizers are spray coating, spray drying, pan coating, and rotary disk atomization. Special equipment include a rotary drum, pan or ribbon or paddle mixer, and fluidized bed.

In situ: The soluble fertilizer is dispersed in a solvent and the solution is mixed with monomers of a polymer. Polymerization will happen and (depending on the method) granules or particles of fertilizers will form.

Simple mixing: The fertilizer granules are simply mixed with the coating at its melting point or with a solution of polymer in a suitable solvent.

16.2.2 Matrix Type

A different mechanism for the controlled release of water-soluble fertilizer comes from matrix-type CRF. This type is less common than coated fertilizers due to its lack of efficiency to precisely control the nutrient content. The release mechanism of matrix type or monolithic dispersion fertilizer involves the diffusion, swelling, and erosion of the fertilizer in a hydrophilic matrix to the soil. The dissolution mechanism of the fertilizer in the matrix materials, whether they are derived from nonbiodegradable polymers or natural materials, depends on the free volume of the structure. In a polymer matrix, diffusion occurs first at the surface of the fertilizer matrix, followed by the partition of the fertilizer between the matrix and the release medium, and then nutrient transport takes place from the surface to the surrounding medium [36]. Figure 16.5 displays the matrix-type fertilizer where the nutrients are embedded into the polymer or other materials.

When the matrix fertilizer is dispersed in the release medium, it is possible that the fertilizer is dissolved immediately before being released to the receiving media such as soil and water. The nutrients given by this burst release are highly concentrated, which may result in plant

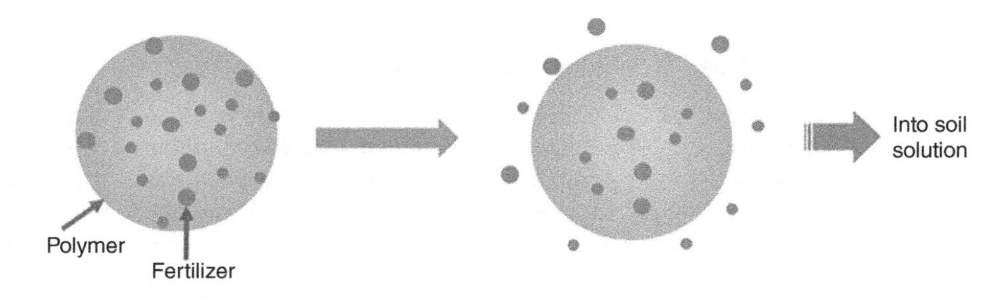

Figure 16.5 Release mechanism of nutrients from matrix-type slow-release fertilizer (SRF).

toxicity. Together with the unpredictable composition, these are seen as the disadvantages that make matrix type fertilizer less favorable in agriculture. However, the preparation process of matrix-type CRF is quite simple. A wide range of fertilizer ingredients can be selected with a less technical effort at a low cost and research investment. Typically, the matrix-based fertilizer is obtained from multiple compositions including inorganic nutrients bound together with minerals, chemicals, bio-based materials, and organic compounds. For example, a biodegradable matrix-based fertilizer has been formulated from starch, chitosan, and lignin with $Al(SO_4)_3H_2O$, water, and/or $Fe_2(SO_4)_3$ at different compactness [4]. In this system, microorganisms were found to take part in degrading the organic materials, with lignin degradation being the slowest. Another preparation process for matrix-based fertilizer in granular form was reported by [37]. In their work, agro-waste like cow dung, rice bran, dried powder of neem leaves, and clay soil were dried and ground into a mixture of 1 : 1 : 1 : 1 ratio. The composite was mixed with biofertilizers (i.e. charcoal mixed with *Azotobacter chroococcum* and *Bacillus subtilis*) in a paste of Acacia gum saresh (25% w/w) in order to make small granules of approximately 5–6 mm diameter. This fertilizer, while being less efficient compared to the conventional fertilizer (urea) in plant growth and productivity, is environmentally friendly as it improves the microorganism colonies that nourish the soil.

Apart from solid matrix-based fertilizers, the nutrients released from fertilizers can be successfully retained in hydrophilic gels. Many biopolymers like alginate or chitosan have been used in the development of hydrogel composites with fertilizer embedded in its network structure, as will be discussed in Section 16.3. This not only gives advantages in delaying the nutrient release but also reduces the frequency of water irrigation due to the decreased evaporation rate [38].

16.2.3 Other Release Mechanisms

Several slightly soluble fertilizers for which the release mechanism depends upon their solubility in water should also be noted here. Urea-formaldehyde is an example where the nitrogen release from the rigid structure is dissolution-limited [39]. Another example can be related to the release of micronutrient boron, which is released over a much longer time due to its slow dissolution from borosilicate glass into the ground [22]. Micronutrients that are essential for plant fertility are only required in small amounts. Although they are generally given as soluble chemical forms, their controlled-release version has also been adopted using an ion-exchange process available via inorganic layered structures such as nanoclay and layered double hydroxides in the release of nitrate [40] and Zn/B [41] or as poorly soluble compounds [16]. High amounts of plant micronutrients Zn and Cu were also loaded on graphene oxide, which is a 2D structure that is available for deposition of various micro- and macronutrients [42].

16.3 Controlled Release Technologies from Bio-based Materials

The reduced nutrient release rate from CRF with incorporated bio-based materials is the focus of this section, whether as coating-based (core-shell) or matrix-based formulations. Bio-based materials are mainly obtained from human activities. The increased population and the resulting accumulation of food supplies throughout the world have been the cause of the global generation of wastes that is estimated to be 3.5 million Mg day^{-1} [43]. This creates pollution and should be reduced as much as possible by recycling [44]. The residual biomass wastes (e.g. postharvest residues, residues from livestock production and food processing) are attractive raw materials for fertilizer production. Recently, biopolymers obtained from post-harvest agricultural products and plant extracts, and carbon-based materials obtained by thermal treatment of waste residues have been explored intensively for CRF applications. The controlled fertilizer release with innovative technologies is required to mitigate global pollution and meanwhile helps effectively sustain the global food supply.

16.3.1 Natural Polymers and Their Fertilizer Applications

Natural polymers have been widely explored as matrix or coating materials for fertilizers in the last decade due to their biodegradability and non-toxicity. Polysaccharides and lignin are the most promising candidates to be used for the CRF system according to their availability, biodegradability, biocompatibility, and non-toxicity to the environment, and they can be simply modified to obtain good coating properties. A traditional polymer-coated fertilizer for controlled-release formation contains a water-soluble nutrient core and a relatively water-insoluble shell as a release-controlling buffer. The release of nutrients depends on the swelling of the hydrophobic shell and the diffusion of the soluble fertilizer nutrient through the shell. In general, the release of nutrients from the polymer coating is composed of three stages including (i) the transportation of water into the capsule, (ii) dissolution of the nutrients, and (iii) gradual release of the nutrients through a semipermeable polymer shell membrane [45]. Not only hydrophobic materials are used as a coating, hydrogels are also among the most common materials for CRF coating due to their ability to reduce water scarcity in growing plants and to improve soil fertility [46]. Nutrients can be released from the hydrogel in a controlled manner, depending on swelling/deswelling transitions of the hydrogel system. Hydrogels can be applied both for matrix and coating-type CRF. The release rate of nutrients from hydrogels generally relates to the concentration gradient of nutrients between inside and outside of the hydrogels. Moreover, "intelligent" or "smart release" fertilizers from natural polymers that exhibit a designable release of nutrients at a specific temperature or pH have received increasing interest [19]. Here, the recent development of natural polymers for CRF will be demonstrated and discussed based on their properties.

16.3.1.1 Polysaccharides

Polysaccharides are long-chain carbohydrate molecules composed of monosaccharide units bound together by glycosidic linkages. Polysaccharides such as starch, alginate, cellulose, chitin, and chitosan are found widely in nature, including in algae (alginate), plants (cellulose and starch) and animals (chitosan). Their ability to form films and coatings with good barrier properties against the release of the nutrients has resulted in their widespread use in slow-release fertilizers. The molecule structures can be linear (such as chitosan), helical, or branched (for example, starch), and can differ in terms of their charge, which can be neutral, positive, or negative. Their chemical structures and properties are shown in Table 16.1. Although polysaccharides are attractive biopolymers, their reasonable biodegradability in an outdoor environment and erosion shorten the duration of the fertilizer effectiveness. The important functional groups in most polysaccharides are the hydroxyl and amino groups, which can be modified to change the character of the polysaccharides, for example, rendering them more hydrophobic or hydrophilic, increasing mechanical strength and modifying biodegradability for further improvement of the controlled-release properties to satisfy the stringent requirements for CRF.

Table 16.1 *Structures and sources of common biopolymers.*

	Polysaccharide					
		Starch				
	Cellulose	Amylose	Amylopectin	Alginate	Chitosan	Lignin
Source	Plant	Plant	Plant	Algae	Animal	Plant
Subunit	β-glucose	α-glucose	α-glucose	α-L-guluronic & β-D-mannuronic	D-glucosamine & N-acetyl-D-glucosamine	Phenyl propanoids
Bonds	1–4	1–4	1–4 & 1–6	1–4	1–4	β-O-4 & α-O-4
Branches	No	No	Yes	No	No	Yes
Water solubility	No	No	No	Yes	No	No
Structure	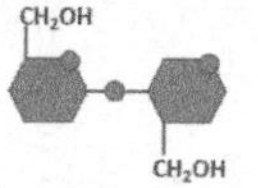 CH_2OH, CH_2OH	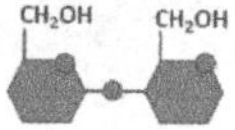 CH_2OH, CH_2OH	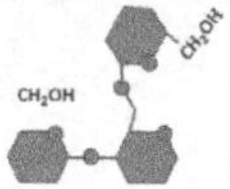 CH_2OH, CH_2OH	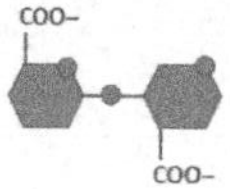 COO–, COO–	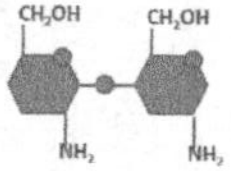 CH_2OH, CH_2OH, NH_2, NH_2	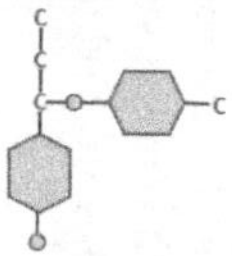 C, C, C, C
Shape	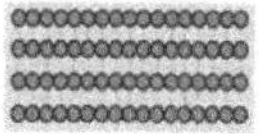		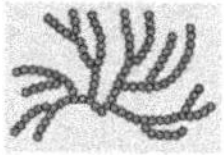			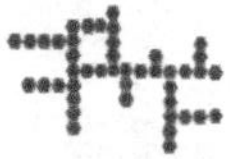

Source: Adapted from Senna and Botaro [47].

16.3.1.1.1 Cellulose

Cellulose is the most abundant natural polymer and a very promising raw material at a low cost for the preparation of various functional polymers to obtain the requirements for CRF. Cellulose can be modified to obtain desired property for CRF through its three hydroxyl groups. As mentioned earlier, hydrophobic materials are commonly used for fertilizer coatings due to their ability to slow down nutrient release. Cellulose acetate (CA), a hydrophobic cellulose ester, is one of the most common cellulose derivatives used as membranes. Nevertheless, due to the instability of the ester bonds under both acidic and alkaline conditions, cellulose ester derivatives are unsuitable to apply in alkaline or acidic soil. On the other hand, a smart-release fertilizer with pH-responsive behavior can be prepared by taking advantage of this feature of cellulose ester [48]. In addition, the degree of substitution of CA can be adjusted to obtain hydrogels with a high swelling property. Recently, hydrogels derived from CA at 2.5 degrees of substitution and grafted with ethylenediaminetetraacetic dianhydride (EDTAD) was introduced for the slow release of NPK fertilizers and the improvement of water retention in soil [47, 49, 50]. Cellulose ethers, such as ethyl cellulose (EC)([51] [52]), and carboxymethyl cellulose (CMC) [53] and their blends with other natural or synthetic polymers have been developed to coat fertilizer granules [54]. These cellulose ether derivatives are produced by replacing the hydrogen atoms of hydroxyl groups in the anhydroglucose units of cellulose with, for example, alkyl groups that act as a hydrophobic coating or with carboxyl groups to increase hydrophilicity. Such modifications enable the slow release of fertilizer nutrients in different manners. To further tailor the CRF properties, a double-coated multifunctional slow-release fertilizer can be beneficial. In this case, a report of EC as a hydrophobic inner coating and cellulose-based superabsorbent polymer (cellulose-SAP) treated with biochemical inhibitor dicyandiamide (DCD) as a hydrophilic outer coating layer was successfully developed for coating of urea granules [55].

Another report focused on the development of environmentally friendly thermo- and pH-responsive cellulose-based hydrogels for controlled-release water-soluble fertilizer [56]. These hydrogel systems were prepared from methylcellulose (MC) and hydroxypropyl methylcellulose (HPMC) blended with potassium sulfate (K_2SO_4). In this system, the K_2SO_4 was used to control the gel temperature of MC and HPMC and to improve the initial swelling rate and mechanical properties of the hydrogels. The SO_4^{2-} can salt out anions and enhance the hydrophobicity of the solution, leading to the formation of hydrogen bonds in the hydrogel network and hydrophobic interactions between the methoxy and hydroxypropyl methyl-substituted segments within the hydrogels. CMC was recently developed to improve its water absorption capacity and pH and salt resistance for the application in saline-alkaline soils by grafted copolymerization with β-cyclodextrin by a combination of inverse emulsion and microfluidic method. The CMC-based microspheres with homogeneous pore structure and high porosity showed good absorbency and retention of water-soluble fertilizer [57]. Silica nanoparticles were also utilized to improve water retention and the slow-release property of CMC-based hydrogels [58]. In this work, a superabsorbent nanocomposite was prepared by in-situ graft polymerization of sulfonated-carboxymethyl cellulose (SCMC) with acrylic acid (AA) in the presence of polyvinylpyrrolidone (PVP), silica nanoparticles, and NPK fertilizer.

Cellulose and its derivatives have not only been developed for the controlled release of granular fertilizers but also for foliar fertilizers. Microcrystalline cellulose (MC), which disperses well in an aqueous medium, was newly designed to prepare a pH-controlled-release ferrous foliar fertilizer with high adhesion capacity on the leaf surface of crops by using attapulgite

(ATP) nanorods as a biocomposite [59]. In this work, MC was first oxidized to form carboxyl cellulose (CC) and later formed a complex with Fe (II) to produce CC–Fe(II) microfibers. After the Fe(II) complex formation, ATP nanorods were coated on the CC–Fe(II) microfibers through hydrogen bonds between –OH groups of ATP and –COOH groups of CC–Fe(II). The release of Fe(II) was efficiently controlled by pH because the chelate and hydrogen bonds are unstable in acidic conditions leading to faster release than in alkaline conditions.

16.3.1.1.2 Starch

Starch is a low-cost, biodegradable, and abundantly available polysaccharide [60]. Starch alone is not feasible to be used as a coating material due to its profound hydrophilic nature, instability at high temperature and pH, and poor mechanical properties like cellulose. Therefore, starch is modified with appropriate additional agents to overcome these discrepancies [61]. For example, a urea-melamine-thermoplastic starch was designed to control the release of nitrogen by using a single-step extrusion process [62]. To decrease the starch degradation rate in the soil to prolong its service time, natural rubber with a lower biodegradability and higher hydrophobicity was grafted onto starch and was applied as fertilizer-coating material [63]. Starch acetate (SA) with a high degree of substitution was developed as a hydrophobic coating for the slow release of urea [64]. In this report, the slow-release urea fertilizer was successfully prepared by a double-layer coating of SA and carboxymethyl starch/xanthan gum. Typically, starch is widely used for hydrogel preparation either in its modified form or as composites/blends. Sulfonated corn starch was proved for its ability to increase the absorbency of water in soil with high salinity [65]. It then was further introduced to form a composite with polyacrylic acid (PAA) for embedding phosphate rock by in-situ solution polymerization to improve the utilization of phosphate rock fertilizer, to enhance water retention and to provide sustained-release P and K-fertilizers [66].

PAA, polyacrylamide (PAAm) [67] and polyvinylalcohol (PVA) [68] are common synthetic hydrogel polymers for agricultural applications. These polymers are frequently blended or chemically cross-linked with (modified) starch to increase the water absorbency of the hydrogel and reduce the toxicity of the hydrogel degradation products [69]. The effect of botanical origins of starch on starch-PAAm superabsorbent (SAP) structures and properties was investigated for slow release of nitrogen by coating the urea granules with EC as the first layer and the obtained starch-PAAm SAP as the second layer as shown in Figure 16.6 [67]. The botanical origins of starch clearly influenced the release behavior of the nitrogen. However, it needed further investigation in terms of fertilizer preparation such as the thickness of the starch-PAAm layer. A controlled release phosphorus fertilizer by phosphorylation of carboxymethyl starch followed by grafting with acrylamide exhibited a high potential for enhanced phosphorus use efficiency [70]. A biodegradable hydrogel based on cassava starch (CSt) prepared via aqueous solution polymerization (CSt-*g*-PAA/natural rubber/polyvinyl alcohol blends) with potential applications in agricultural areas was recently reported by showing the excellent water retention capacity, good reusability, and high biodegradation extent of the gel [71]. Hydrogels can be incorporated with other absorbents or additives to improve their mechanical properties and enhance nutrient capacity. For example, starch-*g*-poly(acrylic acid-*co*-acrylamide) was reinforced by natural char nanoparticles [72]. A one-step process of reactive melt mixing by a modified twin-rotor mixer was developed to prepare starch-based SAPs (SBSAPs) for the slow release of urea fertilizer. The effects of the initiator, cross-linking, and saponification agents under different reaction conditions (time, temperature, and shear intensity)

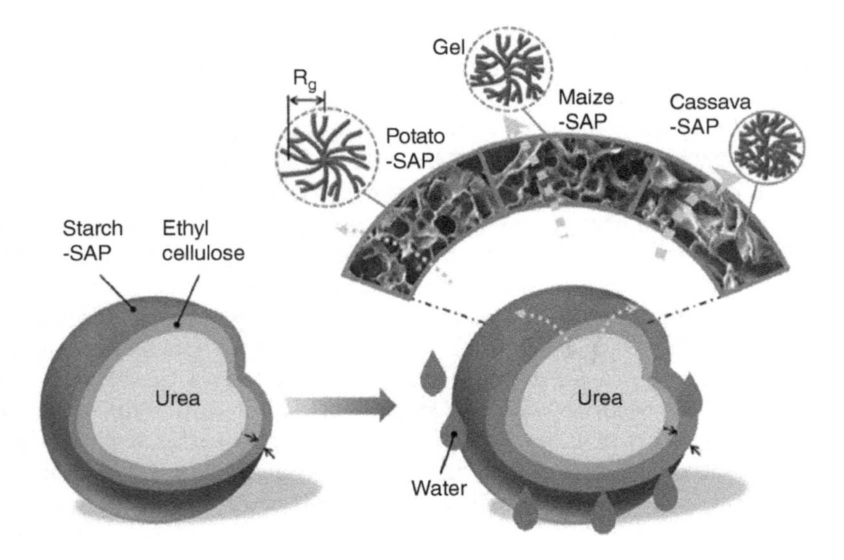

Figure 16.6 Schematic diagram of double-layer slow-release nitrogen fertilizer with ethyl cellulose as the first layer and the starch-SAP as the second layer. Source: Qiao et al. [67]. Reprinted from Qiao et al., Copyright© (2016), with permission from Elsevier.

on the gel strength and microstructure were systematically studied. The embedded urea in the SBSAP gel network showed a typical urea slow-release profile in water and the release rate of urea depended on the gel strength, gel microstructure, and water absorption capacity of the SBSAP gel [73].

16.3.1.1.3 Alginate

Alginate is a natural anionic polymer typically obtained from brown seaweeds. It has been extensively applied in pharmaceutical and biomedical fields due to its biocompatibility, biodegradability, nontoxicity, and hydrophilicity and unique property of gelation with mild conditions by using non-toxic reactants. For example, alginate can form gels by the addition of divalent cationic such as Ca^{2+} to substitute Na^{+} results in cross-linking of the polymer chains, or by decreasing the pH value below the alginate pK_a. Various approaches were used to obtain alginate hydrogels including ionic cross-linking, covalent cross-linking, phase transition with thermal gelation, and free radical polymerization. Hydrogel is a main focus in the technology development for the utilization of alginate because of its hydrophilicity and water solubility. Alginate was usually grafted copolymerization with synthetic monomers like AA or AAm to enhance the degradability of commonly used superabsorbents in various applications [74, 75]. It was later demonstrated for its possible use in agriculture as a soil conditioner and long-term water supply for the plant as well as oligo-alginate growth promoter [76]. Nanocomposites from alginate-based hydrogels with clay minerals were suggested to reduce the final cost of alginate-based hydrogel products for agricultural uses and to improve its properties such as swelling ability, and mechanical and thermal stability [77]. Bentonite [78], montmorillonite [79], and clinoptilolite clays [80] were successfully incorporated in NaAlg-*g*-poly(AA-*co*-AAm) hydrogel via in-situ polymerization in the presence of NPK fertilizer. The presence of the clays in the hydrogel not only exhibited excellent slow-release property but also a good water retention capacity that caused the system to release the nutrients in a more controlled manner compared to the neat hydrogel. On the basis of this study, NaAlg-*g*-poly(AA-*co*-AAm)/clinoptilolite nanocomposite was also applied to coat NPK fertilizer granules by sprinkling with distilled water in a pan coater [81]. A different approach for utilization of alginate to improve fertilizer use efficiency was by design of a double coating on the core urea granule [82]. A composite hydrogel with κ-carrageenan (κC)-*g*-poly(AA)/Celite as the first coating layer instead of blending or polymerizing of monomer with fertilizers could reduce the loss of fertilizer without altering the water absorbency and retention. The hydrophobic second layer was κC–SA (sodium alginate) bead, which was obtained by ionic cross-linking between hydroxyl groups of κC and SA with K^+ and Ca^{2+} ions. Moreover, to improve nutrient retention, calcium alginate (CaAlg) beads were impregnated with ball-milled biochar [83]. The CaAlg-based composite beads showed high kinetic swelling ratios in fertilizer solution and low kinetic swelling ratios in water, indicating that the composite beads of CaAlg-biochar have the potential to retain nutrient-holding capacities, resulting in good controlled release properties. Novel alginate-based cryogel beads loaded with urea and phosphates were recently fabricated via the microinjection and cryo-cross-linking method and coated with ethyl cellulose films to obtain a fully biodegradable SRF. The cryogel beads had high loading capacities of phosphate and urea and showed slow-release behavior for phosphate. However, it could not reduce the release rate of urea [84].

16.3.1.1.4 Chitosan

Chitosan has been extensively explored as a coating material for the manufacture of controlled-release applications in the food, drug, biochemical, and agricultural areas. Chitosan is a natural biopolymer extracted from the exoskeletons of insects and crustaceans. It features good mechanical properties combined with the ability to form fibers, films, gels, and microspheres. Chitosan has a rigid and brittle nature because of inter- and intramolecular hydrogen-bonding interactions that greatly affect its processability. What makes chitosan attractive is its abundant hydroxyl groups and highly reactive amino groups, which can be modified by various chemical reactions, including N-alkylation, N-acylation, and N-carboxyalkylation [85]. The solubility of chitosan in common organic solvents is enhanced by these chemical modifications, which also improve the properties of chitosan for membrane applications. Nevertheless, non-functionalized chitosan as an inner coating and P(AA-*co*-AAm) hydrogel as an outer coating material for controlled release NPK compound fertilizers demonstrated slow nutrient release with excellent water retention capability [86]. The same research team applied PAA and diatomite composite as an outer coating material for urea to improve the controlled release property [87]. Meanwhile, chitosan-based composites have been developed to improve mechanical strength and reduce water permeability, and showed great potential benefits for CRFs [88]. Hydrogels derived from chitosan have also been reported. Chitosan has a low swelling ability when it forms hydrogel due to the slow relaxation rate of polymer chains [89]. Therefore, blending chitosan with other hydrophilic polymers and forming composites with certain absorbents helps improve its water absorbency at gel state and its nutrients' adsorption [90]. Potassium-containing microspheres based on chitosan and montmorillonite clay were prepared by a coagulation method with different proportions of montmorillonite and their in situ release behavior in soil was reported [91]. These results demonstrate that clay provides a porous surface that increases the KNO_3 adsorption. The hybridization process of chitosan with cellulose was accomplished using thiourea formaldehyde resin which cross-links via polycondensation to increase the flexibility of the hybrid material [92]. The hybrid product was then graft-polymerized with acrylic acid in the presence of an initiator to enhance water absorbency. This new approach gave more mechanically robust structures and improved resistance to acidic conditions. A wide range of pH response was acquired due to the presence of chitosan and PAA in one homogeneous entity, which are responsible for the greater water absorbency and enhanced water retention. The in situ hydrogelation of chitosan with salicylaldehyde in the presence of urea fertilizer was designed to improve the water-holding capacity and nitrogen percentage in the soil for addressing both fertilizing and water retention issues as multifunctional soil conditioner [93]. Recently, a novel smart fertilizer with polyelectrolyte layers of cationic chitosan and anionic lignosulfonate deposited on polydopamine-coated ammonium zinc phosphate was successfully developed by layer-by-layer (LBL) electrostatic self-assembly [88]. This report showed excellent pH-responsive release behavior of the system; the nutrient release rate in acidic media was higher than that in the neutral media. In addition, compared with other reported methods for developing environmentally friendly fertilizers, the LBL self-assembly method shows advantages in terms of recyclability, low cost, and renewability.

16.3.1.2 Lignin

Lignin is a good candidate for coating and composite materials due to its abundance in nature and properties such as water impermeability, hydrophobicity, and film-forming ability [94]. Lignin is a complex three-dimensional aromatic copolymer of sinapyl alcohol, coniferyl alcohol, and paracoumaryl alcohol units that are linked by carbon–carbon and carbon–oxygen bonds. It is also well known for its variety of functional groups such as phenolic and aliphatic hydroxyls, carboxylic, methoxyl, and carbonyl groups that offer many possibilities for chemical modification to alter the physicochemical properties of lignin-based materials. The primary approach to use lignin for slow-release nutrients was by facile co-melting of lignin with urea composites followed by pelletization for use as a feed supplement which was patented in 1998 [95]. More recently, a chemical method to trap urea with lignin was prepared by condensation of hydroxymethylated lignin with urea/hydroxymethyl urea/urea-formaldehyde polymers via solution polymerization [96]. Pure lignin granules as fertilizer storage and slow-release systems were developed by polymerization of lignosulfonate with laccase as a catalyst [97]. The obtained water-insoluble lignosulfonate granules were highly brittle and consisted of randomly broken particles with nonhomogeneous size and morphology when dried. To overcome this, incorporation of alginate into the polymerized lignosulfonate could produce perfectly stable granules with regular size. Water-soluble nutrients could be trapped in the lignosulfonate/alginate granules by simple soaking in the fertilizer solutions.

For coating-type fertilizer, lignin was previously used directly without any modification, but plasticizers were needed for improving the elasticity of lignin resulting in better film formation for the coating process. Four commercially available lignins, i.e. two lignosulfonates (Wafex P and Borresperce), a softwood kraft (Indulin AT) and a soda flax lignin (Bioplast), were studied for their potential in urea coating with respect to film-forming properties [98]. Films of Bioplast lignin with plasticizers showed a low water sensitivity and the incorporation of hydrophobic agents or cross-linkers could even further reduce the water sensitivity of Bioplast lignin, retarding the dissolution of urea. To address the availability and heterogeneity of lignin, the extraction of lignin from plants and biomass is one of the key interests for researchers since lignins from different sources have widely varying compositions and structures, which can result in dramatically different properties. Lignin was extracted from olive pomace (OP) biomass and lignin-k-carrageenan composites were formulated to coat triple superphosphate (TSP) fertilizer by pan coating [99]. The influence of polyethylene glycol (PEG) as a plasticizer on the lignin-k-carrageenan film properties was also investigated by testing the mechanical properties of the PEG-containing films. Apart from using plasticizers and hydrophobic agents, chemical modification of lignin can be used to improve the hydrophobicity of the coating and the coating process. One study reported the success of urea coating with modified lignin [100]. The extracted lignins from Kraft and Sulfite black liquors were modified through acetylation reaction, which resulted in an increase in the hydrophobicity of the coating layer and enhanced the adhesion between non-acetylated groups of lignin and urea through hydrogen bonding. The nitrogen release rate was significantly reduced for urea coated with modified lignin. Another report applied PLA blended with modified Kraft lignins for coating of urea. The lignin was modified with acetic anhydride or palmitic anhydride via esterification to increase the hydrophobicity of lignin or it was chemically bonded with nitrogen via Mannich reaction to delay the release of nitrogen [101]. The developed coating materials in this research not only constituted a physical barrier to delay urea dissolution but also offered a chemically controlled slow-release system. However, toxic solvents are still required to dissolve the modified lignin for the coating process and also the price of chemical modification results in lignin to become a costly material. A new design of a slow-release urea formulation from indeed low-cost materials was recently initiated by using wheat straw as the fertilizer carrier and lignin as the binder component [102]. Urea and lignins were impregnated in the wheat straw by the melting and casting process. This study demonstrated the influence of two lignin fractions, which are isolated from pulping black liquor, using acid (acid-lignin) and calcium precipitation (Ca-lignin) on the release of urea from the straw. Calcium-lignin complex from calcium acetate precipitation at high pH of pulping black liquor was a simple and affordable fertilizer system due to its low water solubility and benefited from slow dissolution of urea compared to acid-lignin. This research could lead to further investigation of the polyelectrolyte property of lignin for the new approach of a controlled release system.

16.3.2 Bio-based Modified Polymer Coatings for Controlled-Release Fertilizer

The polymer-coated granular CRFs are usually treated with organic polymers such as polyolefins [103, 104], polyethylene [105], PU [106, 107], polystyrene [108], acrylic acid-*co*-acrylamide [109], super-absorbent hydrogel [110], polysulfone [111], lignin [112], CMC [54, 58], starch [113], and chitosan [114]. In recent years, there has been an increasing interest to use bio-based resin-building blocks to synthesize polymer coatings. Bio-based polymers that are obtained directly from plants and other renewable sources have already been described in Section 16.3.1. However, some bio-based polymers, even though they may not be directly obtained as materials, can be synthesized from natural extracts such as vegetable oils (VOs) or resins. The modification of these synthesis approaches is substantial and provides similar or better products when compared with conventional petroleum-based polymers.

The most commonly used renewable raw materials for polymer synthesis are VOs, but these also include modified polysaccharides, lignocellulose, and proteins [115]. The VOs are composed of triglycerides from fatty acids of aliphatic carboxylic acids' chain with 8–18 carbon atoms and 0–3 carbon–carbon double bonds depending on the plant species and climate [116]. The VOs also possess hydroxyl or epoxy groups that can be modified into useful monomers for the synthesis of bio-based polymers. Modification of VOs by physical, chemical, and biological methods has been studied to improve the usability of VO in the different fields [117–121]. Many types of VOs can be used as monomers to substitute petroleum derivatives. Much attention on the use of VOs is due to their advantages such as supply sources, low cost, renewability, and biodegradability. Examples of polymers obtained from VO monomers including alkyds [122], vinyl resins [123], PUs [124], and epoxy resins [125] have been developed, which have been prepared via olefin metathesis, free radical, cationic, and condensation polymerization [126]. For example, an elastomeric polymer was synthesized via cationic polymerization of linseed oil. In contrast, soybean oil can be used to form monoglyceride maleates which were used to undergo radical copolymerization, resulting in rigid thermoset polymers [127]. Rigid PU foams were also fabricated using rapeseed oil-based polyol and polyols of petrochemical origin which result in the influence of rapeseed oil-based polyol content on the properties of the obtained foams to assist in a cosmetic product [128]. A solvent-free method was proposed to prepare VO-based polyols from different VO sources, e.g. castor, canola, olive, linseed, and grape seed oils, by successive oxidation and ring-opening reactions [129].

16.3.2.1 Bio-based Alkyd Resin

Bio-based resins for coatings are generally referred to as alkyds, which consist of fatty acid-modified polyester resins. Different VOs can be combined as the mixed components of alkyd resin where the weight fraction of each VO affects the alkyd properties [130, 131]. Various VOs have been used to synthesize alkyd resins, such as Tung (*Aleurites fordii*) or Chinese wood oil [122], sacha inchi oil [132], palm fatty acid distillate (PFAD) [133], soya bean fatty acid [134], yellow oleander (*Thevitia peruviana*) seed oil [135], *Jatropha curcas* [136], and castor oil [137]. Alkyd resins are easily produced by condensation of inexpensive materials such as polyhydric alcohols, polybasic acids, and monobasic fatty acids or VOs. The plausible reaction for the preparation of basic alkyd resin is shown in Scheme 16.1 [138].

Nutrient release from Osmocote (an alkyd-resin-coated fertilizer) occurs following the water penetration through the microscopic pores in the coating layer. The osmotic pressure within the pore is developed, which is finally enlarged and allows nutrients to be released. The coating polymer formed as alkyds in this model successfully controls the release rate and timing period.

A drawback of alkyd resins is the relatively high solvent content and economic viability. Changes in the environmental regulation over volatile organic contents (VOCs) permission have also mandated the future of solvent-based coatings. Combined with customer acceptance, in some regions, it is necessary to switch from solvent-based to waterborne technology [139]. Much of the current research is driven toward the development of new approaches to decrease solvent usage and production cost [140]. As a result, most researchers have focused on the development of waterborne alkyd resins. Until now, two primary approaches to obtaining waterborne alkyd resin were disclosed. The first preparation method involves many steps including (i) alkyd formulation to a major chain from several hydrophilic monomers; (ii) neutralization with alkali reagent; and (iii) dispersible alkyd chains in water to obtain water-reducible alkyd resin. In this method, the hydrophilic monomers are allowed to react with other monomers by their −COOH or −OH groups to establish the alkyd resin.

The second method involves hybrid acrylic-alkyd emulsion and mini-emulsion polymerization techniques, which can be carried out while the alkyd resin is dissolved in a solvent or dispersed in water prior to the synthesis step. Acrylate alkyd resin is the modified version of alkyd resin and its derivative that are produced from an emulsion system [141, 142]. This acrylate

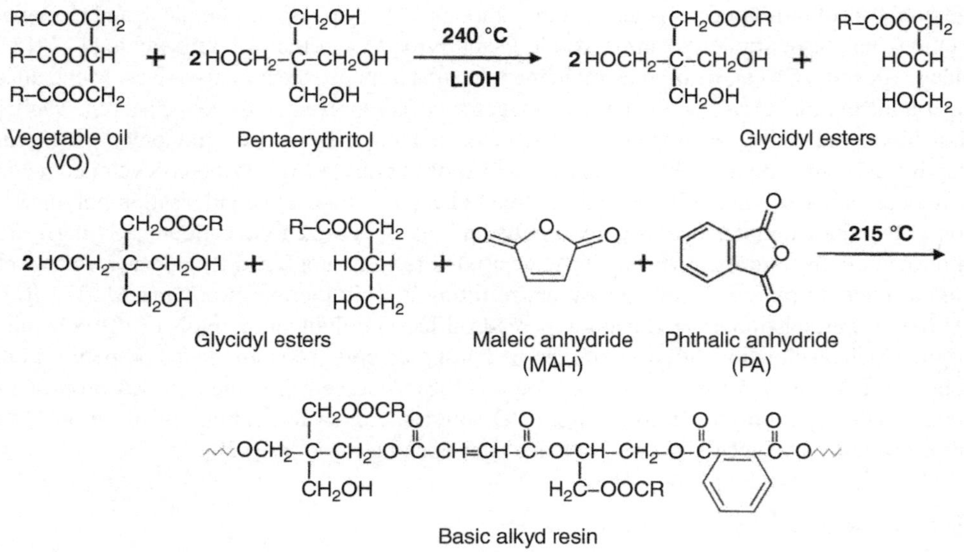

Scheme 16.1 *Plausible reaction scheme for the synthesis of the basic alkyd resin from VO. Source: Liang et al. [138].*

type is also known for its fast curing process [143]. The hybrid formation in acrylate and alkyd hybrid results in splendid compatibility between acrylate and alkyd resins and has excellent mechanical properties including adhesion, pencil hardness, and water resistance tests [144].

The developed waterborne alkyd resins were shown to have good water dispersion and high hydrolytic stability [145, 146]. However, the water resistance in the films remained insufficient and needed further improvement by reducing the excessive carboxyl groups in the side chain of waterborne resins, in addition to the remaining small amounts of unreacted low-molecular-weight polyacrylate, leading to prolonged drying time.

Waterborne alkyd technology has been recently developed [147] and has received much for application in control release urea (CRU). The hydrophobic alkyds, on the other hand, must be reinvestigated and designed to be compatible with polar aqueous environments. The CRU must be coated with natural or seminatural materials that are renewable and environmentally friendly. This retardation of the nutrient release shows benefit since the fertilizer can be supplied with a single application while still meeting the specific nutrient requirements in each crop.

16.3.2.2 Bio-based Polyurethane

In recent years, VO-based polyols have been polymerized into PUs ranging from hard plastic to soft elastomer depending on the desired applications. As a fertilizer-coating material, VO-based polyols containing soft segments of aliphatic chains and hard segments of isocyanates support the elasticity and mechanical strength of the resulting PU. The degradable VO-based PUs are decomposed by urethane bond breakage, leading to the formation of amines and VO-based polyols while releasing CO_2. The VO-based polyols themselves are hydrolyzed into carboxylic acid and alcohol [148, 149]. Thanks to their environmental compatibility, bio-based PUs have been used in polymer-coated CRFs [25, 27, 150–155]). For example, soybean and waste-frying oil-based polyols were prepared through epoxidation and ring opening of propylene glycol. The resulting polyols were subsequently reacted with methylene diphenyl diisocyanate (MDI) to synthesize bio-based PU coatings onto urea granules to form a CRF system. Here, 5 wt% of coating materials achieved a satisfactory N release ranging from 49 to 58 days [153]. A reinforcement of the PU structure can be achieved by adding additional materials such as clays or layered materials. Zhao and coworkers prepared soybean oil-based PUs containing bentonite as a coating layer filler. The effect of fine powder distribution may also reduce the brittleness of the coating since the cracking in the coated fertilizer during transportation and application has been diminished (Scheme 16.2) [157].

The processing of PU-coated CRFs also opens up many possible adjustments. For example, the process can be solvent-free and does not need special equipment. The modification paths are involved with the increased functionalization in order to improve the properties for application as coating materials. A number of systems related to bio-based PU-coated ureas (PCUs) prepared with liquefied locust sawdust or waste oil as polyols have been reported [27, 152, 153]. A biomimetic bio-based PU-CRF was prepared with superhydrophobic properties providing "outerwear" using liquefied biomass (wheat straw, corn stover, and feathers) to reduce the nitrogen release rate [27, 155]. A bio-based thermoset PU coating obtained from polyesteramides and *Gossypium arboreum* (cottonseed) plant oil was also successfully prepared [154, 158]. The usage of castor oil, a typical plant oil that inherently possesses hydroxyl groups, to prepare PU products without prior chemical modification of the oil, was reported [106, 159]. Bio-renewable PUs were reacted with MDI and castor oil at a mass ratio of 40/60 to form a coating layer on the urea surface. The prolonged nitrogen release (up to

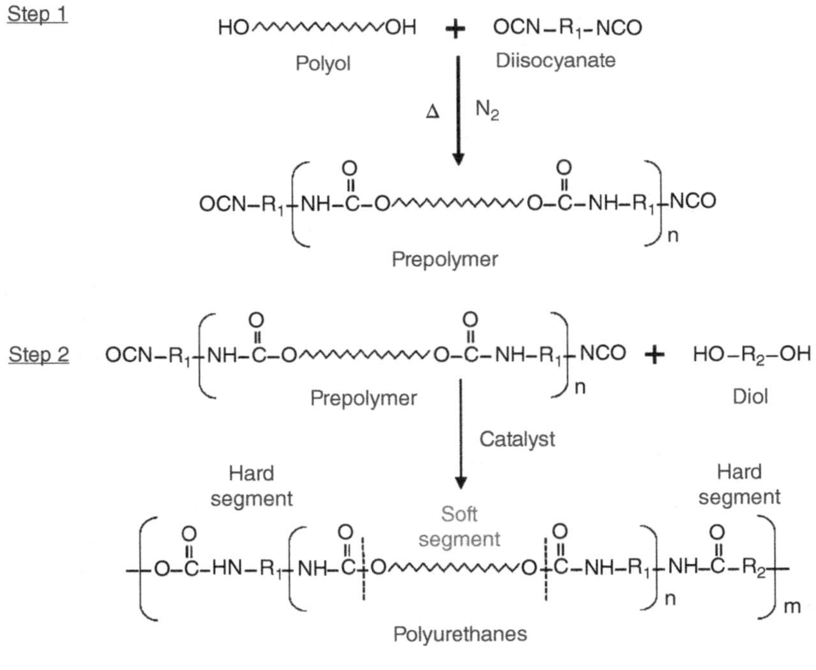

Scheme 16.2 *Schematic diagram of the pure PU synthesis mechanism using the prepolymer technique. Source: Alaa et al. [156].*

33 days) from this polymer-coated fertilizer was successfully obtained, and the performance was superior to other oil-based materials [160]. Bio-based thermoset PU coatings consisting of polyesteramides were obtained from renewable cottonseed plant oil [154]. Bio-based PU coating from VOs can be prepared with additional biodegradable and mechanically resistant properties. PU-like coatings also exhibit well-controlled release properties. Coated granules with 7–9.5% by weight (wt%) of PU show urea release below 50% in water after 60 days [160].

Isocyanate, a key starting material for traditional PU synthesis that is highly reactive and toxic, may react vigorously with water, which requires many safety precautions during storage, transportation, and PU production. Isocyanate also needs to utilize phosgene, which is even more reactive and toxic. As an alternative, non-isocyanate PUs (NIPUs) have been introduced nowadays. One example is the polyaddition reactions of five-membered cyclic carbonates with amines [161, 162]. Scheme 16.3 shows the process comparison between PU and NIPU.

Sustainable resources to produce bio-based NIPUs involving at least three steps were also introduced. Wilkes' group synthesized a cross-linked NIPU with cyclic carbonate-functionalized soybean oil and amines and examined the thermomechanical and tensile properties [162]. Cross-linked NIPU thermosets developed from soybean oil with the highest stress at a break near 6–7 MPa were also reported [164, 165]. Where hybrid systems such as carbonated soybean oil-containing epoxy resins were used, the comparable toughness with unmodified resins after curing with amines was successfully obtained [166]. The cross-linked NIPUs have displayed significant potential in comparison with the mechanical performance of typical PU thermosets.

Despite numerous advantages of bio-based polymer coatings, the use of such polymers as a single-coating material alone could be costly and the polymer mechanical properties could be diminished. Incorporation of organic and mineral fillers in the surface-coating film or matrix bulk can improve the mechanical strength and reduce adhesion between coated granules [167]. Typically, such fillers are less than 20 μm in size and consist of inert inorganic materials like clay, talc, bentonite, kaolin, grinded limestone, barium sulfate, and zeolite. These fillers can be used either as a composite mixture with polymer [168] or formed separately on the polymer surface. The latter is more common and relates to the mixing of fillers with the polymer-coated granules during wetting. Finally, drying of the granules results in filler deposition on the polymer surface that protects the granules from mechanical impact. Zhang and colleagues developed a CRF system utilizing low-cost bio-materials in PUs to form fertilizer with the coating layer, sustaining the release of plant nutrients through micro-holes [25]. In their work, the addition of organosilicon and nano-silica increased the roughness and superhydrophobicity of the bio-based PU composite. Some other fillers such as cellulosic waste have also been used as filler with polymer [169, 170].

16.3.3 Biochar and Other Carbon-based Fertilizers

Carbonaceous bio-based CRFs such as biochars are obtained by converting agricultural residues (e.g. wheat straw, corn, bagasse, chitosan, lignin, and other wastes) using high-temperature processes such as hydrothermal carbonization in aqueous medium and pyrolysis in dry conditions. The difference in raw materials allowed for these two processes is that the hydrothermal treatment accepts wet biomass [171] while pyrolysis requires only dry biomass. Whereas hydrothermal carbonation gives a low amount of hydrochar but a larger amount of carbon with small amounts of oxygen and hydrogen, the pyrolysis products can be solid, liquid, and gaseous phases in different ratios. When the temperature is not that high (~300–700 °C), a higher

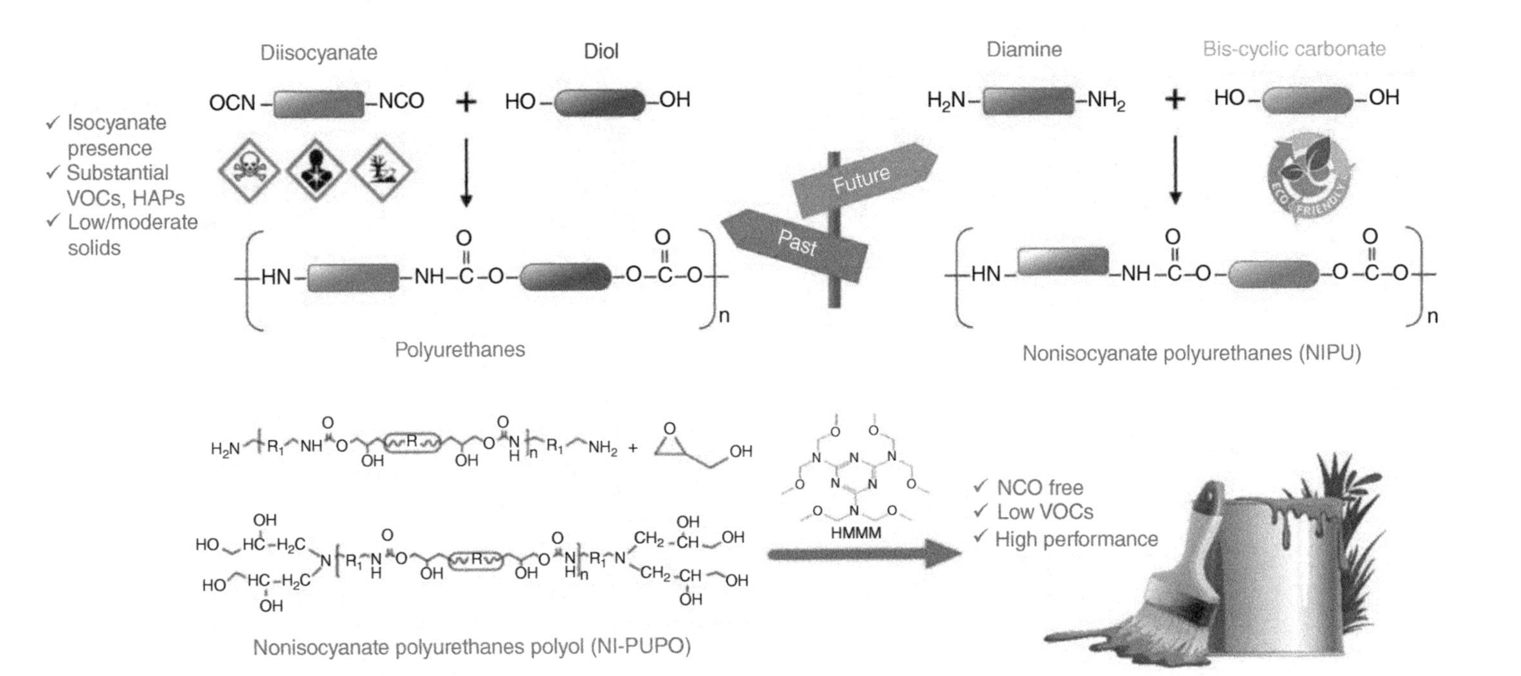

Scheme 16.3 *Schematic comparison of conventional PU and the NIPU. Source: Adapted from Asemani et al. (2019).*

solid or biochar yield is obtained. Biochar is a carbon-rich material obtained predominantly from agricultural residues by pyrolysis [172]. It can be applied in animal farming, manure treatment, and soil amendment [173] due to its ability to sequester carbon while enhancing nutrient contents, water-holding capacity, pH, or microbial activity [174].

The structure of biochar can vary depending on the cellulose content of the raw organic materials. The subsequent pyrolysis will eventually remove water, volatile organic substances, and gaseous residues (CO_2, H_2) as the temperature increases. By this time, alkyl carbons are converted into aromatic carbon comprising the crystalline stacks of graphene sheets and amorphous aromatic structures. Figure 16.7 demonstrates the development of biochar structure upon temperature treatments [176]. The spacing between the planes of biochar turbostratic regions is similar to graphite [177]. Despite the long-range order in two-dimensional graphitic-like layers, the material still has not reached the crystallographic order in the third direction [178]. The biochar-specific surface area is generally higher than that of sandy soil and can also be higher than that of clays ranging from $5\,m^2\,g^{-1}$ for kaolinite to about $750\,m^2\,g^{-1}$ for Na-exchanged montmorillonite. Thus, biochar helps improve soil capacity in water and nutrient content holding as well as the soil aeration through improved pore structure [179].

Previous studies suggest that biochars can be modified before their incorporation into the soil [180]. Organic fertilizer materials such as animal manures and green plants can be mixed with biochar by pelletization or encapsulation to gain better nutrient contents [181–183]. A wide variety of raw materials has not only been used in the making of biochars, such as rice straws, maize straw, oat hull, wood, vegetable, and fruit leaves [180, 184, 185] but also other materials including biosolids from livestock litters [186]. Depending on the feedstock, the

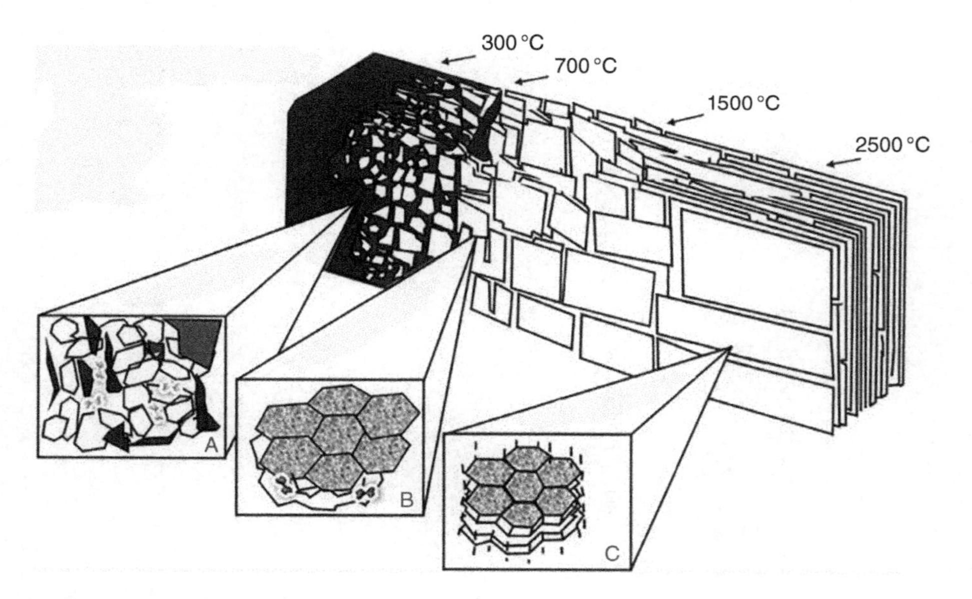

Figure 16.7 *Schematic illustration of organic carbon converted to biochar: structure development with the highest treatment temperature. (A) The pyrolytic process (300–700 °C) increases the degree of aromatic organic carbon, rich in amorphous mass and exposed phytolith particles to the nearby area. (B) Further heating above 500–700 °C results in turbostratic sheet of aromatic carbon and enhances silica crystallization. (C) The silica losts crystallinity and the structure becomes graphitic with order in the third dimension.*
Source: From Li and Delvaux [175], http://creativecommons.org/licenses/by/4.0/.

obtained micro- or mesoporous structures as well as the surface area may be widely different. Normally, a high surface area is more beneficial for the adsorption of nutrients, thus preventing their early release when applied to the soil. Macropores are suggested to be relevant to vital soil functions such as aeration and hydrology [187] and also to the root movement through soil and microbe habitats [176]. Therefore, a higher surface area with a broader volume in biochar has greater benefit than those with low surface area.

The activity of biochar in a controlled nutrient release has been shown when used as an adsorber toward NH_3 and water-soluble ammonium-N (NH_4^+–N) [188]. In the work of Sun et al., biochar from canola straw was employed in the rice paddy field to prevent NH_3 volatilization [189]. The reduction of NH_3 loss up to 39% with N_2O emission improvement up to 19.5% in this work is also consistent with higher rice yield ranging from 8.8 to 28% [82, 190, 191] when nitrogen was utilized efficiently in rice paddy soil. Biochar can also reduce the CH_4 gas emission in rice fields when used in association with a slow-release fertilizer to increase N-use efficiency without compromising crop productivity [192]. Its adsorption capability not only helps capture nitrogen in ammonia but also nitrate form. Kammann and colleagues [193] reported that co-composted biochar particles take up as much as $5.3\,g\,N\,kg^{-1}$ as NO_3^-, which was close to the $3\,g\,NO_3^-\,N\,kg^{-1}$ content uptake by wood biochar from Hagemann et al. [173]. However, this anion instead of cation immobilization by the negatively charged biochar surface has not been fully explained by a conventional cation exchange activity as is typical in most negatively charged surface species.

The nitrogen retention by biochar is more complex than the properties of the material itself. Li and Chen [194] summarized the functions of biochar into four concepts:

1. Biochar mediates soil porosity, aggregation, pH, CEC, etc.
2. Biochar provides more habitats to serve microbial activities.
3. Biochar adsorbs more nutrients (N and P), air (CO_2), and water.
4. A complex organic coating on biochar adds hydrophilicity, redox-active moieties, and additional mesoporosity.

From the usage of biochar as a co-compost with several types of fertilizers introduced earlier in this section, a more complex form, most intensively by organic coating on biochar has been receiving more attention. This is due to its effectiveness in application in the plant field. Unlike traditional CRF, numerous modern controlled release systems use biochar in polymer coatings that present more environmental compatibility with waterborne technology. One of the earliest polyacrylate types obtained from the emulsion process to the CRF form was developed by Zhou et al. [195]. The authors demonstrated that the incorporation of biochar into the waterborne polyacrylate coating layer strengthened the membrane film. Moreover, although the presence of biochar suppressed the microbial activity at the early phase of the study, it gradually recovered after 12 months. This benefits the quality of the CRF product and provides alternative options for the utilization of the crop product. Chen et al. also form the coating material of CRF from the waterborne copolymer of PVA and PVP with biochar received from rice straw, maize straw, and forest litter into CRF beads [184]. The urea core was coated via the rotating technique using a sugar-coating machine and nozzle spray of the biochar-copolymer mixture under 0.7 MPa pressure. Incorporation of biochar in this urea fertilizer system improves the water resistance of the composite film, and the degradability of the coated fertilizer was also assessed by monitoring its weight loss during 120 days soil burial which was reduced to only 34–45% compared to the non-biochar film. Not only N nutrients but also a P fertilizer with effective slow-release features was achieved via the co-pyrolysis of phosphorous from TSP and bone meal with sawdust and switchgrass biomass in biochar preparation [196].

The combination of biochar with SAPs or hydrogel made of acrylic acid and acrylamide monomers in covering NPK fertilizer was introduced by Noordin, Figure 16.8 [197]. It was suggested that the incorporation of biochar into the SAP chain allows the carbon particles to enhance the SAP structure by increasing porosity and thus surface area and capillary effect. This system significantly benefits the water retention in soil and the nutrient (K) release was controllable to be as low as 27% after 30 days.

A study, in which polymers are not involved, but bio-based and natural materials are instead blended with biochar to make a coating layer, was carried out by Dong et al.[198]. The starch adhesive from cassava or corn was used to combine humic acid, bentonite, and biochar in a coating shell to help control the release of NPK nutrients from compound

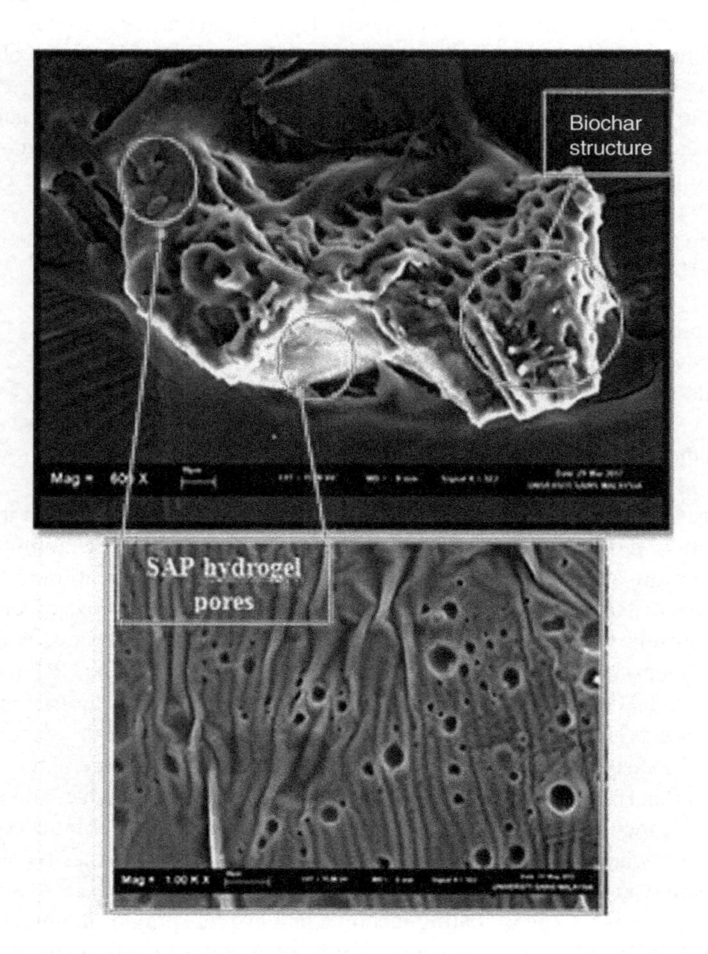

Figure 16.8 FESEM image of the swollen surface of SAP-biochar incorporation hydrogel. Source: Noordin et al. [197]. Reprinted from Noordin et al., Copyright© (2018), with permission from Elsevier.

fertilizer (15% N, 15% P_2O_5, and 15% K_2O). The key factor that affects the N releasing rate is bentonite and follows in order as bentonite > humic acid > starch adhesive > biochar. In spite of this fact, the biochar presence still plays an important role in porous structure formation and N adsorption ability, especially in constraining NH_4^+ beneath the coating layer.

16.4 Conclusion and Foresight

Fertilizer has been a major tool for the agricultural system for centuries. Great developments by scientists and engineers in the early twentieth century made possible the production of the first chemical fertilizers that have been applied to support the food supply needs of the increasing world population. The water-soluble fertilizers are readily dissolved, and the nutrient loss then causes water and soil pollution that is worsened by overuse. Therefore, to prevent its fast release, various petroleum-based polymers have been employed to create controlled release systems for fertilizers that can be categorized in coating or matrix form, designed to exploit the inherent hydrophobic and hydrophilic properties of selected polymers. Recent alerts due to global warming have forced fertilizer development away from dependency on petroleum-based polymers. Abundant bio-based resources can be used as renewable polymer substitutes. These bio-derived materials are categorized here into three types based on how they are received and their technical preparation and application processes. A major group of natural polymers which are obtained directly from plants and biomasses such as starch, cellulose, lignin, alginate, and chitosan, etc. are found the good candidates in such controlled-release applications. These materials, although they may lack hydrophobicity, exhibit hydrophilicity that is employable for water absorbance as hydrogel. The second group of bio-based materials under alkyd and PU groups can serve the coating architecture well by preventing the nutrient release from the inner water-soluble core. Sophisticated synthetic methods are available to give a higher performance for a specific release. Waterborne technology is also focused in order to complete the green preparation loop for environmental purpose. And last, the carbon-based materials that are obtained from thermal technologies such as pyrolysis and the hydrothermal process can give biochar and other carbonaceous porous materials as the products. These materials are predominated in their adsorption capacity to retain H_2O and help controlled-release of plant nutrients.

Although bio-based materials have shown great potential in the development of controlled-release fertilizers, a number of limitations are present. The most challenging issue is related to the reliability of the biomaterials. The natural products collected periodically by crop season possess inconsistencies in physical and chemical properties, which leads to the uncertain quality of the resulting fertilizer. With only a few good results for a predictable controlled-release mechanism on bio-based SRFs as found in literature, it cannot be ensured that their performance on an industrial scale can be the same. That is why more techniques and production developments on bio-based fertilizer will be projected to continue for several years to improve the performance of these fertilizers. However, research cost, production cost, and source of raw bio-based materials have to be economically realistic to the existing and future markets.

Acknowledgments

The authors are grateful for the electric and IT facilities provided during the preparation of this chapter by the National Nanotechnology Center, National Science and Technology Development, Thailand. They also would like to thank Dr. Jesper T.N. Knijnenburg for his valuable suggestions and discussion on the selected topic and outline information.

References

1. Duinen, G.V. (2008). How a century of ammonia synthesis changed the world. *Nature Geoscience* 1: 4.
2. Chen, S., Perathoner, S., Ampelli, C., and Centi, G. (2019). Electrochemical dinitrogen activation: to find a sustainable way to produce ammonia. In: *Studies in Surface Science and Catalysis*. Horizons in Sustainable Industrial Chemistry and Catalyst, vol. 178 (ed. S. Albonetti, S. Perathoner and E.A. Quadrelli), 31–46. Elsevier https://doi.org/10.1016/B978-0-444-64127-4.00002-1.
3. Tilman, D., Cassman, K.G., Matson, P.A. et al. (2002). Agricultural sustainability and intensive production practices. *Nature* 418: 671–677. https://doi.org/10.1038/nature01014.
4. Entry, J.A. and Sojka, R.E. (2008). Matrix based fertilizers reduce nitrogen and phosphorus leaching in three soils. *Journal of Environmental Management* 87: 364–372. https://doi.org/10.1016/j.jenvman.2007.01.044.
5. Giroto, A.S., Guimarães, G.G.F., Foschini, M., and Ribeiro, C. (2017). Role of slow-release nanocomposite fertilizers on nitrogen and phosphate availability in soil. *Scientific Reports* 7: 46032. https://doi.org/10.1038/srep46032.
6. Vitousek, P.M., Aber, J.D., Howarth, R.W. et al. (1997). Human alteration of the global nitrogen cycle: sources and consequences. *Ecological Applications* 7: 737–750. https://doi.org/10.1890/1051-0761(1997)007[0737:HAOTGN]2.0.CO;2.
7. Ramananda Bhat, M., Murthy, D.V.R., and Saidutta, M.B. (2011). Modeling of urea release from briquettes using semi infinite and shrinking core models. *Chemical Product and Process Modeling* 6: 1–19. https://doi.org/10.2202/1934-2659.1548.
8. Vashishtha, M., Dongara, P., and Singh, D. (2010). Improvement in properties of urea by phosphogypsum coating. *International Journal of ChemTech Research* 2: 36–44. https://doi.org/10.1.1.631.6445.
9. Smill, V. (2001). *Enriching the Earth: Fritz Haber, Carl Bosch and the Transformation of World Food Production*. Cambridge, MA: MIT Press.
10. Cabezas, A.R.L., Trivelin, P.C.O., Bendassolli, J.A. et al. (1999). Calibration of a semi-open static collector for determination of ammonia volatilization from nitrogen fertilizers. *Communications in Soil Science and Plant Analysis* 30: 389–406. https://doi.org/10.1080/00103629909370211.
11. Cao, P., Lu, C., and Yu, Z. (2018). Historical nitrogen fertilizer use in agricultural ecosystems of the contiguous United States during 1850–2015: application rate, timing, and fertilizer types. *Earth System Science Data* 10: 969–984. https://doi.org/10.5194/essd-10-969-2018.
12. Alfaro, M.A., Rosolem, C., Goulding, K. et al. (2017). Potassium losses and transfers in agricultural and livestock systems. Proceedings for the Frontiers of Potassium Science Conference. Presented at the Frontiers of Potassium Science Conference, International Plant Nutrition Institute, Rome, Italy (25–27 January2017). https://www.apni.net/k-frontiers/.
13. Uzoh, C.F., Onukwuli, O.D., Ozofor, I.H., and Odera, R.S. (2019). Encapsulation of urea with alkyd resin-starch membranes for controlled N_2 release: synthesis, characterization, morphology and optimum N_2 release. *Process Safety and Environment Protection* 121: 133–142. https://doi.org/10.1016/j.psep.2018.10.015.
14. Uzoma, K.C., Inoue, M., Andry, H. et al. (2011). Effect of cow manure biochar on maize productivity under sandy soil condition: cow manure biochar agronomic effects in sandy soil. *Soil Use and Management* 27: 205–212. https://doi.org/10.1111/j.1475-2743.2011.00340.x.
15. Berber, M.R. (2014). A sustained controlled release formulation of soil nitrogen based on nitrate-layered double hydroxide nanoparticle material. *Journal of Soils and Sediments* 14: 60–66. https://doi.org/10.1007/s11368-013-0766-3.

16. Knijnenburg, J.T.N., Laohhasurayotin, K., Khemthong, P., and Kangwansupamonkon, W. (2019). Structure, dissolution, and plant uptake of ferrous/zinc phosphates. *Chemosphere* 223: 310–318. https://doi.org/10.1016/j.chemosphere.2019.02.024.
17. Morgan, K.T., Cushman, K.E., and Sato, S. (2009). Release mechanisms for slow-and controlled-release fertilizers and strategies for their use in vegetable production. *HortTechnology* 19: 10–12. https://doi.org/10.21273/HORTSCI.19.1.10.
18. Trinh, T.H., Kushaari, K., Shuib, A.S. et al. (2015). Modelling the release of nitrogen from controlled release fertiliser: constant and decay release. *Biosystems Engineering* 130: 34–42. https://doi.org/10.1016/j.biosystemseng.2014.12.004.
19. Calabi-Floody, M., Medina, J., Rumpel, C. et al. (2018). Smart fertilizers as a strategy for sustainable agriculture. In: *Advances in Agronomy*, 119–157. Elsevier https://doi.org/10.1016/bs.agron.2017.10.003.
20. Jarosiewicz, A. and Tomaszewska, M. (2003). Controlled-release NPK fertilizer encapsulated by polymeric membranes. *Journal of Agricultural and Food Chemistry* 51: 413–417. https://doi.org/10.1021/jf020800o.
21. Trenkel, M.E. (2010). *Slow- and Controlled-Release and Stabilized Fertilizers: An Option for Enhancing Nutrient Use Efficiency in Agriculture*, 2nde. Paris, France: International Fertilizer Industry Association (IFA).
22. Gowariker, V., Krishnamurthy, S., Gowariker, S. et al. (2009). *The Fertilizer Encyclopedia*. Hoboken, NJ: Wiley.
23. Azeem, B., KuShaari, K., Man, Z.B. et al. (2014). Review on materials & methods to produce controlled release coated urea fertilizer. *Journal of Controlled Release* 181: 11–21. https://doi.org/10.1016/j.jconrel.2014.02.020.
24. Shaviv, A. (2001). Advances in controlled-release fertilizers. In: *Advances in Agronomy* (ed. D.L. Sparks), 1–49. Academic Press https://doi.org/10.1016/S0065-2113(01)71011-5.
25. Zhang, S., Yang, Y., Gao, B. et al. (2017). Superhydrophobic controlled-release fertilizers coated with bio-based polymers with organosilicon and nano-silica modifications. *Journal of Materials Chemistry A* 5: 19943–19953. https://doi.org/10.1039/C7TA06014A.
26. Xie, L., Liu, M., Ni, B., and Wang, Y. (2012). Utilization of wheat straw for the preparation of coated controlled-release fertilizer with the function of water retention. *Journal of Agricultural and Food Chemistry* 60: 6921–6928. https://doi.org/10.1021/jf3001235.
27. Zhang, S., Yang, Y., Gao, B. et al. (2016). Bio-based interpenetrating network polymer composites from locust sawdust as coating material for environmentally friendly controlled-release urea fertilizers. *Journal of Agricultural and Food Chemistry* 64: 5692–5700. https://doi.org/10.1021/acs.jafc.6b01688.
28. França, D., Medina, Â.F., Messa, L.L. et al. (2018). Chitosan spray-dried microcapsule and microsphere as fertilizer host for swellable – controlled release materials. *Carbohydrate Polymers* 196: 47–55. https://doi.org/10.1016/j.carbpol.2018.05.014.
29. Sempeho, S.I., Kim, H.T., Mubofu, E., and Hilonga, A. (2014). Meticulous overview on the controlled release fertilizers. *Advances in Chemotherapy* 2014: 1–16. https://doi.org/10.1155/2014/363071.
30. Liu, L., Kost, J., Fishman, M.L., and Hicks, K.B. (2008). A review: controlled release systems for agricultural and food applications. In: *New Delivery Systems for Controlled Drug Release from Naturally Occurring Materials*, ACS Symposium Series (ed. N. Parris, L. Liu, C. Song and V.P. Shastri), 265–281. Washington, DC: American Chemical Society https://doi.org/10.1021/bk-2008-0992.ch014.
31. Shaviv, A. (2005). Controlled release fertilizers. The IFA International Workshop on Enhanced-Efficiency Fertilizers. Presented at the IFA International Workshop on Enhanced-Efficiency Fertilizers, International Fertilizer Industry Association, Frankfurt, Germany.
32. Ransom, C.J., Jolley, V.D., Blair, T.A. et al. (2020). Nitrogen release rates from slow- and controlled-release fertilizers influenced by placement and temperature. *PLoS One* 15: e0234544. https://doi.org/10.1371/journal.pone.0234544.
33. Yang, Y., Tong, Z., Geng, Y. et al. (2013). Biobased polymer composites derived from corn stover and feather meals as double-coating materials for controlled-release and water-retention urea fertilizers. *Journal of Agricultural and Food Chemistry* 61: 8166–8174. https://doi.org/10.1021/jf402519t.

34. Tian, H., Liu, Z., Zhang, M. et al. (2019). Biobased polyurethane, epoxy resin, and polyolefin wax composite coating for controlled-release fertilizer. *ACS Applied Materials & Interfaces* 11: 5380–5392. https://doi.org/10.1021/acsami.8b16030.
35. Avelino, F., Miranda, I.P., Moreira, T.D. et al. (2019). The influence of the structural features of lignin-based polyurethane coatings on ammonium sulfate release: kinetics and thermodynamics of the process. *Journal of Coating Technology and Research* 16: 449–463. https://doi.org/10.1007/s11998-018-0123-y.
36. Melaj, M.A. and Gimenez, R.B. (2019). Hydrogels: an effective tool to improve nitrogen use efficiency in crops. In: *Polymers for Agri-Food Applications* (ed. T.J. Gutiérrez), 127–141. Cham, Switzerland: Springer International Publishing https://doi.org/10.1007/978-3-030-19416-1_8.
37. Kumar, M., Bauddh, K., Sainger, M. et al. (2015). Increase in growth, productivity and nutritional status of wheat (*Triticum aestivum* L.) and enrichment in soil microbial population applied with biofertilizers entrapped with organic matrix. *Journal of Plant Nutrition* 38: 260–276. https://doi.org/10.1080/01904167.2014.957391.
38. Montesano, F.F., Parente, A., Santamaria, P. et al. (2015). Biodegradable superabsorbent hydrogel increases water retention properties of growing media and plant growth. *Agriculture and Agricultural Science Procedia* 4: 451–458. https://doi.org/10.1016/j.aaspro.2015.03.052.
39. Giroto, A.S., Guimarães, G.G.F., and Ribeiro, C. (2018). A novel, simple route to produce urea : urea–formaldehyde composites for controlled release of fertilizers. *Journal of Polymers and the Environment* 26: 2448–2458. https://doi.org/10.1007/s10924-017-1141-z.
40. López-Rayo, S., Imran, A., Hansen, H.C.B. et al. (2017). Layered double hydroxides: potential release-on-demand fertilizers for plant zinc nutrition. *Journal of Agricultural and Food Chemistry* 65: 8779–8789. https://doi.org/10.1021/acs.jafc.7b02604.
41. Songkhum, P., Wuttikhun, T., Chanlek, N. et al. (2018). Controlled release studies of boron and zinc from layered double hydroxides as the micronutrient hosts for agricultural application. *Applied Clay Science* 152: 311–322. https://doi.org/10.1016/j.clay.2017.11.028.
42. Kabiri, S., Degryse, F., Tran, D.N.H. et al. (2017). Graphene oxide: a new carrier for slow release of plant micronutrients. *ACS Applied Materials & Interfaces* 9: 43325–43335. https://doi.org/10.1021/acsami.7b07890.
43. Chojnacka, K., Moustakas, K., and Witek-Krowiak, A. (2020). Bio-based fertilizers: a practical approach towards circular economy. *Bioresource Technology* 295: 122223. https://doi.org/10.1016/j.biortech.2019.122223.
44. Hoornweg, D., Bhada-Tata, P., and Kennedy, C. (2013). Environment: waste production must peak this century. *Nature* 502: 615–617. https://doi.org/10.1038/502615a.
45. Mikula, K., Izydorczyk, G., Skrzypczak, D. et al. (2020). Controlled release micronutrient fertilizers for precision agriculture – a review. *Science of the Total Environment* 712: 136365. https://doi.org/10.1016/j.scitotenv.2019.136365.
46. Ramli, R.A. (2019). Slow release fertilizer hydrogels: a review. *Polymer Chemistry* 10: 6073–6090. https://doi.org/10.1039/C9PY01036J.
47. Senna, A.M. and Botaro, V.R. (2017). Biodegradable hydrogel derived from cellulose acetate and EDTA as a reduction substrate of leaching NPK compound fertilizer and water retention in soil. *Journal of Controlled Release* 260: 194–201. https://doi.org/10.1016/j.jconrel.2017.06.009.
48. Pang, L., Gao, Z., Feng, H. et al. (2019). Cellulose based materials for controlled release formulations of agrochemicals: a review of modifications and applications. *Journal of Controlled Release* 316: 105–115. https://doi.org/10.1016/j.jconrel.2019.11.004.
49. Senna, A.M., Braga do Carmo, J., Santana da Silva, J.M., and Botaro, V.R. (2015). Synthesis, characterization and application of hydrogel derived from cellulose acetate as a substrate for slow-release NPK fertilizer and water retention in soil. *Journal of Environmental Chemical Engineering* 3: 996–1002. https://doi.org/10.1016/j.jece.2015.03.008.
50. Senna, A.M., Novack, K.M., and Botaro, V.R. (2014). Synthesis and characterization of hydrogels from cellulose acetate by esterification crosslinking with EDTA dianhydride. *Carbohydrate Polymers* 114: 260–268. https://doi.org/10.1016/j.carbpol.2014.08.017.

51. Costa, M.M.E., Cabral-Albuquerque, E.C.M., Alves, T.L.M. et al. (2013). Use of polyhydroxybutyrate and ethyl cellulose for coating of urea granules. *Journal of Agricultural and Food Chemistry* 61: 9984–9991. https://doi.org/10.1021/jf401185y.
52. Li, Y., Zhen, D., Liao, S.J. et al. (2018). Controlled-release urea encapsulated by ethyl cellulose/butyl acrylate/vinyl acetate hybrid latex. *Polish Journal of Chemical Technology* 20: 108–112. https://doi.org/10.2478/pjct-2018-0062.
53. Kenawy, E., Azaam, M.M., El-nshar, E.M., (2018). Preparation of carboxymethyl cellulose-*g*-poly (acrylamide)/montmorillonite superabsorbent composite as a slow-release urea fertilizer. *Polymers for Advanced Technologies.* 29, 2072–2079. https://doi.org/10.1002/pat.4315.
54. Ni, B., Liu, M., Lü, S. et al. (2011). Environmentally friendly slow-release nitrogen fertilizer. *Journal of Agricultural and Food Chemistry* 59: 10169–10175. https://doi.org/10.1021/jf202131z.
55. Zhang, M. and Yang, J. (2020). Preparation and characterization of multifunctional slow release fertilizer coated with cellulose derivatives. *International Journal of Polymeric Materials and Polymeric Biomaterials* 1–8: 774–781. https://doi.org/10.1080/00914037.2020.1765352.
56. Chen, Y.-C. and Chen, Y.-H. (2019). Thermo and pH-responsive methylcellulose and hydroxypropyl methylcellulose hydrogels containing K_2SO_4 for water retention and a controlled-release water-soluble fertilizer. *Science of the Total Environment* 655: 958–967. https://doi.org/10.1016/j.scitotenv.2018.11.264.
57. Qi, H., Ma, R., Shi, C. et al. (2019). Novel low-cost carboxymethyl cellulose microspheres with excellent fertilizer absorbency and release behavior for saline-alkali soil. *International Journal of Biological Macromolecules* 131: 412–419. https://doi.org/10.1016/j.ijbiomac.2019.03.047.
58. Olad, A., Zebhi, H., Salari, D. et al. (2018). Slow-release NPK fertilizer encapsulated by carboxymethyl cellulose-based nanocomposite with the function of water retention in soil. *Materials Science and Engineering: C* 90: 333–340. https://doi.org/10.1016/j.msec.2018.04.083.
59. Wang, M., Zhang, G., Zhou, L. et al. (2016). Fabrication of pH-controlled-release ferrous foliar fertilizer with high adhesion capacity based on nanobiomaterial. *ACS Sustainable Chemistry & Engineering.* 4: 6800–6808. https://doi.org/10.1021/acssuschemeng.6b01761.
60. Guilherme, M.R., Aouada, F.A., Fajardo, A.R. et al. (2015). Superabsorbent hydrogels based on polysaccharides for application in agriculture as soil conditioner and nutrient carrier: a review. *European Polymer Journal* 72: 365–385. https://doi.org/10.1016/j.eurpolymj.2015.04.017.
61. Tang, X. (2011). Recent advances in starch, polyvinyl alcohol based polymer blends, nanocomposites and their biodegradability. *Carbohydrate Polymers* 85 (1): 7–16. https://doi.org/10.1016/j.carbpol.2011.01.030.
62. Giroto, A.S., Guimaraes, G.G., Colnago, L.A. et al. (2019). Controlled release of nitrogen using urea-melamine-starch composites. *Journal of Cleaner Production* 217: 448–455. https://doi.org/10.1016/j.jclepro.2019.01.275.
63. Vudjung, C. and Saengsuwan, S. (2018). Biodegradable IPN hydrogels based on pre-vulcanized natural rubber and cassava starch as coating membrane for environment-friendly slow-release urea fertilizer. *Journal of Polymers and the Environment* 26: 3967–3980. https://doi.org/10.1007/s10924-018-1274-8.
64. Lü, S., Gao, C., Wang, X. et al. (2014). Synthesis of a starch derivative and its application in fertilizer for slow nutrient release and water-holding. *RSC Advances* 4: 51208–51214. https://doi.org/10.1039/C4RA06006G.
65. Lim, D.W., Whang, H.S., Yoon, K.J., and Ko, S.W. (2001). Synthesis and absorbency of a superabsorbent from sodium starch sulfate-g-polyacrylonitrile. *Journal of Applied Polymer Science* 79: 1423–1430. https://doi.org/10.1002/1097-4628(20010222)79:8<1423::aid-app90>3.0.co;2-v.
66. Zhong, K., Lin, Z.-T., Zheng, X.-L. et al. (2013). Starch derivative-based superabsorbent with integration of water-retaining and controlled-release fertilizers. *Carbohydrate Polymers* 92: 1367–1376. https://doi.org/10.1016/j.carbpol.2012.10.030.
67. Qiao, D., Liu, H., Yu, L. et al. (2016). Preparation and characterization of slow-release fertilizer encapsulated by starch-based superabsorbent polymer. *Carbohydrate Polymers* 147: 146–154. https://doi.org/10.1016/j.carbpol.2016.04.010.
68. Noppakundilograt, S., Pheatcharat, N., and Kiatkamjornwong, S. (2015). Multilayer-coated NPK compound fertilizer hydrogel with controlled nutrient release and water absorbency. *Journal of Applied Polymer Science* 132: https://doi.org/10.1002/app.41249.

69. Holliman, P.J., Clark, J.A., Williamson, J.C., and Jones, D.L. (2005). Model and field studies of the degradation of cross-linked polyacrylamide gels used during the revegetation of slate waste. *Science of the Total Environment* 336: 13–24. https://doi.org/10.1016/j.scitotenv.2004.06.006.
70. Alharbi, K., Ghoneim, A., Ebid, A. et al. (2018). Controlled release of phosphorous fertilizer bound to carboxymethyl starch-g-polyacrylamide and maintaining a hydration level for the plant. *International Journal of Biological Macromolecules* 116: 224–231. https://doi.org/10.1016/j.ijbiomac.2018.04.182.
71. Tanan, W., Panichpakdee, J., and Saengsuwan, S. (2019). Novel biodegradable hydrogel based on natural polymers: synthesis, characterization, swelling/reswelling and biodegradability. *European Polymer Journal* 112: 678–687. https://doi.org/10.1016/j.eurpolymj.2018.10.033.
72. Salimi, M., Motamedi, E., Motesharezedeh, B. et al. (2020). Starch-*g*-poly(acrylic acid-*co*-acrylamide) composites reinforced with natural char nanoparticles toward environmentally benign slow-release urea fertilizers. *Journal of Environmental Chemical Engineering* 8: 103765. https://doi.org/10.1016/j.jece.2020.103765.
73. Xiao, X.M., Yu, L., Xie, F.W. et al. (2017). One-step method to prepare starch-based superabsorbent polymer for slow release of fertilizer. *Chemical Engineering Journal* 309: 607–616. https://doi.org/10.1016/j.cej.2016.10.101.
74. Marandi, G.B., Sharifnia, N., and Hosseinzadeh, H. (2006). Synthesis of an alginate-poly(sodium acrylate-*co*-acrylamide) superabsorbent hydrogel with low salt sensitivity and high pH sensitivity. *Journal of Applied Polymer Science* 101: 2927–2937. https://doi.org/10.1002/app.23373.
75. Yin, Y.H., Ji, X.M., Dong, H. et al. (2008). Study of the swelling dynamics with overshooting effect of hydrogels based on sodium alginate-*g*-acrylic acid. *Carbohydrate Polymers* 71: 682–689. https://doi.org/10.1016/j.carbpol.2007.07.012.
76. Abd El-Rehim, H.A. (2006). Characterization and possible agricultural application of polyacrylamide/sodium alginate crosslinked hydrogels prepared by ionizing radiation. *Journal of Applied Polymer Science.* 101: 3572–3580. https://doi.org/10.1002/app.22487.
77. Schexnailder, P. and Schmidt, G. (2009). Nanocomposite polymer hydrogels. *Colloid and Polymer Science.* 287: 1–11. https://doi.org/10.1007/s00396-008-1949-0.
78. Wen, P., Han, Y., Wu, Z. et al. (2017). Rapid synthesis of a corncob-based semi-interpenetrating polymer network slow-release nitrogen fertilizer by microwave irradiation to control water and nutrient losses. *Arabian Journal of Chemistry* 10: 922–934. https://doi.org/10.1016/j.arabjc.2017.03.002.
79. Rashidzadeh, A. and Olad, A. (2014). Slow-released NPK fertilizer encapsulated by NaAlg-*g*-poly(AA-co-AAm)/MMT superabsorbent nanocomposite. *Carbohydrate Polymers* 114: 269–278. https://doi.org/10.1016/j.carbpol.2014.08.010.
80. Rashidzadeh, A., Olad, A., Salari, D., and Reyhanitabar, A. (2014). On the preparation and swelling properties of hydrogel nanocomposite based on sodium alginate-*g*-poly (acrylic acid-*co*-acrylamide)/clinoptilolite and its application as slow release fertilizer. *Journal of Polymer Research* 21: 1–15. https://doi.org/10.1007/s10965-013-0344-9.
81. Rashidzadeh, A., Olad, A., and Reyhanitabar, A. (2015). Hydrogel/clinoptilolite nanocomposite-coated fertilizer: swelling, water-retention and slow-release fertilizer properties. *Polymer Bulletin* 72: 2667–2684. https://doi.org/10.1007/s00289-015-1428-y.
82. Wang, J., Pan, X., Liu, Y. et al. (2012). Effects of biochar amendment in two soils on greenhouse gas emissions and crop production. *Plant and Soil* 360: 287–298. https://doi.org/10.1007/s11104-012-1250-3.
83. Wang, B., Gao, B., Zimmerman, A.R. et al. (2018). Novel biochar-impregnated calcium alginate beads with improved water holding and nutrient retention properties. *Journal of Environmental Management.* 209: 105–111. https://doi.org/10.1016/j.jenvman.2017.12.041.
84. Shan, L., Gao, Y., Zhang, Y. et al. (2016). Fabrication and use of alginate-based cryogel delivery beads loaded with urea and phosphates as potential carriers for bioremediation. *Industrial and Engineering Chemistry Research* 55: 7655–7660. https://doi.org/10.1021/acs.iecr.6b01256.
85. Chen, C., Gao, Z., Qiu, X., and Hu, S. (2013). Enhancement of the controlled-release properties of chitosan membranes by crosslinking with suberoyl chloride. *Molecules* 18: 7239–7252. https://doi.org/10.3390/molecules18067239.

86. Wu, L. and Liu, M. (2008). Preparation and properties of chitosan-coated NPK compound fertilizer with controlled-release and water-retention. *Carbohydrate Polymers* 72: 240–247. https://doi.org/10.1016/j.carbpol.2007.08.020.
87. Wu, L., Liu, M.Z., and Liang, R. (2008). Preparation and properties of a double-coated slow-release NPK compound fertilizer with superabsorbent and water-retention. *Bioresource Technology* 99: 547–554. https://doi.org/10.1016/j.biortech.2006.12.027.
88. Li, T., Gao, B., Tong, Z. et al. (2019). Chitosan and graphene oxide nanocomposites as coatings for controlled-release fertilizer. *Water, Air, and Soil Pollution* 230: 146. https://doi.org/10.1007/s11270-019-4173-2.
89. Mi, F.-L., Kuan, C.-Y., Shyu, S.-S. et al. (2000). The study of gelation kinetics and chain-relaxation properties of glutaraldehyde-cross-linked chitosan gel and their effects on microspheres preparation and drug release. *Carbohydrate Polymers* 41: 389–396. https://doi.org/10.1016/S0144-8617(99)00104-6.
90. Jamnongkan, T. and Kaewpirom, S. (2010). Potassium release kinetics and water retention of controlled-release fertilizers based on chitosan hydrogels. *Journal of Polymers and the Environment* 18: 413–421. https://doi.org/10.1007/s10924-010-0228-6.
91. dos Santos, B.R., Bacalhau, F.B., dos Pereira, T., S. et al. (2015). Chitosan-montmorillonite microspheres: a sustainable fertilizer delivery system. *Carbohydrate Polymers* 127: 340–346. https://doi.org/10.1016/j.carbpol.2015.03.064.
92. Essawy, H.A., Ghazy, M.B., El-Hai, F.A., and Mohamed, M.F. (2016). Superabsorbent hydrogels via graft polymerization of acrylic acid from chitosan-cellulose hybrid and their potential in controlled release of soil nutrients. *International Journal of Biological Macromolecules* 89: 144–151. https://doi.org/10.1016/j.ijbiomac.2016.04.071.
93. Iftime, M.M., Ailiesei, G.L., Ungureanu, E., and Marin, L. (2019). Designing chitosan based eco-friendly multifunctional soil conditioner systems with urea controlled release and water retention. *Carbohydrate Polymers* 223: 115040. https://doi.org/10.1016/j.carbpol.2019.115040.
94. Dastpak, A., Yliniemi, K., Monteiro, M.C.D. et al. (2018). From waste to valuable resource: lignin as a sustainable anti-corrosion coating. *Coatings* 8: 454. https://doi.org/10.3390/coatings8120454.
95. Castro, F.B.D., 1998. Controlled release urea product, method for its production and use of said product as feed supplement. WO1998027830A1.
96. Jiao, G.J., Xu, Q., Cao, S.L. et al. (2018). Controlled-release fertilizer with lignin used to trap urea/hydroxymethylurea/urea-formaldehyde polymers. *BioResources* 13: 1711–1728. https://doi.org/10.15376/biores.13.1.1711-1728.
97. Legras-Lecarpentier, D., Stadler, K., Weiss, R. et al. (2019). Enzymatic synthesis of 100% lignin biobased granules as fertilizer storage and controlled slow release systems. *ACS Sustainable Chemistry & Engineering* 7: 12621–12628. https://doi.org/10.1021/acssuschemeng.9b02689.
98. Mulder, W.J., Gosselink, R.J.A., Vingerhoeds, M.H. et al. (2011). Lignin based controlled release coatings. *Industrial Crops and Products* 34: 915–920. https://doi.org/10.1016/j.indcrop.2011.02.011.
99. Fertahi, S., Bertrand, I., Amjoud, M. et al. (2019). Properties of coated slow-release triple superphosphate (TSP) fertilizers based on lignin and carrageenan formulations. *ACS Sustainable Chemistry & Engineering* 7: 10371–10382. https://doi.org/10.1021/acssuschemeng.9b00433.
100. Behin, J. and Sadeghi, N. (2016). Utilization of waste lignin to prepare controlled-slow release urea. *International Journal of Recycling of Organic Waste in Agriculture* 5: 289–299. https://doi.org/10.1007/s40093-016-0139-1.
101. Li, J., Wang, M., She, D., and Zhao, Y. (2017). Structural functionalization of industrial softwood kraft lignin for simple dip-coating of urea as highly efficient nitrogen fertilizer. *Industrial Crops and Products* 109: 255–265. https://doi.org/10.1016/j.indcrop.2017.08.011.
102. Sipponen, M.H., Rojas, O.J., Pihlajaniemi, V. et al. (2017). Calcium chelation of lignin from pulping spent liquor for water-resistant slow-release urea fertilizer systems. *ACS Sustainable Chemistry & Engineering* 5: 1054–1061. https://doi.org/10.1021/acssuschemeng.6b02348.
103. Shoji, S. and Kanno, H. (1994). Use of polyolefin-coated fertilizers for increasing fertilizer efficiency and reducing nitrate leaching and nitrous oxide emissions. *Fertilizer Research* 39: 147–152. https://doi.org/10.1007/BF00750913.

104. Xu, M., Li, D., Li, J. et al. (2013). Polyolefin-coated urea decreases ammonia volatilization in a double rice system of southern china. *Agronomy Journal* 105: 277–284. https://doi.org/10.2134/agronj2012.0222.
105. Wei, Y., Li, J., Li, Y. et al. (2017). Research on permeability coefficient of a polyethylene controlled-release film coating for urea and relevant nutrient release pathways. *Polymer Testing* 59: 90–98. https://doi.org/10.1016/j.polymertesting.2017.01.019.
106. Liang, D., Zhang, Q., Zhang, W. et al. (2019). Tunable thermo-physical performance of castor oil-based polyurethanes with tailored release of coated fertilizers. *Journal of Cleaner Production* 210: 1207–1215. https://doi.org/10.1016/j.jclepro.2018.11.047.
107. Liu, J., Yang, Y., Gao, B. et al. (2019). Bio-based elastic polyurethane for controlled-release urea fertilizer: fabrication, properties, swelling and nitrogen release characteristics. *Journal of Cleaner Production* 209: 528–537. https://doi.org/10.1016/j.jclepro.2018.10.263.
108. Yang, Y., Zhang, M., Li, Y. et al. (2012). Improving the quality of polymer-coated urea with recycled plastic, proper additives, and large tablets. *Journal of Agricultural and Food Chemistry* 60: 11229–11237. https://doi.org/10.1021/jf302813g.
109. Tao, S., Liu, J., Jin, K. et al. (2011). Preparation and characterization of triple polymer-coated controlled-release urea with water-retention property and enhanced durability. *Journal of Applied Polymer Science* 120: 2103–2111. https://doi.org/10.1002/app.33366.
110. Zhang, Y., Gao, P., Zhao, L., and Chen, Y. (2016). Preparation and swelling properties of a starch-*g*-poly(acrylic acid)/organo-mordenite hydrogel composite. *Frontiers of Chemical Science and Engineering* 10: 147–161. https://doi.org/10.1007/s11705-015-1546-y.
111. Tomaszewska, M. and Jarosiewicz, A. (2002). Use of polysulfone in controlled-release NPK fertilizer formulations. *Journal of Agricultural and Food Chemistry* 50: 4634–4639. https://doi.org/10.1021/jf0116808.
112. Majeed, Z., Mansor, N., Ajab, Z., and Man, Z. (2017). Lignin macromolecule's implication in slowing the biodegradability of urea-crosslinked starch films applied as slow-release fertilizer: lignin macromolecule's implication in biodegradability of UcS films. *Starch – Stärke* 69: 1600362. https://doi.org/10.1002/star.201600362.
113. Chen, L., Xie, Z., Zhuang, X. et al. (2008). Controlled release of urea encapsulated by starch-*g*-poly(l-lactide). *Carbohydrate Polymers* 72: 342–348. https://doi.org/10.1016/j.carbpol.2007.09.003.
114. Majeed, Z., Ramli, N.K., Mansor, N., and Man, Z. (2015). A comprehensive review on biodegradable polymers and their blends used in controlled-release fertilizer processes. *Reviews in Chemical Engineering* 31: 69–95. https://doi.org/10.1515/revce-2014-0021.
115. Zhang, C., Garrison, T.F., Madbouly, S.A., and Kessler, M.R. (2017). Recent advances in vegetable oil-based polymers and their composites. *Progress in Polymer Science* 71: 91–143. https://doi.org/10.1016/j.progpolymsci.2016.12.009.
116. Lligadas, G., Ronda, J.C., Galià, M., and Cádiz, V. (2010). Plant oils as platform chemicals for polyurethane synthesis: current state-of-the-art. *Biomacromolecules* 11: 2825–2835. https://doi.org/10.1021/bm100839x.
117. Cruz, M.V., Araújo, D., Alves, V.D. et al. (2016). Characterization of medium chain length polyhydroxyalkanoate produced from olive oil deodorizer distillate. *International Journal of Biological Macromolecules* 82: 243–248. https://doi.org/10.1016/j.ijbiomac.2015.10.043.
118. Ionescu, M., Radojčić, D., Wan, X. et al. (2015). Functionalized vegetable oils as precursors for polymers by thiol-ene reaction. *European Polymer Journal* 67: 439–448. https://doi.org/10.1016/j.eurpolymj.2014.12.037.
119. Meiorin, C., Muraca, D., Pirota, K.R. et al. (2014). Nanocomposites with superparamagnetic behavior based on a vegetable oil and magnetite nanoparticles. *European Polymer Journal* 53: 90–99. https://doi.org/10.1016/j.eurpolymj.2014.01.018.
120. Sharmin, E., Zafar, F., Akram, D. et al. (2015). Recent advances in vegetable oils based environment friendly coatings: a review. *Industrial Crops and Products* 76: 215–229. https://doi.org/10.1016/j.indcrop.2015.06.022.
121. Stemmelen, M., Lapinte, V., Habas, J.-P., and Robin, J.-J. (2015). Plant oil-based epoxy resins from fatty diamines and epoxidized vegetable oil. *European Polymer Journal* 68: 536–545. https://doi.org/10.1016/j.eurpolymj.2015.03.062.

122. Thanamongkollit, N., Miller, K.R., and Soucek, M.D. (2012). Synthesis of UV-curable tung oil and UV-curable tung oil based alkyd. *Progress in Organic Coating* 73: 425–434. https://doi.org/10.1016/j.porgcoat.2011.02.003.
123. Black, M. and Rawlins, J.W. (2009). Thiol–ene UV-curable coatings using vegetable oil macromonomers. *European Polymer Journal* 45: 1433–1441. https://doi.org/10.1016/j.eurpolymj.2009.02.007.
124. Desroches, M., Caillol, S., Lapinte, V. et al. (2011). Synthesis of biobased polyols by thiol-ene coupling from vegetable oils. *Macromolecules* 44: 2489–2500. https://doi.org/10.1021/ma102884w.
125. Stemmelen, M., Pessel, F., Lapinte, V. et al. (2011). A fully biobased epoxy resin from vegetable oils: from the synthesis of the precursors by thiol-ene reaction to the study of the final material. *Journal of Polymer Science Part A: Polymer Chemistry* 49: 2434–2444. https://doi.org/10.1002/pola.24674.
126. Xia, Y. and Larock, R.C. (2010). Vegetable oil-based polymeric materials: synthesis, properties, and applications. *Green Chemistry* 12: 1893–1909. https://doi.org/10.1039/c0gc00264j.
127. Can, E., Küsefoğlu, S., and Wool, R.P. (2001). Rigid, thermosetting liquid molding resins from renewable resources. I. Synthesis and polymerization of soy oil monoglyceride maleates. *Journal of Applied Polymer Science* 81: 69–77. https://doi.org/10.1002/app.1414.
128. Zieleniewska, M., Leszczyński, M.K., Kurańska, M. et al. (2015). Preparation and characterisation of rigid polyurethane foams using a rapeseed oil-based polyol. *Industrial Crops and Products* 74: 887–897. https://doi.org/10.1016/j.indcrop.2015.05.081.
129. Zhang, C., Madbouly, S.A., and Kessler, M.R. (2015). Biobased polyurethanes prepared from different vegetable oils. *ACS Applied Materials & Interfaces* 7: 1226–1233. https://doi.org/10.1021/am5071333.
130. Dubrulle, L., Lebeuf, R., Fressancourt-Collinet, M., and Nardello-Rataj, V. (2017). Optimization of the vegetable oil composition in alkyd resins: a kinetic approach based on FAMEs autoxidation. *Progress in Organic Coating* 112: 288–294. https://doi.org/10.1016/j.porgcoat.2017.06.021.
131. Otabor, G.O., Ifijen, I.H., Mohammed, F.U. et al. (2019). Alkyd resin from rubber seed oil/linseed oil blend: a comparative study of the physiochemical properties. *Heliyon* 5: e01621. https://doi.org/10.1016/j.heliyon.2019.e01621.
132. Flores, S., Obregón, D., and Flores, A. (2018). Preparation and evaluation of paints fabricated from alkyd resin based on sacha inchi oil. Eurocoat 2018 Conference, Paris Expo-Porte de Versailles, France (27–29 March 2018). https://doi.org/10.13140/RG.2.2.24184.75528.
133. Chiplunkar, P.P., Shinde, V.V., and Pratap, A.P. (2017). Synthesis and application of palm fatty acid distillate based alkyd resin in liquid detergent. *Journal of Surfactants and Detergents* 20: 137–149. https://doi.org/10.1007/s11743-016-1905-9.
134. Yin, X., Duan, H., Wang, X. et al. (2014). An investigation on synthesis of alkyd resin with sorbitol. *Progress in Organic Coating* 77: 674–678. https://doi.org/10.1016/j.porgcoat.2013.12.005.
135. Bora, M.M., Gogoi, P., Deka, D.C., and Kakati, D.K. (2014). Synthesis and characterization of yellow oleander (*Thevetia peruviana*) seed oil-based alkyd resin. *Industrial Crops and Products* 52: 721–728. https://doi.org/10.1016/j.indcrop.2013.11.012.
136. Boruah, M., Gogoi, P., Adhikari, B., and Dolui, S.K. (2012). Preparation and characterization of *Jatropha curcas* oil based alkyd resin suitable for surface coating. *Progress in Organic Coating* 74: 596–602. https://doi.org/10.1016/j.porgcoat.2012.02.007.
137. Hlaing, N.H. and Oo, M.M. (2008). Manufacture of alkyd resin from castor oil. *World Academy of Science, Engineering and Technology* 48: 155–161.
138. Liang, L., Liu, C., Xiao, X. et al. (2014). Optimized synthesis and properties of surfactant-free water-reducible acrylate-alkyd resin emulsion. *Progress in Organic Coating* 77: 1715–1723. https://doi.org/10.1016/j.porgcoat.2014.05.015.
139. Arendt, J. and Kim, J. (2015). Zero-VOC, surfactant-free alkyd dispersion. *PCI – Paint and Coatings Industry*.
140. Gooch, J.W. (2002). *Emulsification and Polymerization of Alkyd Resins*. New York: Kluwer Academic/Plenum Publishers.
141. Clark, M.D. and Helmer, B.J. (2001). Acrylic modified waterborne alkyd dispersions. US Patent 6,242,528 B1.

142. Wang, S.T., Schork, F.J., Poehlein, G.W., and Gooch, J.W. (1996). Emulsion and miniemulsion copolymerization of acrylic monomers in the presence of alkyd resin. *Journal of Applied Polymer Science* 60: 2069–2076. https://doi.org/10.1002/(SICI)1097–4628(19960620)60:12<2069::AID-APP4>3.0.CO;2-K
143. Athawale, V.D. and Nimbalkar, R.V. (2011). Waterborne coatings based on renewable oil resources: an overview. *Journal of the American Oil Chemists' Society* 88: 159–185. https://doi.org/10.1007/s11746-010-1668-9.
144. Nabuurs, T., Baijards, R.A., and German, A.L. (1996). Alkyd-acrylic hybrid systems for use as binders in waterborne paints. *Progress in Organic Coating* 27: 163–172. https://doi.org/10.1016/0300-9440(95)00533-1.
145. Akbarinezhad, E., Ebrahimi, M., Kassiriha, S.M., and Khorasani, M. (2009). Synthesis and evaluation of water-reducible acrylic–alkyd resins with high hydrolytic stability. *Progress in Organic Coating* 65: 217–221. https://doi.org/10.1016/j.porgcoat.2008.11.012.
146. Wang, C., Chin, C.-K., and Gianfagna, T. (2000). Relationship between cutin monomers and tomato resistance to powdery mildew infection. *Physiological and Molecular Plant Pathology* 57: 55–61. https://doi.org/10.1006/pmpp.2000.0279.
147. Uzoh, C.F., Obele, M.C., Umennabuife, U.M., and Onukwuli, O.D. (2019). Synthesis of rubber seed oil waterborne alkyd resin from glucitol. *Journal of Coating Technology and Research* 16: 1727–1735. https://doi.org/10.1007/s11998-019-00235-0.
148. Barrioni, B.R., de Carvalho, S.M., Oréfice, R.L. et al. (2015). Synthesis and characterization of biodegradable polyurethane films based on HDI with hydrolyzable crosslinked bonds and a homogeneous structure for biomedical applications. *Materials Science and Engineering: C* 52: 22–30. https://doi.org/10.1016/j.msec.2015.03.027.
149. Lu, P., Zhang, Y., Jia, C. et al. (2016). Degradation of polyurethane coating materials from liquefied wheat straw for controlled release fertilizers. *Journal of Applied Polymer Science* 133: https://doi.org/10.1002/app.44021.
150. Irfan, S.A., Razali, R., KuShaari, K. et al. (2018). A review of mathematical modeling and simulation of controlled-release fertilizers. *Journal of Controlled Release* 271: 45–54. https://doi.org/10.1016/j.jconrel.2017.12.017.
151. Li, L., Sun, Y., Cao, B. et al. (2016). Preparation and performance of polyurethane/mesoporous silica composites for coated urea. *Materials and Design* 99: 21–25. https://doi.org/10.1016/j.matdes.2016.03.043.
152. Li, Y., Jia, C., Zhang, X. et al. (2018). Synthesis and performance of bio-based epoxy coated urea as controlled release fertilizer. *Progress in Organic Coating* 119: 50–56. https://doi.org/10.1016/j.porgcoat.2018.02.013.
153. Liu, X., Yang, Y., Gao, B. et al. (2017). Environmentally friendly slow-release urea fertilizers based on waste frying oil for sustained nutrient release. *ACS Sustainable Chemistry & Engineering* 5: 6036–6045. https://doi.org/10.1021/acssuschemeng.7b00882.
154. Patil, C.K., Rajput, S.D., Marathe, R.J. et al. (2017). Synthesis of bio-based polyurethane coatings from vegetable oil and dicarboxylic acids. *Progress in Organic Coating* 106: 87–95. https://doi.org/10.1016/j.porgcoat.2016.11.024.
155. Xie, J., Yang, Y., Gao, B. et al. (2017). Biomimetic superhydrophobic biobased polyurethane-coated fertilizer with atmosphere "outerwear.". *ACS Applied Materials & Interfaces* 9: 15868–15879. https://doi.org/10.1021/acsami.7b02244.
156. Alaa, M.A., Yusoh, K., and Hasany, S.F. (2015). Pure polyurethane and castor oil based polyurethane: synthesis and characterization. *Journal of Mechanical Engineering Science* 8: 1507–1515. https://doi.org/10.15282/jmes.8.2015.25.0147.
157. Zhao, M., Wang, Y., Liu, L. et al. (2018). Green coatings from renewable modified bentonite and vegetable oil based polyurethane for slow release fertilizers. *Polymer Composites* 39: 4355–4363. https://doi.org/10.1002/pc.24519.
158. Miao, S., Wang, P., Su, Z., and Zhang, S. (2014). Vegetable-oil-based polymers as future polymeric biomaterials. *Acta Biomaterialia* 10: 1692–1704. https://doi.org/10.1016/j.actbio.2013.08.040.
159. Ogunniyi, D. (2006). Castor oil: a vital industrial raw material. *Bioresource Technology* 97: 1086–1091. https://doi.org/10.1016/j.biortech.2005.03.028.

160. Bortoletto-Santos, R., Ribeiro, C., and Polito, W.L. (2016). Controlled release of nitrogen-source fertilizers by natural-oil-based poly(urethane) coatings: the kinetic aspects of urea release. *Journal of Applied Polymer Science* 133: https://doi.org/10.1002/app.43790.
161. Ochiai, B., Sato, S.-I., and Endo, T. (2007). Synthesis and properties of polyurethanes bearing urethane moieties in the side chain. *Journal of Polymer Science Part A: Polymer Chemistry* 45: 3408–3414. https://doi.org/10.1002/pola.22093.
162. Tamami, B., Sohn, S., and Wilkes, G.L. (2004). Incorporation of carbon dioxide into soybean oil and subsequent preparation and studies of nonisocyanate polyurethane networks. *Journal of Applied Polymer Science* 92: 883–891. https://doi.org/10.1002/app.20049.
163. Asemani, H.R. and Mannari, V. (2019). Synthesis and evaluation of non-isocyanate polyurethane polyols for heat-cured thermoset coatings. *Progress in Organic Coating* 131: 247–258. https://doi.org/10.1016/j.porgcoat.2019.02.036.
164. Javni, I., Hong, D.P., and Petrović, Z.S. (2013). Polyurethanes from soybean oil, aromatic, and cycloaliphatic diamines by nonisocyanate route. *Journal of Applied Polymer Science* 128: 566–571. https://doi.org/10.1002/app.38215.
165. Javni, I., Hong, D.P., and Petrović, Z.S. (2008). Soy-based polyurethanes by nonisocyanate route. *Journal of Applied Polymer Science* 108: 3867–3875. https://doi.org/10.1002/app.27995.
166. Parzuchowski, P.G., Jurczyk-Kowalska, M., Ryszkowska, J., and Rokicki, G. (2006). Epoxy resin modified with soybean oil containing cyclic carbonate groups. *Journal of Applied Polymer Science* 102: 2904–2914. https://doi.org/10.1002/app.24795.
167. Kalia, A., Sharma, S.P., Kaur, H., and Kaur, H. (2020). Novel nanocomposite-based controlled-release fertilizer and pesticide formulations: prospects and challenges. In: *Multifunctional Hybrid Nanomaterials for Sustainable Agri-Food and Ecosystems* (ed. K.A. Abd-Elsalam), 99–134. (Micro and Nano Technologies). Elsevier https://linkinghub.elsevier.com/retrieve/pii/B9780128213544000054.
168. Basak, B.B., Pal, S., and Datta, S.C. (2012). Use of modified clays for retention and supply of water and nutrients. *Current Science* 102: 1272–1278.
169. Elhassani, C.E., Essamlali, Y., Aqlil, M. et al. (2019). Urea-impregnated HAP encapsulated by lignocellulosic biomass-extruded composites: a novel slow-release fertilizer. *Environmental Technology and Innovation* 15: 100403. https://doi.org/10.1016/j.eti.2019.100403.
170. Guo, H., White, J.C., Wang, Z., and Xing, B. (2018). Nano-enabled fertilizers to control the release and use efficiency of nutrients. *Current Opinion in Environmental Science & Health* 6: 77–83. https://doi.org/10.1016/j.coesh.2018.07.009.
171. Bisinoti, M.C., Moreira, A.B., Melo, C.A., Fregolente, L.G., Bento, L.R., Vitor dos Santos, J., Ferreira, O.P. (2019). Application of carbon-based nanomaterials as fertilizers in soils, in: Nascimento R, Ferreira OP, Olveira Sousa Neto Vde *Nanomaterials Applications for Environmental Matrices*. Elsevier, pp. 305–333. https://linkinghub.elsevier.com/retrieve/pii/B9780128148297000082.
172. Yuan, J.-H., Xu, R.-K., and Zhang, H. (2011). The forms of alkalis in the biochar produced from crop residues at different temperatures. *Bioresource Technology* 102: 3488–3497. https://doi.org/10.1016/j.biortech.2010.11.018.
173. Hagemann, N., Kammann, C.I., Schmidt, H.-P. et al. (2017). Nitrate capture and slow release in biochar amended compost and soil. *PLoS One* 12: e0171214. https://doi.org/10.1371/journal.pone.0171214.
174. Glaser, B., Wiedner, K., Seelig, S. et al. (2015). Biochar organic fertilizers from natural resources as substitute for mineral fertilizers. *Agronomy for Sustainable Development* 35: 667–678. https://doi.org/10.1007/s13593-014-0251-4.
175. Li, Z. and Delvaux, B. (2019). Phytolith-rich biochar: a potential Si fertilizer in desilicated soils. *GCB Bioenergy* 11: 1264–1282. https://doi.org/10.1111/gcbb.12635.
176. Downie, A. (2011). Biochar production and use: environmental risks and rewards. [Dissertation on the internet]. Sydney (AU): The University of New South Wales. https://unsworks.unsw.edu.au/fapi/datastream/unsworks:10236/SOURCE02.
177. Pusceddu, E., Santilli, S.F., Fioravanti, G. et al. (2019). Chemical-physical analysis and exfoliation of biochar-carbon matter: from agriculture soil improver to starting material for advanced nanotechnologies. *Materials Research Express* 6: 115612. https://doi.org/10.1088/2053-1591/ab4ba8.

178. Emmerich, F.G. and Luengo, C.A. (1996). Babassu charcoal: a sulfurless renewable thermo-reducing feedstock for steelmaking. *Biomass and Bioenergy* 10: 41–44. https://doi.org/10.1016/0961-9534(95)00060-7.
179. Kolb, S. (2007). Understanding the mechanisms by which a manure-based charcoal product affects microbial biomass and activity. PhD thesis. University of Wisconsin, Green Bay.
180. González, M.E., Cea, M., Medina, J. et al. (2015). Evaluation of biodegradable polymers as encapsulating agents for the development of a urea controlled-release fertilizer using biochar as support material. *Science of the Total Environment* 505: 446–453. https://doi.org/10.1016/j.scitotenv.2014.10.014.
181. Kotaka, K. (2005). Process for producing nitrogenous fertilizer and apparatus for producing nitrogenous fertilizer. WO2005054154.
182. Magrini-Bair, K.A., Czernik, S., Pilath, H. et al. (2009). Biomass derived, carbon sequestering, designed fertilizers. *Annals of Environmental Science* 3: 217–225.
183. Radlein, D., Piskorz, J., Majerski, P. (n.d.). Method of producing slow-release nitrogenous organic fertilizer from biomass. US Patent 5,676,727.
184. Chen, S., Yang, M., Ba, C. et al. (2018). Preparation and characterization of slow-release fertilizer encapsulated by biochar-based waterborne copolymers. *Science of the Total Environment* 615: 431–437. https://doi.org/10.1016/j.scitotenv.2017.09.209.
185. Yao, Y., Gao, B., Chen, J., and Yang, L. (2013). Engineered biochar reclaiming phosphate from aqueous solutions: mechanisms and potential application as a slow-release fertilizer. *Environmental Science & Technology* 47: 8700–8708. https://doi.org/10.1021/es4012977.
186. Dzvene, A.R., Chiduza, C., Mnkeni, P.N., and Peter, P.C. (2019). Characterisation of livestock biochars and their effect on selected soil properties and maize early growth stage in soils of Eastern Cape province, South Africa. *South African Journal of Plant and Soil* 36: 199–209. https://doi.org/10.1080/02571862.2018.1536930.
187. Troeh, F.R. and Thompson, L.M. (2005). *Soils and Soil Fertility*. Ames, IA: Blackwell Publishing.
188. Taghizadeh-Toosi, A., Clough, T.J., Sherlock, R.R., and Condron, L.M. (2012). A wood based low-temperature biochar captures NH_3–N generated from ruminant urine-N, retaining its bioavailability. *Plant and Soil* 353: 73–84. https://doi.org/10.1007/s11104-011-1010-9.
189. Sun, H., Zhang, H., Min, J. et al. (2016). Controlled-release fertilizer, floating duckweed, and biochar affect ammonia volatilization and nitrous oxide emission from rice paddy fields irrigated with nitrogen-rich wastewater. *Paddy and Water Environment* 14: 105–111. https://doi.org/10.1007/s10333-015-0482-2.
190. Zhang, A., Bian, R., Pan, G. et al. (2012). Effects of biochar amendment on soil quality, crop yield and greenhouse gas emission in a Chinese rice paddy: a field study of 2 consecutive rice growing cycles. *Field Crops Research* 127: 153–160. https://doi.org/10.1016/j.fcr.2011.11.020.
191. Zhang, A., Cui, L., Pan, G. et al. (2010). Effect of biochar amendment on yield and methane and nitrous oxide emissions from a rice paddy from Tai Lake plain. *Agriculture, Ecosystems & Environment* 139: 469–475. https://doi.org/10.1016/j.agee.2010.09.003.
192. Kim, J., Yoo, G., Kim, D. et al. (2017). Combined application of biochar and slow-release fertilizer reduces methane emission but enhances rice yield by different mechanisms. *Applied Soil Ecology* 117–118: 57–62. https://doi.org/10.1016/j.apsoil.2017.05.006.
193. Kammann, C.I., Schmidt, H.-P., Messerschmidt, N. et al. (2015). Plant growth improvement mediated by nitrate capture in co-composted biochar. *Scientific Reports* 5: 11080. https://doi.org/10.1038/srep11080.
194. Li, S. and Chen, G. (2019). Contemporary strategies for enhancing nitrogen retention and mitigating nitrous oxide emission in agricultural soils: present and future. *Environment, Development and Sustainability* https://doi.org/10.1007/s10668-019-00327-2.
195. Zhou, Z., Du, C., Li, T. et al. (2015). Biodegradation of a biochar-modified waterborne polyacrylate membrane coating for controlled-release fertilizer and its effects on soil bacterial community profiles. *Environmental Science and Pollution Research* 22: 8672–8682. https://doi.org/10.1007/s11356-014-4040-z.
196. Zhao, L., Cao, X., Zheng, W. et al. (2016). Copyrolysis of biomass with phosphate fertilizers to improve biochar carbon retention, slow nutrient release, and stabilize heavy metals in soil. *ACS Sustainable Chemistry & Engineering* 4: 1630–1636. https://doi.org/10.1021/acssuschemeng.5b01570.

197. Noordin, N., Ghazali, S., and Adnan, N. (2018). Impact of sap-biochar incorporation on controlled release water retention fertilizer (CRWR) towards growth of okras (*Abelmoschus esculentus*). *Materials Today: Proceedings* 5: 21911–21918. https://doi.org/10.1016/j.matpr.2018.07.050.
198. Dong, D., Wang, C., Van Zwieten, L. et al. (2020). An effective biochar-based slow-release fertilizer for reducing nitrogen loss in paddy fields. *Journal of Soils and Sediments*. 20: 3027–3040. https://doi.org/10.1007/s11368-019-02401-8.

Index

High-Performance Materials from Bio-based Feedstocks, First Edition. Edited by Andrew J. Hunt, Nontipa Supanchaiyamat, Kaewta Jetsrisuparb and Jesper T.N. Knijnenburg.